New Approaches to Nonlinear Problems in Dynamics

Proceedings of a Conference
Asilomar Conference Grounds, Pacific Grove, California
December 9-14, 1979

The conference was sponsored by the
Engineering Foundation
345 East 47th Street, New York, New York 10017

CONFERENCE ORGANIZING COMMITTEE

Okan Gurel, *IBM, Yorktown Heights*
J. Karl Hedrick, *Massachusetts Institute of Technology*
Philip J. Holmes (Chairman), *Cornell University*
Hans Othmer, *Rutgers University*

The Organizers and the Engineering Foundation gratefully
acknowledge the additional support provided by the
National Science Foundation (Solid Mechanics Division)

NEW APPROACHES TO NONLINEAR PROBLEMS IN DYNAMICS

edited by
Philip J. Holmes
Cornell University

siam
Philadelphia
1980

Library of Congress Catalog Card Number: 80-52593
ISBN: 0-89871-167-3

Table of Contents

Introduction vii

List of Participants xi

Section One. Mathematical Methods 1

Dynamical Systems and Invariant Manifolds
Philip J. Holmes and Jerrold E. Marsden 3

Computation of Invariant Manifolds .... *Brian D. Hassard* 27

A Qualitative Approach to Steady-State Bifurcation Theory
Martin Golubitsky and David Schaeffer 43

Solution Branching—A Constructive Technique
D. W. Decker and H. B. Keller 53

Patterns of Bifurcation .... *John Guckenheimer* 71

Section Two. Aerospace and Mechanical Engineering 105

Nonlinear Attitude Dynamics of Satellites
Rudrapatna V. Ramnath and Yee-Chee Tao 107

Bifurcation Analysis of Aircraft High Angle-of-Attack Flight Dynamics
Raman K. Mehra and James V. Carroll 127

Nonlinear Aeroelasticity .... *Earl H. Dowell* 147

Nonlinear Behavior in Rail Vehicle Dynamics .... *Neil K. Cooperrider* 173

Accuracy of Statistical Linearization .... *J. J. Beaman* 195

Section Three. Chemical Engineering 209

Bifurcations of a Model Diffusion-Reaction System .. *C. R. Kennedy and R. Aris* 211

Bifurcation Phenomena in Stirred Tanks and Catalytic Reactors
W. H. Ray and K. F. Jensen 235

A Singularity Theory Approach to Qualitative Behavior of Complex Chemical Systems .... *Martin Golubitsky, Barbara L. Keyfitz and David Schaeffer* 257

Waves in Distributed Chemical Systems: Experiments and Computations
Pavel Raschman, Milan Kubicek and Milos Marek 271

Computer-Aided Analysis of Nonlinear Problems in Transport Phenomena
Robert A. Brown, L. E. Scriven and William J. Silliman 289

Section Four. Electrical and Civil Engineering 309

Steady Motions Exhibited by Duffing's Equation: A Picture Book of Regular and Chaotic Motions .... *Yoshisuke Ueda* 311

Periodically Forced Relaxation Oscillations .... *Mark Levi* 323

A Discrete Dynamical System with Subtly Wild Behavior
D. G. Aronson, M. A. Chory, G. R. Hall and R. P. McGehee 339

Buckling of Shallow Elastic Structures .... *Raymond H. Plaut* 361

Section Five. Review: Mathematical Methods and Mechanics 377

Remarks on Bifurcation Theory in Differential Equations .... *Jack K. Hale* 379

Bifurcation Theory and Averaging in Mechanical Systems .... *P. R. Sethna* 392

On Some Global Results of Point Mapping Dynamical Systems .... *C. S. Hsu* 405

Jets and Geneticity in Qualitative Dynamics .... *Lawrence Markus* 418

Section Six. Bifurcation with Symmetry 431

An Application of Singularity Theory to Convection in a Spherical Shell
David Schaeffer and Martin Golubitsky 433

The Analysis of Nonlinear Equations with Symmetries .... *Hans G. Othmer* 437

Section Seven. Stochastic Problems 449
Statistical Performance Analysis of Nonlinear Stochastic Systems
James H. Taylor 451
Experimentally Derived Reformulation of the Wheelset Nonlinear Hunting Problem .... *Larry M. Sweet* 463
Bifurcation in Nonlinear Stochastic Systems .... *S. T. Ariaratnam* 470

Section Eight. Strange Attractors 475
Chaos and Bijections Across Dimensions .... *Otto E. Rossler* 477
Experimental Models for Strange Attractor Vibrations in Elastic Systems
Francis C. Moon 487
On Coupled Cells .... *Igor Schreiber, Milan Kubicek and Milos Marek* 496

Section Nine. Large Scale and Distributed Systems 509
Optimization of Distributed Parameter Structures with Repeated Eigenvalues
Edward J. Haug 511
A General Limit Cycle Analysis Method for Multivariable Systems
James H. Taylor 521

INTRODUCTION

The Engineering Foundation sponsored and organized a conference on New Approaches to Nonlinear Problems in Dynamics, which was held December 9-14, 1979, at the Asilomar Conference Grounds, Pacific Grove, California. Fifty-three engineers and mathematicians met for five days in an informal atmosphere to discuss recent developments in the solution of nonlinear problems.

This volume of proceedings reflects the organization of the meeting into five morning sessions, each with four or five forty-minute presentations and followed by less formal afternoon and evening sessions with shorter presentations. Almost all the attendees participated in the latter sessions, but it has been possible to include only a limited number of these short contributions in this volume.

Three of the morning sessions were directly concerned with problems arising in engineering: mechanical and aerospace (Tuesday), chemical (Wednesday), and electrical and civil (Thursday). The opening session was devoted to introductory surveys of mathematical methods and the closing session to proven applications of the techniques and reviews of their usefulness. The absence of studies of hydrodynamic stability and turbulence is notable here. The organizing committee felt that the fluid mechanics community had already been fairly well served by conferences similar to this one, while workers in the areas of structural engineering, solid mechanics, and chemical engineering had had fewer opportunities to learn about recent mathematical developments. It was therefore agreed to limit the scope of the meeting, with the hope that what was lost in breadth might be gained in the depth of understanding developed among a relatively small group.

In recent years the topics of qualitative analysis of dynamical systems, bifurcation theory, and global analysis have aroused considerable interest among mathematicians. Much of this work has its origins in Poincaré's studies of celestial mechanics [14], and in the work of Birkhoff [5], Liapunov [12], and others of the Russian School [2],[3]. Hale gives an outline of this fundamental work in his contribution to these proceedings. In the past fifteen or twenty years there has been a veritable explosion in the research in this area. Smale, in a classic paper [16], outlined a number of outstanding problems, and thereby stimulated much of this work. Until the mid-1970's, however, these new analytical tools were largely in the hands of pure mathematicians, although a number of potential applications had been sketched, notably by Ruelle and Takens [15], who suggested the importance of "strange attractors" in the study of turbulence. (The mathematical ideas behind these developments are ably presented in this volume by the contributions of Marsden, Hassard, Golubitsky, Decker, Guckenheimer, Hale, and Markus.)

A number of engineers and physicists soon realized that the geometric ideas and other concepts from the study of qualitative structures in the phase spaces of dynamical systems might be of help in their work. In mechanical engineering, the work of Mehra, Hsu, and Sethna (all represented in this volume) began to demonstrate the value of these new developments. Holmes and Marsden [7],[8] also brought new approaches to well-known problems in mechanics, including the panel flutter studies of Dowell [6].

The topic of "strange attractors," irregular bounded motions which arise naturally in deterministic dynamical systems of dimension $\geqq 3$, also began to attract more general interest. Ueda demonstrated their existence in a class of nonlinear oscillators with periodic forcing which occur in electrical circuit theory. A summary of some of his work appears in this volume. Also in these proceedings, Moon discusses his experimental measurements on the forced vibrations of a magneto-elastic beam. It is interesting to compare the Poincaré maps produced in the two cases. The existence of Smale horseshoes in the attractor seems important in both problems (cf. [9],[10]).

Similar promising starts had been made in studies of chemical reactor kinetics. The paper of Uppal, Ray, and Poore on the continuous stirred tank reactor [17] established it as an archetypical system in nonlinear chemical kinetics. (See Ray's contribution in this volume.) Aris' monograph on mathematical modelling [4] provides a nice introduction to this area. Applications were also being proposed in other engineering fields, and all in all it seemed that the time was ripe for a general discussion. In this respect the interdisciplinary and informal nature of the meeting was felt to be important, since it would permit research workers in several areas to exchange their ideas and problems, and to discover their common ground in the mathematical techniques. An example which occurred to the organizing committee at the outset was the area of tracked vehicle dynamics, including rail-wheel interactions, in which aggressively nonlinear forces occur. This was an obvious candidate for the application of new techniques. The papers of Cooperrider, Beaman, and Sweet show clearly the magnitude and complexity of these problems, and hopefully the ideas exchanged at Asilomar will lead to new applications and progress in this area.

The noncontinuous or piecewise linear force functions encountered in the above studies do not lend themselves to the analytical methods of local bifurcation theory, but some of the global methods might be usefully applied. One area which was mentioned in discussions was the relationship between equivalent linearization or describing function techniques (see the papers of Beaman and Taylor) and the Hopf bifurcation theorem and other techniques for analyzing limit cycles. Allwright [1] has made some contributions in this area.

There emerged from the meeting the general feeling that qualitative analysis is no longer abstract, and that the complaint that "qualitative is but poor quantitative" can justifiably (if humorously) be countered by "who needs quantity without quality?" Knowledge of the generic or typical properties of dynamical systems and their bifurcations, at the very least, allows one to organize one's expectations in the analysis of a nonlinear problem. It is worthwhile to know what to look for before starting out. In this respect the work (presented herein) of Keyfitz, Golubitsky, and Schaeffer on the H_2–O_2 reaction and of the latter two on convection in a spherical shell provide nice examples of the power of unfolding theory and of the role of symmetries in the analysis of practical problems. Plaut's work on the stability of arches, also contained in this volume, explores imperfection sensitivity with respect to asymmetries—a phenomenon well known to civil engineers.

In 1962 Smale proposed the famous horseshoe as an example of a two-dimensional mapping with a complex invariant set [16]. He suggested that something of this type occurred in the forced van der Pol oscillator, a piecewise linear version of which had been studied by Levinson [11]. In this volume Levi sketches his study of the oscillator in which he demonstrated the existence of horseshoes. In both this work and the work of Holmes and Marsden [9], [10] on periodically forced systems with chaotic solutions, the results of classical perturbation methods are interpreted geometrically to yield new information.

Not only are the new methods being related to more traditional techniques of nonlinear analysis, such as perturbation theory and the averaging method, but, in addition, computational realizations of many of the once abstract theorems are now appearing. Hassard's work on approximating invariant manifolds (see pp. 27–42) and Decker's work on the computation of bifurcation diagrams (see pp. 53-70) are notable here. Also, in a presentation not included in these proceedings, Meyer described an algorithm for computing normal forms of vector fields [13].

Acknowledgment. I would like to thank Okan Gurel for taking detailed notes during the conference. His notes were of great help in the preparation of this introduction.

PHILIP J. HOLMES
Cornell University
February, 1980

REFERENCES

[1] D. J. Allwright, *Harmonic balance and the Hopf bifurcation*, Math. Proc. Camb. Phil. Soc., 82 (1977), 453-467.

[2] A. A. Andronov, E. A. Vitt, and S. E. Khaiken, *Theory of Oscillators*, (trans. F. Immirzi), Pergamon Press, Oxford, 1966.

[3] A. A. Andronov and L. Pontryagin, *Systems Grossières*, Dokl. Akad. Nauk. SSSR, 14 (1937), 247-251.

[4] R. Aris, *Mathematical Modelling Techniques*, Pitman, London, 1978.

[5] G. D. Birkhoff, *Dynamical Systems*, Amer. Math. Soc. Colloq. Publ., 9, American Mathematical Society, Providence, RI, 1927.

[6] E. H. Dowell, *Aeroelasticity of Plates and Shells*, Noordhoff International Publishing, Leiden, The Netherlands, 1975.

[7] P. J. Holmes, *Bifurcations to divergence and flutter in flow-induced oscillations: ~~An infinite~~ A finite dimensional analysis*, J. Sound Vib., 53 (1977), 471-503.

[8] P. J. Holmes and J. E. Marsden, *Bifurcations to divergence and flutter in flow-induced oscillations: An infinite dimensional analysis*, Automatica, 14 (1978), 367-384.

[9] P. J. Holmes, *Averaging and chaotic motions in forced oscillations*, SIAM J. Appl. Math., 38 (1980), 65-80.

[10] P. J. Holmes and J. E. Marsden, *A partial differential equation with infinitely many periodic orbits: Chaotic oscillations of a forced beam*, 1980, submitted for publication.
[11] N. Levinson, *A second order differential equation with singular solution*, Annals. of Math., 50 (1949), 127-153.
[12] A. M. Liapunov, *Problème Général de la Stabilité du Mouvement*, Princeton University Press, Princeton, NJ, 1949.
[13] K. R. Meyer and D. S. Schmidt, *Entrainment domains*, Funkcialaj Ekvacioj, 20 (1977), 171-192.
[14] H. Poincaré, *Les Methods Nouvelles de la Mécanique Celeste*, Vols. I, II, III, Gauthier-Villars, Paris, 1893-1899.
[15] D. Ruelle and F. Takens, *On the nature of turbulence*, Comm. Math. Phys., 20 (1971), 167-192.
[16] S. Smale, *Differentiable dynamical systems*, Bull. Amer. Math. Soc., 73 (1967), 747-817.
[17] A. Uppal, W. H. Ray, and A. Poore, *On the dynamical behavior of continuous stirred tank reactors*, Chem. Eng. Sci., 29 (1974), 967.

LIST OF PARTICIPANTS

David Anderson, Associate Professor, University of Texas Health Center and Department of Mathematics, Southern Methodist University, Dallas, TX 75275.

S. T. Ariaratnam, Professor, Solid Mechanics Division, University of Waterloo, Waterloo, Ontario N2L 3G1, Canada.

Rutherford Aris, Professor, Department of Chemical Engineering, University of Minnesota, Minneapolis, MN 55455.

D. Aronson, Professor, School of Mathematics, University of Minnesota, Minneapolis, MN 55455.

Anil Kumar Bajaj, Graduate Student, Department of Aerospace Engineering, University of Minnesota, Minneapolis, MN 55455.

John Ball, Visiting Professor, Department of Mathematics, University of California, Berkeley, CA 94720.

Joe Beaman, Department of Mechanical Engineering, University of Texas, Austin, TX 78705.

Robert A. Brown, Assistant Professor, Department of Chemical Engineering, Massachusetts Institute of Technology, Cambridge, MA 02139.

Shui-Nee Chow, Mathematics Department, Michigan State University, East Lansing, MI 48824.

Peter L. Christiansen, Laboratory of Applied Mathematics, The Technical University of Denmark, DK-2800 Lynby, Denmark.

Sandford S. Cole, Director, Engineering Foundation, 345 East 47th St., New York City, NY 10017.

Neil K. Cooperrider, Mechanical Engineering Department, Arizona State University, Tempe, AZ 85281.

Dwight Decker, Mathematics Department, North Carolina State University, Raleigh, NC 27650.

Earl H. Dowell, Professor, Department of Mechanical and Aerospace Engineering, Princeton University, Princeton, NJ 08544.

Martin Golubitsky, Department of Mathematics, Arizona State University, Tempe, AZ 85281.

John Guckenheimer, Professor, Division of Natural Sciences, University of California, Santa Cruz, CA 95064.

Okan Gurel, IBM, 1133 Westchester Ave., White Plains, NY 10604.

Jack K. Hale, Professor, Division of Applied Mathematics, Brown University, Providence, RI 02912.

Brian Hassard, Assistant Professor, Department of Mathematics, State University of New York, Buffalo, NY 14216.

E. Haug, Professor, Division of Materials Engineering, University of Iowa, Iowa City, IA 52242.

Philip J. Holmes, Assistant Professor, Department of Theoretical and Applied Mechanics, Cornell University, Ithaca, NY 14853.

C. S. Hsu, Professor, Department of Mechanical Engineering, University of California, Berkeley, CA 94720.

Barbara L. Keyfitz, Assistant Professor, Department of Mathematics, Arizona State University, Tempe, AZ 85281.

Mark Levi, Assistant Professor, Mathematics Department, Northwestern University, Evanston, IL 60201.

Milos Marek, Visiting Professor, Department of Chemical Engineering, University of Wisconsin, 1415 Johnson Drive, Madison, WI 53706.

Lawrence Markus, Professor, Mathematics Department, University of Minnesota, Minneapolis, MN 55455.

Jerrold E. Marsden, Professor, Department of Mathematics, University of California, Berkeley, CA 94720.

Raman K. Mehra, President, Scientific Systems, Inc., 186 Alewife Dr., Parkway, Cambridge, MA 02138.

Kenneth R. Meyer, Professor, Department of Mathematics, University of Cincinnati, Cincinnati, OH 45208.

F. C. Moon, Professor, Department of Theoretical and Applied Mechanics, Cornell University, Ithaca, NY 14853.

Syhar H. Mufti, Division of Mechanical Engineering, National Research Council of Canada, Ottawa, Ontario, K1A OR6, Canada.

Douglas Muster, Brown and Root Professor, Department of Mechanical Engineering, University of Houston, Houston, TX 77035.

Tyre A. Newton, Professor, Department of Pure and Applied Mathematics, Washington State University, Pullman, WA 99164.

Hans G. Othmer, Department of Mathematics, Rutgers University, New Brunswick, NJ 08903.

Norman H. Packard, Physics Board, University of California, Santa Cruz, CA 95064.

Daniel Y. Perng, Staff Engineering, Exxon Research and Engineering Co., P.O. Box 45, Linden, NJ 07036.

J. Carl Pirkle, Jr., Staff Engineering, Exxon Research and Engineering Co., P.O. Box 45, Linden, NJ 07036.

Richard E. Plant, Associate Professor, Department of Mathematics, University of California, Davis, CA 95616.

Raymond H. Plaut, Professor, Department of Civil Engineering, Virginia Polytechnic Institute and State University, Blacksburg, VA 24061.

Aubrey B. Poore, Associate Professor, Mathematics Research Center, 610 Walnut Street, Madison, WI 53705.

R. V. Ramnath, Leader, Division of Education, C. S. Draper Laboratory, 555 Technology Square, Cambridge, MA 02173.

W. Harmon Ray, Professor, Department of Chemical Engineering, University of Wisconsin, Madison, WI 53706.

Otto E. Rossler, Professor, Institute for Physical and Theoretical Chemistry, University of Tubingen, Auf der Morgenstelle 8, D-7400 Tubingen, West Germany.

S. Shankar Sastry, Research Assistant, Electronics Research Laboratories, University of California, Berkeley, CA 94720.

David G. Schaeffer, Professor, Department of Mathematics, Duke University, Durham, NC 22706.

L Scriven, Professor, Department of Chemical Engineering and Materials Science, University of Minnesota, Minneapolis, MN 55455.

P. R. Sethna, Professor, Department of Aeronautical Engineering, University of Minnesota, Minneapolis, MN 55455.

Michael Shearer, Visiting Professor, Mathematics Research Center, University of Wisconsin, Madison, WI 53706.

Larry M. Sweet, Associate Professor, Department of Mechanical and Aerospace Engineering, Princeton University, Princeton, NJ 08540.

D. L. Taylor, Assistant Professor, Department of Mechanical and Aerospace Engineering, Cornell University, Ithaca, NY 14853.

James H. Taylor, Associate Professor, School of Mechanical and Aerospace Engineering, Oklahoma State University, Stillwater, OK 74074.

Hans Troger, Professor, University of Technology, Karlsplatz 13, Vienna A-1040, Austria.

Yoshisuke Ueda, Department of Electrical Engineering, Kyoto University, Yoshida-nonmachi, Kyoto 606, Japan.

Enol Utku, Professor, School of Engineering, Duke University, Durham, NC 27706.

Paul K. C. Wang, Professor, University of California, Boelter Hall, Los Angeles, CA 90024.

SECTION ONE
MATHEMATICAL MODELS

Dynamical Systems and Invariant Manifolds

Philip J. Holmes* and Jerrold E. Marsden**

Abstract. We review some basic terminology in dynamical systems with the purpose of bridging some of the communication gaps that may exist between mathematicians and engineers at this conference. Recent results on panel flutter and on the existence of horseshoes in the dynamics of a forced beam are briefly sketched to illustrate some of the concepts of interest to both groups.

1. Dynamical Systems on Manifolds. A vector field X on a manifold M is a (smooth) mapping from M to TM, the tangent bundle of M, that assigns to each point $x \in M$ a vector tangent to M at x. Often, M is Euclidean n space $\mathbb{R}^n$, so X is a vector field in the sense of advanced calculus: $X(x^1, \ldots, x^n) = (X^1(x^1, \ldots, x^n), X^2(x^1,\ldots, x^n), \ldots, X^n(x^1,\ldots, x^n))$. A vector field X may be thought of as the right hand side of a system of first order differential equations in the large, that is, a dynamical system. In $\mathbb{R}^n$, this simply corresponds to the system of n-ordinary differential equations

$$\frac{dx^i}{dt} = X^i(x^1(t), \ldots, x^n(t)), \quad i = 1, \ldots, n.$$

or, abstractly,

$$\dot{x} = X(x(t)).$$

One might ask whether it is worthwhile to engineers to invest in the machinery of dynamical systems on manifolds. This question may be answered in the affirmative on two grounds as follows:

---

*Department of Theoretical and Applied Mechanics, Cornell University, Ithaca, N.Y. 14853.

**Department of Mathematics, University of California, Berkeley, CA 94720.

†Notes by Shankar Sastry.

(i) Quite often in practice, the systems that one encounters have state spaces which result from imposing smooth constraints on vector spaces. For instance, in a circuit the dynamics of the capacitor voltages and the inductor currents are constrained along a manifold specified by Kirchoff's laws and the static non-linear resistor characteristics.

(ii) It often helps to identify the state space of a physical object with a more abstract mathematical object with a manifold structure and exploit the convenient geometric intuition of manifolds. For instance in rigid body rotation the state space may be identified with SO(3), the space of all proper orthogonal 3×3 matrices, which is a compact 3-dimensional manifold.

We now state some of the simplest results of dynamical systems culminating in invariant manifold theory (the general reference used here is Abraham-Marsden [1]). First we give some definitions and preliminaries.

Let X be a (time-independent) vector field on M. An <u>integral curve</u> or <u>trajectory</u> of X at $x_0 \in M$ is a curve $x(t)$ in M such that $\dot{x}(t) = X(x(t))$ for each t in an open interval I and $x(0) = x_0$. If X is smooth (or locally Lipschitz will do), then the integral curve of X at x_0 exists and is unique. The vector field X is said to be <u>complete</u> if the domain of each integral curve can be extended to all of $\mathbb{R}$ (i.e. the open interval I can be chosen to be $\mathbb{R}$). If M is compact, any (smooth) vector field X on M is complete; or if the support of a vector field X is compact, it is complete. We assume hence forward that the vector field X is complete, for simplicity. In this case we can associate with X a one-parameter family of diffeomorphism on M called the <u>flow</u> of X denoted F_t and defined by

$$F_0(x_0) = x_0$$

and

$$\frac{d}{dt} F_t(x_0) = X(F_t(x_0))$$

A point x_0 is called a <u>critical point</u> (or <u>singular point</u> or <u>equilibrium point</u>) of X if $X(x_0) = 0$. This is equivalent to x_0

being a fixed point of the flow: $F_t(x_0) = x_0$. The linearization of X at x_0 is the linear map

$$X'(x_0) : T_{x_0}M \to T_{x_0}M \quad \text{defined by}$$

$$X'(x_0)v = \frac{d}{d\lambda}\left(TF_\lambda(x_0)\cdot v\right)\Big|_{\lambda=0}$$

In $\mathbb{R}^n$, $X'(x_0)$ is the matrix $\partial X^i/\partial x^j$ evaluated at x_0. The eigenvalues of $X'(x_0)$ are called the characteristic exponents of X at x_0 and their exponentials are called the characteristic multipliers of X at x_0.

Of basic interest to engineers is the stability of a critical point, defined as follows: If x_0 is a critical point of X then x_0 is stable (in the sense of Liapounov) if for any neighbourhood U of x_0, there is a neighbourhood V of x_0, such that $F_t(x) \in U$ for all $x \in V$ and all $t \geq 0$. The point x_0 is said to be asymptotically stable if there exists a neighbourhood V of x_0 such that

$$F_t(V) \subset F_s(V) \text{ if } \quad t > s \quad \text{and} \quad \lim_{t\to\infty} F_t(V) = \{x_0\}.$$

An important sufficient condition for checking stability is the following theorem of Liapounov:

Liapounov's Theorem. Let x_0 be a critical point of X and let the characteristic exponents of X at x_0 have strictly negative real parts. Then x_0 is asymptotically stable (similarly, if the characteristic exponents of X have strictly positive real parts, then x_0 is asymptotically unstable i.e. asymptotically stable as $t \to -\infty$).

A critical point x_0 is said to be hyperbolic if none of its characteristic exponents has zero real part. A result of Hartman shows that near a hyperbolic critical point the flow looks like that of its linearization (i.e. is conjugate to the flow of its linearization). Thus, in the plane we have (upto diffeomorphism) the hyperbolic flows shown in Figure 1.

The case when the critical point is not hyperbolic is of obvious interest, for instance in Hamiltonian dynamics, and will be discussed

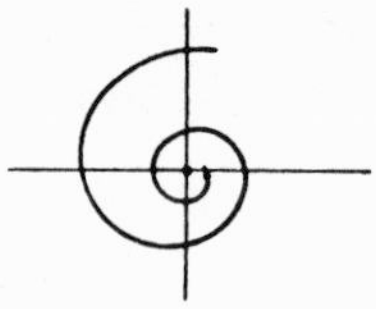

(a) $Re\,\mu_1 = Re\,\mu_2 < 0$
μ_1, μ_2 not real. (stable focus)

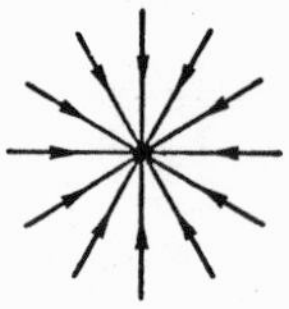

(b) $\mu_1 < 0, \mu_2 < 0$ (stable node)

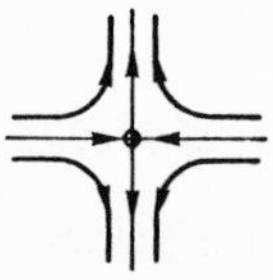

(c) $\mu_1 > 0, \mu_2 < 0$ (saddle)

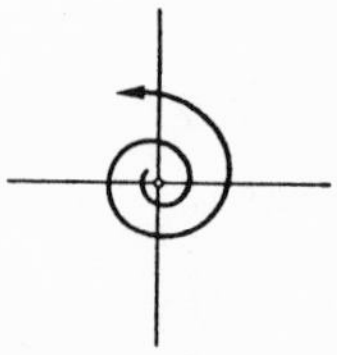

(d) $Re\mu_1 = Re\mu_2 > 0$
μ_1, μ_2 not real
(unstable focus)

(e) $\mu_1 > 0, \mu_2 > 0$
(unstable node).

Figure 1 Hyperbolic equilibria with characteristic exponents. (a) $Re\,\mu_1 = Re\,\mu_2 < 0$, with μ_1, μ_2 not real (stable focus). (b) $\mu_1 < 0$, $\mu_2 < 0$ (stable node). (c) $\mu_1 < 0$, $\mu_2 > 0$ (saddle). (d) $Re\,\mu_1 = Re\,\mu_2 > 0$, with μ_1, μ_2 not real (unstable focus). (e) $\mu_1 > 0$, $\mu_2 > 0$ (unstable node).

below.

We next discuss another possible critical element of the vector field X, namely a closed orbit. A periodic point of X is a point $x \in M$ such that for some $\tau > 0$, $F_{t+\tau}(x) = F_t(x)$ for all $t \in \mathbb{R}$, and the period of x is the smallest $\tau > 0$ satisfying this condition. A closed orbit is the orbit of a periodic non-equilibrium point. We have seen how the linearization, $X'(x_0)$ of the vector field X at an equilibrium point x_0 approximates the flow of X near x_0. We now discuss the asymptotic behavior of orbits close to a closed orbit using the Poincaré map on a local-transversal section. This is defined as follows: A local transversal section of X at $x \in M$ is a submanifold $S \subset M$ of codimension one with $x \in S$ and with $X(s)$ not contained in (transversal to) T_sS for all $s \in S$. Then, if γ is a closed orbit of X with period τ and S a local transversal section of X at $x \in \gamma$ then a Poincaré map of γ is a mapping

$\Theta: W_0 \to W_1$ where

(i) W_0, W_1 are open neighbourhoods on S of $x \in S$ and Θ is a diffeomorphism,

(ii) There is a continuous function $\delta: W_0 \to \mathbb{R}$, such that $\Theta(s) = F(s, \tau - \delta(s))$; and

(iii) If $t \in (0, \tau - \delta(s))$ then $F(s,t) \notin W_0$.

This definition is visualized in Figure 2.

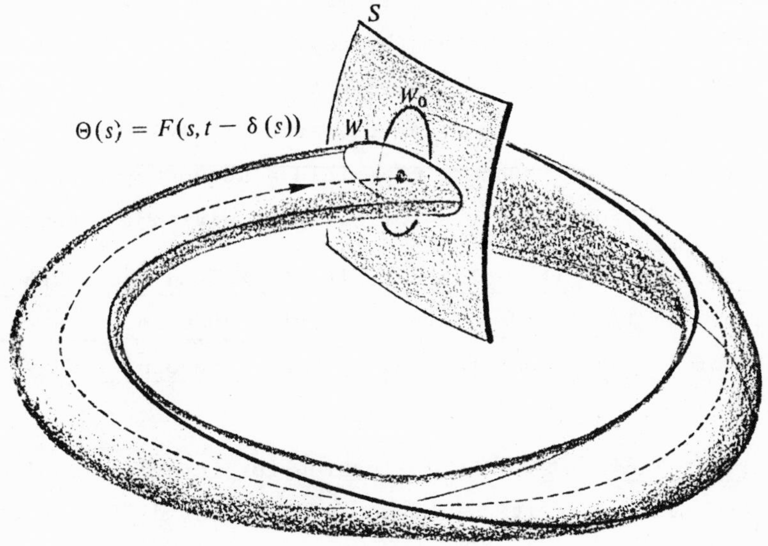

Figure 2. Visualization of the Poincaré map.

It is a basic result that Poincaré maps exist. Also, they are unique upto configuration -- i.e. if S' is another local transverse section at $x' \in \gamma$ with associated Poincaré map Θ' then there are open neighbourhoods $W_2 \subset S$, $W_2' \subset S'$, $W_2 \subset W_0 \cap W_1$, $W_2' \subset W_0' \cap W_1'$ and a diffeomorphism $H: W_2 \to W_2'$ such that the diagram

$$\begin{array}{ccc} \Theta^{-1}(W_2) \cap W_2 & \xrightarrow{\ \Theta\ } & W_2 \cap \Theta(W_2) \\ H \downarrow & & \downarrow H \\ W_2' & \xrightarrow{\ \Theta'\ } & S' \end{array}$$

commutes.

The linear approximation to Θ at x is $T_x\Theta: T_xS \to T_xS$ and the uniqueness of the Poincaré map upto configuration makes $T_x\Theta'$ similar to $T_x\Theta$ so that the eigenvalues of $T_x\Theta$ are independent of $x \in \gamma$ and the specific transverse section. These eigenvalues of $T_x\Theta$ are referred to as the characteristic multipliers of X at γ. Another linear approximation to the flow near γ is given by $T_xF_\tau : T_xM \to T_xM$. It is clear that T_xF_τ has an eigenvalue 1 corresponding to the eigenvector $X(x)$ (since τ is the period of the closed orbit). The remaining eigenvalues are the characteristic multipliers of X at γ.

We now define what we understand by asymptotic stability of a closed orbit. An orbit $F_t(y)$ is said to wind toward γ if for any transversal S to X at $x \in \gamma$ there is a t_0 such $F_{t_0}(y) \in S$ and successive applications of the Poincaré map yield a sequence of points that converge to x. We then have the following condition for this stability.

Proposition. *If γ is a closed orbit of X and the characteristic multipliers of γ lie inside the unit circle, then there is a neighbourhood U of γ such that for any $y \in U$, the orbit $F_t(y)$ winds towards γ.*

2. Invariant Manifolds. The motivation for invariant manifolds comes from the study of critical elements of linear differential equations of the form

$$\dot{x} = Ax, \qquad x \in \mathbb{R}^n .$$

Let W_s, W_c and W_u be the (generalized) real eigenspaces of A associated with eigenvalues of A lying the open left half plane, the imaginary axes and open right half plane respectively. Each of these spaces is invariant under the flow of $\dot{x} = Ax$ and represents respectively a stable, center and unstable manifold.

To return to the non-linear case, a submanifold $S \subset M$ is said to be invariant under the flow of X if for $x \in S$, $F_t(x) \in S$ for small $t > 0$; i.e. X is tangent to S. Invariant manifolds are, then, "non-linear eigenspaces". We may define invariant manifolds S of a critical element γ to be stable or unstable depending on whether they are comprised of orbits in S that wind toward γ with increasing time, or wind toward γ with decreasing time. In a neighbourhood of x in the critical element γ, the tangent spaces to the stable and unstable manifolds are provided by the eigenspaces in T_xM of characteristic multipliers of modulus <1 and modulus >1 respectively. The eigenspace corresponding to eigenvalues of modulus = 1 (not including $T_x\gamma$) is tangent to the center manifold of γ. We state this result formally as the local center stable manifold (for proof, see for example Kelley's appendix in Abraham-Robbin [2]).

Local-Center Stable Manifold Theorem. *If* $\gamma \subset M$ *is a critical element of* X, *there exist submanifolds* S_γ, CS_γ, C_γ, CUS_γ, US_γ *of* M (also denoted $W^s(\gamma)$, $W^{cs}(\gamma)$, $W^c(\gamma)$, $W^{cu}(\gamma)$ and $W^u(\gamma)$) such that

(i) *Each sub-manifold is locally invariant under* X *and contains* γ

(ii) *For* $x \in \gamma, T_x(S_\gamma)$ [resp. $T_x(CS_\gamma)$, $T_x(C_\gamma)$, $T_x(CUS_\gamma)$, $T_x(US_\gamma)$] *is the sum of the eigenspace in* T_xM *of characteristic multipliers of modulus* <1 [resp. ≤ 1, $= 1$, ≥ 1, >1] *and the subspace* $T_x\gamma$.

(iii) *If* $x \in S_\gamma$, *then* $\bigcap_{n\in\mathbb{Z}} F_{(n,\infty)}(x) = \gamma$; *and if* $x \in US_\gamma$, *then* $\bigcap_{n\in\mathbb{Z}} F_{(-\infty,n)}(x) = \gamma$.

(iv) S_γ *and* US_γ *are (locally) unique*.

Comments: (i) The configuration of these manifolds is slightly different in the cases covered $\gamma = \{x\}$, a critical point in which case $T_x\gamma = \{0\}$ or γ is a closed orbit in which case $T_x\gamma$ is the subspace generated by $X(x)$. These two cases are shown in Figure 3.

(ii) The stable and unstable manifolds are unique; but the center manifold is not unique (see Kelley's article cited above, Marsden and McCracken [10] and Wan [17]).

(iii) The theorem says in addition that if γ is hyperbolic then locally, the orbits behave qualitatively (actually, up to diffeomorphism) like the linear case.

(iv) If γ is hyperbolic we only have the locally unique manifolds S_γ and US_γ. These can be extended to *globally unique, immersed submanifolds* by means of the integral of X. This is the *global stable manifold theorem of Smale*. (There is also a global center manifold theorem due to Fenichel).

(v) From the stability (instability) of the center manifold C we can conclude the stability (instability) of the center stable CS (center-unstable CUS) manifold. This is an important theorem of Pliss and

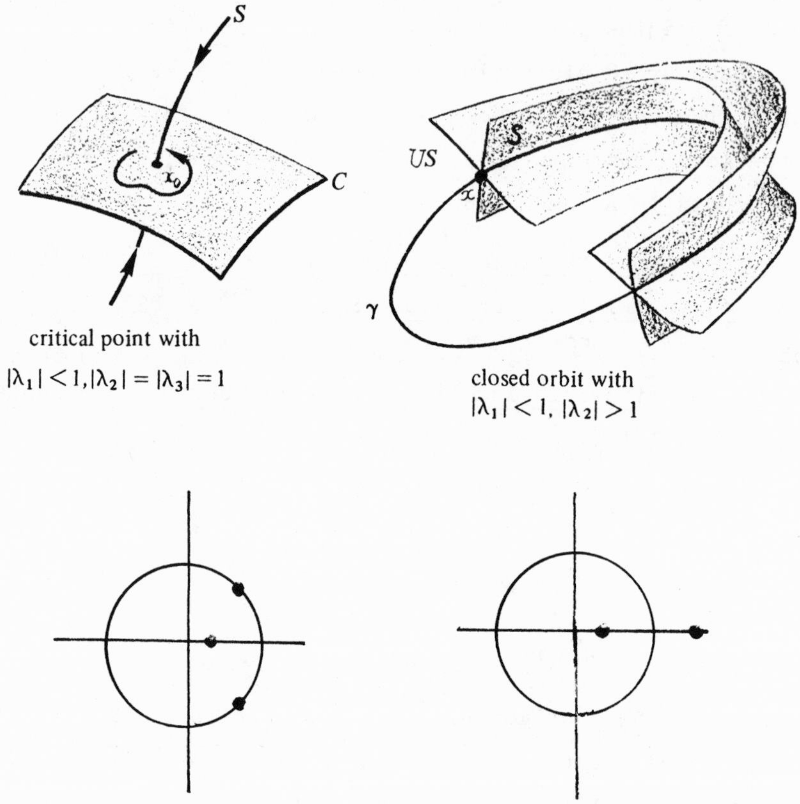

Figure 3. Invariant Manifolds.

Kelley.

It may be mentioned here in conclusion that the theory of invariant manifolds can be generalized in two important directions: (i) to maps rather than to dynamical systems and (ii) to arbitrary non-wandering sets of the flow rather than elementary critical elements. We make a few comments on (i) and (ii).

(i) Mappings rather than flows arise in at least 3 basic ways:

(a) Many systems are directly described by discrete dynamics: $x_{n+1} = f(x_n)$. For example, many population problems are best understood this way. Delay and difference equations can be viewed in this category as well.

(b) The Poincaré map of a closed orbit has already been mentioned.

(c) Suppose we are interested in non-autonomous systems of the form $\dot{x} = f(x,t)$ where f is T-periodic in t. Then the map P that advances solutions by time T, also called the Poincaré map, is very basic to a qualitative study of the orbits. (See Figure 4). This map is often used in forced oscillations as we shall see in Section 5.

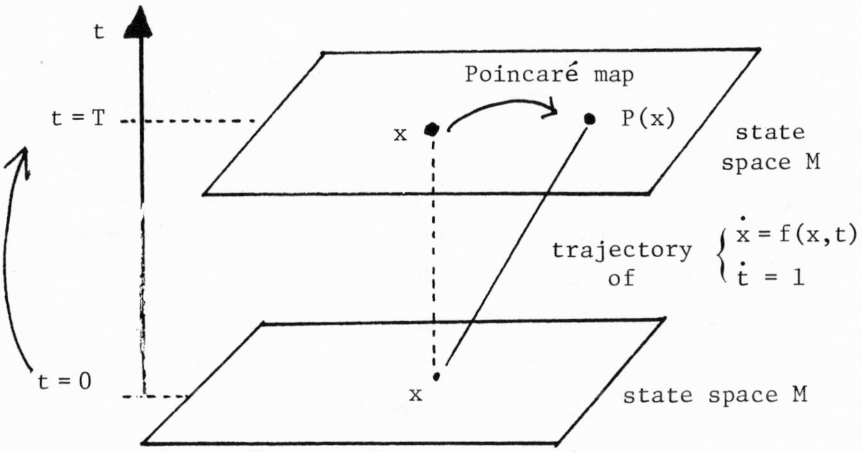

Figure 4. The Poincaré map for forced oscillations.

(ii) The generalization of the invariant manifold theory to arbitrary nonwandering sets is significant; see for instance, Hirsch, Pugh and Shub [4]). The main problem here is the lack of a spectrum (characteristic multipliers) to be able to define a hyperbolic property. The definitions are rather in terms of contractions and expansions under the flow of the norm of the tangent to the flow (with the norm induced by an appropriate Riemannian metric). In the study of chaotic dynamics such complex invariant sets can arise in very concrete systems as many of the other lectures will demonstrate. In Section 5 we shall briefly sketch an example of a complex invariant set, namely the horseshoe.

3. Invariant Manifold Theory for Partial Differential Equations. In applications to partial differential equations a useful assumption is that the semi-flow F_t defined by the equations be smooth for each $t \geq 0$, i.e. $F_t: Z \to Z$ is smooth, where Z is a suitably chosen Banach space of functions and F_t takes Cauchy data at $t = 0$ to the solution

at time t. This enables the invariant manifold machinery to go through along with bifurcation theorems (for an example of applications to the Hopf bifurcation, see ([10]). General conditions for checking smoothness are technical, but the following special instances are easily understood:

(i) F_t is smooth for semilinear p.d.e.s.

$$\frac{\partial u}{\partial t} = \Delta u + f(u) \quad \text{with} \quad u \in H_0^2(\Omega) = Z$$

or more abstractly $\dot{x} = Ax + F(x)$

where A is a generator and F is a smooth function of Z to Z. This is due to Segal [15].

(ii) F_t is smooth for the Navier-Stokes equation

$$\dot{u} = +(u\cdot\nabla)u = \text{grad } p; \quad \text{div } u = 0; \quad Z = H_0^2(\Omega)$$

See [10]. This is essentially due to Kato and Fujita.

(iii) F_t is smooth for the Euler-equations for a fluid in Lagrangian coordinates. This case is due to Ebin and Marsden; see [9]; (F_t is not, however, smooth in Eulerian coordinates).

(iv) F_t is not smooth for the Korteweg de Vries equation (but is in a kind of Lagrangian coordinates); cf. Ratiu [14].

(v) F_t seems not to be smooth for quasilinear hyperbolic equations, for instance in 3 dimensional (conservative) elasticity.

One now assumes that the spectra of $DF_t(x_0)$ or $D\Theta(x)$ split into 3 pieces, one inside the unit circle (at a non zero distance from the unit circle), the second on the unit circle and the third outside the unit circle (at a non-zero distance from it). Then there are corresponding invariant manifolds. The idea is to apply the invariant manifold theorems for smooth maps to each F_t separately; since $F_t \circ F_s = F_{t+s} = F_s \circ F_t$, these manifolds can be chosen common to all the F_t.

4. Applications of Invariant Manifold Theory to Bifurcations. Invariant manifolds are useful in qualitative investigations and in bifurcation theory. To give a specific example we will show the application of the center-manifold theory to reduce the dimension of a bifurcation problem; this method is due to Ruelle and Takens (for details see Marsden and McCracken [10]). Let F_t be a flow on a Banach space Z depending on a bifurcation parameter $\lambda \in \mathbb{R}^p$. The idea is to apply

the invariant manifold theorems to the <u>suspended flow</u>

$$F_t : Z \times \mathbb{R}^p \to Z \times \mathbb{R}^p$$
$$(x,\lambda) \to (F_t(x),\lambda)$$

The invariant manifold theorem shows that if the spectrum of the linearlization of $DF_t(z_0,\lambda_0)$ at a fixed point (z_0,λ_0) splits into $\sigma_s \cup \sigma_c$ where σ_s lies inside the unit disc (at a non-zero distance) and σ_c is on the unit disc, then the flow F_t leaves invariant manifolds S and C tangent to the eigenspaces corresponding to σ_s and σ_c respectively; S is the stable and C is the center manifold respectively. For suspended systems we always have $1 \in \sigma_c$. We now state the center manifold theorem for flows in this context:

<u>Center Manifold Theorem for Suspended Flows</u>. <u>Let</u> Z <u>be a Banach space (or manifold) and let</u> ψ <u>be the time one map of the suspended flow defined in a neighbourhood of</u> (z_0,λ_0). <u>Assume that</u> $\psi(z_0,\lambda_0) = (z_0,\lambda_0)$, <u>that</u> ψ <u>has</u> k <u>continuous derivatives, that</u> $d\psi(z_0,\lambda_0)$ <u>has spectral radius</u> 1 <u>and that the spectrum of</u> $d\psi(z_0,\lambda_0)$ <u>splits into a part on the unit circle and a part at a nonzero distance from the unit circle. Let</u> Y <u>denote the generalized eigenspace of</u> $d\psi(z_0,\lambda_0)$ <u>belonging to the part of the spectrum on the unit circle and that</u> Y <u>has dimension</u> $d < \infty$. <u>Then, there exists a submanifold</u> M <u>defined in a neighbourhood</u> V <u>of</u> (z_0,λ_0) <u>in</u> $Z \times \mathbb{R}^p$ <u>passing through</u> (z_0,λ_0) <u>and tangent to</u> Y <u>at</u> (z_0,λ_0) <u>such that</u>

(i) <u>If</u> $x \in M$ <u>and</u> $\psi(x) \in V$, <u>then</u> $\psi(x) \in M$.

(ii) <u>If</u> $\psi^n(x) \in V$ <u>for</u> $n = 0,1,2,$ <u>then as</u> $n \to \infty$, $\psi^n(x) \to M$.

For dynamical bifurcations the center manifold theorem plays the same role as the Lyapunov-Schmidt procedure for static bifurcation -- namely, it reduces the bifurcation problem to a finite dimensional one. In the instance of the Hopf bifurcation with a single parameter $(\lambda \in \mathbb{R})$, we obtain as center manifold for the suspended flow a 3-manifold tangent to the eigenspace of the two simple, purely imaginary eigenvalues crossing the imaginary axis at $\lambda = \lambda_0$ and tangent to the λ-axis at $\lambda = \lambda_0$. By looking at λ = constant sections, the problem is now reduced to that of a vector-field in two dimensions.

This general method has been useful in a number of specific problems.

We illustrate briefly by sketching one used by Holmes and Marsden [6] for the two parameter panel flutter problem (see Dowell [3] for background).

We consider the one dimensional thin panel shown in Figure 5 and are interested in bifurcations near the zero solution

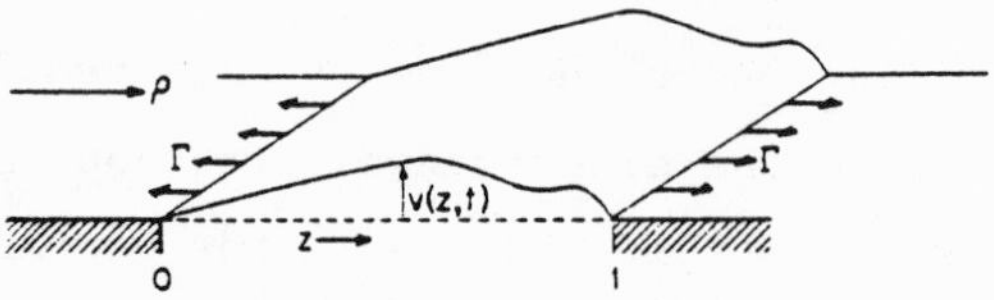

FIG. 5. The panel flutter problem.

written as

$$\alpha\dot{v}'''' + v'''' - \{\Gamma + \kappa\int_0^1 (v'(z))^2\,dz + \sigma\int_0^1 (v'(z)\cdot\dot{v}'(z))\,dz\}v'' + \rho v' + \sqrt{\rho}\delta\dot{v} + \ddot{v} = 0 \tag{4.1}$$

Here $\cdot \equiv \partial/\partial t$, $' \equiv \partial/\partial z$; viscoelastic structural damping terms are α, σ; aerodynamic damping terms are $\sqrt{\rho}\delta$; κ is the non-linear (membrane) stiffness; ρ the dynamic pressure and Γ an in-plane tensile load. All quantities are assumed non-dimensionalized and boundary conditions at $z = 0,1$ are typically simply supported $(v = v'' = 0)$ or clamped $(v = v' = 0)$. The control parameter is $\lambda = (\rho,\Gamma)$, $\rho > 0$. We redefine (4.1) as an ordinary differential equation on a Banach space $Z = H_0^2([0,1]) \times L^2([0,1])$ where H_0^2 denotes H^2 functions on $[0,1]$ vanishing at 0 and 1. Define the norm on Z by $\|(v,\dot{v})\|_Z = (|\dot{v}|^2 + |v''|^2)^{1/2}$ with $|\cdot|$ denoting the L^2 norm and define the linear operator.

$$A_\mu = \begin{pmatrix} 0 & I \\ C_\mu & D_\mu \end{pmatrix}; \quad \begin{aligned} C_\mu v &= -v'''' + \Gamma v'' - \rho v' \\ D_\mu v &= \alpha v'''' - \sqrt{\rho}\delta v \end{aligned}$$

The basic domain of A_μ consists of $\{v,\dot{v}\} \in Z$ such that $\dot{v} \in H_0^2$ and $v + \alpha\dot{v} \in H^4$; specific boundary conditions necessitate further restrictions.

Also, define the nonlinear operator

$$B(v,\dot{v}) = \begin{pmatrix} 0 \\ \\ [\kappa|v'|^2 + \sigma\langle v',\dot{v}'\rangle]\, v'' \end{pmatrix}$$

where $\langle\cdot,\cdot\rangle$ denotes the L^2 inner product. Then we can write (4.1) in the form

$$\dot{x} = A_\mu x + B(x) \quad \text{with} \quad x = \{v,\dot{v}\} \in \text{the domain of definition of } A_\mu .$$

Using ideas of Segal [15] and Parks [13] it can be shown that (4.1) does define (globally) a smooth flow on the domain of $A_\mu \subset Z$, and for $\alpha = 0.005$ and $\delta = 0.1$ the bifurcations take place in the vicinity of $\rho = 108$ and $\Gamma = -22$ when a double zero eigenvalue occurs. The center manifold theorem again reduces this to a 2-dimensional problem. This leads to the bifurcation diagram of Figure 6 (which is the Andronov-Takens normal form for the two-dimensional flow $\ddot{x} = -\nu_2 x - \nu_1 \dot{x} - x^3 - x^2\dot{x}$ with parameters ν_1, ν_2).

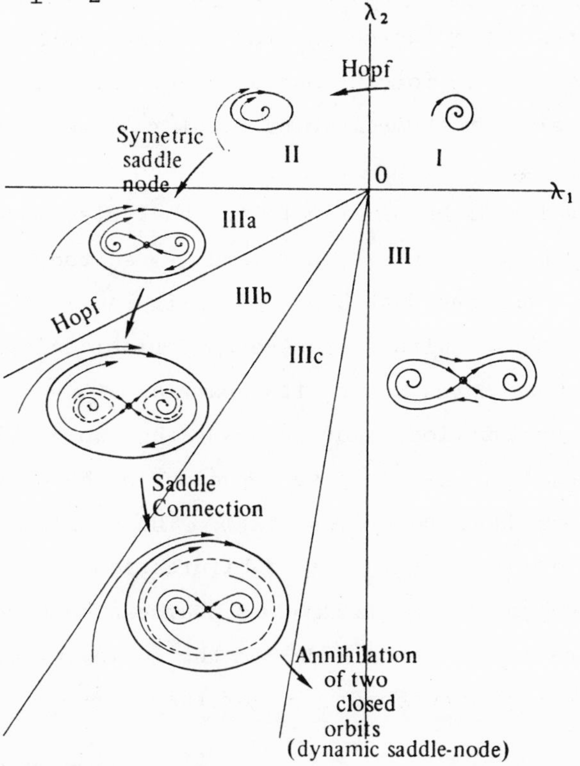

Figure 6 Takens' (2,—) normal form showing the local phase portrait in each region of parameter space.

In particular, there is a supercritical Hopf bifurcation on crossing from region I to region II, and a symmetric saddle node bifurcation on crossing from region I to region III. On crossing from region II to region IIIa the stable closed orbit persists and the unstable critical point bifurcates to a saddle and two unstable fixed points. This bifurcation diagram is actually a structurally stable (symmetric) bifurcation in the sense that it persists under any symmetric perturbation.

The transition from region IIIb to IIIc is especially interesting because a homoclinic loop occurs at the instant of bifurcation; i.e. there is an orbit from the saddle point to itself.

As we shall see in the next section, small (unsymmetric) perturbations of homoclinic loops generally invite horeshoes and "chaos". In fact, this may provide an explanation for the "chaos" discussed in Dowell's lecture.

5. <u>A Horseshoe in the Dynamics of a Forced Beam</u>. In this section we describe a situation in which complex dynamics arises by perturbing a Hamiltonian system with forcing and damping. Several other lectures will be on similar themes; in particular we refer to those of Hale and Levi.

A physical model will help motivate the analysis. One considers a beam that is buckled by an external load Γ so there are two stable equilibrium states and one unstable. (See Figure 7). The whole structure is then shaken with a transverse periodic displacement $f \cos \omega t$ and the beam moves due to its inertia. In a (related) experiment one observes periodic motion about the two stable equilibria for small f but as f is increased, the motion becomes aperiodic or "chaotic". (See Moon and Holmes [12]). The mathematical problem is to develop theorems to explain this bifurcation.

There are a number of specific models that can be used to describe the beam in Figure 7. One such model is the following p.d.e. for the deflection $w(z,t)$ of the center line of the beam:

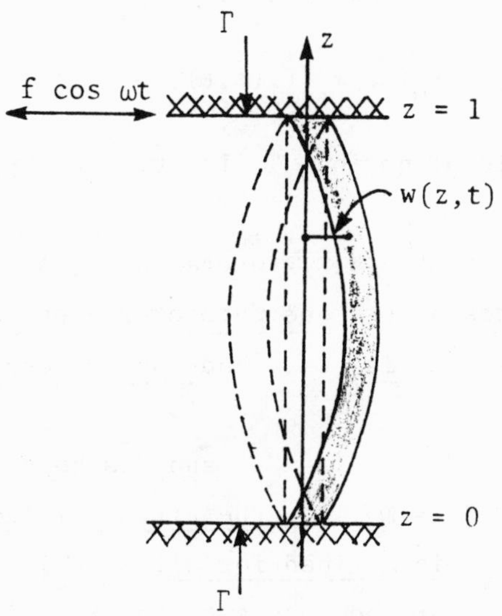

Figure 7. A buckled beam undergoing periodic forcing.

$$\ddot{w} + w'''' + \Gamma w'' - \kappa\left(\int_0^1 [w']^2 d\zeta\right) w'' = \varepsilon(f \cos \omega t - \delta\dot{w}) \tag{5.1}$$

where $\cdot = \partial/\partial t$, $' = \partial/\partial z$, Γ = external load, κ = stiffness due to "membrane" effects, δ = damping, and ε is a small parameter used to measure the relative size of f and δ. We use hinged boundary conditions: $w = w'' = 0$ at $z = 0, 1$. We assume the beam is in its first buckled state: $\pi^2 < \Gamma < 4\pi^2$.

A simpler model is obtained by looking for "lowest mode" solutions of the form $w(z,t) = x(t) \sin(\pi z)$. Substituting into (5.1), one finds the following Duffing type equation for x:

$$\ddot{x} - \beta x + \alpha x^3 = \varepsilon(\gamma \cos \omega t - \delta\dot{x}), \tag{5.2}$$

where

$$\beta = \pi^2(\Gamma - \pi^2) > 0, \quad \alpha = \kappa\pi^4/2, \; \gamma = 4f/\pi.$$

The methods used are inspired by Melnikov [11]; see Holmes [5] for an account. We shall set it up in an abstract fashion that applies to the above p.d.e.

We consider an evolution equation in a Banach space X of the form

$$\dot{x} = f_0(x) + \varepsilon f_1(x,t), \tag{5.3}$$

where f_1 is periodic of period T in t. Our hypothesis on (5.3) are as follows:

1. (a) Assume $f_0(x) = Ax + B(x)$ where A is an (unbounded) linear operator that generates a C^0 one parameter group of transformations on X and where $B: X \to X$ is C^∞, and has bounded derivatives on bounded sets.

(b) Assume $f_1: X \times S^1 \to X$ is C^∞ and has bounded derivatives on bounded sets, where $S^1 = \mathbb{R}/(T)$, the circle of length T.

(c) Assume that F_t^ε is defined for all $t \in \mathbb{R}$ for $\varepsilon > 0$ sufficiently small and F_t^ε maps bounded sets in $X \times S^1$ to bounded sets in $X \times S^1$ uniformly for small $\varepsilon \geq 0$ and t in bounded time-intervals.

Assumption 1 implies that the associated suspended autonomous system on $X \times S^1$,

$$\begin{cases} \dot{x} = f_0(x) + \varepsilon f_1(x,\theta), \\ \dot{\theta} = 1, \end{cases} \tag{5.4}$$

has a smooth local flow, F_t^ε, which can be extended globally in time, i.e. solutions do not escape to infinity in finite time. Energy estimates suffice to prove this for equation (5.1).

2. (a) Assume that the system $\dot{x} = f_0(x)$ (the unperturbed system) is Hamiltonian with energy $H_0: X \to \mathbb{R}$.

(b) Assume there is a symplectic 2-manifold $\Sigma \subset X$ invariant under the flow F_t^0 and that on Σ there is a fixed point p_0 and a homoclinic orbit $x_0(t)$, i.e.,

$$f_0(p_0) = 0, \ \dot{x}_0(t) = f_0(x_0(t))$$

and

$$p_0 = \lim_{t \to +\infty} x_0(t) = \lim_{t \to -\infty} x_0(t)$$

This means that X carries a skew symmetric continuous bilinear map $\Omega: X \times X \to \mathbb{R}$ that is weakly non-degenerate (i.e., $\Omega(u,v) = 0$ for all v implies $u = 0$) called the symplectic form and there is a smooth function $H_0 : X \to \mathbb{R}$ such that

$$\Omega(f_0(x), u) = dH_0(x) \cdot u$$

for all x in D_A, the domain of A. Consult Abraham and Marsden [1] for details about Hamiltonian systems.

The next assumption states that the homoclinic orbit through p_0 arises from a hyperbolic saddle.

3. Assume that $\sigma(Df_0(p_0))$, the spectrum of $Df_0(p_0)$, consists of two nonzero real eigenvalues $\pm\lambda$, with the remainder of the spectrum on the imaginary axis, strictly bounded away from 0. Assume that $\sigma(e^{tDf_0(p_0)})$, the spectrum of $e^{tDf_0(p_0)}$, equals the closure of $e^{t\sigma(Df_0(p_0))}$.

Consider the suspended system (5.4) with its flow $F_t^\varepsilon : X \times S^1 \to X \times S^1$. Let $P : X \to X$ be defined by

$$P^\varepsilon(x) = \pi_1 \cdot (F_T^\varepsilon(x,0))$$

where $\pi_1 : X \times S^1 \to X$ is the projection onto the first factor. The map P^ε is the Poincaré map for the flow F_t^ε. Note that $P^0(p_0) = p_0$, and that fixed points of P^ε correspond to periodic orbits of F_t^ε. One can prove that for $\varepsilon > 0$ small, there is a unique fixed point p_ε for P^ε near p_0; moreover p_ε is a smooth function of ε.

Our final hypothesis means in effect that the perturbation $f_1(x,t)$ is Hamiltonian plus damping. Using an assumption like 3, above, this condition can be stated either in terms of the spectrum of the linearization of equation (5.4) or in terms of the Poincaré map.

4. Assume that for $\varepsilon > 0$ the spectrum of $DP^\varepsilon(p_\varepsilon)$ lies strictly inside the unit circle with the exception of a single real eigenvalue $e^{T\lambda_\varepsilon^+} > 1$.

We remark that the fixed point p_0 perturbs to another fixed point p_ε for the perturbed system. The same is true for the local invariant manifolds [4] of the map P^ε, $W_\varepsilon^{ss}(p_\varepsilon)$ and $W_\varepsilon^u(p_\varepsilon)$, which remains C^r close to the unperturbed manifolds $W_0^s(p_0)$ and $W_0^u(p_0)$.

Here $W_\varepsilon^{ss}(p_\varepsilon) \subset W_\varepsilon^s(p_\varepsilon)$ and the superscript ss denotes the strong stable manifold Assumptions 3 and 4 guarantee that the center-stable manifold $W_0^{sc}(p_0)$ of the unperturbed system and the perturbed stable manifold $W_\varepsilon^s(p_\varepsilon)$ are codimension one, while the unstable manifolds are one dimensional. The flow in $X \times S^1$ similarly has a periodic orbit γ_ε, C^r close to $\{p_0\} \times S^1$, with invariant manifolds close to $W_0^s(p_0) \times S^1$, etc. One now proceeds to calculate the separation of the perturbed manifolds $W_\varepsilon^s(p_\varepsilon)$ and $W_\varepsilon^u(p_\varepsilon)$, by calculating the $O(\varepsilon)$ components of perturbed solution curves of equation (5.3) from the first variation equation of (5.3):

$$\frac{d}{dt} x_1^s(t,t_0) = Df_0(x_0(t-t_0))\, x_1^s(t,t_0) + f_1(x_0(t-t_0),t) \tag{5.5}$$

Here we have expanded solution curves in $W_\varepsilon^s(\gamma_\varepsilon)$; a similar expression holds for those in $W_\varepsilon^u(\gamma_\varepsilon)$. Points in $W_\varepsilon^s(p_\varepsilon)$ are obtained by intersecting $W_\varepsilon^s(\gamma_\varepsilon)$ with the section $X \times \{0\}$. This can also be done on general sections $X \times \{t_0\}$ and equation (5.5) contains t_0 as an initial starting time.

It is then possible to compute a function $M(t_0)$ which acts as a measure of the separation of the perturbed manifolds $W_\varepsilon^s(p_\varepsilon)$, $W_\varepsilon^u(p_\varepsilon)$ on different Poincaré sections $X \times \{t_0\}$.

Theorem. *Let hypotheses 1-4 hold. Let*

$$M(t_0) = \int_{-\infty}^{\infty} \Omega(f_0(x_0(t-t_0)),\ f_1(x_0(t-t_0),t))\, dt \tag{5.6}$$

Suppose that $M(t_0)$ has a simple zero as a function of t_0. Then for $\varepsilon > 0$ sufficiently small, the stable manifold $W_\varepsilon^s(p_\varepsilon(t_0))$ of p_ε for $P_{t_0}^\varepsilon$ and the unstable manifold $W_\varepsilon^u(p_\varepsilon(t_0))$ intersect transversally.

We refer to Holmes and Marsden [7] for the proof. We also have:

Theorem. *If the diffeomorphism $P_{t_0}^\varepsilon : X \to X$ possesses a hyperbolic saddle point p_ε and an associated transverse homoclinic point $q \in W_\varepsilon^u(p_\varepsilon) \pitchfork W_\varepsilon^s(p_\varepsilon)$, with $W_\varepsilon^u(p_\varepsilon)$ of dimension 1 and $W_\varepsilon^s(p_\varepsilon)$ of codimension 1, then some power of $P_{t_0}^\varepsilon$ possesses an invariant zero*

<u>dimensional hyperbolic set</u> Λ <u>homeomorphic to a Cantor set on which a power of</u> $P^{\varepsilon}_{t_0}$ <u>is conjugate to a shift on two symbols</u>.

This implies the following:

<u>Corollary</u>. <u>A power of</u> $P^{\varepsilon}_{t_0}$ <u>restricted to</u> Λ <u>possesses a dense set of periodic points, there are points of arbitrarily high period and there is a non-periodic orbit dense in</u> Λ.

We now briefly sketch some intuition behind this result.

If the hypotheses above hold, we end up with a Poincaré map $P^{\varepsilon}: X \to X$ that has a hyperbolic saddle point x'_{ε} which has a 1 dimensional unstable manifold intersecting a codimension 1 stable manifold transversally. For $X = \mathbb{R}^2$, this situation implies that the dynamics contains a Smale horseshoe. Figure 8 shows the situation in $\mathbb{R}^2$. The rectangle R gets squashed horizontally, stretched vertically and laid down as shown. A little thought shows that $\bigcap_{-\infty}^{\infty} (P^{\varepsilon})^N(R) = \Lambda$ is locally an interval × a cantor set. This structure is responsible for the complex dynamics. The account given in Smale [16] is very readable.

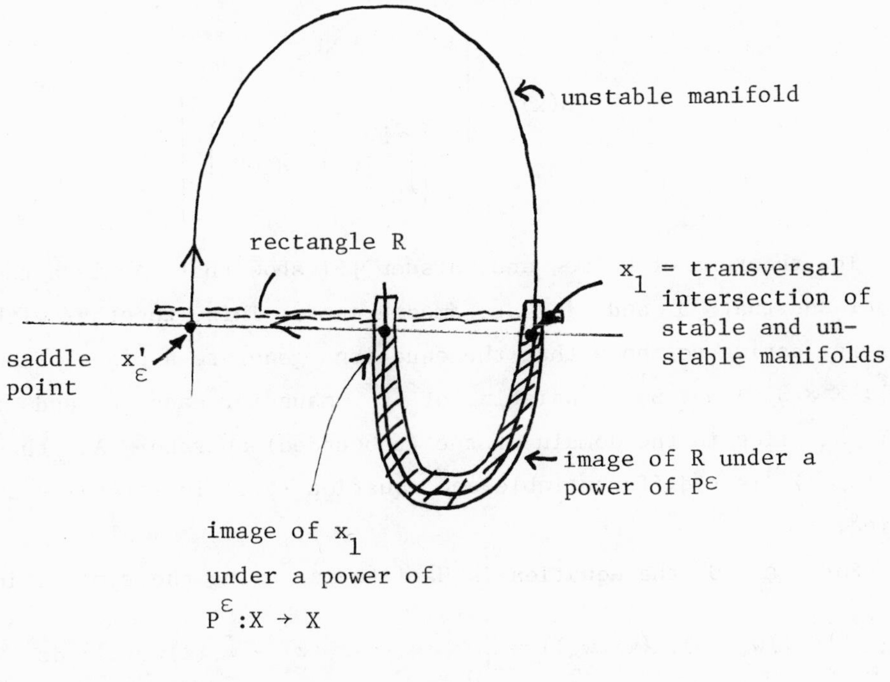

Figure 8. A Smale horseshoe

One can use these results to show that the beam equation (5.1) possesses horseshoes if the force, f, exceeds a certain critical level, dependent upon the damping, δ. The basic space is $X = H_0^2 \times L^2$ where H_0^2 denotes the set of all H^2 functions on $[0,1]$ satisfying the boundary condition $w = 0$ at $z = 0,1$. For $x = (w,\dot{w}) \in X$, the X-norm is the "energy" norm $\|x\|^2 = |w''|^2 + |w|^2$ where $|\cdot|$ denotes the L_2 norm. We write (5.1) in the form (5.3): where

$$f_0(x) = Ax + B(x) \quad \text{and} \quad f_1(x,t) = \begin{pmatrix} 0 \\ f\cos\omega t - \delta\dot{w} \end{pmatrix}.$$

Here A is the linear operator

$$A\begin{pmatrix} w \\ \dot{w} \end{pmatrix} = \begin{pmatrix} \dot{w} \\ -w'''' - \Gamma w'' \end{pmatrix},$$

with domain

$$D(A) = \{(w,\dot{w}) \in H^4 \times H^2 \,|\, w = w'' = 0 \text{ and } \dot{w} = 0 \text{ at } z = 0,1\}$$

and B is the nonlinear mapping of X to X given by

$$B(x) = \begin{pmatrix} 0 \\ \kappa\left(\int_0^1 [w']^2\, d\zeta w''\right) \end{pmatrix}.$$

The theorems of Holmes and Marsden [6] show that A is a generator and that B and f_1 are smooth maps. This, together with energy estimates shows that the equations generate a global flow $F_t^\varepsilon : X \times S^1 \to X \times S^1$ consisting of C^∞ maps for each ε and t. If x_0 lies in the domain of the (unbounded) operator A, then $F_t^\varepsilon(x_0,s)$ is t-differentiable and equation (5.1) is literally satisfied.

For $\varepsilon = 0$ the equation is Hamiltonian using the symplectic form

$$\Omega((w_1,\dot{w}_1),(w_2,\dot{w}_2)) = \int_0^1 \{\dot{w}_2(z)w_1(z) - \dot{w}_1(z)w_2(z)\}\, dz$$

and

$$H(w,\dot{w}) = \frac{1}{2}|\dot{w}|^2 = \frac{\Gamma}{2}|w'|^2 + \frac{1}{2}|w''|^2 + \frac{\kappa}{4}|w'|^4$$

The invariant symplectic 2 manifold Σ is the plane in X spanned by the functions $(a \sin \pi z, b \sin \pi z)$ and the homoclinic loop is given by

$$w_0(z,t) = \frac{2}{\pi}\sqrt{\frac{\Gamma-\pi^2}{\kappa}} \sin(\pi z) \operatorname{sech}(t\pi\sqrt{\Gamma-\pi^2})$$

For $\varepsilon = 0$ one finds by direct calculation that the spectrum of $Df_0(p_0)$, where $p_0 = (0,0)$, is discrete with two real eigenvalues

$$\pm\lambda = \pm\pi\sqrt{\Gamma-\pi^2}$$

and the remainder pure imaginary (since $\Gamma < 4\pi^2$) at

$$\lambda_n = \pm n\pi\sqrt{\Gamma-n^2\pi^2}, \quad n = 2,3, \ldots$$

The Melnikov function is given by

$$M(t_0) = \int_{\infty}^{\infty} \Omega\left(\begin{matrix} \dot{w} \\ -w'''' + \kappa|w'|^2 w'' - \Gamma w' \end{matrix} , \begin{matrix} 0 \\ f\cos\omega t - \delta\dot{w} \end{matrix}\right) dt$$

$$= \int_{\infty}^{\infty}\left(\int_0^1 f\cos\omega t\, \dot{w}(z,t-t_0) - \delta\dot{w}(z,t-t_0)\dot{w}(z,t-t_0)\,dz\right) dt$$

Substituting the expressions for w, $\dot{w}$ along the homoclinic orbit, the integral can be evaluated by contour methods to give

$$-M(t_0) = \frac{4\omega}{\pi}\sqrt{\frac{\Gamma-\pi^2}{\kappa}}\, f \frac{\sin(\omega t_0)}{\cosh\left(\frac{\omega}{2\sqrt{\Gamma-\pi^2}}\right)} + \frac{4\delta(\Gamma-\pi^2)^{3/2}}{3\pi\kappa}$$

If

$$|f| > \frac{4\delta(\Gamma-\pi^2)}{3\omega\delta\sqrt{K}} \left\{ \cosh\left(\frac{\omega}{2\sqrt{\Gamma-\pi^2}} \right) \right\}$$

Then $M(t_0)$ has simple zeros, so the stable and unstable manifolds intereact transversally. Note that in [5] the integral was given incorrectly.

This shows that there is a complicated invariant hyperbolic Cantor set Λ embedded in the Poincaré map of equation 5.1 for a calculable open set of parameter values. Although the dynamics near Λ is complex, we do not assert that Λ is a strange attractor i.e. that Λ is a structurally stable attractor. In fact, Λ is unstable in the sense that its generalized unstable manifold (or outset), $W^u(\Lambda)$ is non-empty (it is one dimensional and thus points starting near Λ may wander, remaining near Λ for a relatively long time, but eventually leaving a neighborhood of Λ and approaching an attractor. This kind of behavior has been referred to as transient chaos (or pre-turbulence). In two dimensions, Λ can coexist with two simple sinks of period one or with a strange attractor, depending on the parameter values. There is experimental evidence for transient chaos in the magnetic cantilever problem (Moon and Holmes [12]).

We close with a comment on the bifurcations in which the transversal intersections are created. Since the Melnikov function $M(t_0)$ has nondegenerate maxima and minima, it can be shown that, near the parameter values at which $M(t_0) = M'(t_0) = 0$, but $M''(t_0) \neq 0$, the stable and unstable manifolds $W_\varepsilon^s(p_\varepsilon(t_0))$. $W_\varepsilon^u(p_\varepsilon(t_0))$ have quadratic tangencies. This "Newhouse" mechanism implies that $P_{t_0}^\varepsilon$ can have infinitely many stable periodic orbits of arbitrarily high periods near the bifurcation point, at least in the finite dimensional examples . In practice it may be difficult to distinguish these long period stable periodic points from transient chaos and from "true" chaos itself. In fact it is not yet understood what role the Newhouse sinks play in the experimental and computer generated chaotic motions.

REFERENCES

[1] R. Abraham and J. Marsden [1978], Foundations of Mechanics, 2nd ed., Benjamin; New York.

[2] R. Abraham and J. Robbin [1967], Transversal Mappings and Flows, Benjamin, New York, (Kelley's proof of the "Stable, Center-Stable, Center, Center Unstable, and Unstable Manifolds" appears in Appendix C, pp. 134-154.)

[3] Dowell, E.H. [1975], "Aeroelasticity of Plates and Shells" Noordhoff.

[4] Hirsch, M.W., Pugh, C. and Shub, M. [1977], Invariant Manifolds Lec. Notes Math., 583, Springer, New York.

[5] P. Holmes [1980], "Averaging and chaotic motions in forced oscillations," SIAM J. Appl. Math., 38, pp. 65-80.

[6] Holmes P.J. and Marsden, J.E. [1978], "Bifurcation to divergence and flutter in flow induced oscillations.... An infinite dimensional analysis," Automatica 14, 367-384.

[7] ______________. "A partial differential equation with infinitely many periodic orbits: chaotic oscillations of a forced beam," (preprint).

[8] Marsden, J.E. [1978], "Qualitative Methods in Bifurcation Theory; Bull. Am. Math. Soc., 84, pp. 1125-1148.

[9] Marsden, J.E., Ebin, D., and Fischer, A., [1970], "Diffeomorphism groups, hydrodynamics and relativity," Proc. 13th Biennial seminar of Canadian Math. Congress, ed., J.R. Transtone, Montreal, pp. 135-275.

[10] Marsden, J.E. and McCracken, M., [1976], The Hopf Bifurcation and its Applications, Springer, New York.

[11] Melnikov, V.K., [1963],"On the stability of the center for time periodic perturbations." Trans. Moscow Math. Soc. 12, 1-57.

[12] Moon, F.C. and Holmes, P.J., [1979], "A magnetoelastic strange attractor," J. Sound. and Vibration , 65, 275-296.

[13] Parks, P.C., [1966], "A stability criterion for a panel flutter problem via the second method of Lyapunov, Differential Equations and Dynamical Systems," J.K. Hale and J. P. LaSalle (Eds.) Academic Press, New York.

[14] T. Ratiu [1979], On the smoothness of the time t map of the KdV equation and the bifurcation of the eigenvalues of Hill's operator, Springer lecture notes 755, 248-294.

[15] Segal, I. [1962],"Nonlinear semigroups," Ann. of Math. (2) 78, pp. 339-364.

[16] Smale, S. [1967],"Differentiable dynamical systems," Bull. Am. Math. Soc. 73: 747-817.

[17] Y.H. Wan [1977] ,"On the uniqueness of invariant manifolds," J. Diff. Equations, 24, 265-273.

Computation of Invariant Manifolds

Brian D. Hassard*

Abstract. In the center manifold approach to Hopf bifurcation theory, quadratic terms in the Taylor expansion of the center manifold are involved in determining the amplitude, stability and period of the bifurcating family of periodic solutions. A portion of the code BIFOR2, which analyzes the Hopf bifurcation in an autonomous system of ordinary differential equations, "computes" a center manifold in the sense of generating a partial Taylor expansion. A more general version of this portion of BIFOR2 generates up to cubic terms in the Taylor expansions of the stable and unstable manifolds associated with isolated equilibrium points of such systems. An application of these latter expansions is described: homoclinic orbits are approximated for large positive (resp. negative) t by solutions of the lower order systems obtained by restricting the original system to the approximate stable (resp. unstable) manifolds, and for intermediate t by solutions of a two-point boundary value problem. The advantage of using higher order than linear approximation to the manifolds is discussed in the context of an example, the computation of an accurate traveling wave solution of the Hodgkin-Huxley model nerve conduction system.

1. Introduction. The title "Computation of Invariant Manifolds" could cover a vast range of topics, so let me first narrow down the material to be discussed. Taylor expansions of center, stable, and unstable manifolds associated with isolated equilibrium points of autonomous ordinary differential systems will provide the common theme of this talk. Coefficients of these expansions are, in fact, computed in the programs to be described. However, the objects of primary interest are periodic, homoclinic and heteroclinic orbits of the systems of o.d.e.'s, rather than the invariant manifolds themselves. This talk

*Department of Mathematics, 106 Diefendorf Hall, State University of New York at Buffalo, Buffalo, N. Y. 14214. This work was partially supported by grant MCS-7905790 from the National Science Foundation.

will outline some of the ways in which expansions of invariant manifolds contribute to the computation of these orbits.

Y. H. Wan introduced me to center manifolds and to Kelley's theorem [1] as a tool for reducing the analysis of the Hopf bifurcation in R^n to study of a two-dimensional system. At the time, I was looking at the periodic solutions that occur in the Hodgkin-Huxley model current clamped nerve conduction equations [2]. The appeal of the center manifold approach was largely the fact that it helped organize the analysis into a number of straightforward steps, each easily checked, so that one could have confidence in the results. The computer program written for the study [2] and based upon the derivation of bifurcation formulae [3] has since evolved, dropping symbolic manipulation in favor of numerical differentiation [4], then becoming more efficient in execution and sparing in the use of array space [5]. For those interested in computing the Hopf bifurcation, the present version is called BIFOR2, is in ANSI X3.9 (1966) FORTRAN, uses $2n^2 + 15n$ locations of array storage space, and is available.

One of the stages in the center manifold approach to bifurcation theory is expansion of a center manifold. For the Hopf bifurcation, it usually suffices to expand to quadratic terms, and this is the sense in which center manifolds are computed in BIFOR2. The quadratic terms in the expansion contribute to the cubic terms in the system obtained by restricting the n-dimensional system to the center manifold, and hence play a role in determining the amplitude, stability and period of the bifurcating family of periodic solutions.

Recently, that portion of the bifurcation analysis code devoted to center manifold expansion has been adapted to compute up to cubic terms in the Taylor expansions of the stable and unstable manifolds associated with isolated equilibrium points of autonomous ordinary differential systems, and for much of this talk we will discuss the role of such expansions in the computation of homoclinic and heteroclinic orbits. As an example, we shall describe the computation of an accurate homoclinic orbit (traveling wave profile) for the Hodgkin-Huxley model nerve conduction equations.

2. The Hopf Bifurcation and Center Manifold Expansion. The Hopf bifurcation for an autonomous system

$$\dot{x} = f(x; \nu), \quad x \in R^n$$

occurs when an isolated equilibrium point $x = x_*(\nu)$ loses linear stability due to a complex conjugate pair of eigenvalues

$$\lambda_1(\nu) = \bar{\lambda}_2(\nu) = \alpha(\nu) + i\,\omega(\nu)$$

of the Jacobian matrix $A(\nu) = \partial f/\partial x(x_*(\nu);\, \nu)$. That is, $x_*(\nu)$ becomes unstable as ν crosses ν_c, a critical value of the bifurcation parameter ν , because $\alpha(\nu)$ becomes positive. The conditions are,

$$\alpha(\nu_c) = 0 \ , \quad \alpha'(\nu_c) \neq 0 \ , \quad \omega(\nu_c) > 0 \ ,$$

$$\mathrm{Re}\ \lambda_j(\nu_c) < 0 \quad (j=3,\ldots,n) \ .$$

Under certain smoothness conditions (f jointly C^4 in x and ν will suffice, [5]) there exists a family of periodic solutions $p_\epsilon(t)$, one for each ϵ in an interval $(0, \epsilon_1)$, such that $p_\epsilon(t)$ occurs for $\nu = \nu_c + \mu(\epsilon) = \nu_c + \mu_2\epsilon^2 + O(\epsilon^3)$. Explicitly,

$$p_\epsilon(t) = x_*(\nu) + \mathrm{Re}\,\{\, v_1 z + w_{20} z^2 + w_{11} z\bar{z} \,\} + O(|z|^3)$$

where

$$z = \xi + \chi_{20}\, \xi^2/2 + \chi_{11}\, \xi\bar{\xi} + \chi_{02}\bar{\xi}^2/2 + O(|\xi|^3)$$

$$\xi = \epsilon\ \exp(2\pi i t/T(\epsilon))$$

$$T(\epsilon) = (2\pi/\omega(\nu_c))(\,1 + \tau_2\epsilon^2 + O(\epsilon^3)) \ .$$

One of the characteristic exponents associated with $p_\epsilon(t)$ is given by

$$\beta(\epsilon) = \beta_2\epsilon^2 + O(\epsilon^3) \ .$$

The most important number is the coefficient μ_2 , for if nonzero it determines the direction of bifurcation, i.e. whether the periodic solutions occur for $\nu > \nu_c$ or for $\nu < \nu_c$. The next most important coefficient is $\beta_2 = -\,2\alpha'(\nu_c)\mu_2$, for if $\beta_2 > 0$, $p_\epsilon(t)$ is unstable whereas if $\beta_2 < 0$, $p_\epsilon(t)$ is orbitally, asymptotically stable with asymptotic phase.

In the center manifold approach to Hopf bifurcation theory, the family of periodic solutions is seen as sitting in a family of two-dimensional manifolds within R^n, having Taylor expansions

$$x_*(\nu) + \mathrm{Re}\{ v_1(\nu)z + w_{20}(\nu)z^2 + w_{11}(\nu)z\bar{z} \} + O(|z|^3)$$

where $z = y_1 + iy_2$. (If desired, the expansions may be written strictly in terms of the real variables y_1 and y_2.) The family of two-dimensional manifolds collectively represents a three-dimensional center manifold for the suspended system

$$\dot{x} = f(x; \nu), \qquad \dot{\nu} = 0 \ .$$

Existence and smoothness of a center manifold for the suspended system follow from the Center Manifold Theorem [1, 8]. The vector $v_1(\nu)$ is the eigenvector of $A(\nu)$ for eigenvalue $\lambda_1(\nu)$, while the vectors $w_{20}(\nu) \in C^n$ and $w_{11}(\nu) \in R^n$ are found as solutions of a certain pair of linear algebraic systems.

The program BIFOR2 first locates ν_c by solving the algebraic equation $\alpha(\nu) = 0$. Then BIFOR2 computes all of the coefficients in the approximation to the family of periodic solutions shown above, for $\nu = \nu_c$. In particular, BIFOR2 computes the quadratic terms in the expansion of the slice $\nu = \nu_c$ of the center manifold, which contribute to the cubic terms in the two-dimensional o.d.e. obtained by restricting the original system to the center manifold, and hence are involved in computing μ_2, τ_2, and β_2.

For further discussions of center manifolds and bifurcation theory, see the papers by Marsden and by Guckenheimer in the present volume, as well as the books [5] and [6]. Numerous examples and exercises appear in [5]: examples worked by hand are given in Chapter 2 and applications of BIFOR2 are described in Chapter 3. Also in [5], the same approach used for o.d.e.'s is carried through for differential-delay equations (Chapter 4), for certain reaction-diffusion equations (Chapter 5), and a proof of the Center Manifold Theorem is included (Appendix A).

Several interesting physical examples of Hopf bifurcation appear in the present volume. For a simple experimental setup which clearly exhibits the phenomena, see the paper by Sethna. The only equipment required is a water faucet and a short piece of flexible hose! (In the Asilomar talk, at this point work in progress by I. M. El Henawy was shown. Mr. El Henawy has used BIFOR2 to analyze the Hopf bifurcation that occurs in a catalyst particle system as the Lewis number is decreased past a critical value. For the parameter values considered, stable, small amplitude periodic solutions appear. To apply BIFOR2, the partial differential system was first reduced to an approximating ordinary differential system of order $n = 50$, by discretizing the space variable.)

3. Stable and Unstable Manifolds, and Traveling Waves. Many of the ideas we shall discuss may be found in recent work by Miura [7], and we have little doubt that the ideas have also occurred to others. In the past, however, there have been technical obstacles to the practical application of manifold expansion techniques in all but very simple systems. In developing a code for analysis of the Hopf bifurcation, we met and overcame these obstacles: one is no longer restricted to low order systems with relatively simple nonlinearities. (See Appendix E of [5] for a brief history of BIFOR2.)

As motivation for considering stable and unstable manifolds, consider first the problem of computing traveling wave (hereinafter TW) solutions of one dimensional partial differential equations of reaction-diffusion type. A TW variable $t = \xi + c\tau$ is introduced, where τ is time, ξ is the space coordinate and c the assumed wavespeed. The profile of the TW is then obtained as either a homoclinic orbit, or a heteroclinic orbit, of an autonomous system

$$\dot{x} = f(x; c) . \tag{1}$$

Whether a homoclinic or heteroclinic orbit is sought depends upon whether the TW is to meet the same or different conditions "upstream" or "downstream", i.e. as $\xi \to -\infty$ and as $\xi \to \infty$. For simplicity in the following, we shall consider just the homoclinic case, so the orbit $x(t; c)$ sought must obey $x \to x_*(c)$ as $|t| \to \infty$, where $x_*(c)$ is an equilibrium point for the system (1).

The stable manifold $S(c)$ for (1) at $x_*(c)$ may be defined as the set of initial conditions x_0 at $t = 0$ for which the corresponding trajectories of (1) exist for all positive t and tend to $x_*(c)$ as $t \to \infty$. Similarly, the unstable manifold $U(c)$ for (1) at $x_*(c)$ may be defined as the set of initial conditions x_0 at $t = 0$ for which the corresponding trajectories of (1) exist for all negative t and tend to $x_*(c)$ as $t \to -\infty$. The homoclinic orbit problem is then equivalent to determining a value of $c = c_0$ such that $S(c)$ and $U(c)$ intersect in more than just the single point $x_*(c)$.

4. Expansion of Stable and Unstable Manifolds. The characterization of stable and unstable manifolds given above is simple and geometric, however does not lend itself to computations as readily as the algebraic characterization of these same manifolds which we shall now outline.

Suppose that the eigenvalues of the Jacobian matrix for the system (1) at the equilibrium point $x_*(c)$ obey

$$\text{Re}\,\lambda_1 \geq \text{Re}\,\lambda_2 \geq \cdots \geq \text{Re}\,\lambda_m > 0 > \text{Re}\,\lambda_{m+1} \geq \cdots \geq \text{Re}\,\lambda_n. \tag{2}$$

Let $P(c)$ denote a real matrix, the first m columns of which span the union of the generalized eigenspaces for λ_1 through λ_m and the last $n-m$ columns of which span the union of the generalized eigenspaces for λ_{m+1} through λ_n. Upon changing variables according to

$$x = x_*(c) + P(c)\begin{vmatrix} y \\ z \end{vmatrix} \tag{3}$$

where $y \in R^m$ and $z \in R^{n-m}$, (1) becomes

$$\begin{aligned} \dot{y} &= A(c)\,y + g(y, z; c) \\ \dot{z} &= B(c)\,z + h(y, z; c) \end{aligned} \tag{4}$$

where $g = h = g_y = g_z = h_y = h_z = 0$ at $y = z = 0$. Under smoothness conditions on the system (4) (for example, C^{L+1} jointly in y, z and c for some $L \geq 1$) there now exists [1,8] a vector-valued function $Y(z; c)$ which represents the stable manifold in some neighborhood of $z = 0$. The function is defined locally, is unique, and inherits smoothness properties from the system (4). Furthermore, $Y(0; c) = 0$, $Y_z(0; c) = 0$, and any orbit of (4) which tends to the origin as $t \to \infty$ may be written as $(Y(z(t); c), z(t))$ for all sufficiently large t, where $z(t)$ satisfies the $(n-m)$th order system

(5) $$\dot{z} = B(c)\, z + h(Y(z;c), z; c) .$$

The eigenvalues of $B(c)$ are λ_{m+1} through λ_n, and all have negative real parts. Note that the stable manifold may be understood intuitively as follows: the variable z "wants" to decay, while y "wants" to grow. The solutions $(y(t), z(t))$ of interest are the decaying ones. To obtain these solutions, y is "forced" to decay by being expressed as a function of z.

If trajectories of (1) with the property $x(t; c) \to x_*(c)$ as $t \to \infty$ are desired, the large t portion of these trajectories may, in principle, be computed by integrating (5) forwards in t. The practical difficulty is that the function $Y(z; c)$ is not known very well, in general. The best one can usually do is the first few terms in the Taylor expansion.

Since $y(t; c) = Y(z(t;c); c)$ satisfies (4),

$$\dot{y} = Y_z\, \dot{z} = Y_z\, [\, B(c)\, z + h(Y(z; c), z; c)\,]$$

and so $Y(z; c)$ must satisfy the partial differential equation

(6) $$A\,Y - Y_z\, B\, z = Y_z\, h(Y, z; c) - g(Y, z; c) .$$

Consider the Taylor expansion

(7) $$Y(z; c) = \Sigma_2^M\, Y_k(z; c) + O(\|z\|^{M+1})$$

where $M \le L$ and each function $Y_k(z; c)$ is an homogeneous polynomial of combined degree k in the variables $z_1, z_2, \ldots z_{n-m}$. The coefficients of the polynomials are real m-vectors which may be obtained by solving the systems of linear algebraic equations which result when the expansion (7) is set in (6). The solution scheme must proceed by solving first for the coefficients of $Y_2(z; c)$, then for those of Y_3 and so forth because the forcing terms (the right hand sides) in the algebraic systems for the coefficients of Y_k, $(k \ge 3)$, involve the coefficients of the polynomials Y_p for $2 \le p \le k$. The total dimension of the linear system to be solved for the coefficients of Y_p is $(m)[(p+n-m-1)!]/[p!(n-m-1)!]$ and the key to keeping the amount of linear algebra reasonable lies in choosing the matrix P so that the matrices A and B are in real canonical form. This breaks the linear system down into a set of decoupled linear systems of low dimension. Although straightforward, the procedure just outlined tends to be tedious if performed by hand. To use our FORTRAN code which carries the expansion to cubic terms, one need only supply a subroutine to evaluate the function $f(x; c)$ and the Jacobian matrix $\partial f/\partial x(x; c)$.

5. Approximation by Trajectories of the Restriction of the Original System to the Approximate Stable Manifold. Let

$$Y^M(z; c) = \Sigma_2^M Y_k(z; c) . \tag{8}$$

Then, one may integrate the system

$$\dot{z} = B(c) z + h(Y^M(z; c), z; c) \tag{9}$$

in order to approximate trajectories of (5). For any vector $v \in R^{n-m}$, define the (Lyapunov) norm

$$\|v\|_B^2 = \int_0^\infty \|e^{\tau B} v\|^2 \, d\tau . \tag{10}$$

(This norm is actually easy to compute, provided B is in real canonical form and λ_{m+1} through λ_n are distinct. If so, the norm (10) is nothing more than a weighted Euclidean norm.) One result concerning the approximation of trajectories of (5) by means of those of (9) is then, for all sufficiently small $\|z_0\|_B$, the initial value problems consisting of (5) and (9) with the same initial conditions $z(t_0) = z_0$, each have unique solutions, say $z_5(t)$ and $z_9(t)$, which exist for all $t \geq t_0$, tend to 0 as $t \to \infty$ and furthermore obey

$$\max_{t \geq t_0} \| z_5(t) - z_9(t) \|_B = O(\|z_0\|^{M+1}) . \tag{11}$$

(Much more precise results are easily obtained, but (11) will suffice for present purposes.)

The advantage of integrating (9) rather than (1) (or equivalently, (4)) is that $z = 0$ is a stable equilibrium point for (9), whereas $x = x_*(c)$ has an associated unstable manifold. Thus one can compute the portion, t large and positive, of a homoclinic orbit for (1), without having to integrate "against" the unstable manifold. The restriction to large t is necessary, because the expansion (7) of the stable manifold is valid only in a neighborhood of the equilibrium point. As more terms are included in the expansion, the expectation is that the neighborhood within which the expansion provides an adequately accurate approximation to the stable manifold, will become larger. However, practical constraints (the growth in the number of distinct coefficients and the requirement of approximating higher order partial derivatives, for two) limit the expansion to a few terms.

6. Joining Trajectories of (1) to the Approximate Stable Manifold.

Away from the equilibrium point, then, trajectories must be computed as solutions of (1). The question arises, how to patch these trajectories to the approximate stable manifold (8) . It is a result from Lyapunov's theory [8] , that as $t \to \infty$, $\|z(t)\|_B$ decays monotonically to 0 , where z is a solution either of (5) or of (9) . (Indeed, this property is the reason for the choice of norm (10).) Thus, for all sufficiently small δ , if $x(t; c)$ is any trajectory of (1) which tends to $x_*(c)$ as $t \to \infty$, the trajectory necessarily crosses through the set

$$\{ x_*(c) + P(c) \begin{vmatrix} Y(z; c) \\ z \end{vmatrix} , \|z\|_B = \delta \}$$

in exactly one point. The set

$$B^s(\delta, M; c) = \{ x_*(c) + P(c) \begin{vmatrix} Y^M(z; c) \\ z \end{vmatrix} , \|z\|_B = \delta \} \quad (12)$$

is therefore an appropriate place to join numerical solutions of (1) to the approximate stable manifold.

We discussed the stable manifold first only because in Lyapunov's theory, the limit $t \to \infty$ is more familiar than the limit $t \to -\infty$. Under the same condition (2) on the eigenvalues, there also exists a vector-valued function $Z(y; c)$ which represents the unstable manifold for (4) in a neighborhood of 0 , and hence the unstable manifold for (1) in a neighborhood of $x_*(c)$. Z may be expanded as

$$Z(y; c) = \Sigma_2^N Z_k(y; c) + O(\|y\|^{N+1}) \quad (13)$$

where $N \le L$ and $Z_k(y; c)$ is a homogeneous polynomial of degree k in the variables $y_1, y_2, \ldots y_m$ with coefficients in R^{n-m} . Let

(14) $$Z^N(y;\ c) = \Sigma_2^N Z_k(y;\ c)$$

and for any vector u in R^m, let

(15) $$\|u\|_A^2 = \int_{-\infty}^0 \|e^{\tau A} u\|^2 \, d\tau \ .$$

Then, for all sufficiently small ϵ , if $x(t;\ c) \to x_*(c)$ as $t \to -\infty$, the trajectory must cross the set

$$\{\ x_*(c) + P(c) \begin{vmatrix} y \\ Z(y;\ c) \end{vmatrix} ,\ \|y\|_A = \epsilon\ \}$$

in exactly one point, and the set

(16) $$B^u(\epsilon,\ N;\ c) = \{\ x_*(c) + P(c) \begin{vmatrix} y \\ Z^N(y;\ c) \end{vmatrix} ,\ \|y\|_A = \epsilon\ \}$$

is an appropriate place to join the numerical solutions of (1) to the approximate unstable manifold.

7. Reduction of the Homoclinic Orbit Problem to a Two-Point Boundary Value Problem. The entire homoclinic orbit may thus be approximated in 3 parts,

i) for $t \le t_A$, by $x_*(c) + P(c) \begin{vmatrix} y \\ Z^N(y;\ c) \end{vmatrix}$, where $y = y(t;\ c)$ satisfies

(17) $$\dot{y} = A(c)\ y + g(y,\ Z^N(y;\ c);\ c) \ ,$$

ii) for $t_A \le t \le t_B$, by a solution of $\dot{x} = f(x;\ c)$,

iii) for $t \ge t_B$, by $x_*(c) + P(c) \begin{vmatrix} Y^M(z;\ c) \\ z \end{vmatrix}$, where $z = z(t;\ c)$ satisfies

(18) $$\dot{z} = B(c)\ z + h(Y^M(z;\ c),\ z;\ c) \ .$$

The joining conditions are, $\|y(t_A;\ c)\|_A = \epsilon$, $x(t_A;\ c) \in B^u(\epsilon,\ N;\ c)$, $x(t_B;\ c) \in B^s(\delta,\ M;\ c)$, $\|z(t_B;\ c)\|_B = \delta$. Since the systems are autonomous, the origin in t may be shifted and one or other of the pair t_A , t_B may be defined arbitrarily, say be setting $t_A = 0$.

The portion (ii) of the homoclinic orbit is to be obtained by solving a two-point boundary value problem, a task for which several codes are available [9] . The condition on x at t_A represents $n + 1 - m$ distinct boundary conditions and the condition at t_B represents $m + 1$ distinct boundary conditions. When the parameter c and the length $L = t_B - t_A$ of the interval are taken as additional unknowns, the order of the augmented system $\dot{x} = f(x; c)$, $\dot{c} = 0$, $\dot{L} = 0$ agrees with the number of independent boundary conditions. Once the two-point boundary value problem has been solved, the initial conditions for integrating the ends i) and iii) of the homoclinic orbit are determined.

For simplicity in the following discussion, we shall set $\delta = \varepsilon$ and $M = N$. Assume that the two-point BVP can be solved for all sufficiently small ε , and let $x = x_\varepsilon(t)$, $c = c_\varepsilon$, $L = L_\varepsilon$ denote the solution. Assume further that (1) has a homoclinic orbit $x_0(t)$ which occurs for $c = c_0$, and that $c_\varepsilon \to c_0$, $L_\varepsilon \to \infty$, and

$$\inf_{\phi} \sup_{0 \le t \le L_\varepsilon} \| x_\varepsilon(t) - x_0(t - \phi) \| \to 0$$

in the limit $\varepsilon \to 0$. We now conjecture, that in fact

$$c_\varepsilon = c_0 + O(\varepsilon^{N+1}) ,$$

$$\inf_{\phi} \sup_{0 \le t \le L_\varepsilon} \| x_\varepsilon(t) - x_0(t - \phi) \| = O(\varepsilon^{N+1}) .$$

The power ε^{N+1} in these conjectures arises from the power $\|z_0\|_B^{N+1}$ in the local approximation (11) and is supported by numerical evidence as we shall see below. When the tangent space approximations to the stable and unstable manifolds are employed, our conjecture is that the global truncation error is $O(\varepsilon^2)$. When our code for expansion of these manifolds is employed, $N = 3$ and the global truncation error is expected to be $O(\varepsilon^4)$.

8. Application to Nerve Conduction Theory. In a class of homoclinic orbit problems derived from nerve conduction theory [10,11] , the unstable manifold is one-dimensional and the traditional means [12,13] for computing the desired orbit has been simple shooting forwards in t , with initial conditions taken as

$$x_*(c) \pm \varepsilon\, v_1(c) \tag{19}$$

where $v_1(c)$ is the eigenvector of the Jacobian matrix for (1) at $x_*(c)$ which corresponds to the positive eigenvalue $\lambda_1(c)$ and c is an estimate of the wavespeed c_0 for which the homoclinic orbit exists. (Strictly speaking, the "traditional means" uses an ad hoc approximation to $v_1(c)$.) The eigenvector is normalized in some fashion, and ϵ taken small. The sign $\pm$ is chosen so that the orbit proceeds away from the equilibrium point in the direction dictated by the physical model. The basis for the shooting procedure is the observation that, if c is too large, the computed trajectory diverges in one fashion, while if c is too small, the trajectory diverges in a distinct fashion. A bisection scheme for the wavespeed is thus obtained. It is a property of the nerve conduction systems, that an approximate wavespeed may be computed with bisection as outlined, to machine precision on currently available machines, without the computed trajectories necessarily returning at some $t = t_B$ to a "small" neighborhood of $x_*(c)$. The truncation error in the computed wavespeed is then due entirely to approximating the unstable manifold. (We are ignoring the truncation error associated with the use of numerical integration.) Since (19) is the tangent space approximation to the unstable manifold, we expect that $c_\epsilon = c_0 + O(\epsilon^2)$. We further expect that when the initial condition (19) is replaced with $x(0; c) \in B^u(\epsilon, N; c)$, the computed wavespeed $c = c(\epsilon, N)$ will satisfy $c(\epsilon, N) = c_0 + O(\epsilon^{N+1})$.

Miura [7] recently computed an accurate homoclinic orbit for the Fitzhugh-Nagumo model nerve conduction system, including "nonlinear correction terms" in the initial condition. In the present terminology, he expanded the unstable manifold and investigated the truncation error in the computed wavespeed $c(\epsilon, N)$ as ϵ and N are varied. The results he obtained, table 4.1 of [7], show clearly an $O(\epsilon^{N+1})$ truncation error.

The Fitzhugh-Nagumo system is simple enough that one can carry through the manifold expansion procedure by hand. For the Hodgkin-Huxley system, however, hand calculations are excessively tedious, (as we found out in the bifurcation analysis of the study [2]). To demonstrate our code for manifold expansions, we computed a homoclinic orbit for the Hodgkin-Huxley system and investigated the truncation error due to the initial conditions in a manner analogous to Miura's study of the Fitzhugh-Nagumo system. We took the system for $T = 6.3^\circ$ C , and used the same normalization as Evans and Feroe [14] . Gears method, IMSL subroutine DVOGER, was used to integrate (1) . The single step error was adjusted so that the truncation error from the numerical integration does not affect any of the digits in the results given below. The initial conditions used were $x(0) \in B^u(\epsilon, N; c)$, where the matrix $P(c)$ was formed from the real and imaginary parts of the eigenvectors of the Jacobian matrix for (1) at $x_*(c)$, the eigenvectors being normalized to have first component 1 . The computed wavespeeds $c(\epsilon, N)$ are given in the following table. Correct digits in the wavespeeds are underlined.

Computed Wavespeed $c(\epsilon, N)$ for the Hodgkin-Huxley System

ϵ	$c(\epsilon, 1)$	$c(\epsilon, 2)$	$c(\epsilon, 3)$
10	2.03698707	2.10627786	2.11859117
1	2.12340178	2.12384432	2.12385678
.1	2.12385546	2.12385727	2.12385728
.01	2.12385727	2.12385728	2.12385728

From this table, we infer that the wavespeed in $c_0 = 2.12385728$, with an uncertainty of one digit in the last place. The wavespeed result in [14], although quoted to 23 significant figures, is only accurate to 6 figures; the discrepancy may be due to truncation error from using initial conditions of the form (19) .

The table above exhibits an $O(\epsilon^{N+1})$ truncation error. Note that for $N = 3$, the computation is more efficient than for $N = 1$ or $N = 2$ because comparably accurate wavespeeds are obtained with larger values of ϵ , i.e. for lesser amounts of numerical integration, and in this particular computation, the numerical integration is far more expensive than the manifold expansion. The reduction in the amount of numerical integration also diminishes the contribution of the numerical integration to the overall error, and thereby improves the attainable accuracy. For this particular computation, the increase in attainable accuracy is not especially important, however it would be if (for example) the unstable manifold were two-dimensional and one of the eigenvalues of $A(c)$ greatly exceeds the other in magnitude.

If the profile of the TW is desired as well, manifold expansions help resolve a difficulty that has plagued studies of this TW to date. The difficulty is the computation of the return to the rest state. In [14] for example, the stability analysis of the TW is not fully convincing because the basic TW profile was not known beyond 7 units of t past the voltage maximum. The decay to rest is evidently "slow", and the strong influence of the unstable manifold makes the integrations forwards in t very sensitive to perturbations. The ability of the cubic expansion of the stable manifold to represent this manifold in a larger neighborhood than is possible with the tangent space approximation, thus produces a considerable saving in the amount of computational effort required to generate trajectories which return to a "small" neighborhood of the rest state. We have computed an accurate single-pulse TW solution in this manner, and intend to analyze the stability of this solution and of a double pulse solution recently found.

9. Partial Rectification Maps. In addition to constructing manifold expansions, our code also constructs what we have been calling "partial rectification maps". These maps effect transformations of the systems (9) and (17) so that in the new variables, the nonlinear terms only begin to appear at higher order. The idea of such nonlinear transformations of systems of o.d.e.'s in a neighborhood of an equilibrium point may be traced back to Poincaré, and Arnol'd [15] attributes the basic theorem to Poincaré and DuLac. The order to which the nonlinearity can be moved depends upon conditions on the eigenvalues, conditions which are checked by the code. Technically, the construction of "partial rectification maps" is similar to the construction of manifold expansions, and much of the same computer code is used for both purposes. Approximate solutions of the partially rectified systems may then be obtained by solving the linear part of these systems alone, and the approximate solutions, when expressed in terms of the original variables provide analytic approximations to solutions of (9) and (17). Ideally, the partial rectification maps are constructed to the same order as the corresponding manifold expansions, so that the analytic approximation obtained has the same approximation property (11) . In particular, for the Hodgkin-Huxley TW system, our code will construct

approximations with error terms $O(\varepsilon^4)$ for the "ends" of the TW profile. One use for these approximations is the computation of the end portions of Miura's "wave integrals" [7] in order to verify the accuracy of the overall solution.

10. Summary. In summary, expansions of center manifolds have a secure place in the analysis of the periodic solutions which arise by the mechanism of Hopf bifurcation, while expansions of stable and unstable manifolds have a largely unexplored role to play in the analysis and computation of homoclinic and heteroclinic orbits. Mathematically, these expansions are an elegant way to proceed. We are confident that as more and more problems are solved using such techniques, they will become accepted as standard tools.

REFERENCES

[1] A. KELLEY, The stable, center-stable, center, center-unstable, and unstable manifolds, J. Diff. Eqns. 3(1967), pp. 546-570.

[2] B. D. HASSARD, Bifurcation of periodic solutions of the Hodgkin-Huxley model for the squid giant axon, J. Theoretical Biology 71 (1978), pp. 401-422.

[3] B. D. HASSARD and Y. H. WAN, Bifurcation formulae derived from center manifold theory, J. Math. Anal. Appl. 63(1978),pp. 297-312.

[4] B. D. HASSARD, The numerical evaluation of Hopf bifurcation formulae, in Information Linkage Between Applied Mathematics and Industry, P. C. C. Wang (ed.), Academic Press, N. Y. (1979)

[5] B. D. HASSARD, N. D. KAZARINOFF, and Y. H. WAN, Theory and Applications of Hopf Bifurcation, to appear, London Math. Society Lecture Note Series, Cambridge Univ. Press (1980)

[6] J. E. MARSDEN and M. MCCRACKEN, The Hopf Bifurcation and Its Applications, Applied Math. Sciences vol. 19, Springer-Verlag, N. Y. (1976)

[7] R. MIURA, Accurate solution of travelling wave equations I, preprint .

[8] P. HARTMAN, Ordinary Differential Equations, Baltimore (1973)

[9] B. CHILDS, M. SCOTT, J. W. DANIEL, E. DENMAN and P. NELSON, (eds.) Codes for Boundary Value Problems in Ordinary Differential Equations, Lecture Notes in Computer Science 76, Springer-Verlag N. Y. (1979)

[10] A. L. HODGKIN and A. F. HUXLEY, A quantitative description of membrane current and its application to conduction and excitation in nerve, J. Physiol. 117(1952), pp. 500-544.

[11] R. FITZHUGH, Mathematical models of excitation and propogation in nerve, in Biological Engineering, H. P. Schwan, (ed.), McGraw-Hill, N. Y. (1969)

[12] K. S. COLE, H. A. ANTOSIEWICZ, and P. RABINOWITZ, Automatic computation of nerve excitation, J. SIAM 3(1955), pp. 153-172

[13] J. W. COOLEY and F. A. DODGE, Digital computer solutions for excitation and propogation of the nerve impulse, Biophys. J. 6 (1966), pp. 583-599.

[14] J. W. EVANS and J. FEROE, Local stability theory of the nerve impulse, Math. Biosciences 37(1977), pp. 23-50.

[15] V. I. ARNOL'D, Supplementary Chapters of the Theory of Ordinary Differential Equations, Nauka, Moscow (1978) (in Russian)

A Qualitative Approach to Steady-State Bifurcation Theory

Martin Golubitsky† and David Schaeffer††

Abstract. We describe how the singularity theory of smooth mappings may be applied to problems in steady-state bifurcation theory. We restrict our attention to one state variable and present the first part of a classification theorem for such problems.

We describe here the approach [3,4] that we have used to analyse steady-state bifurcation problems in terms of a simple example from elasticity - the buckling of an Euler strut. Additional lectures by Barbara Keyfitz, David Schaeffer, and myself will illustrate more difficult (and interesting) applications.

For the most part, the mathematical content of our theory is obtained by adapting results from the singularity theory of smooth mappings as developed by R. Thom, J. Mather, V. Arnold, and others to the specific situation found in bifurcation theory. The somewhat controversial Catastrophe Theory is a branch of singularity theory; thus the way we plan to use our theory should be put in perspective. Applications of Catastrophe Theory have been attempted in two distinct ways.

The first is to analyse specific models given by the application and the second is to let the mathematics suggest models in situations where no commonly agreed upon model exists. It is this latter approach which has surely led to the notoriety surrounding Catastrophe Theory. The problems we have studied are based exclusively upon the first approach. To be specific, these methods have proved useful in the study of

(1) mode jumping in the buckling of a rectangular plate [7],

†Department of Mathematics, Arizona State University, Tempe, AZ 85281. This work was sponsored in part by N.S.F. Grant MCS-79-5799.
††Department of Mathematics, Duke University, Durham, NC 27706. This work was sponsored in part by N.S.F. Grant MCS-7902010.

(2) steady state behavior in a continuous flow stirred tank chemical reactor [1],

(3) the explosion peninsula in H_2-O_2 combustion [2] ,

(4) thermal convection in spherical geometry [5] ,

(5) boundary effects in the Taylor problem in hydrodynamic stability [6], and

(6) bifurcation behavior near a double eigenvalue in a model reaction-diffusion equation known as the "Brusselator" [8].

We consider two theoretical issues :

(A) Solve $G(x,\lambda) = 0$ where $x = (x_1,\ldots,x_n) \in \mathbb{R}^n$ and $\lambda \in \mathbb{R}$ are near 0. Here, $G(x,\lambda) = (g_1(x,\lambda),\ldots,g_n(x,\lambda))$ is C^∞ and $G(0,0) = 0$.

(B) Find all small C^∞ perturbations of the zero set $\{G = 0\}$.

We think of λ as some externally controlled parameter varied quasi-statically by an experimenter and those values x for which $G(x,\lambda)=0$ as measuring the possible equilibrium states of the given system.

For example, consider the finite element analogue of the Euler strut illustrated in Figure 1.

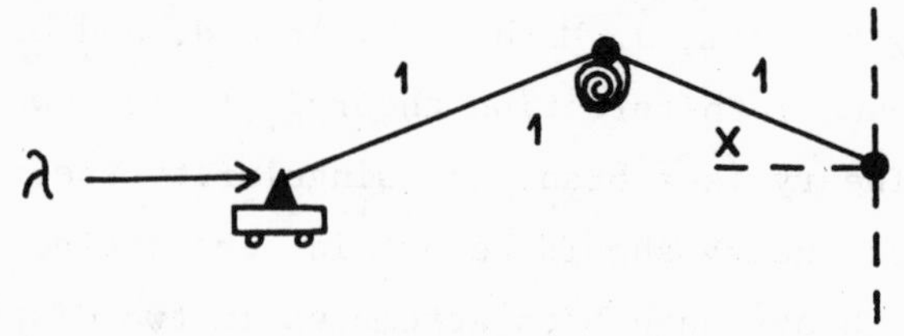

Figure 1

This system, consisting of two rigid rods of unit length connected by frictionless pins, is subjected to a compressive force λ which is resisted by a torsional spring of unit strength. The angular deviation x of the rod from the horizontal measures the single state of the system. The potential energy function for this system is

$$(1) \quad V(x,\lambda) = \frac{x^2}{2} + 2\lambda \cos x$$

with the steady state configurations being described by

(2) $G(x,\lambda) = \frac{\partial V}{\partial x} = x - 2\lambda \sin x = 0.$

A simple calculation shows that when $\lambda = \lambda_o = \frac{1}{2}$ this system undergoes bifurcation from the trivial undeflected solution $x = 0$.

We now consider (A) above. Our main interest is in knowing how many solutions exist for $\lambda < \lambda_o$ and $\lambda > \lambda_o$. This information is preserved by the following change of coordinates.

Definition 3. Let $G,H : \mathbb{R}^n \times \mathbb{R} \to \mathbb{R}^n$ be bifurcation problems. G and H are contact equivalent if

$$G(x,\lambda) = T(x,\lambda).H(X(x,\lambda), \Lambda(\lambda))$$

where T is an invertible nxn matrix, $\det(d_x X) > 0$, and $\Lambda' > 0$.

A typical result of our theory (which may be applied to the above example) is :

Proposition 4. Let $G : \mathbb{R}\times\mathbb{R} \to \mathbb{R}$ be a bifurcation problem. Assume

(a) $G = G_x = G_\lambda = G_{xx} = 0$ at (x_o,λ_o)

and (b) $G_{xxx}.G_{x\lambda} < 0$ at (x_o,λ_o).

Then G is contact equivalent on a neighborhood of (x_o,λ_o) to $H(x,\lambda) = (x-x_o)^3 - (\lambda-\lambda_o)(x-x_o) = 0$.

Note : We call (a) the defining conditions and (b) the non-degeneracy conditions.

In particular the bifurcation diagram is the pitchfork pictured in Figure 2.

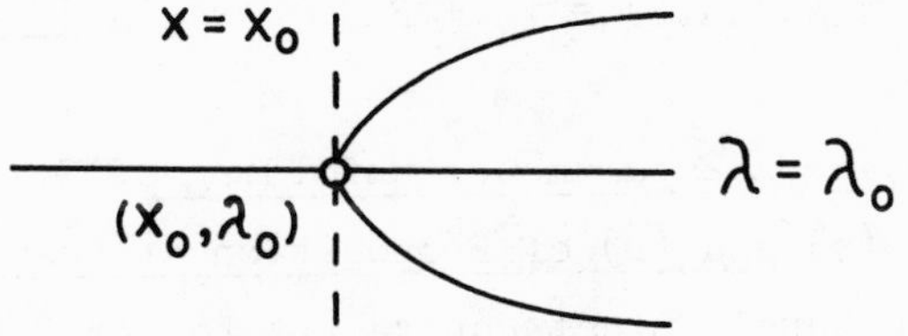

Figure 2.

It is straight-forward to apply this proposition to G in (2) to determine that there is exactly one solution x for $\lambda < \lambda_o$ and three solutions for $\lambda > \lambda_o$. (Of course, in this situation the use of these methods is not a necessity).

The method for finding all small perturbations ((B) above) to a given problem G is broken into two parts - one analytic and one geometric. The first part uses the most basic central result in Singularity Theory ; the existence of universal unfoldings.

Definition 5. (a) We call $F : \mathbb{R}^n \times \mathbb{R} \times \mathbb{R}^k \to \mathbb{R}^n$ a k-parameter unfolding of G if $F(x,\lambda,0) = G(x,\lambda)$.
(b) Given $F(x,\lambda,\beta)$ and another unfolding $E(x,\lambda,\beta)$ we say that E factors through F if

(6) $E(x,\lambda,\beta) = T(x,\lambda,\beta).F(X(x,\lambda,\beta),\Lambda(\lambda,\beta)\alpha(\beta))$.

Equation (6) means that for each β there exists an α depending smoothly on β such that $E(.,.,\beta)$ is contact equivalent to $F(.,.,\alpha)$.
(c) F is a universal unfolding if every unfolding E factors through F.

Note that the factoring of E through F implies that the bifurcation diagrams $E(.,.,\beta) = 0$ as β varies are qualitatively the same (in the sense of (A)) as some subset of the bifurcation diagrams $F(.,.,\alpha) = 0$ as α varies. Thus, a universal unfolding gives an analytic description of the qualitatively different small perturbations of G. However, it is by no means clear that under quite general circumstances these universal unfoldings do exists. It is beyond the scope of this lecture to describe how one finds universal unfoldings for a given G ; however, we can describe typical results for our model problem.

Proposition 7. (i) $F(x,\lambda,\alpha,\beta) = x^3 + \beta x^2 - \lambda x + \alpha$ is a universal unfolding of $x^3 - \lambda x$.
(ii) Let $F(x,\lambda,a,b)$ be a 2-parameter unfolding of $G(x,\lambda)$ where G satisfies conditions (a) and (b) of Proposition 4. Then F is a universal unfolding of G (and thus contact equivalent to the F in (i)) if

$$\det \begin{pmatrix} 0 & 0 & G_{x\lambda} & G_{xxx} \\ 0 & G_{\lambda x} & G_{\lambda\lambda} & G_{\lambda xx} \\ F_a & F_{ax} & F_{a\lambda} & F_{axx} \\ F_b & F_{bx} & F_{b\lambda} & F_{bxx} \end{pmatrix} \neq 0$$

where the various derivatives are computed at $(x,\lambda,\alpha,\beta) = (x_o,\lambda_o,0,0)$.

One can now realize physically the universal unfolding parameters α and β in (i) for our model problem. Consider the following imperfections : let a be the angle at which the torsional spring exerts no torque and let b measure a vertical load applied to the center pin of the system. Then the potential energy becomes

(8) $V(x,\lambda,a,b) = \frac{1}{2}(x-a)^2 + 2\lambda \cos x + b \sin x.$

The steady states are now given by a 2-parameter unfolding of the idealized problem

(9) $F(x,\lambda,a,b) = \frac{\partial V}{\partial x} = x - a - 2\lambda \sin x + b \cos x.$

It is easy to check that the determinant in Proposition 7 (ii) is non-zero at $(x_o,\lambda_o,a,b) = (0,\frac{1}{2},0,0)$ so that the given imperfections are universal unfolding parameters for the idealized problem.

To complete our analysis of (B) one needs an algorithm for enumerating the qualitatively different bifurcation diagrams occurring as small perturbations of $\{G = 0\}$. What follows is a first step. For ease of exposition we restrict to n = 1. Let $G : \mathbb{R}\times\mathbb{R} \to \mathbb{R}$ be a bifurcation problem with $F : \mathbb{R}\times\mathbb{R}\times\mathbb{R}^k \to \mathbb{R}$ its universal unfolding. Our first goal is to enumerate the stable diagrams in $\{F = 0\}$. By stable we mean those diagrams which do not change their qualitative type under small perturbations. A usual ploy, which we follow, is to list the ways in which a bifurcation diagram can fail to be stable. There are three such ways which are shown in Figure 3 along with the perturbations which change the qualitative nature of the given diagram.

	Ideal	Defining Conditions	Perturbation
Bifurcation (B)		$F = F_x = F_\lambda = 0$ at (x_o, λ_o)	
Hysteresis (H)		$F = F_x = F_{xx} = 0$ at (x_o, λ_o)	
Double limit (DL)	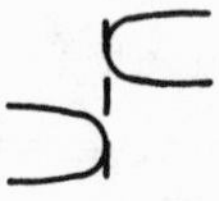	$F = F_x = 0$ at (x_1, λ_o) and (x_2, λ_o)	

Figure 3

Proposition 10. Let Σ be the set of those $\alpha \in \mathbb{R}^k$ such that the bifurcation diagram $F(.,.,\alpha) = 0$ exhibits either bifurcation, hysteresis, or double limit behavior at some point (x_o, λ_o). If α_1 and α_2 are in the same connected component of $\mathbb{R}^k \sim \Sigma$ then $F(.,.,\alpha_1)$ and $F(.,.,\alpha_2)$ are contact equivalent.

For our sample problem $F(x,\lambda,\alpha,\beta) = x^3 + \beta x^2 - \lambda x + \alpha$ we have the results pictured in Figure 4. There are

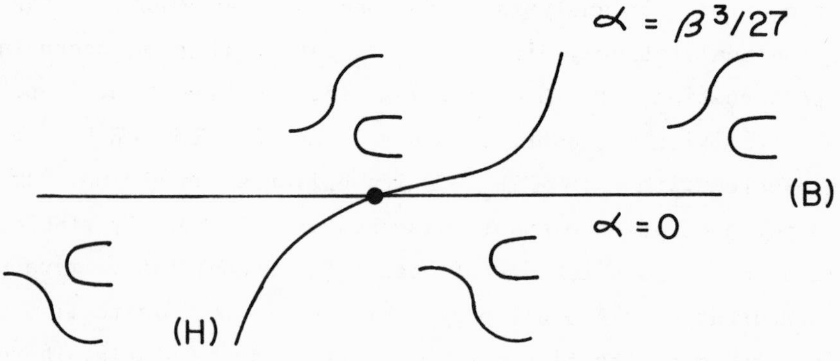

Figure 4

four connected components to the complement of Σ ($DL = \emptyset$) ; and pictured within each component is a qualitative representation of the stable bifurcation diagram associated with that component. The information given by Figure 4 is really the only new information obtained about this problem by the infusion of singularity theory.

For the remainder of this lecture we will describe a partial classification of bifurcation problems with one state variable.

Definition 11. (a) Consider $G : \mathbb{R}\times\mathbb{R} \to \mathbb{R}$ with $G(x,0) = a_\ell x^\ell + \ldots$ with $a_\ell \neq 0$. We call ℓ the rank of G.
(b) The codimension of G is the minimum number of parameters necessary in a universal unfolding of G.

The pitchfork $x^3 - \lambda x$ discussed above has rank 3 and codimension 2.

Theorem 12. *Let $G : \mathbb{R}\times\mathbb{R} \to \mathbb{R}$ be a bifurcation problem with rank ≤ 3 and codimension ≤ 5, then G is contact equivalent to one of the following thirteen normal forms :*

	Normal Form	Universal Unfolding	Codimension	Name
(i)	x	x	0	Non-singular
(ii)	$x^2 \pm \lambda$	$x^2 \pm \lambda$	0	Limit Point
(iii)	$x^2 - \lambda^2$	$x^2 - \lambda^2 + \alpha$	1	Simple bifurcation
(iv)	$x^2 - \lambda^2$	$x^2 + \lambda^2 + \alpha$	1	Isola
(v)	$x^3 \pm \lambda$	$x^3 + \alpha x \pm \lambda$	1	Hysteresis
(vi)	$x^2 \pm \lambda^3$	$x^2 \pm \lambda^3 + \beta\lambda + \alpha$	2	Asymmetric cusp
(vii)	$x^3 \pm \lambda x$	$x^3 + \beta x^2 \pm \lambda x + \alpha$	2	Pitchfork
(viii)	$x^2 \pm \lambda^4$	$x^2 \pm \lambda^4 + \gamma\lambda^2 + \beta\lambda + \alpha$	3	
(ix)	$x^3 \pm \lambda^2$	$x^3 + \gamma x\lambda + \beta x + \alpha \pm \lambda^2$	3	Asymmetric (or winged) cusp
(x)	$x^2 \pm \lambda^5$	$x^2 \pm \lambda^5 + \delta\lambda^3 + \gamma\lambda^2 + \beta\lambda + \alpha$	4	
(xi)	$x^3 - 3mx\lambda^2 \pm 2\lambda^3$, $m < 1$	$x^3 - 3\gamma(\lambda^2 + \delta\lambda + \varepsilon)x \pm 2(\lambda^3 + \beta\lambda + \alpha)$	5	London Underground
(xii)	$x^3 - 3mx\lambda^2 \pm 2\lambda^3$, $m > 1$	$x^3 - 3\gamma(\lambda^2 + \delta\lambda + \varepsilon)x \pm 2(\lambda^3 + \beta\lambda + \alpha)$	5	
(xiii)	$x^3 - 3x\lambda^2 \pm 2(\lambda^3 \pm \lambda^4)$	$x^3 - 3\gamma(\lambda^2 + \delta\lambda + \varepsilon)x \pm 2(\lambda^3 \pm \lambda^4 + \beta\lambda + \alpha)$	5	

As a last example we consider the winged cusp.

Proposition 13. (A) Let $G : \mathbb{R}\times\mathbb{R} \to R$ satisfy

(a) $G = G_x = G_\lambda = G_{xx} = G_{x\lambda} = 0$ at $(0,0)$

and

(b) $G_{xxx} \cdot G_{\lambda\lambda} \neq 0$ at $(0,0)$.

Then G is contact equivalent to $x^3 \pm \lambda^2$

(B) There are seven stable diagrams which appear in the universal unfolding of $x^3 + \lambda^2$. They are pictured in Figure 5.

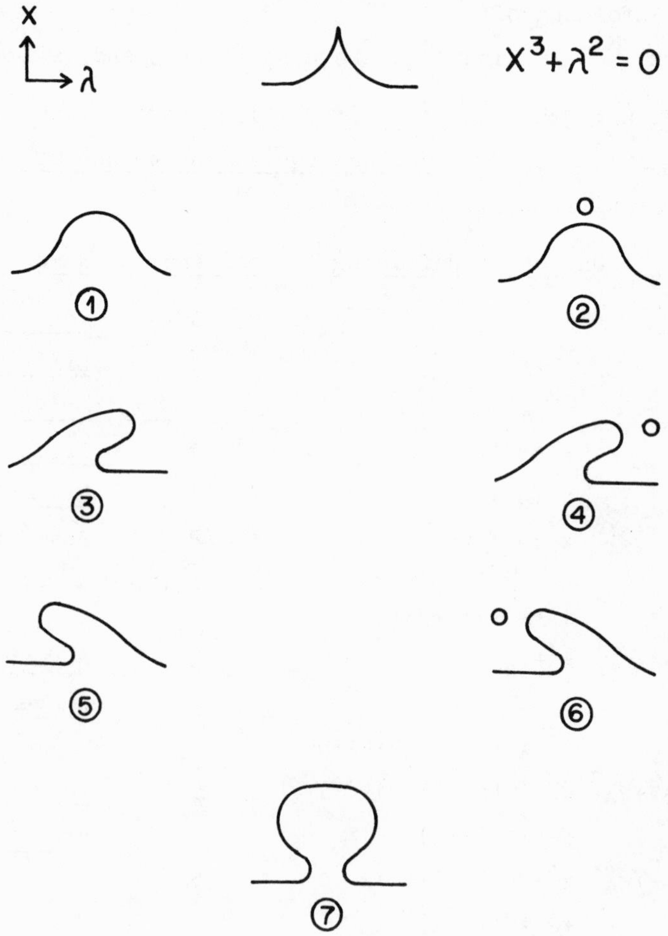

Figure 5.

REFERENCES

1 M. GOLUBITSKY and B.L. KEYFITZ, "A qualitative study of the steady state solutions for a continuous flow stirred tank chemical reactor", SIAM J. Math. Anal. To appear.

2 M. GOLUBITSKY, B.L. KEYFITZ and D. SCHAEFFER, "A singularity theory analysis of the thermal-chainbranching model for the explosion peninsula". Preprint.

3 M. GOLUBITSKY and D. SCHAEFFER, "A theory for imperfect bifurcation via singularity theory", Comm. Pure Appl. Math. 32 (1979) 21-98.

4 ———————, "Imperfect bifurcation in the presence of symmetry", Commun. Math. Phys. 67 (1979) 205-232.

5 ———————, "The Bénard problem in spherical geometry ; an example with O(3) symmetry". In preparation.

6 D. SCHAEFFER, "Qualitative analysis of a model for boundary effects in the Taylor problem", Proc. Camb. Phil. Soc. To appear.

7 D. SCHAEFFER and M. GOLUBITSKY, "Boundary conditions and mode jumping in the buckling of a rectangular plate", Commun. Math. Phys. 69 (1979) 209-236.

8 ———————, "Bifurcation analysis near a double eigenvalue of a model chemical reaction", ARMA. To appear.

Solution Branching—A Constructive Technique

D. W. Decker* and H. B. Keller**

Abstract. The employment of Euler-Chord and Euler-Newton continuation methods for the generation of solution branches of non-linear equilibrium problems is described. The continuation parameter is chosen to be an approximation to arclength along the solution branch. Step-size estimates along arcs composed of regular and simple limit points are derived. Convergence of Euler-Newton (Chord) stepping from simple bifurcation points is demonstrated. The problem of the approach toward a simple bifurcation point is considered and limiting approach rates determined. In both cases, the choice of approximate arclength as the continuation parameter improves convergence behavior along branches with a "vertical" tangent at bifurcation.

1. Introduction. Bifurcation theory deals with the branching of solutions to nonlinear functional equations. Although time dependent solution branching may occur, only equilibrium problems are treated here, of the general form;

$$G(u, \lambda) = 0 \tag{1.1}$$

Here $G(.,.)$ is a nonlinear mapping between Banach spaces depending on some, usually physically relevant, real control parameter λ. We seek arcs or branches of solutions for a range of the parameter λ, say of the form depicted in Figure (1). In this presentation we report on continuation techniques that will allow the generation of much of the solution structure of this Figure. The methods and results are briefly outlined with the details and proofs to appear elsewhere [5].

*Mathematics Department, North Carolina State University, Raleigh, N.C. 27650.
**Applied Mathematics Department, California Institute of Technology, Pasadena, Ca. 91125. This work was supported under U. S. Army Research Office contract DAAG 29-78-C-0011.

Section (2) contains the basic formulation of the problem and determines the structure of a solution arc in the neighborhood of certain singular points, that is, solution points where the linearized operator associated with (1.1) is singular. These allowable singular points are either, (i) simple limit points, of which the points A and B are representatives or, (ii) simple bifurcation points such as C or D. In order to allow continuation through simple limit points, Section (3) defines an inflated system [8] in which solution parametrization is an approximation to arc-length. In this formulation, denoted

$$F(x(s), s) = 0 \tag{1.2}$$

the linearized operator F_x remains non-singular at simple limit points. Section (4) introduces Euler-Newton and Euler-Chord continuation and the basic convergence and step-size estimates for these methods are indicated. These estimates immediately allow convergence through regular and simple limit points.

At simple bifurcation points, the linearized inflated operator differentiates between the points C and D, the distinguishing feature being the solution arc $x(s) \equiv (u(s), \lambda(s))$ for which $d\lambda/ds = 0$ at C.

Section (5) determines the behavior of the linearized inflated operator in the neighborhood of both these types of bifurcation points. The convergence of the Euler-Newton (Chord) method shooting from a simple bifurcation point is demonstrated in Section (6). The behavior

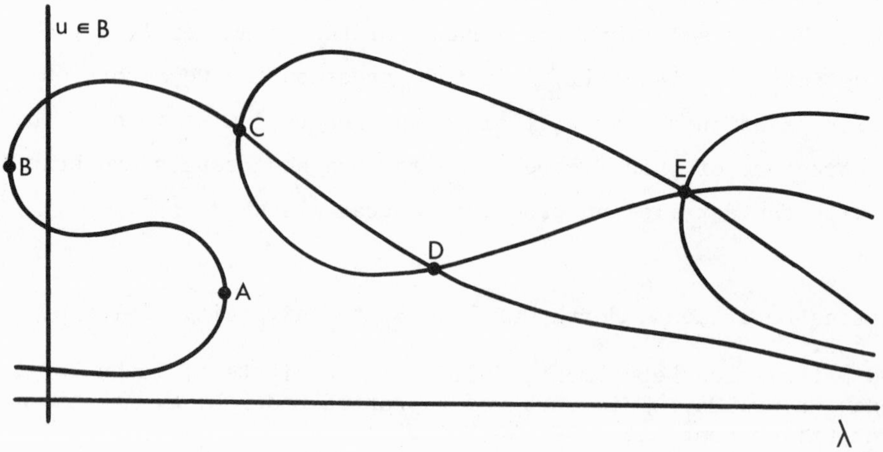

Figure 1

of the linearized inflated operator along a solution arc with a "vertical" tangent at bifurcation is shown to be an aid to convergence. Section (7) is concerned with the problem of the approach to a simple bifurcation point. The growth of $\|F_x^{-1}(x(s), s)\|$ is shown to limit the possible approach rates. These "optimal" rates are determined for both the Chord and Newton methods. Once again, a branch with $d\lambda/ds = 0$ at bifurcation is seen to have a convergence advantage.

We note that although only a simple bifurcation point is considered, many of the results apply to multiple bifurcation, such as that occurring at the point E.

2. Regular, Simple Bifurcation and Limit Points. The abstract formulation of the problem to be considered is of the form

(2.1) $$G(u, \lambda) = 0$$

where $G(.,.)$ is a nonlinear mapping from $B_1 \times \mathbb{R}$ into B_2. Here B_1 and B_2 are Banach spaces with $B_1 \subseteq B_2$. The desire is to find solutions of (2.1) for some range of the physical parameter λ.

First, suppose one knows a solution point (u_o, λ_o). Then provided the Frechet derivative of G with respect to u at (u_o, λ_o), denoted $G_u^o \equiv G_u(u_o, \lambda_o)$, is nonsingular, the Implicit Function Theorem [9] guarantees a locally unique solution arc $(u(\lambda), \lambda)$. Thus for continuation through such regular points, parametrization by λ appears appropriate. However, in the neighborhood of singular points of G_u^o it may be unsatisfactory, and in the next section a proper choice of parametrization will be seen to be important.

For this reason, consider a smooth solution arc $(u(s), \lambda(s))$ with the parameter as yet unspecified, and suppose G_u is singular at some point $(u(s_o), \lambda(s_o))$. We shall assume the singularity of G_u^o to be of the following form;

(2.2) $$N(G_u^o) = \text{span } \{\phi_1\}, \; \|\phi_1\| = 1$$
$$R(G_u^o) \text{ is closed and codim } R(G_u^o) = 1$$

That is, G_u^o is a Fredholm operator of index zero and its adjoint operator $G_u^{o*} : B_2^* \to B_1^*$ satisfies;

(2.3) $$N(G_u^{o*}) = \text{span } \{\phi_1^*\} \text{ and } R(G_u^o) = \{x \varepsilon B_2 | \phi_1^* x = 0\}$$

In addition, we assume the zero eigenvalue of G_u^o to be simple (algebraic multiplicity one), and so we may take

$$\phi_1^* \phi_1 = 1 \tag{2.4}$$

With these assumptions we proceed to find conditions that must be satisfied by any solution arc through $(u(s_o), \lambda(s_o)) \equiv (u_o, \lambda_o)$. Differentiating

$$G(u(s), \lambda(s)) = 0 \tag{2.5}$$

and evaluating at s_o we find;

$$G_u^o \dot{u}_o + G_\lambda^o \dot{\lambda}_o = 0 \tag{2.6}$$

$$G_u^o \ddot{u}_o + G_\lambda^o \ddot{\lambda}_o = -(G_{uu}^o \dot{u}_o \dot{u}_o + 2G_{u\lambda}^o \dot{u}_o \dot{\lambda}_o + G_{\lambda\lambda}^o \dot{\lambda}_o^2) \tag{2.7}$$

The existence of a solution $\dot{u}_o$ of (2.6) requires $\dot{\lambda}_o G_\lambda^o \varepsilon R(G_u^o)$ or

$$\dot{\lambda}_o d = o, \text{ where } d \equiv \phi_1^* G_\lambda^o \tag{2.8}$$

This presents two possibilities (i) $d = 0$ or (ii) $d \neq 0$ but $\dot{\lambda}_o = 0$. The second case we shall call a simple limit point, and it will later be shown that no bifurcation may occur at such a point. Returning to the first case we consider

$$G_u^o \phi_o + G_\lambda^o = o \text{ with } \phi_1^* \phi_o = 0 \tag{2.9}$$

This has a unique solution and therefore $\dot{u}_o = \xi_o \phi_o + \xi_1 \phi_1$ where $\xi_o = \dot{\lambda}_o$ and ξ_1 are at present arbitrary. Placing this expression into (2.7), the existence of a solution $\ddot{u}_o$ requires

$$a\xi_1^2 + 2b\xi_1\xi_o + c\xi_o^2 = 0 \tag{2.10}$$

where we have introduced

$$a = \phi_1^* G_{uu}^o \phi_1 \phi_1 \qquad b = \phi_1^*(G_{uu}^o \phi_o + G_{u\lambda}^o)\phi_1 \tag{2.11}$$
$$c = \phi_1^*(G_{uu}^o \phi_o \phi_o + 2G_{u\lambda}^o \phi_o + G_{\lambda\lambda}^o)$$

Adding to (2.10) an equation allowing a unique determination of the parameter s results in the Algebraic Bifurcation Equations (ABE)

$$\text{(2.12)} \qquad a\xi_1^2 + 2b\xi_1\xi_o + c\xi_o^2 = 0 \qquad \text{(a)}$$
$$2\xi_o^2 + \xi_1^2 = 1 \qquad \text{(b)}$$

with Jacobian

$$J = \left| \begin{pmatrix} a\xi_1 + b\xi_o & b\xi_1 + c\xi_o \\ \xi_1 & 2\xi_o \end{pmatrix} \right| \tag{2.13}$$

It can be shown [3, 4, 8] that each isolated ($J \neq 0$) root (ξ_o^*, ξ_1^*) of (2.12) generates a smooth solution arc $(u(s), \lambda(s))$ for s near s_o of the form

$$(2.14)\quad \begin{aligned} u(s) &= u_o + (s - s_o)(\xi_o(s)\phi_o + \xi_1(s)\phi_1) + (s - s_o)^2 v(s) \\ \lambda(s) &= \lambda_o + (s - s_o)\xi_o(s) \end{aligned}$$

where $\phi_1^*(s)v(s) = 0$ and $\xi_o(s_o) = \xi_o^*$, $\xi_1(s_o) = \xi_1^*$. This result is well known in other forms [1, 2, 9, 10].

Now equations (2.12) can have at most two isolated roots and hence at most two non-tangential solution arcs intersecting at (u_o, λ_o).

The main purpose of this work is to determine a continuation method for following a solution arc up to such a bifurcation point and to determine the subsequent branching behavior.

We note that in a completely analogous fashion one may derive ABE's and prove existence results when the eigenspace of G_u^o is of dimension $m > 1$. However in this case multiple limit point bifurcation may also occur [4]. Many of the results that follow can be carried over to these situations.

3. Inflated System at Regular, Simple Limit and Bifurcation Points. We now describe a technique due to Keller [8] in which the parameter is chosen to be an approximation to arc-length along a solution curve. Suppose (u_o, λ_o) is a known solution point of (2.1), we set $(u(s_o), \lambda(s_o)) \equiv (u_o, \lambda_o)$ and $(\dot{u}(s_o), \dot{\lambda}(s_o)) \equiv (\dot{u}_o, \dot{\lambda}_o)$ where

$$(3.1)\qquad G_u^o\dot{u}_o + G_\lambda^o\dot{\lambda}_o = 0 \text{ and } \|\dot{u}_o\|^2 + \dot{\lambda}_o^2 = 1$$

Then for s near s_o we require $(u(s), \lambda(s))$ to satisfy

$$(3.2)\qquad N(u, \lambda, s) = \dot{u}^*(s_o)(u(s)-u(s_o)) + \dot{\lambda}(s_o)(\lambda(s)-\lambda_o) - (s-s_o) = 0$$

Here $\dot{u}(s_o)^* \epsilon B_1^*$ is chosen such that $\dot{u}^*(s_o)\dot{u}(s_o) = \|\dot{u}(s_o)\|^2$. This is a local parametrization which approximates arc-length, and results in the "inflated" system of equations

$$(3.3)\qquad \begin{pmatrix} G(u, \lambda) \\ N(u, \lambda, s) \end{pmatrix} \equiv F(x, s) = 0, \text{ where } x \equiv (u, \lambda)$$

Thus along any smooth solution arc through (u_o, λ_o) we have

$$(3.4)\qquad F_x(s)\dot{x}(s) = -F_s(s) = \binom{0}{1}$$

that is

$$(3.5)\qquad \begin{aligned} G_u(s)\dot{u}(s) + G_\lambda(s)\dot{\lambda}(s) &= 0 \\ \dot{u}(s_o)^*\dot{u}(s) + \dot{\lambda}(s_o)\dot{\lambda}(s) &= 1 \end{aligned}$$

The advantage of this arc-length formulation is apparent from [8];

Theorem (3.1). Let (u_o, λ_o) be a regular or simple limit point of

$G(u, \lambda)$ and let G be C^2 near (u_o, λ_o). Then there exists a unique solution arc $(x(s), s)$ of (3.3) through (u_o, λ_o) for $|s - s_o| < \delta$, some $\delta > o$, and along this arc the Frechet derivative

$$(3.6) \qquad F_x(x(s), s) = \begin{pmatrix} G_u(s) & G_\lambda(s) \\ \dot{u}^*(s_o) & \dot{\lambda}(s_o) \end{pmatrix}$$

is non-singular. //

Hence regular points of $G(u, \lambda)$ remain regular points of $F(x, s)$ and simple limit points of G become regular points of F. As a consequence, no bifurcation may occur in either case.

Now suppose the solution point (u_o, λ_o) is a simple bifurcation point of G and $(\dot{u}_o, \dot{\lambda}_o) = (\xi_o\phi_o + \xi_1\phi_1, \xi_o)$ is a solution of (2.6) and the ABE (2.12). Then, with an equivalent choice of norm in B_1 such that $\|\dot{u}_o\|^2 = \xi_o^2 + \xi_1^2$, (3.1) is satisfied and $x(s_o) \equiv (u_o, \lambda_o)$, $\dot{x}(s_o) \equiv (\dot{u}_o, \dot{\lambda}_o)$ satisfies the inflated system (3.3-5) at $s = s_o$. This choice of norm also allows one to take

$$(3.7) \qquad \dot{u}^*(s_o) = \xi_o\phi_o^* + \xi_1\phi_1^*$$

where ϕ_1^* is as before and ϕ_o^* is chosen such that $\phi_i^*\phi_j = \delta_{ij}$ $i, j = 0, 1$. The behavior of the inflated linear operator at s_o is contained in [3, 5];

<u>Theorem (3.2)</u>. Let (u_o, λ_o) be a simple bifurcation point of $G(u, \lambda)$ and (ξ_o, ξ_1) a root of the ABE (2.12). Then the Frechet derivative

$$(3.8) \qquad F_x(x_o, s_o) = \begin{pmatrix} G_u^o & G_\lambda^o \\ \xi_o\phi_o^* + \xi_1\phi_1^* & \xi_o \end{pmatrix}$$

is a Fredholm operator of index zero with

$$(3.9) \qquad N(F_x^o) = \text{span}\,\{\Phi_1\} \text{ where } \Phi_1 \equiv \begin{pmatrix} \xi_1\phi_o - 2\xi_o\phi_1 \\ \xi_1 \end{pmatrix}$$

$$(3.10) \quad R(F_x^o) = \{x \varepsilon B_2 \times \mathbb{R} \,|\, \Phi_1^* x = 0\} \quad \text{where} \quad \Phi_1^* \equiv (\phi_1^*, 0)$$

For $\xi_o \neq 0$ the zero eigenvalue is simple, for $\xi_o = 0$ it is of algebraic multiplicity two. //

Hence provided $\xi_o \neq 0$, the inflated system inherits the structure of $G(u, \lambda)$ at a simple bifurcation point. The inflated structure when $\xi_o = 0$ will later be seen to hold a distinct advantage over that of G in regard to the convergence of iterative schemes.

Now any smooth solution arc through $x_o = (u_o, \lambda_o)$ must satisfy (since $F_{xs} = F_{ss} \equiv 0$)

(3.11) $$F_x^o \dot{x}(s_o) = -F_s^o$$

(3.12) $$F_x^o \ddot{x}(s_o) = -F_{xx}^o \dot{x}(s_o)\dot{x}(s_o)$$

Since $\dot{x}(s_o) = (\xi_o\phi_o + \xi_1\phi_1, \xi_o)$ satisfies (3.11) we place this in (3.12) and require the right hand side to be in $R(F_x^o)$ resulting in

(3.13) $$(\phi_1^*, 0)F_{xx}^o \dot{x}(s_o)\dot{x}(s_o) = a\xi_1^2 + 2b\xi_1\xi_o + c\xi_o^2 = 0$$

and from (3.1) $$2\xi_o^2 + \xi_1^2 = 1.$$

Thus the requirements for the existence of branching in F(x, s) are identical with those of G(u, λ), and bifurcation occurs simultaneously.

4. Euler-Newton (Chord) Continuation. Now suppose x(s) is a smooth solution arc through $x(s_o)$, a regular, simple limit or bifurcation point. Under what conditions can one construct solutions for s near s_o? The continuation procedure to be used will be a single Euler step, coupled with either the Chord or Newton's method. That is, from the known solution point $(x(s_o), s_o)$, one constructs an initial guess for the solution at $s = s_o + \Delta s$ from

(4.1) $$F_x(x(s_o), s_o)\dot{x}(s_o) = -F_s(x(s_o), s_o)$$

(4.2) $$x^o(s) = x(s_o) + (s - s_o)\dot{x}(s_o)$$

in order to start

(4.3) (Chord) $$F_x(x^o(s),s)(x^{\nu+1}(s)-x^\nu(s)) = -F(x^\nu(s),s)$$

$$\nu = 0, 1, \ldots$$

or

(4.4) (Newton) $$F_x(x^\nu(s),s)(x^{\nu+1}(s)-x^\nu(s)) = -F(x^\nu(s),s)$$

Now if we define

(4.5)
(i) $$M(s) = \|F_x^{-1}(x(s), s)\| \qquad o < |s - s_o| < \delta$$
(ii) $$\max_{s_o \le t \le s} \|\ddot{x}(t)\| = \kappa(s)$$
(iii) $$\|x^o(s) - x(s)\| = r(s)$$

and assume

(4.6) $$\|F_x(y, s) - F_x(x(s), s)\| \le K(s)\|y - x(s)\| \quad \text{for } \|y - x(s)\| \le r(s)$$

then geometric convergence of the Chord Method can be guaranteed provided [8];

(4.7) $$M(s)K(s)r(s) < \frac{1}{3}$$

Here one views (4.6-7) as equations determining the allowable first iterate error $r(s)$. Since $r(s) \leq \frac{1}{2}(s - s_o)^2\kappa(s)$, convergence is assured provided $M(s)$ and $K(s)$ are well-behaved as $s \to s_o$. Indeed a maximum step-size may be estimated from

(4.8) $$K(s)M(s)\kappa(s)(s - s_o)^2 < \frac{2}{3}$$

Now for Newton's method, well known conditions sufficient for convergence [6], may be stated as follows, if

(4.9) $$\begin{aligned} &\|F_x^{-1}(x^o(s), s)\| \leq a \\ &\|F_x^{-1}(x^o(s), s)\ F(x^o(s), s)\| \leq b \\ &\|F_x(y_1, s) - F_x(y_2, s)\| \leq c\|y_1 - y_2\| \\ &\text{for } \|y_i - x^o(s)\| \leq 2b \quad i = 1, 2 \end{aligned}$$

then quadratic convergence is assured provided $abc < \frac{1}{2}$. In our case defining

(4.10) $$M_o(s) = \|F_x(x^o(s), s)^{-1}\|$$

and assuming

(4.11) $$\|F_x(y, s)\| \leq H(s) \text{ for } \|y - x(s)\| \leq r(s)$$

we have convergence provided

(4.12) $$M_o^2(s)H(s)L(s)r(s) < \frac{1}{2}$$

where

(4.13) $$\begin{aligned} &\|F_x(y_1, s) - F_x(y_2, s)\| \leq L(s)\|y_1 - y_2\| \\ &\text{for } \|y_i - x^o(s)\| \leq 2M_o(s)H(s)r(s) \qquad i = 1, 2 \end{aligned}$$

Once again (4.12-13) may be viewed as equations determining (a lower bound for) the maximum allowable $r(s)$, and hence step-size from

(4.14) $$\kappa(s)H(s)L(s)M_o^2(s)(s - s_o)^2 < 1 \quad .$$

Often the bounds (4.7) and (4.12) are not restrictive and since $M(s)$ (and hence $M_o(s)$) is bounded in the neighborhood of regular or simple limit points, solution arcs composed of such points may be constructively determined. In this situation one may view (4.8) and (4.14) as defining convergence tubes about the solution arc $x(s)$, which determine the allowable Euler step sizes (Figure 2). Problems may arise, however, in the neighborhood of simple bifurcation points where $M(s) \to \infty$ and hence $r(s) \to o$. The size of the "shrinking" convergence tubes will be determined by the rate of growth of $M(s)$ which we now determine.

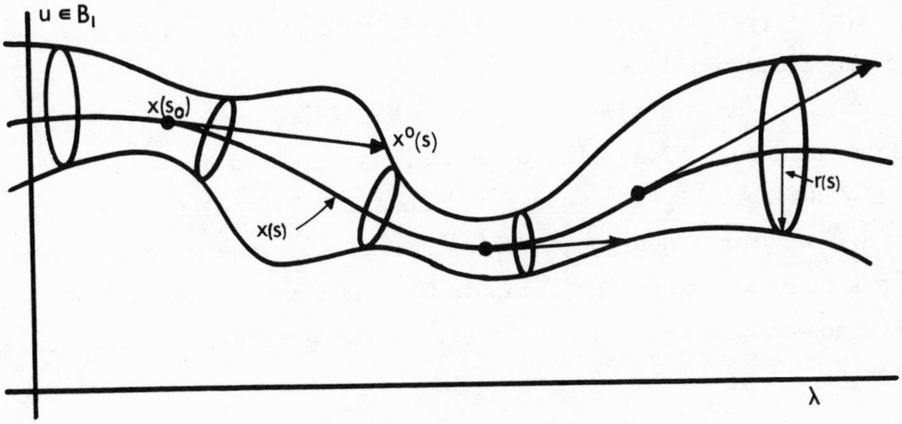

Figure 2 Convergence Tube

5. Inflated Linear Operator Near a Simple Bifurcation Point. The behavior of $F_x(x(s), s)$ near a simple bifurcation point $(x(s_o), s_o)$ will depend on whether the zero eigenvalue is simple ($\xi_o \neq 0$) or of multiplicity two ($\xi_o = 0$).

For the simple case, with $F(.,.)$ sufficiently smooth, and $F_x(x(s_o), s_o)$ a Fredholm operator of index zero, one can prove [5, 7] the existence of a smooth eigenvalue-eigenvector pair such that

$$F_x(x(s), s)\Phi(s) = \alpha(s)\Phi(s) \tag{5.1}$$

$$\Phi(s_o) = \Phi_1, \quad \alpha(s_o) = 0 \tag{5.2}$$

It is the approach to zero of $\alpha(s)$ which will determine the growth of $M(s)$. Differentiating (5.1), evaluating at $s = s_o$, and acting with Φ_1^* we see

$$\Phi_1^* F_x^o \dot{\Phi}(s_o) + \Phi_1^* F_{xx}^o \dot{x}(s_o)\Phi_1 = \alpha(s_o)\Phi_1^* \dot{\Phi}(s_o) + \dot{\alpha}(s_o)\Phi_1^*\Phi_1 \tag{5.3}$$

Recalling (2.3-4) (3.9-10) and (5.2) we see

$$-2\xi_o \dot{\alpha}(s_o) = \Phi_1^* F_{xx}^o \dot{x}(s_o)\Phi_1 \tag{5.4}$$

Using the definition of $F(x, s)$ and the ABE (5.4) becomes

$$\dot{\alpha}(s_o) = J/2\xi_o \tag{5.5}$$

Thus along a branch generated by an isolated root of the ABE with $\xi_o \neq 0$ the eigenvalue approaches zero like $(s - s_o)$.

For later use we note the adjoint-operator $F_x^*(x(s), s)$ also has an eigenvalue-eigenvector pair $(\alpha(s), \Phi^*(s))$, with $\alpha(s_o) = 0$, $\Phi^*(s_o) = \Phi_1^*$. Defining $B_1 = B_1 \times \mathbb{R}$, $B_2 = B_2 \times \mathbb{R}$ and the subspaces

$$(5.6) \qquad N(s) = \text{span}\,\{\Phi(s)\}$$

$$R(s) = \{x\varepsilon B_2 \mid \Phi^*(s)x = 0\} \qquad X(s) = \{x\varepsilon B_1 \mid \Phi^*(s)x = 0\}$$

we have the decompositions

$$(5.7) \qquad B_1 = N(s) \oplus X(s) \qquad B_2 = N(s) \oplus R(s)$$

and define the projections $P(s)$ onto $N(s)$ parallel to $R(s)$ and $Q(s) = I - P(s)$. Now $F_x(s_o)$ maps $X(s_o)$ in a 1 - 1 fashion onto $R(s_o) = R(F_x^o)$. For s near s_o, $F_x : X(s) \to R(s)$ does the same and therefore has a bounded inverse as a mapping between these spaces.

In the non-simple case we may construct the generalized eigenspace for which

$$(5.8) \qquad F_x^o \Phi_1 = 0 \quad \text{and} \quad F_x^o \Phi_2 = \Phi_1$$

One may then prove the existence of a smooth two-dimensional invariant subspace satisfying [3, 5]

$$(5.9) \qquad \begin{pmatrix} F_x(s) & 0 \\ 0 & F_x(s) \end{pmatrix} \begin{pmatrix} \Phi_1(s) \\ \Phi_2(s) \end{pmatrix} = (\Phi_1(s), \Phi_2(s))B(s)$$

where

$$B(s) = \begin{pmatrix} b_{11}(s) & b_{12}(s) \\ b_{21}(s) & b_{22}(s) \end{pmatrix} \quad \text{and} \quad B(s_o) = \begin{pmatrix} 0 & 0 \\ 1 & 0 \end{pmatrix}$$

The eigenvalues of $F_x(s)$ restricted to $N(s)$ are those of $B(s)$. Provided $\dot{b}_{12}(s_o) \neq 0$ these eigenvalues behave like $(s - s_o)^{1/2}$. With an appropriate choice of Φ_1 and Φ_2 one finds $\dot{b}_{12}(s_o) = -J$. In the same fashion $F_x^*(s)$ has a two dimensional invariant subspace spanned by $(\Phi_1^*(s), \Phi_2^*(s))$ and defining

$$(5.10) \quad N(s) = \text{span}\{\Phi_1(s), \Phi_2(s)\} \quad R(s) = \{x\varepsilon B_2 \mid \Phi_i^*(s)x = 0 \quad i = 1, 2\}$$

$$X(s) = \{x\varepsilon B_1 \mid \Phi_i^*(s)x = 0 \quad i = 1, 2\}$$

we have the decompositions

$$(5.11) \qquad B_2 = N(s) \oplus R(s) \qquad B_1 = N(s) \oplus X(s)$$

with corresponding projections $P(s)$ and $Q(s)$. Also $F_x(s) : X(s) \to R(s)$ is 1 - 1 and onto for s near s_o.

From these considerations we may state [5]

<u>Lemma (5.1)</u>. Let $x(s)$ be a smooth solution arc through $x(s_o)$, a simple bifurcation point, leading to an isolated root of the ABE (2.12). Then for $o < |s - s_o| < \delta$ some $\delta > o$ there exists a constant $N > o$ such that

$$(5.12)\qquad \begin{array}{lll} \text{(i)} & \text{if } \dot{\lambda}_o \equiv \xi_o \neq 0 & \|F_x^{-1}(x(s),s)\| \le \left|\dfrac{N}{s-s_o}\right| \\ \text{(ii)} & \text{if } \dot{\lambda}_o = 0 & \|F_x^{-1}(x(s),s)\| \le \left|\dfrac{N}{s-s_o}\right|^{1/2} \end{array}$$

We note that for $\dot{\lambda}_o = \xi_o = 0$ the behavior of the inflated system is, for our purpose, more satisfactory than that of the original system $G(u, \lambda)$. The eigenvalue of G remains simple for $\xi_o = 0$ and, near $s = s_o$ we have

$$(5.13)\qquad G_u(s)\phi(s) = \beta(s)\phi(s)$$
$$\phi(s_o) = \phi_1 \qquad \beta(s_o) = 0$$

Then as before we have

$$(5.14)\qquad \phi_1^* G_u(s_o)\dot{\phi}(s_o) + \phi_1^*(G^o_{uu}\dot{u}_o + G^o_{u\lambda}\dot{\lambda}_o)\phi_1 = \dot{\beta}(s_o)\phi_1^*\phi_1 + \beta(s_o)\phi_1^*\dot{\phi}(s_o)$$

and thus $\dot{\beta}(s_o) = \phi_1^* G^o_{uu}\dot{u}_o\phi_1 = \xi_1 a$. From (2.12) with root $(\xi_o, \xi_1) = (0, 1)$ we see $a = o$ and hence $\dot{\beta}(s_o) = 0$. Hence $\beta(s)$ vanishes at least like $(s - s_o)^2$ and thus $\|G^{-1}(u(s), \lambda(s))\|$ will grow at least as fast as $(s - s_o)^{-2}$. Inflation is therefore an aid to convergence in the neighborhood of such bifurcation points.

6. Shooting From a Bifurcation Point. In this section we assume $x(s_o)$ is a known solution at a simple bifurcation point. Further we assume (ξ_o, ξ_1) is an isolated root of the ABE and construct the tangent $\dot{x}(s_o) = (\xi_o\phi_o + \xi_1\phi_1, \xi_o)$. Neighboring solution points on the arc determined by this root are desired. Using the estimates of the two previous sections we establish convergence criteria for Euler-Newton (Chord) continuation.

First, for the Chord method, (4.8) and (5.12) provide

$$(6.1)\qquad \kappa(s)K(s)N|s - s_o|^\alpha < \frac{2}{3}$$

where $\alpha = 1$ for $\dot{\lambda}_o = \xi_o \neq 0$, and $\alpha = \frac{3}{2}$ for $\xi_o = 0$.
Provided $\kappa(s)$, the solution arc curvature, and $K(s)$ are well behaved for s near s_o, (6.1) assures convergence. Branches with $\dot{\lambda}_o = 0$ will in general allow a larger step-size (Figure 3).

For Newton's method, the estimates for $M_o(s)$ may be taken to be of the same form as those for $M(s)$, since only first order information was required in their derivation. In the same way we define the subspaces and projections $N(s)$, $R(s)$, $P(s)$, $Q(s)$ for $F_x(x^o(s), s)$ rather

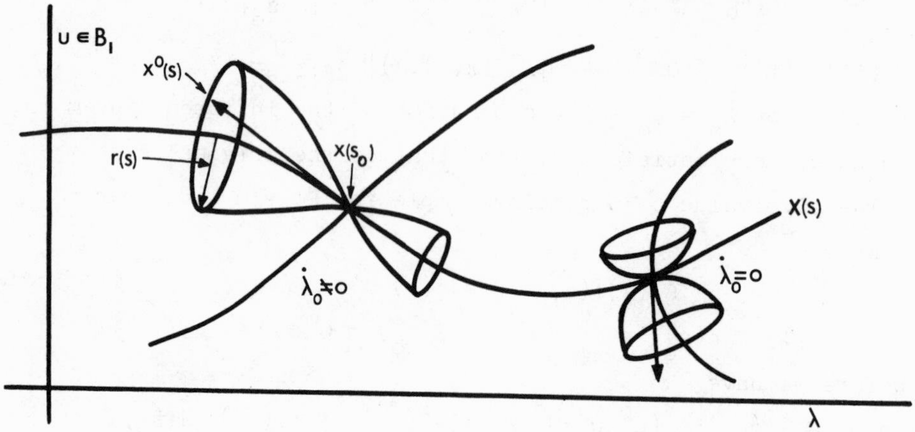

Figure 3

than $F_x(x(s), s)$. For a branch with $\dot{\lambda}_0 = 0$, (4.14) becomes (using (5.12))

(6.2) $$\kappa(s)L(s)H(s)N^2|s - s_0| < 1$$

where $L(s)$ is a Lipschitz bound on $F_x(y)$ for $\|y - x^0(s)\|$ of order $|s - s_0|^{3/2}$.

Convergence can still be assured, but with generally smaller steps than the Chord method allows. However turning to a branch with $\dot{\lambda}_0 = 0$, (4.14) would demand $\kappa(s)L(s)H(s)N^2 < 1$, which sufficiently small steps cannot assure. Hence, although the estimate (4.14) is adequate near regular points of F, it proves unsatisfactory at bifurcation points of this type. Sharper estimates are now derived by making use of the ABE.

Recalling the projections of the previous section

(6.3) $$\|F_x^{-1}(x^0(s),s)F(x^0(s),s)\| = \|F_x^{-1}(x^0(s),s)(P(s)+Q(s))F(x^0(s),s)\| \le \|F_x^{-1}(x^0(s),s)\|\,\|P(s)F(x^0(s),s)\| + \|F_x^{-1}(x^0(s),s)Q(s)F(x^0(s),s)\|$$

Now assuming the required smoothness, expanding $F(x^0(s), s)$ about $s = s_0$ with $x^0(s) = x(s_0) + (s - s_0)\dot{x}(s_0)$, $F_{xs} = F_{ss} = 0$, yields

(6.4) $$F(x^0(s),s) = F_{xx}(x(s_0),s_0)\dot{x}(s_0)\dot{x}(s_0)(s-s_0)^2/2 + O((s-s_0)^3)$$

For the simple eigenvalue situation $P(s_0)$ projects onto Φ_1, but $\Phi_1^*\Phi_1 = -2\xi_0 \neq 0$ and $\Phi_1^*F_{xx}(x(s_0), s_0)\dot{x}(s_0)\dot{x}(s_0) = 0$ from the ABE and (3.13). Hence

(6.5) $$\|P(s)F(x^o(s), s)\| = O((s - s_o)^3)$$

Since $F_x^{-1}(x(s), s)$ is bounded from $R(s)$ to $X(s)$ and $\|F(x^o(s), s)\| = O((s - s_o)^2)$ we have

(6.6) $$\|F_x^{-1}(x^o(s), s)Q(s)F(x^o(s), s)\| = O((s - s_o)^2)$$

Finally, using (5.12), the estimate (6.3) becomes

(6.7) $$\|F_x^{-1}(x^o(s), s)F(x^o(s), s)\| \le C_o(s - s_o)^2, \quad |s - s_o| < \delta$$

for some constants C_o, $\delta > o$.

Using this revised estimate, the convergence requirement ($abc < \frac{1}{2}$) becomes

(6.8) $$\kappa(s)L(s)NC_o|s - s_o| < 1$$

where $L(s)$ is a Lipschitz bound on $F_x(y)$ for $\|y - x^o(s)\| \le 2C_o(s-s_o)^2$. For this simple eigenvalue case, (6.8) would allow a Newton step-size of the same order as that of the Chord method (6.1).

Returning to the situation with $\dot{\lambda}_o = 0$, $P(s)$ now projects onto a two dimensional subspace, however (6.5) can still be shown to hold. Hence, although now

(6.9) $$\|F_x^{-1}(x^o(s), s)\| \|P(s)F(x^o(s), s)\| = O(|s - s_o|^{5/2})$$

equation (6.6) remains unaltered, and the overall estimate (6.7) is unchanged. Convergence will now require

(6.10) $$\kappa(s)L(s)NC_o|s - s_o|^{3/2} < 1$$

For both the Chord and Newton methods, the calculation of a branch with $\dot{\lambda}_o = 0$ proves easier than when $\dot{\lambda}_o \neq 0$. This is just the opposite for the uninflated system, where (Section (5))

(6.11) $$\|G_u^{-1}(u(s), \lambda(s)\| \le \frac{N}{|s - s_o|^\alpha}$$

with α at least 2 is the behavior to be expected.

We conclude that if $x(s_o)$ is a known simple bifurcation point, and the solution of the ABE results in two isolated roots, then nearby points on each branch may be computed using an Euler-Newton scheme. For this case the Chord method has no step-size advantage.

7. Location of a Simple Bifurcation Point. We now consider the more difficult problem of using a continuation method to approach and locate a simple bifurcation point. The previous estimates will be

seen to limit the possible approach rates of the Euler-Newton (Chord) methods, and these rates will depend on $d\lambda/ds$ at the bifurcation point.

Suppose Euler-Newton or Euler-Chord continuation has been used to follow a solution arc up to some position $(x(s_o), s_o)$. Further suppose this solution point $x(s_o)$ is nearby a simple bifurcation point which is <u>known</u> to exist at the position $s = s_*$. We wish to determine a sequence of allowable Euler steps (to be followed at each stage by Chord or Newton iteration) in order to approach and compute the bifurcation point $x(s_*)$. Within the context of the estimates derived in the previous sections, the largest allowable approach steps will be determined. Now provided $\dot{\lambda}(s_*) \neq 0$ on the branch being followed, the estimates (4.8) and (5.12) for the Chord method become

(7.1) $$(s - s_o)^2 \leq \sigma|s_* - s|$$

(7.2) where $$\sigma^{-1} \equiv \max_{s_o \leq t \leq s_*} 3/2\kappa(t)K(t)N$$

Now assuming σ is also known we solve for the maximum step-size $s_1 - s_o$ allowing convergence of the Chord iterates. Using this value, $x(s_1)$ is computed with the Chord method and the process repeated. In this way an approach sequence $s_1, s_2, \ldots$ is generated with

(7.3) $$(s_k - s_{k-1})^2 = \sigma|s_* - s_k|$$

and depicted in Figure (4). Defining $e_k = s_* - s_k$, and solving (7.3) yields

(7.4) $$e_k \leq (4/\sigma)e_{k-1}^2 \qquad k = 1, 2, \ldots$$

giving a quadratic approach to s_* for e_o sufficiently small. With the present estimates, this provides the best or "optimal" approach rate for the Chord method.

Turning to Newton's method we still have the estimate (6.3) and

(7.5) $$F(x^o(s),s) = F_{xx}(x(s_o),s_o)\dot{x}(s_o)\dot{x}(s_o)\frac{(s-s_o)^2}{2} + O((s-s_o)^3)$$

However

(7.6) $$F_{xx}(x(s_o),s_o)\dot{x}(s_o)\dot{x}(s_o)-F_{xx}(x(s_*),s_*)\dot{x}(s_*)\dot{x}(s_*)=O((s_*-s_o))$$

as $s_o \to s_*$. Thus (6.5) becomes

$$(7.7) \qquad \|P(s)F(x^o(s), s)\| = R(s; s_o)(s - s_o)^2(s_* - s_o)$$

with $R(s; s_o)$ bounded for s, s_o near s_*. This results in the bound

$$(7.8) \qquad \|F_x^{-1}(x^o(s),s)F(x^o(s),s)\| \leq C_o \frac{(s-s_o)^2(s_*-s_o)}{(s_*-s)} \qquad 0<|s-s_*|<\delta$$

and the convergence estimate $abc < 1/2$ becomes

$$(7.9) \qquad (s - s_o)^2(s_* - s_o) \leq \Sigma^2(s_* - s)^2$$

Here Σ is a constant difficult to estimate, but assuming it to be known we may solve for the maximum step-size $s_1 - s_o$. Using this procedure to generate the sequence $s_1, s_2, \ldots$ results in

$$(7.10) \qquad e_k \leq (1/\Sigma)e_{k-1}^{3/2} \qquad k = 1, 2, \ldots$$

Although super-linear, this approach is slower than the Chord method. This must be balanced, however, with the linear convergence of the Chord method in the calculation of $x(s_k)$ at each stage.

For the approach to a singular point along a branch with $\dot{\lambda}(s_*) = 0$, the Chord estimate (7.1) is modified to (Figure (5))

$$(7.11) \qquad (s - s_o)^2 \leq \sigma|s_* - s|^{1/2}$$

This convergence requirement allows an approach with

$$(7.12) \qquad e_k \leq (4/\sigma)^2 e_{k-1}^4 \qquad k = 1, 2, \ldots$$

For Newton's method the estimate (7.8) becomes

$$(7.13) \qquad \|F_x^{-1}(x^o(s),s)F(x^o(s),s)\| \leq C_o \frac{(s-s_o)^2(s_*-s_o)}{|s_*-s|^{1/2}}$$

resulting in the convergence requirement

$$(7.14) \qquad (s - s_o)^2(s_* - s_o) \leq \Sigma(s_* - s)$$

This allows an approach rate satisfying

$$(7.15) \qquad e_k \leq (4/\Sigma)e_{k-1}^3 \qquad k = 1, 2, \ldots$$

Just as a branch with $\dot{\lambda}(s_*) = 0$ allowed a larger step in shooting from a bifurcation point, it now allows a faster approach to the bifurcation point than a branch with $\dot{\lambda}(s_*) \neq 0$.

The above results are of limited practical application since it is doubtful the position s_* of a bifurcation point would be known in advance. However they do describe the nature of an optimal approach to the singular point, thereby limiting the behavior to be expected

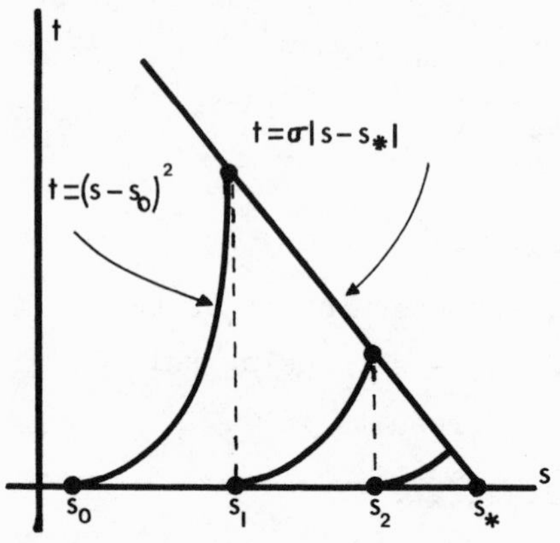

Figure 4 Chord with $\dot{\lambda}(s_*) \neq 0$

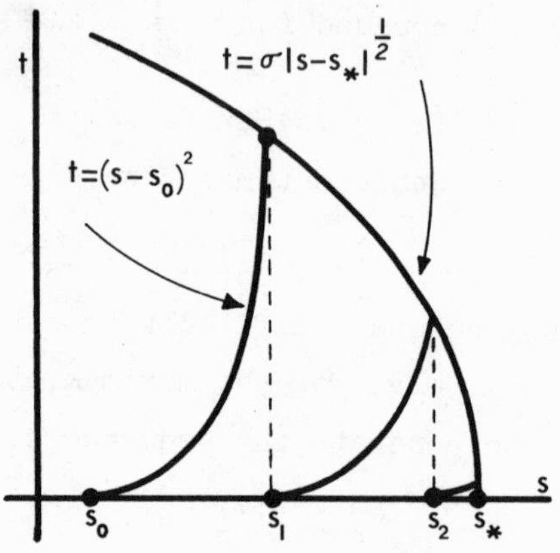

Figure 5 Chord with $\dot{\lambda}(s_*) = 0$

in numerical computations. Some of the considerations involved in the design of practical approach algorithms are the object of present study and will be presented elsewhere.

REFERENCES

[1] S. N. CHOW, J. K. HALE, and J. MALLET-PARET, Applications of generic bifurcation, I, Arch. Rational Mech. Anal., 59(1975), pp. 159-188.

[2] M. G. CRANDALL and P. H. RABINOWITZ, Bifurcation from simple eigenvalues, J. Func. Anal., 8(1971), pp. 321-340.

[3] D. W. DECKER, Topics in Bifurcation Theory, Thesis, California Institute of Technology, Pasadena, California 1978.

[4] D. W. DECKER and H. B. KELLER, Multiple limit point bifurcation, to appear in J. Math. Anal. Appl.

[5] D. W. DECKER and H. B. KELLER, Continuation methods for bifurcation problems, in preparation.

[6] L. V. KANTOROVICH, Functional Analysis in Normed Spaces, MacMillan, New York 1964.

[7] H. B. KELLER, Bifurcation Theory and Nonlinear Eigenvalue Problems, unpublished lecture notes, California Institute of Technology, Pasadena, California (1976).

[8] H. B. KELLER, Numerical solution of bifurcation and nonlinear eigenvalue problems, in Applications of Bifurcation Theory, Academic Press, New York 1977, pp. 359-384.

[9] L. NIRENBERG, Topics in Nonlinear Functional Analysis, Courant Institute of Mathematical Sciences, New York 1974.

[10] D. H. SATTINGER, Group representation theory and branch points of nonlinear functional equations, SIAM J. Math. Anal., 8(1977), pp. 179-201.

Patterns of Bifurcation

John Guckenheimer*

Abstract. This paper summarizes basic results about the bifurcations of dynamical systems. It emphasizes results which relate different elementary bifurcations in one parameter families of flows. There are two principal techniques for developing this relationship: the study of codimension two bifurcations of equilibrium points and the study of iterating one dimensional mappings.

This paper is intended to be a user's guide to recent developments in the theory of bifurcations of differential equations. We explain new results and then discuss ways in which they might be helpful in the numerical investigations of specific problems. The reader is warned, however, that there is still little practical experience in these areas. The recipes we provide for studying the bifurcations of differential equations have not been thoroughly tested numerically and surprises can be expected until more problems have been treated. The structure of the paper is as follows. In this introduction we describe the nature of the problems dealt with and the strategy to be employed in solving them. Section §1 reviews standard results of bifurcation theory from our perspective, thereby establishing the language used in the rest of the paper. Section §2 discusses dynamical phenomena in

*Board of Studies in Mathematics, University of California, Santa Cruz, California 95064. Research partially supported by National Science Foundation grant MCS 79-02002.

systems depending upon two parameters. Section §3 introduces one dimensional difference equations and explains their relevance to global patterns of bifurcation which seem to occur in certain problems.

Our central concern is with systems of ordinary differential equations. Much of bifurcation theory has concerned itself with partial differential equations where the issue of existence and uniqueness of solutions of evolution equations arises. There are a variety of techniques for coping with the problems of an infinite number of degrees of freedom, but we shall not concern ourselves with them. This is not to say that the theory we describe is not applicable but only that some method (such as a center manifold theorem [25]) has reduced the problem to one of a finite dimensional system of ordinary differential equations. We shall also have occasion to work with different equations and discuss connections between the bifurcation theories for differential and difference equations. In all cases, we work with systems which depend upon parameters—usually one or two.

The strategy which we employ makes heavy use of transversality and the concept of genericity. The aim of bifurcation theory is the description of qualitative changes which take place in the solutions of differential equations as parameters are varied. We shall always assume that these changes are occurring in a way which is robust. By this we mean roughly that if the entire system with its parameters is changed slightly then the phenomena being studied remain. This does not mean that we are uninterested in all degenerate situations. Faced with a mildly degenerate situation, we study it by trying to find a context in which it does become robust. In a number of settings, there are only a few different types of degeneracy which can occur. By studying these in specific examples, one can compile a dictionary (or "zoo") of the bifurcations which one expects to encounter in typical problems. By allowing the degree of degeneracy to grow, one can build a picture of how different bifurcations usually interact with one another. Since numerical computation of qualitative properties near degenerate situations is difficult, this can be of great benefit in understanding numerical solutions of complicated problems.

The general picture with which we work is stratified in the spirit of Thom's catastrophe theory [40]. One starts with some space of systems, usually a space of smooth vector fields (= differential equations), diffeomorphisms (= invertible difference equations), or maps (= difference equations). This space should have some nice topological structure. In this space, one introduces a concept of "structural stability"* or "robustness" which gives the technical interpretation of when all systems close to a given one are qualitatively similar. In nice situations, these robust systems will be dense, and the ones which are not will have a stratified structure. This means that there should be a partition of the non-robust systems into strata. Each stratum should be a submanifold; i.e., it should be defined (locally) as the set of zeros of k independent, smooth functions where k is the codimension of the stratum. A typical family of systems which depends upon k parameters will contain only systems whose codimension is at most k.

*Structural stability usually has a particular technical meaning [33]. Structurally stable vectorfields are dense only in one and two dimensions.

This scheme does not adapt itself well to systems of ordinary differential equations when one takes a concept of robustness which requires that solutions be followed for arbitrarily large times. The higher codimension bifurcations of dynamical systems do not seem to be amenable to an elegant mathematical characterization. The theory is a mess and seems inherently messy. Nevertheless, there is a collection of results and techniques which can be useful in applications as an alternative to commonly employed asymptotic expansions. The framework of this theory has the virtues of conceptual clarity and a reasonable foundation of non-trivial analysis upon which to base examination of geometrically complicated phenomena.

§1. Elementary Codimension 1 Bifurcations

In this section we describe the elementary codimension one bifurcations of equilibria and periodic orbits which provide the language upon which later sections are based. These elementary bifurcations give the simplest ways in which the qualitative structure of the solutions of a system of differential equations change. We begin by recalling some facts about equilibria and periodic solutions of differential equations.

A system of differential equations $\dot{x}_i = f_i(x_1,\ldots,x_n)$, $i = 1,\ldots,n$, or $\dot{x} = F(x)$ defines a vector field X on $\mathbb{R}^n$ and its flow $\Phi : \mathbb{R}^n \to \mathbb{R}$. The flow is the map which advances points along their solutions: $\Phi(x,t) = x(t)$ where $x(t)$ is the solution with initial condition $x = x(0)$. There are two kinds of distinguished solutions of a system of differential equations. An equilibrium solution has $x(t) \equiv x(0)$ and $F(x) = 0$. A periodic solution of (least) period τ is a solution with

$x(\tau) = x(0)$ but $x(t) \neq x(0)$ for $0 < t < \tau$. Our primary concern is the way in which the equilibrium and periodic solutions change as parameters in a system are varied.

The technique which we use for studying equilibrium solutions depends upon looking at the Taylor series expansion of the vector field there. If p is an equilibrium solution for the equations $\dot{x} = F(x)$ and if $F(x) = L(x-p) + o(x-p)$, then the first part of our analysis is an examination of the linearized equations $\dot{\xi} = L\xi$ with $\xi = (x-p)$. The behavior of this equation depends upon the eigenvalues of L. If L has no eigenvalues which are zero or pure imaginary, then p is called hyperbolic. Hyperbolicity has two important consequences. First, if L has no zero eigenvalue, then p is a simple (transverse) zero of the equation $F(x) = 0$. The implicit function theorem implies that if F depends upon parameters μ, then there will be a smooth function $p(\mu)$ of equilibria for the equation $\dot{x} = F_{\mu}(x)$. No changes in the number of equilibria occur at p. The second conclusion is that the qualitative structure of the flow Φ for the original system of equations is the same as that for the linearized system near $\xi = 0$. This is a consequence of Hartman's Theorem and the Stable Manifold Theorem [33]. In particular, if s eigenvalues of L (with multiplicities) have negative real parts and $n-s = u$ eigenvalues of L have positive real parts, then there are manifolds W^s and W^u of dimensions s and u, respectively, which contain all solutions asymptotic to p as $t \to \infty$ or $t \to -\infty$. These are called the stable and unstable manifolds of p. No qualitative changes in the flow occur near a hyperbolic equilibrium when the vectorfield is perturbed.

There is a corresponding concept of hyperbolicity for periodic orbits. Let γ be a periodic orbit of period τ and let $\phi_\tau: \mathbb{R}^n \to \mathbb{R}^n$ be the transformation which advances all solutions τ units of time. Then the Jacobian derivative at $p \in \gamma$ of ϕ_τ automatically has one eigenvalue 1 with eigenvector tangent to γ. Thus one says that γ is <u>hyperbolic</u> if 1 is a simple eigenvalue and there are no other eigenvalues of modulus 1. (Pure imaginary eigenvalues of a linear vector field give eigenvalues of modulus 1 for a time t map ϕ_t upon integration.) Once again hyperbolic periodic orbits behave nicely with respect to parameter changes and have stable and unstable manifolds.

We next consider the simplest ways in which equilibria and periodic orbits can fail to be hyperbolic. For each of these, we embed the corresponding vector field in a one parameter family of vector fields which "unfolds" the degeneracy. These one parameter families have the property that when they are perturbed there will still be a parameter value at which a degeneracy occurs and this bifurcation is qualitatively similar to the original one. These elementary codimension one bifurcations comprise the dictionary upon which the study of more complicated phenomena is based.

The first case we consider is an equilibrium at which the linearization has a simple zero eigenvalue and no pure imaginary eigenvalues. To ensure that this occurs in the simplest possible way, it is necessary to place a condition on the quadratic terms in the Taylor series expansion of F at p. If L is the Jacobian matrix of F at p, v is the right zero eigenvector of L, and w is the left zero eigenvector of L, then the condition is that $w \cdot D^2F(v,v) \neq 0$. Here D^2F is the array of second partial derivatives of F at p. This condition means that there is a non-zero quadratic part to F in the direction of the degeneracy of its linear part. With this assumption, we have a saddle-node. The invariant manifolds of the flow near a saddle-node are depicted in Figure 1 for the case n = 3 with L having one positive and one negative eigenvalue. Topologically, these invariant manifolds are half-spaces for a saddle-node.

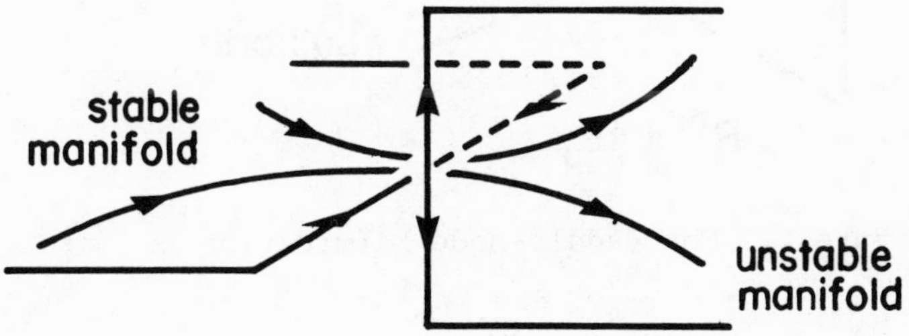

Figure 1. Stable and unstable manifolds of a saddle-node.

Of additional interest for the saddle-node is its bifurcation behavior. To understand this, we embed $\dot{x} = F(x)$ in a one parameter family of differential equations $\dot{x} = F(x)$ so that the eigenvalue $\lambda(\mu)$ which satisfies $\lambda(\mu_o) = 0$ at the saddle-node also satisfies $\frac{d\lambda}{d\mu}(\mu_o) \neq 0$. With this assumption, the locus of equilibria in $\mathbb{R}^n \times \mathbb{R}$ is a curve, with a quadradic tangency to $\mathbb{R}^n \times \{\mu_o\}$. This is shown in Figure 2. Depending upon the signs of $\frac{d\lambda}{d\mu}(\mu_o)$ and $w \cdot D^2F(v,v)$, this bifurcation represents the birth or coalescence of a pair of equilibria. For each parameter value on the appropriate side of μ_o, there is a pair of

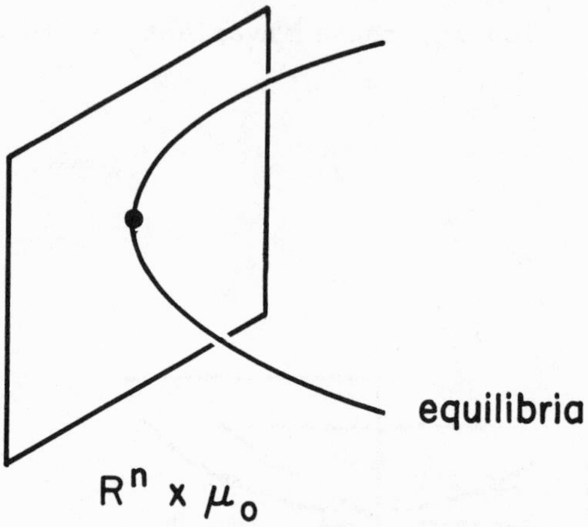

Figure 2. The saddle-node bifurcation

hyperbolic equilibria, the dimensions of whose stable manifolds differs by 1. The simplest example of a saddle-node bifurcation is given by the single equation $\dot{x} = \mu - x^2$. For $\mu < 0$, the flow of this equation is always to the left. For $\mu > 0$, there are equilibria at $+\sqrt{\mu}$ (a sink) and at $-\sqrt{\mu}$ (a source). All other saddle-node bifurcations look like this example with hyperbolic behavior in transverse directions.

The second elementary bifurcation involving an equilibrium p occurs when the matrix of the linearized equations at p has a pair of pure imaginary eigenvalues. This is the case of Hopf bifurcation. In a one parameter family the equilibrium varies smoothly, but the dimensions of its stable and unstable manifolds changes. This happens in association with a family of periodic solutions originating or coalescing at the bifurcating equilibrium. The Hopf bifurcation has been the subject of enormous interest the past few years as a mechanism whereby stable equilibria give way to oscillatory behavior with changes in parameters.

Computations involving the Hopf bifurcation are done most easily when special coordinates are introduced. These special coordinates are called <u>normal forms</u>. Let $\dot{x} = F_\mu(x)$ be a one parameter family such that when $\mu = \mu_o$, there is an equilibrium p at which the Jacobian derivative L of $F\mu_o$ has a pair of pure imaginary eigenvalues $\pm i\lambda$ and no other eigenvalues which are pure imaginary or zero. We shall make a series of reductions and coordinate transformations to better study the bifurcation at p. In the first of these reductions, we note that coordinates (u,v,w) can be chosen so that L has the form $\left(\begin{array}{c|c|c} L_1 & 0 & 0 \\ \hline 0 & L_2 & 0 \\ \hline 0 & 0 & L_3 \end{array}\right)$

where L_1, L_2, L_3 are square matrices with the eigenvalues of L_1 in the left half plane, the eigenvalues of L_2 in the right half plane, and the eigenvalue of L_3 being $\pm i\lambda$. Now the behavior of the (u,v) coordinates is approximately like that in the hyperbolic case, so we restrict our attention to the w coordinates in studying the bifurcation. This process of restricting our attention to a 2-dimensional system of equations can be rigorously justified, even in many infinite dimensional cases, by the use of the <u>Center Manifold Theorem</u> [25].

Having reduced the problem to a 2-dimensional system, we next observe that the coordinates in this place can be chosen so that the Jacobian derivative at p is the matrix of an infinitesimal rotation:

$L = \begin{pmatrix} 0 & -\lambda \\ \lambda & 0 \end{pmatrix}$. In polar coordinates the linearized equations are $\dot{r} = 0$, $\dot{\phi} = \lambda$. By further coordinate transformations of the form identity + non-linear terms, some of the non linear terms in the equations can be eliminated. To see how this works, consider the quadratic equations $\dot{x}_1 = -x_2 + P_1(x_1,x_2)$ and $\dot{x}_2 = x_1 + P_2(x_1,x_2)$ where P_1 and P_2 are homogeneous quadratic polynomials. Introduce the coordinate transformation $y_1 = x_1 + Q_1(x_1,x_2)$, $y_2 = x_2 + Q_2(x_1,x_2)$ with Q_1 and Q_2 quadratic. Then $x_i = y_i - Q_i(y_1 y_2)$ + higher order terms and $\dot{y}_i = \dot{x}_i +$

$$\frac{\partial Q_i}{\partial x_i}\dot{x}_i + \frac{\partial Q_i}{\partial x_2}\dot{x}_2 = \lambda\left[(-1)^j x_j - \frac{\partial Q_i}{\partial x_i}x_2 + \frac{\partial Q_i}{\partial x_2}x_1\right] + P_i + \text{higher order terms} =$$

$$\lambda\left[(-1)^j y_j + (-i)^j Q_j - \frac{\partial Q_i}{\partial x_1}y_2 - \frac{\partial Q_i}{\partial x_i}y_1\right] + P_i + \text{higher order terms.}$$ Here

$i \neq j$. Let us compute the quadratic terms in parentheses. If

$Q_i(x_1,x_2) = \alpha_{i1}x_1^2 + \alpha_{i2}x_1x_2 + \alpha_{i3}x_2^2$, then $+ Q_2 - \frac{\partial Q_1}{\partial x_1}x_2 + \frac{\partial Q_1}{\partial x_2}x_1 =$

$(\alpha_{12} + \alpha_{21})x_1^2 + (-2\alpha_{11} + 2\alpha_{13} + \alpha_{22})x_1x_2 + (-\alpha_{12} + \alpha_{23})x_2^2$ and $-Q_1 - \frac{\partial Q_2}{\partial x_1}x_2 + \frac{\partial Q_2}{\partial x_2}x_1 = (-\alpha_{11} + \alpha_{22})x_1^2 + (-\alpha_{12} - 2\alpha_{21} + 2\alpha_{23})x_1x_2 +$

$(-\alpha_{13} - \alpha_{22})x_2^2$. A bit of linear algebra shows that we can always

choose Q_1 and Q_2 so that $P_i + \left((-1)^j Q_j - \frac{\partial Q_i}{\partial x_i}x_2 + \frac{\partial Q_i}{\partial x_2}x_1\right) = 0$. Thus

it is possible to make a coordinate transformation after which the equations have no quadratic terms.

These computations are relatively long and tedious. A systematic exposition involves the introduction of Lie algebras [38] and will not be given here. Applied to the Hopf bifurcation, the result of these computations is that coordinates can always be found for which our system of equations takes the form

$$\dot{x}_1 = -x_2 + (ax_1 - bx_2)(x_1^2 + x_2^2) + \text{higher order terms}$$
$$\dot{x}_2 = x_1 + (ax_2 + bx_1)(x_1^2 + x_2^2) + \text{higher order terms}$$

or in polar coordinates

$$\dot{r} = ar^3 + \text{higher order terms}$$
$$\dot{\phi} = \lambda + br^2 + \text{higher order terms}$$

Now one condition that the Hopf bifurcation occurs in the simplest possible way is that the coefficient a in these last equations is non-zero. Its sign determines whether the singular point at the origin is weakly attracting (-) or weakly repelling (+). A computation of the coefficient for a general vector field in Cartesian coordinates is in [25]. The other condition placed upon a one parameter family is that the derivative of the real part of the eigenvalues at the bifurcation point with respect to the parameter should not vanish. Thus a normal form for the Hopf bifurcation in polar coordinates is given by the equation

$$\dot{r} = \mu r + ar^3 + \text{higher order terms}$$
$$\dot{\phi} = \lambda + \text{higher order terms}$$

These equations are separable (ignoring the higher order terms), so it is easy to study their solutions. If $\mu a < 0$, then there is a periodic orbit defined by $r = \sqrt{-\mu/a}$. It is stable if $\mu > 0$ and repelling if $\mu < 0$. The flow is depicted in Figure 3. This completes our description of the Hopf bifurcation. In many examples, reducing a system of equations to normal form, thereby calculating the coefficient of r^3, is a lengthy task when it can be accomplished. The paper by Hassard in this volume describes computational procedures for these calculations.

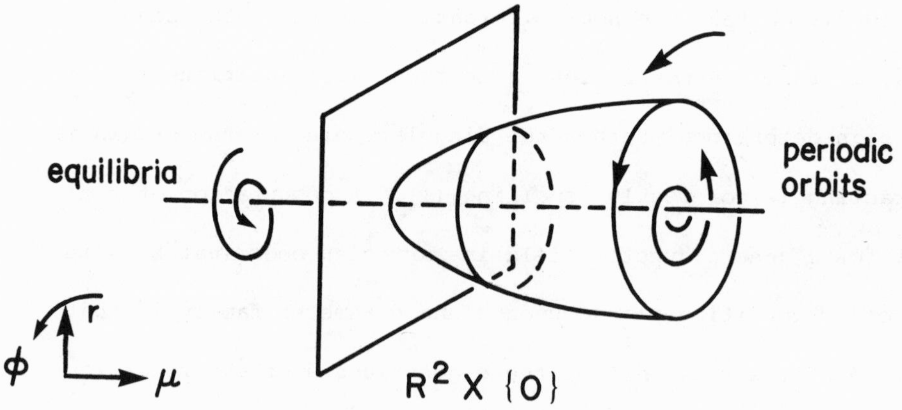

Figure 3. The Hopf bifurcation

The Hopf bifurcation theorem gives us an effective way of locating systems of equations with periodic orbits. Periodic orbits are subject to other elementary bifurcations. Pictorially, these are most easily studied via a <u>return map</u> for the periodic orbit. If γ is a periodic orbit in $\mathbb{R}^n$, a <u>cross-section</u> Σ is an (n-1) dimensional surface intersecting γ transversally at a point $p \in \gamma$. Solutions with initial conditions in Σ near p return to Σ in a time near the period of γ. The return map $\mathcal{O}:\Sigma \to \Sigma$ is defined near p by setting $\mathcal{O}(x)$ to be the first point of intersection with Σ of the solution with initial condition x. The map $\mathcal{O}$ has a fixed point at p. By using return maps, there is an equivalence between the bifurcation of fixed points for (invertible) maps of $\mathbb{R}^{n-1}$ and of periodic orbits for differential equations on $\mathbb{R}^n$. Thus we shall describe the bifurcations of fixed points for discrete time systems defined by a map $f:\mathbb{R}^{n-1} \to \mathbb{R}^{n-1}$ and translate the results back to information about flows.

The condition that a fixed point p of a map f be hyperbolic is that no eigenvalues of the Jacobian derivative Df(p) have absolute value 1. There are analogues of Hartman's Theorem and the Stable Manifold Theorem described above for fixed points. Near a hyperbolic fixed point, the orbits of a map look like those of its linearization, and there are stable and unstable manifolds tangent to the eigenspaces belonging to eigenvalues inside and outside the unit circle. To study the elementary codimension one bifurcations, there are three cases to consider: eigenvalue 1, eigenvalue -1, and a pair of complex eigenvalues of absolute value 1 for the Jacobian derivative at a fixed point. These give rise to bifurcations which are called the saddle-node, flip, and secondary Hopf bifurcations, respectively.

The analysis of the saddle-node bifurcation proceeds just as in the case of equilibria. In the direction of the degeneracy, one assumes that the quadratic part of the map does not vanish. The normal form for a one parameter family having a saddle node bifurcation is $\theta_\mu(x) = \mu + x - x^2$. For $\mu < 0$, we always have $\phi_\mu(x) < x$. For $\mu > 0$, there are two fixed points, an attractor at $x = \sqrt{\mu}$ and a repellor at $x = -\sqrt{\mu}$. The fixed points and the direction of trajectories look like those for the flow case illustrated in Figure 2.

The flip bifurcation has no counterpart for equilibria. The first thing to note is that a map with a negative eigenvalue reverses orientation in the direction of its eigenvector. Thus the center manifold of a periodic orbit with eigenvalue -1 is a Möbius band with the periodic orbit along its center line. Secondly, note that if $\theta : \mathbb{R} \to \mathbb{R}$ satisfies $\theta(p) = p$, $\theta'(p) = -1$, then $(\theta \circ \theta)'(p) = 1$ and $(\theta \circ \theta)''(p) = 0$. Thus $\theta \circ \theta$ would be a candidate for a saddle-node bifurcation at p, but the condition that the quadratic part be non-zero cannot hold. This produces a different picture at this bifurcation. The normal form is $\theta_\mu(x) = (-1 + \mu)x + x^3$. For $\mu < 0$, the fixed point is stable and all points are asymptotic to it. For $\mu > 0$, the fixed point is unstable, but $(\theta_\mu \circ \theta_\mu)(x) = (-1 + \mu)^2 x + 2(-1 + \mu)x^3$ + higher order terms has fixed points near $x = \pm\sqrt{(1-\mu)/2}$ in addition to $x = 0$. These two points are a stable periodic orbit of period 2 for the map θ_μ. For a corresponding flow, they represent a periodic orbit with approximately twice the period of the original one from which it bifurcated. See Figure 4 for a diagram.

The secondary Hopf bifurcation is more complex than the bifurcations we have studied up to this point. Let $\theta : \mathbb{R}^2 \to \mathbb{R}^2$ have a fixed point at which there are a pair of complex eigenvalues of absolute value one. We can introduce coordinates so that the fixed point is the origin and the derivative of $\theta(0)$ is a rotation $\begin{pmatrix} \cos\lambda & -\sin\lambda \\ \sin\lambda & \cos\lambda \end{pmatrix}$. We

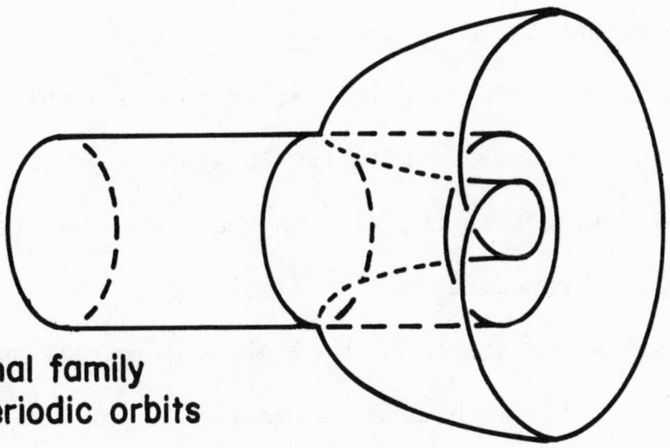

Figure 4. The flip bifurcation

would like to make further coordinate changes to remove non-radial dependence in the quartic terms of the Taylor series of θ. This computation is a normal form computation similar to the one outlined for the Hopf bifurcation of equilibria. Here we need to assume that the eigenvalues are not third or fourth roots of unity to obtain the desired normal form. The special cases of <u>low order resonance</u> when the eigenvalues are $\pm e^{2\pi i/3}$ or $\pm e^{\pi i/2}$ must be treated separately, and some details of the fourth order resonance case are still not clear [5]. If there are no low order resonances, then the fourth order Taylor polynomial of θ at 0 can be assumed to be $\theta(r,\phi) = (r + ar^3, \lambda + \phi + br^2)$ in polar coordinates.

The lack of ϕ dependence in the r coordinate of θ leads to invariant curves which bifurcate from the fixed point in a one-parameter family, but the existence of these curves is substantially harder to prove than the analysis of the previous bifurcations. The invariant curve is the cross-section of an invariant two-dimensional torus for the vector field. It is composed of many trajectories rather than a single one and this accounts for the increased difficulty in its analysis. There is the additional complication that the flow on the torus can itself have different qualitative properties.

The study of differential equations on torii is a classical subject initiated by Poincaré. A flow without equilibria on a two dimensional torus has a rotation number [3] which measures the ratio of the asymptotic rates at which trajectories wind around the torus relative to the two angular variables. This rotation number is rational if and only if the flow has periodic orbits. If the flow is sufficiently smooth (Lipschitz first derivative) and the rotation number is irrational, then all trajectories are dense in the torus and the motion is quasiperiodic [3]. Typically in a one parameter family of flows on a torus, the set of parameter values for which the rotation number is rational is open and dense, but its complement has positive measure [17]. Thus, the details of the qualitative structure of flows on a torus are sensitive to small changes. It is often difficult to determine numerically whether one has a rational or irrational flow.

This completes our dictionary of elementary codimension one bifurcations of periodic orbits. Before passing to codimension 2 phenomena, there is one final codimension one bifurcation phenomenon we want to describe. If p and q are equilibria for a system of differential equations, then the stable manifold of p and the unstable manifold of q generally meet transversally. However, in a generic one parameter family, there may be parameter values for which there are non-transverse intersections. The simplest example occurs in the plane with a common separatrix of two saddle points. These two saddle-points could be the same, in which case we have a saddle-loop. This is a trajectory which is asymptotic to a saddle point equilibrium as $t \to \pm\infty$. A common bifurcation involving periodic orbits is that they disappear by their period becoming larger and larger as they become saddle loops. We encounter examples in the next section.

§2. Codimension 2 Bifurcations

In the last section we compiled a list of elementary bifurcation phenomena which occur in one parameter families of systems of differential equations. There are many problems in which several bifurcations occur and one would like to find some pattern or rationale for the relationship between different bifurcations. In this section and the next we describe two approaches to discerning such patterns. We reach the "state of the art" quickly, and the reader is alerted that there is little experience in applying these theories. Thus the prevalence of the patterns we describe is yet to be determined.

Consider now a system of equations which depends upon two parameters. In the parameter plane one expects to find curves along which various of the elementary codimension one bifurcations take place. If these curves end inside the parameter plane, they do so at places where the corresponding system is more degenerate. Here we reverse this procedure and examine the "morphology" of bifurcation curves which meet at degenerate equilibria occurring in generic two parameter families. By passing near one of these codimension two bifurcations in a one parameter family, several bifurcations may occur in a way which is determined by the morphology we describe.

There are five distinct ways in which an equilibrium can have a degeneracy of codimension two. In the first two, the linear part is the same as for the saddle and Hopf bifurcations, but there is an additional degeneracy in a higher order term. In the other three cases, there is additional degeneracy in the linear terms. These three cases correspond to a double eigenvalue zero with a corresponding block $\begin{pmatrix} 0 & 1 \\ 0 & 0 \end{pmatrix}$ in the Jordan normal form, a zero eigenvalue together with a pair of pure imaginary eigenvalues, and a pair of pure imaginary eigenvalues. Let us consider the first four cases in turn. The fifth has not been analyzed yet.

The first example corresponds to the saddle-node. The non-hyperbolic behavior occurs in one-dimension, and the bifurcation is described by Thom's cusp catastrophe [40]. A normal form for a two parameter family is given by $\dot{x} = \mu + \varepsilon x - x^3$. Here saddle-nodes occur along the curve $27\mu^2 - 4\varepsilon^3 = 0$. The second example with the degenerate Hopf bifurcation is also easy to analyze [36]. In normal form polar coordinates, one studies the equation $r = \mu r + \varepsilon r^3 - r^5$. When $\mu = 0$, $\varepsilon \neq 0$, ordinary Hopf bifurcation occurs. When $\varepsilon > 0$, $\varepsilon^2 + 4\mu = 0$, there is a saddle-node in this amplitude equation for r. This indicates that a saddle-node of periodic orbits occurs. The bifurcation diagrams for these two cases are shown in Figures 5 and 6.

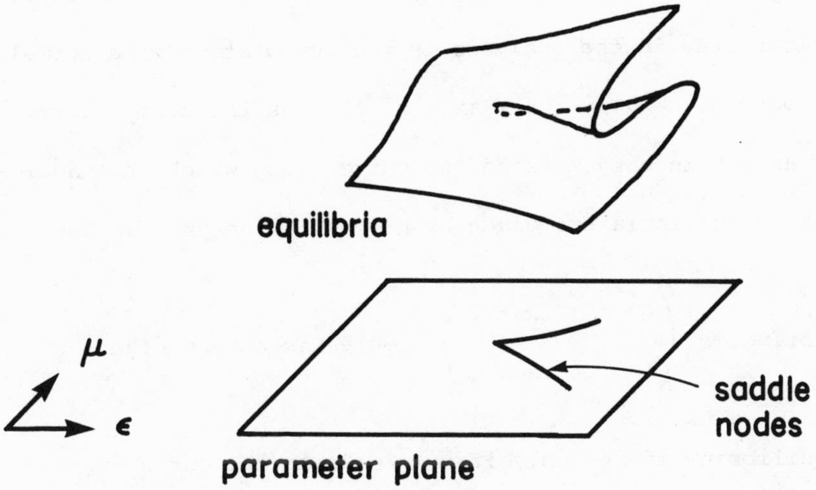

Figure 5. Thom's cusp catastrophe of equilibria

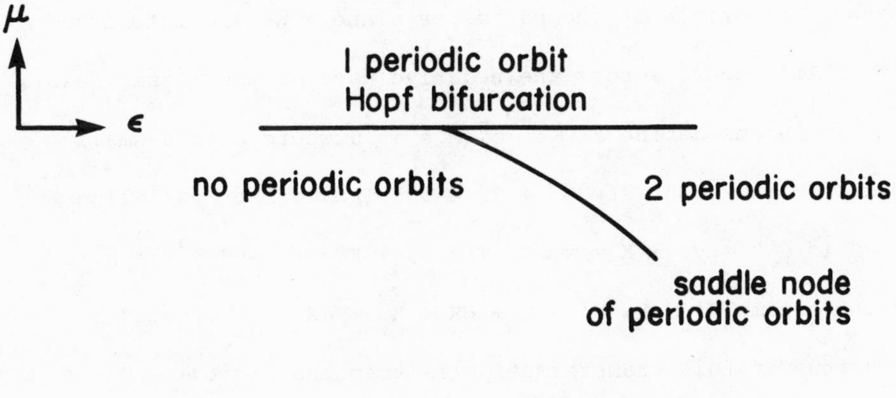

Figure 6. The bifurcation diagram of a degenerate Hopf bifurcation.

The study of the remaining cases requires new techniques beyond those which we have already introduced. Let us examine the case of two zero eigenvalues first. We assume that X is a vector field on the plane which has an equilibrium at the origin with Jacobian derivative $\begin{pmatrix} 0 & 1 \\ 0 & 0 \end{pmatrix}$. The first step in the analysis is the computation of a normal form. Here we obtain $\dot{x} = y$, $\dot{y} = \mu + \varepsilon x + x^2 \pm xy$ as the normal form. In the (μ,ε) plane it is easy to find the curve along which the saddle-nodes occur. The equilibria are given by $\mu + \varepsilon x + x^2 = y = 0$. The Jacobian derivative is $\begin{pmatrix} 0 & \varepsilon + 2x \pm y \\ 1 & \pm x \end{pmatrix}$. This has a zero eigenvalue at an equilibrium if and only if $x = -\varepsilon/2$, so the curve of saddle-nodes is given by $4\mu = \varepsilon^2$. Note also that pure imaginary eigenvalues giving Hopf bifurcations occur when $\mu = 0$ and $\varepsilon < 0$.

It is not easy to determine the fate of periodic orbits which arise through Hopf bifurcation in this system. The appropriate technique [39] is to introduce a rescaling of the equation so that the size of observed periodic behavior remains approximately constant as one approaches the origin of the parameter plane. We are interested in behavior which occurs across the negative part of the ε axis near the origin, so let us define $E = \delta^{-2}\varepsilon$, $M = \delta^{-4}\mu$ where δ is a small parameter, and M and E satisfy $M^2 + E^2 = 1$. The correct rescaling for our purposes is $\delta^3 Y = y$, $\delta^2 X = x$, $\delta^{-1} T = t$, $\delta^2 E = \varepsilon$, and $\delta^4 M = \mu$. The equations become $X' = Y$, $Y' = M + EX + X^2 \pm \delta XY$, $\frac{d}{dT}$ denoted by "'". The virtue of this transformation is that the limit $\delta \to 0$ gives us a system of equations which we can analyze.

The system of equations $X' = Y$, $Y' = M + EX + X^2$ is a Hamiltonian system whose solutions are constant along the level curves of $H(X,Y) = Y^2 - (MX + \frac{1}{2}EX^2 + \frac{1}{3}X^3)$. For $E = -1$, $M = 0$, there is a family of periodic orbits for $H(X,Y) \in (0,\frac{1}{6})$ ending in a center when $H = 0$ and a saddle loop when $H = \frac{1}{6}$. These periodic orbits and saddle loop represent the rescaled limiting behavior of our original equations as $\delta \to 0$. Next, we regard δ as a parameter and apply a perturbation argument to determine whether a given periodic orbit remains closed to first order in δ [1]. The criterion for this to happen is that the integral of the divergence of the vector field over the interior of the periodic orbit should vanish. Here the divergence is just $\pm\delta X$. Thus for each θ near $-\frac{\pi}{2}$ we need to compute whether there are closed level curves γ of $H_\theta(X,Y) = Y^2 - ((\cos\theta)X + \frac{1}{2}(\sin\theta)X^2 + \frac{1}{3}X^3)$ for which $\int_{\text{interior}} x\, dA = 0$. The result of these computations is that there is a range of $\theta < -\frac{\pi}{2}$ for which there are such γ. In this range, there is exactly one γ for which the integral vanishes and the size of the curve γ is a decreasing function of θ. The solutions end with a saddle-loop.

The interpretation of these results is that if one traverses the elongated ellipse $\delta^2\varepsilon + \mu = \delta^4$ in the parameter plane in the clockwise direction, then a Hopf bifurcation is encountered when $\varepsilon = \delta$ and $\mu = 0$. This Hopf bifurcation gives rise to a limit cycle which grows until it disappears as a saddle loop. In this transition, the relative position of two of the saddle separatrices changes. The bifurcation behavior of this example is illustrated in Figure 7. It is worthwhile to remark here that the nonlinear interaction of different bifurcating "modes" produces periodic behavior even though each mode is itself not the kind which leads to nonstationary asymptotic behavior. However, this periodic behavior occurs in a thin wedge of the parameter plane and thus may be difficult to locate in an example.

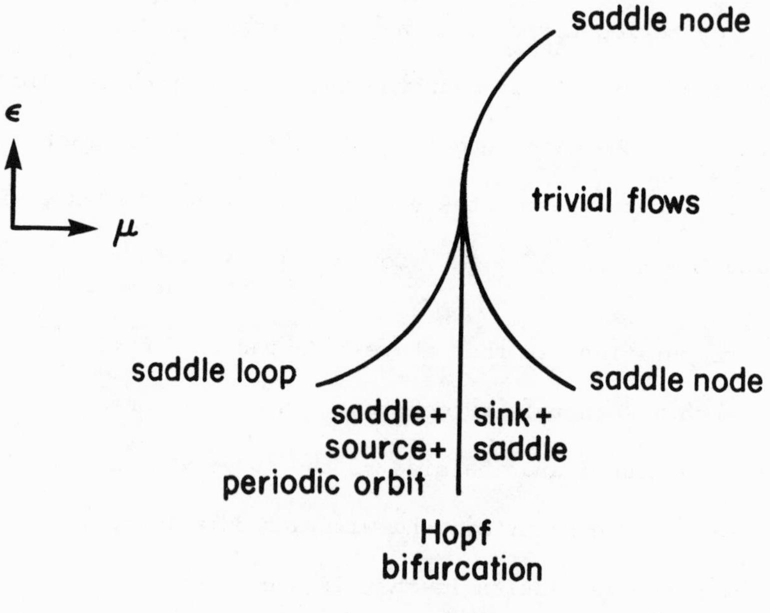

Figure 7. The bifurcation diagram of an equilibrium with a double zero eigenvalue.

The last bifurcation we examine in this section involves an equilibrium at which the linearization has a pair of pure imaginary eigenvalues and a zero eigenvalue. The pattern of analysis resembles that of the previous case, but there are additional complications to contend with. The equations of interest occur as three dimensional systems and this means that much more complex asymptotic behavior is possible than even with the flows on torii discussed in the previous section. In some cases such aperiodic solutions are part of typical two parameter families and cannot be avoided without essential restrictions on the equations of interest. Their presence lends a certain fuzziness to the detailed quantitative description of the bifurcation behavior. Further exploration of these details is warranted.

One begins the study of these bifurcations with a normal form calculation. The flow of a linear vectorfield on $\mathbb{R}^3$ with a pair of pure imaginary eigenvalues and a zero eigenvalue is rotation around an axis. In cylindrical coordinates (r,θ,z), the linear equations are $\dot{r} = 0$, $\dot{\theta} = \lambda$, $\dot{z} = 0$. In computing normal forms for these equations, the first observation one makes is that all terms involving θ on the right hand sides of the equations can be eliminated. This makes it natural to consider first the class of systems which are equivariant with respect to rotations around the z-axis. Such equations have no θ dependence at all in the equations for r and z, so that one can study the two dimensional system for (r,z) first. This reduction returns one to a two dimensional space and the kind of analysis used in the previous example. For these reduced equations, the equations for $\dot{r}$ is odd in r and the equation for $\dot{z}$ is even in r. Thus the lowest order terms in the equations are

$$\dot{r} = r(\varepsilon + a_1 r^2 + a_2 z)$$
$$\dot{z} = \mu + a_3 r^2 + a_4 z^2 .$$

We have included the unfolding parameters (ε,μ) in these equations.

Linear rescaling in (r,z,t,ε,μ) allows one to fix most of the constants (a_1,a_2,a_3,a_4) in these equations, but not all of them. After allowing for reversals of the direction of time, there are still four qualitatively different cases which depend upon the signs of these coefficients. In each of these cases, one can study the bifurcation of equilibria following the procedures we have already outlined. In two of the four cases, Hopf bifurcations of the (r,z) equations do not occur and the analysis is complete. In the other two cases, Hopf bifurcation does occur and additional rescaling like that used for the case of two zero eigenvalues is necessary to study the periodic orbits which occur in (r,z) space. The bifurcation diagrams for all these different cases are shown in Figure 8.

There are two general comments to be made about aperiodic asymptotic behavior. The first is that a periodic orbit in the (r,z) plane represents an invariant torus in (r,θ,z) space. Thus all of the complications associated with the secondary Hopf bifurcation described in the previous section occur here. The second comment relates to the disappearance of these invariant torii as parameters are varied. In one case, the torus becomes fat for suitable parameters, approaching a set which is a sphere together with its axis. (The picture looks like an innertube blown up inside a basketball.) For systems which preserve a rotational symmetry, one passes through this transition in a simple fashion. On the other hand, non-symmetric systems will not usually have the property that the stable and unstable manifolds of the two equilibria on the z-axis coincide with the sphere. They will generally intersect transversally and be associated with complicated "homoclinic" orbits.

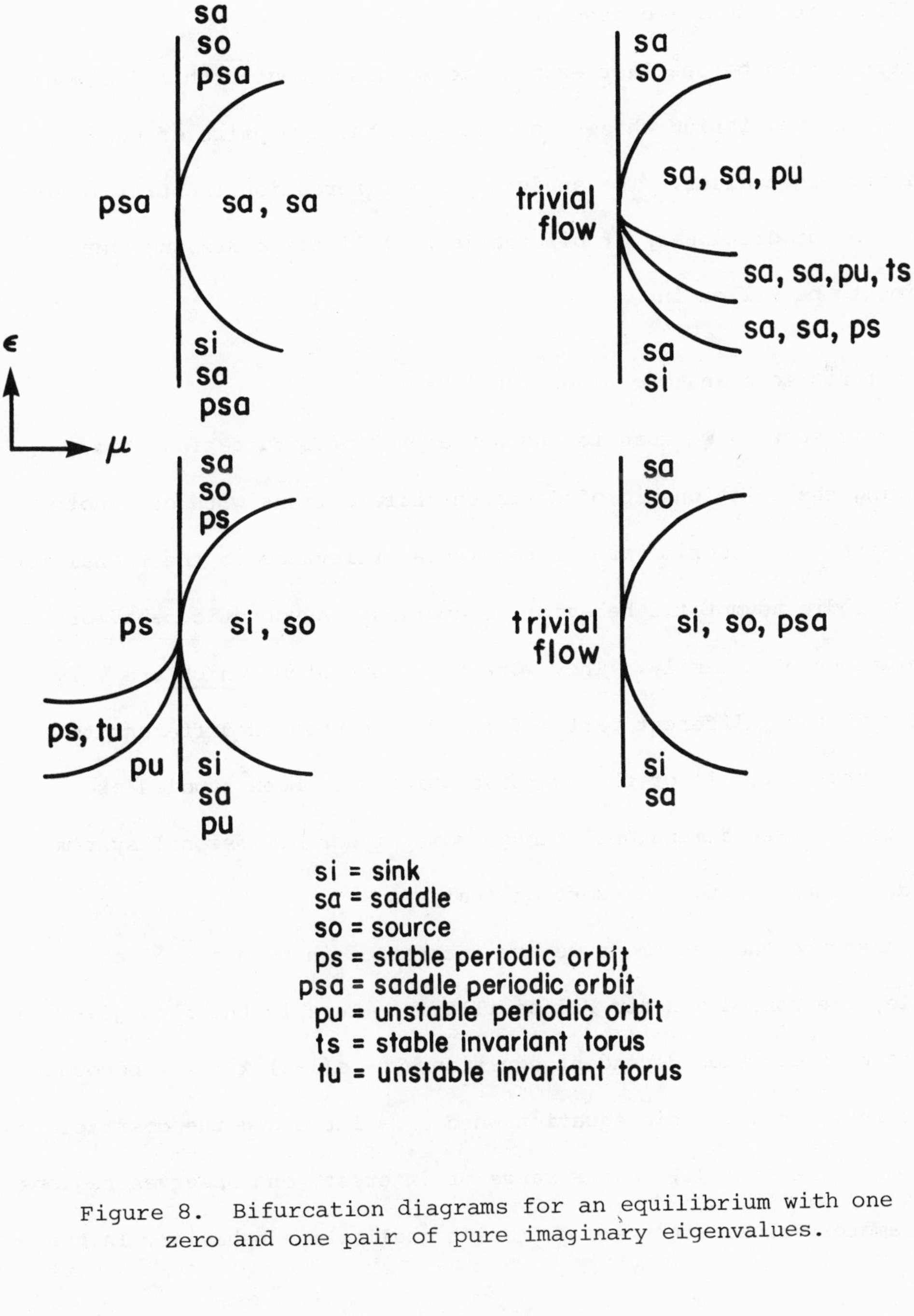

Figure 8. Bifurcation diagrams for an equilibrium with one zero and one pair of pure imaginary eigenvalues.

There will be a thin wedge of values in the parameter plane where there are solutions with complicated, aperiodic asymptotic behavior. We refrain from a further description of these flows here and refer to [15] for a more detailed description.

There should be one last example to be described in this section: that of an equilibrium whose linearization has two pairs of pure imaginary eigenvalues. The study of this bifurcation is still incomplete, so our dictionary of bifurcations still has a missing entry waiting to be filled in.

§3. Bifurcations and One Dimensional Maps

In this section we turn to another aspect of bifurcation theory involving the relationship of different bifurcations with one another. The features we describe are of particular relevance to the transition from periodic asymptotic behavior to aperiodic asymptotic behavior in solutions of a family. They work best in systems which are very "stiff" in that different parts of the flow happen on different time scale. Thus far, all of the examples which have been studied are essentially three dimensional though some higher dimensional systems should be amenable to this sort of analysis.

We describe the phenomena we are interested in by means of an example, the forced van der Pol oscillator. This is the flow given by the second order time dependent equation $\ddot{x} - \varepsilon(1-x)^2\dot{x} + x = b\varepsilon\cos\omega t$. We are interested in this equation when ε is large and the coefficients b and/or ω are varied. In the range of interest, one observes regions of parameter values with very different qualitative behavior. In the first type of region, all solutions are asymptotically periodic with all solutions (but one) periodic with least period $(2n-1)\ 2\pi/\omega$, n an odd integer. In the second type of region, there are stable periodic orbits of periods $(2n\pm 1)\ 2\pi/\omega$ as well as a large family which includes aperiodic orbits as well as periodic orbits of arbitrarily long periods. We want to describe the relationship between the different bifurcations which occur in the transition from one parameter region to the other. Since there are an infinite number of periodic orbits which qualitatively differ from one another in the second type of parameter region but only a finite number of periodic orbits in the first type, there are an infinite number of bifurcations in this transition. We look for patterns in the way these bifurcations occur.

One feature of this example is that the equation is very stiff when ε is large. This means that there are families of solutions which approach each other so closely that their trajectories are indistinguishable after a certain period of time. In this example, throughout much of the phase space there are curves of initial conditions whose solutions approach each other to a distance of order $e^{-\varepsilon t}$ after time t. Almost all trajectories eventually spend most of their time in this region. Taking account of this fact, there will be a region R of $(x,\dot{x})$ phase space such that integration of the van der Pol equations yields a map $F:R \to R$ whose image is so thin that we can think of it as a curve. By choosing appropriate coordinates in R, we may arrange that F depends only on the first coordinate. The iteration of F is then described entirely by a function which tells us what happens to the first coordinate under F. In the van der Pol example, this coordinate should be taken as an angular variable and the major feature of the dynamics are reduced to studying a map $f:S^1 \to S^1$ of the circle to itself.

This procedure of reducing the study of a flow to a one dimensional map is suitable in situations where solutions rapidly approach each other in all but two directions (the flow direction and one other). At least approximately, one can then project along these directions of strong contraction to obtain a (semi) flow on some two dimensional object (called a branched manifold [41]). If this has a cross-section, then its return map gives a map of some one dimensional object (which may have branches). There are theoretical difficulties with this process at those places where doubling back along this one dimensional structure takes place, and these are the places where bifurcation of long periodic orbits takes place. Thus there is a breakdown in the strategy we outline for investigating the relationship between different bifurcations, but it is not clear how serious it is. We do know that some aspects of the one dimensional analysis we now describe briefly carry over to the analysis of flows and some do not.

For one dimensional maps there are restrictions on which types of periodic orbits can occur in conjunction with which other orbits. The situation which has been studied most intensively is that of a function $f:I \to I$ on an interval which has a single critical point and maps the boundary of the interval to itself [13]. The quadratic functions $Q_a(x) = ax(1-x)$, $a \in [0,4]$ are a prototype of this class. Here there is an ordering of periodic orbits and bifurcations which has an intermediate value property: between a pair of bifurcations of periodic orbits, other bifurcations of intermediate periodic orbits must occur. The calculations which establish the ordering of periodic orbits are essentially combinatorial in nature.

Of particular interest in this study of one dimensional maps is the sequence of bifurcations which leads from maps having a finite number of periodic orbits to maps with an infinite number of periodic orbits. This sequence consists of flip bifurcations, each producing a periodic orbit of twice the period of the previous one. Sequences of flips such as this do occur also for diffeomorphisms of the plane and hence for systems of differential equations themselves. There are features of passage through these sequences of bifurcations which are reminiscent of critical phenomena in physical systems [9]. They provide one general mechanism by which systems acquire aperiodic solutions as parameters are varied.

There have been several numerical studies of systems whose bifurcations are approximately described by families of one dimensional maps. At the end of these sets of bifurcations, one has aperiodic solutions which may not be stable. This happens with the van der Pol equation, for example. Nonetheless, there do seem to be substantial sets of parameter values in these studies for which large sets of solutions have aperiodic asymptotic behavior. There are a variety of current efforts which seek to understand this phenomenon for one dimensional maps, but there are hardly any results which are directly applicable to differential equations.

The one example of a system of differential equations whose bifurcations have been thoroughly analyzed from this point of view are the "Lorenz equations." This is a three dimensional system of equations obtained as a truncation of the Navier-Stokes equations for a convective layer heated below. Here a convincing description (still not rigorously established) has been given for the sequence of bifurcations which occur in a range that includes the transitions producing aperiodic solutions and sets of aperiodic solutions which are stable ("strange attractors"). The one dimensional maps upon which the analysis rests have a discontinuity caused by the presence of an equilibrium with a two dimensional stable manifold and a one dimensional unstable manifold. See [16] for details.

4. Concluding Remarks

We have provided a survey of the different patterns of bifurcation which have been observed for systems of differential equations. The difficulties in applying these results have been slighted. Demonstrating that particular bifurcations occur in problems arising outside mathematics is often a substantial task. Nonetheless, the geometric and dynamical approach we have adopted here offers some advantages over alternative methods. Calculations of Taylor series at equilibria lead to strong conclusions about non-equilibrium behavior. There are examples which provide prototypes for the kinds of bifurcation sequences (secondary bifurcation) one can expect to find in the transition to aperiodic asymptotic behavior. Whether this approach will produce an exhaustive understanding of such mechanisms depends upon additional experience with a variety of examples.

We should also point to one direction for extending the theory which is important for applications. Systems encountered in applications often have symmetries or constraints, and these can force additional degeneracy in a bifurcating vector field. The dictionary of bifurcations which we have presented in this paper should be supplemented to include the analysis of generic bifurcations of systems which are equivariant with respect to simple symmetry groups. This process has begun for the study of bifurcations of equilibria, but there are few results about bifurcations involving non-equilibrium behavior in the presence of symmetry groups.

REFERENCES

[1] A. ANDRONOV, E. LEONTOVICH, I. GORDON, and A. MAIER, The Theory of Bifurcation of Plane Dynamical Systems, 1971.

[2] V. I. ARNOLD, Lectures of bifurcations in versal families, Russian Math Surveys 27(]972), pp. 54-123.

[3] V. I. ARNOLD, Small denominators I, mappings of the circle onto itself, Izv. Akad. Nauk. SSSR Ser. Mat. 25(1961), pp. 21-86.

[4] V. I. ARNOLD, On matrices depending upon a parameter, Russian Math Surveys 26(1971), no. 2, pp. 29-43.

[5] V. I. ARNOLD, Loss of stability of self oscillations close to resonance and versal deformations of equivariant vectorfields, Funct. Anal. Appl. 10(1976), pp. 316-317.

[6] R. I. BOGDANOV, Orbital equivalence of singular points of vectorfields on the plane, Funct. Anal. Appl. 10 (]976), pp. 316-317.

[7] R. I. BOGDANOV, Versal deformations of a singular point of a vectorfield on a plane in the case of zero eigenvalues, Proceedings of the I.G. Petrovskii Seminar 2(1976), pp. 37-65.

[8] F. DUMORTIER, Singularities of vector fields in the plane, Journal of Differential Equations, 23(1977), pp. 53-106.

[9] M. FEIGENBAUM, Quantitative universality for a class of nonlinear transformations, J. Stat. Phys. 19(1978), pp. 25-

[10] N. K. GAVRILOV and L. P. S. SIL'NIKOV, On three dimensional dynamical systems close to systems with a structurally unstable homoclinic curve, Math. USSR, Sb 17(1972), pp. 467-485 and 19(1973), pp. 139-156.

[11] M. GOLUBITSKY and D. SCHAEFFER, A theory for imperfect bifurcation via singularity theory, Comm. Pure Appl. Math. 32(1979), pp. 21-98.

[12] M. GOLUBITSKY and D. SCHAEFFER, Imperfect bifurcation in the presence of symmetry, mimeographed.

[13] J. GUCKENHEIMER, On the bifurcation of maps of the interval, Inventiones Mathematicae, 39(1977), pp. 165-178.

[14] J. GUCKENHEIMER, Lectures on bifurcation theory, C.I.M.E. Lecture Notes, to appear.

[15] J. GUCKENHEIMER, On a codimension two bifurcation, mimeographed, 1979.

[16] J. GUCKENHEIMER and R. F. WILLIAMS, Structural stability of Lorenz attractors, to appear.

[17] M. R. HERMAN, Mesure de Lebesque et nombre de rotation, Geometry and Toplogy, Springer Lecture Notes in Mathematics 597, ed. J. Palis and M. do Carmo, 1977, pp. 271-293.

[18] P. HOLMES, A nonlinear oscillator with a strange attractor, preprint, Ithaca, 1977.

[19] P. HOLMES and J. MARSDEN, Bifurcation to divergence and flutter in flow-induced oscillations, an infinite dimensional analysis, Automatica 14(1978), pp. 367-384.

[20] G. IOOSS and D. JOSEPH, Bifurcation and stability of nT-periodic solutions at points of resonance, preprint, 1978.

[21] W. LANGFORD, Periodic and steady-state mode interactions lead to tori, SIAM J. Appl. Math. 37(1979), pp. 22-47.

[22] M. LEVI, Thesis, Courant Institute of New York University, 1978.

[23] N. KOPELL and L. HOWARD, Target patterns and horseshoes from a perturbed central force problem: some temporally periodic solutions to reaction diffusion equations, preprint, 1979.

[24] J. E. MARSDEN, Qualitative methods in bifurcation theory, Bull. Am. Math. Soc. 84(1978), pp. 1125-1148.

[25] J. MARSDEN and H. MCCRACKEN, The Hopf Bifurcation Theorem and Its Applications, Springer Verlag, 1976.

[26] S. NEWHOUSE, Diffeomorphisms with infinitely many sinks, Topology 12(1974), pp. 9-18.

[27] S. E. NEWHOUSE, The abundance of wild hyperbolic sets and non-smooth stable sets for diffeomorphisms, mimeographed, IHES, 1977.

[28] S. NEWHOUSE and J. PALIS, Cycles and bifurcation theory, Asterisque, 31(1976), pp. 43-140.

[29] S. NEWHOUSE, J. PALIS, and F. TAKENS, Stable arcs of diffeomorphisms, Bull. Am. Math. Soc. 82(1976), pp. 499-502.

[30] J. PALIS and F. TAKENS, Topological equivalence of normally hyperbolic dynamics systems, Topology 16(1977), pp. 335-345.

[31] L. SILNIKOV, On a new type of bifurcation of multidimensional dynamical systems, Soc. Math. Dokl. 10(1969), pp. 1368-1371.

[32] L. P. SIL'NIKOV, A contribution to the problem of the structure of an extended neighborhood of a structurally stable equilibrium of saddle-focus type, Math. USSR Sb. 10(1970), pp. 91-102.

[33] S. SMALE, Differentiable dynamical systems, Bull. Am. Math. Soc., 73(1967), pp. 747-817.

[34] J. SOTOMAYOR, Bifurcations of vector fields on two dimensional manifolds, Publ. IHES 43(1973), pp. 1-46.

[35] J. SOTOMAYOR, Generic bifurcations of dynamical systems, Dynamical Systems, ed. M. Peixoto, Academic Press, 1973, pp. 561-582.

[36] F. TAKENS, Unfoldings of certain singularities of vectorfields: generalized Hopf bifurcations, J. Diff. Eq., 14(1973), pp. 476-493.

[37] F. TAKENS, Integral curves near mildly degenerate singular points of vectorfields, Dynamical Systems, ed. M. Peixoto, Academic Press, 1973, pp. 599-617.

[38] F. TAKENS, Singularities of vector fields, I.H.E.S. 43(1973), pp. 47-100.

[39] F. TAKENS, Forced oscillations and bifurcations, applications of global analysis, Communications of Maths Institute, Rijkswuniversiteit, Utrecht, 3(1947), pp. 1-59.

[40] THOM, R., Structural Stability and Morphogenesis, W.A. Benjamin, Inc., Reading, Mass., 1975.

[41] WILLIAMS, R. F., Expanding attractors, Publ. Math. I.H.E.S. 43(1974).

SECTION TWO
AEROSPACE AND MECHANICAL ENGINEERING

Nonlinear Attitude Dynamics of Satellites

Rudrapatna V. Ramnath* and Yee-Chee Tao*

Abstract. The problem of predicting the rotational motion of a rigid-body satellite orbiting the earth under the influence of external torques is studied. The body can have an asymmetric mass distribution. The attitude motion in response to the external torques is formulated as a perturbation from a nominal torque-free case. The torques considered are due to geogravity-gradient and geomagnetic effects. The equations of motion are parameterized by the ratio (which is small) of the orbital and attitude frequencies. Asymptotic solutions are developed for the perturbation equations by the multiple scales technique.

The independent variable, time, is extended into a space of higher dimension by means of new scales, fast, slow, etc. Integration of the equations is carried out separately in the new variables. The rapid and slow aspects of the attitude dynamics are systematically separated, resulting in a more efficient computer implementation and enhanced physical insight. The theory is applied to predict the attitude dynamics of an asymmetric rigid-body satellite. A comparison is made of the maximum errors as the step size increases as predicted by the multiple scales solution and direct numerical integration. An improvement in computational speed of an order of magnitude is demonstrated.

1. Introduction. The rotational motion of a rigid body in inertial space has been a subject of classical interest and can be seen in the works of Euler, Poinsot, Cayley, Klein [1-3] and others. However, in recent times interest has been enhanced with the advent of artificial satellites and the need to know their orientation more precisely. The motivation towards a more accurate prediction of rotational motion

---

* The Charles Stark Draper Laboratory, Inc., Cambridge, Massachusetts 02139. Dr. Ramnath is also lecturer at the Massachusetts Institute of Technology in the Departments of Aeronautics and Astronautics and Mechanical Engineering. This work was supported by NASA Goddard Space Flight Center, Contract NAS5-20848.

stems from earth resource technology using satellites as well as the more obvious military applications. There is thus a strong need to understand the angular motions of satellites and to calculate them accurately for long time intervals.

Determination of the rotational motion of an orbiting body under the influence of disturbing torques is not easy because of the nonlinear nature of the problem. As the equations of motion are highly nonlinear, exact analytical solutions cannot be obtained except in very special cases. Therefore, the main recourse is to use approximations and numerical solutions. However, with the direct numerical approach, the ability to construct solutions that are valid for long time intervals is usually beset with such difficulties as possible round-off and truncation errors and often prohibitive computational time and cost, even with the use of modern high-speed computers.

The problem of determining the rotational motion of a body in the absence of external torques - i.e., the Euler-Poinsot problem - was solved in classical times analytically by G. Kirchhoff [2] in terms of Jacobean elliptic functions. The problem of orientation of the body with respect to an inertial frame of reference was solved by Cayley, Klein, and Sommerfeld [3] for the torque-free case. These solutions employed the Euler symmetric parameters in terms of the unitary quaternion, which renders the representation free from singularities at certain orientations.

When there is an external torque, the problem becomes considerably more difficult, and exact analytical solutions are not available. One has to resort to approximations. In addition to numerical approaches, asymptotic solutions have frequently been used. A common approach involves the averaging method of Bogoliubov and Mitropolsky [4], and a number of applications have been presented. They include the works of Chernous′ko [5] on the variation of parameters for the angular momentum vector and the rotational kinetic energy for an asymmetric satellite; Holland and Sperling [6] on the slow evolution of the angular momentum vector due to gravity-gradient torque, and others [7-9]. The two-variable method has also been successfully applied to some nonlinear problems, primarily in orbital mechanics by Kevorkian [10] and others. The multiple scales method has been applied to predict the satellite attitude motion due to gravity-gradient and geomagnetic torques by Tao and Ramnath [11,12] for both rigid conventional and dual-spin satellites.

In this paper, we present a general formulation of the problem of rotational motion of rigid bodies in orbit in which the mass distribution can be asymmetric. We approach the problem from the point of asymptotic approximations and multiple scales technique. The equations are parameterized by means of the fast and slow motions. The rationale is that satellite attitude dynamics exhibit an intimate mixture of fast and slow motions. For instance, the attitude motions are much faster than the motion around the orbit. A systematic separation of the fast and slow variations is achieved by the multiple scales technique and leads to insight and efficient computation. The general formulation enables us to treat several problems, including conventional and dual-spin satellites as influenced by geogravity-gradient and geomagnetic torques.

2. The Problem. The attitude dynamics of a satellite can be studied in two parts: (i) the rotational motion with respect to a set of axes fixed in the body, and (ii) the orientation of the body axes with respect to an inertial frame of reference. The first part, i.e., motion in the body axes, is described by Euler's equations for a rigid body. The orientation with respect to an inertial frame can be represented by either Euler angles $\{\theta_i\}$ or Euler's symmetric parameters $\{\beta_i\}$. We will use the Euler parameter representation instead of the Euler angles because of its inherent advantages. The rotational motion in body axes is formulated as an Encke perturbation problem, using the torque-free case as the nominal solution. The perturbation equations, which are also nonlinear, are then solved asymptotically by the mutliple scales method.

2.1 Rotational Equations. By Newton's Second Law, the rotational motion of a rigid body about a moving frame which coincides with the principal axes (x,y,z) of the body, is represented by Euler's equation:

$$I\dot{\bar{\omega}} + (\bar{\omega}\times)I\bar{\omega} = \bar{M} \tag{1}$$

where $\bar{\omega}$ is a 3-vector of angular velocities, I is the principal moment of inertia matrix, and $\bar{M}$ is the external torque.

2.2 Euler Symmetric Parameters. These are similar to the Euler angles, which relate the orientations of two coordinate systems. A transformation matrix can be constructed using either system to transform a vector defined in one system to the other. A number of useful differences exist between the two systems, some of which can be enumerated as shown in Table 1.

TABLE 1

Differences between Euler Angles and Euler Symmetric Parameters

EULER ANGLES θ_i for i = 1,2,3	EULER SYMMETRIC PARAMETERS β_i for i = 0,1,2,3
Three dimensions	Four dimensions with a constraint
Singular at certain orientations	Free from singularities
Propagated by a nonlinear differential equation	Propagated by a linear differential equation
Easy physical interpretation i.e., nutation, precession, rotation	Physical interpretation not direct

In view of the information given in Table 1, we choose Euler symmetric parameters for our formulation. They are derived from "Euler's Theorem", which states that a completely general rotation or a sequence of rotations of a rigid body is equivalent to a single rotation ψ about a unit vector $\hat{\ell} = (\ell_1, \ell_2, \ell_3)$, which is fixed in both the body and reference frames. The Euler symmetric parameters are then defined as:

$$\beta_0 = \cos(\psi/2), \qquad \beta_i = \ell_i \sin(\psi/2), \qquad i=1,2,3 \tag{2}$$

together with a constraint $\sum_{i=0}^{3} \beta_i^2 = 1$. The β_i, $i = 0,1,2,3$, form a unitary quaternion and satisfy the linear equation

$$(3) \qquad \frac{d\bar{\beta}}{dt} = \psi[\bar{\omega}(t)]\bar{\beta}$$

where $\bar{\beta}$ is the 4-vector of Euler parameters and $\psi(\bar{\omega})$ is a skew symmetric 4×4 matrix function of angular velocities. The problem addressed is embodied in Eq. (1) and (3), which describe the attitude motion in body axes and the orientation with respect to inertial space, respectively. These equations are studied by the multiple scales theory [13-15].

2.3 Euler-Poinsot Problem and Kirchhoff's Solution. Closed form solutions of Euler's equation are usually not possible because of their nonlinear characteristics. However, Euler's equation can be solved if the external torque is zero, and this problem is called the Euler-Poinsot problem. A complete analytical solution for the Euler-Poinsot problem, attributed to Kirchhoff [2], in terms of an elliptic integral of the first kind can be written as:

$$(4) \quad \omega_x = a(1 - k^2 \sin^2 \phi)^{1/2}; \quad \omega_y = b \sin \phi; \quad \omega_z = c \cos \phi$$

where $\lambda(t - \tau) = \int_0^{\phi} d\phi/(1 - k^2 \sin^2 \phi)^{1/2} \equiv F(k,\phi)$. $F(k,\phi)$ is the incomplete elliptic integral of the first kind. The constants a, b, c, λ, k, and τ are all determined in terms of the kinetic energy T, the angular momentum H, the inertia, and initial conditions. Kirchhoff's solution seems to be less popular than the Poinsot construction in the engineering world, perhaps because it involves the elliptic integral. However, for the long-term attitude prediction problem, Kirchhoff's solution is useful.

3. Multiple Scales Theory. This method has been developed relatively recently and has been applied to a number of complex problems in diverse fields. Here, we will present only a very brief review, as many publications on the subject can be found. However, instead of treating the subject in a somewhat *ad hoc* manner as is done sometimes, we will use

it in the context of extensions as developed by Sandri and Ramnath [1ᴊ-15]. This provides a unified framework connecting many apparently disparate concepts and techniques into a systematic unified theory. The foundations of the general approach rest on concepts of mappings and their composition and are beyond the scope of this paper.

The method is particularly useful when a phenomenon exhibits a mixture of rapid and slow motions. The existence of the fast and slow motions is used to parameterize a system in terms of a small parameter. The domain of the independent variable, time, is extended into a space of higher dimension by means of "clocks" or scale functions. Mathematically, the differential equations are extended into a set of partial differential equations which are then solved asymptotically, order by order. The solutions of the partial differential equations are then restricted such that their solutions coincide with the solutions in the original domain along certain lines in the extended space - "trajectories" - which correspond to the scale functions. The scale functions are not restricted to be real. Indeed, in many nonlinear and nonautonomous problems, it is essential that the scale functions be nonlinear and complex quantities. The degrees of freedom available, - i.e., of the trajectories and the extension itself - are utilized to obtain a simpler description of the phenomenon.

Physically, the approach is tantamount to employing a number of independent observers who follow a complex phenomenon by performing readings on appropriate clocks. Thus, "slow" and "fast" observers who use slow and fast clocks to measure time will perceive different aspects of the phenomenon. A combination of the different behaviors will provide a composite description of how the phenomenon transpires.

In applications, we find the following definition of an extension useful: Given a function $f(t)$ and a function $F(\tau_i)$ of the n independent variables $\tau_1, \tau_2, \ldots, \tau_n$ (each of which can be an N-dimensional vector), F is said to be an extension of f

$$f(t) \rightarrow F(\tau_1, \tau_2, \ldots, \tau_n)$$

if and only if there exists a set of relations $\tau_i = \tau_i(t)$ which, when inserted into F yield

$$F(\tau_1(t), \tau_2(t), \ldots, \tau_n(t)) = f(t)$$

In applications, the trajectories $\tau_i(t, \varepsilon)$ turn out to be nonlinear and complex quantities. For these and other related issues, the reader is referred to the literature [13-15].

4. Development of the Solution. In deriving the approximate solution by the multiple scales technique, we treat the total problem in two parts: Euler's equation with zero torque is treated as the nominal case; then this solution is modified via the perturbation equations to accommodate for the presence of torque.

4.1 Solution of Euler's Equation. We can write Euler's equation in the parameterized form:

$$\text{(5)} \quad I\dot{\bar{\omega}} + (\bar{\omega} \times)I\bar{\omega} = \bar{M}(\varepsilon) = \varepsilon^2\bar{M}_1 + \varepsilon^3\bar{M}_2 + \ldots; \; \bar{\omega}(t_0) = \bar{\omega}_0; \; 0 < \varepsilon << 1$$

where $\bar{M}_1$, $\bar{M}_2$,... are terms in the expansion of the disturbing torque, and the small parameter ε is the ratio of the orbital and attitude frequencies. In general, $\bar{M}_i$ would depend on the angular velocities and orientation variables $\bar{\beta}$. To the leading order in ε^2, $\bar{M}_1$ depends only on the nominal values of these variables. Let $\bar{\omega}_N$ be the torque-free Kirchhoff's solution given by Eq. (4), satisfying the particular initial condition. Further, we express $\bar{\omega}(\varepsilon)$ in the perturbation form

$$\text{(6)} \qquad \bar{\omega}(t) = \bar{\omega}_N(t) + \varepsilon\, \delta\bar{\omega}(t)$$

Substituting Eq. (6) into Eq. (5) we have:

$$I\, \delta\dot{\bar{\omega}} + (\bar{\omega}_N \times)I\, \delta\bar{\omega} + (\delta\bar{\omega} \times)I\bar{\omega}_N + \varepsilon(\delta\bar{\omega} \times)I\, \delta\bar{\omega} = \varepsilon M_1 + \varepsilon^2 M_2 + \ldots$$

with $\delta\bar{\omega}(t_0) = 0$.

We note that so far there has been no approximation. Further, by computing in this perturbation form, one has the advantage of reducing round-off errors. For notational simplicity, the periodic matrix operator A(t) is defined as:

(7) $$A(t) = -I^{-1}[(\bar{\omega}_N \times)I - (I\bar{\omega}_N \times)]$$

The perturbed motion is now described by:

(8) $$\delta\dot{\bar{\omega}} - A(t)\delta\bar{\omega} + \varepsilon I^{-1}(\delta\bar{\omega} \times)I\,\delta\bar{\omega} = \varepsilon I^{-1}\bar{M}_1 + \varepsilon^2 I^{-1}\bar{M}_2 + \ldots$$

We see that this equation is weakly nonlinear. We now invoke Floquet theory [16], which is suggested by the periodicity of A(t), and proceed as follows. Let us define matrices R_A and P_A by means of the relations

(9) $$R_A = \frac{1}{T_\omega}\,\ell n[\Phi_A(T_\omega,0)]; \quad P_A^{-1} = \Phi_A(t,0)\exp(-R_A t)$$

where $\Phi_A(t,0)$ is the transition matrix for A(t), and R_A is a constant matrix. Using these results, we separate the periodic part of the non-linear dynamics by means of the transformation

(10) $$\delta\bar{\omega}(t) = P_A^{-1}(t)\bar{u}(t)$$

The nonperiodic part of the nonlinear attitude dynamics in Eq. (8) can be shown to satisfy the equation:

(11) $$\dot{\bar{u}} = R_A\bar{u} - \varepsilon P_A I^{-1}(P_A^{-1}\bar{u} \times)I(P_A^{-1}\bar{u}) + \varepsilon P_A I^{-1}\bar{M}_1 + \varepsilon^2 P_A I^{-1} M_2 + \ldots$$

Further, R_A is proportional to $(1/T_\omega)$ and hence $R_A = O(\varepsilon)$. For a satellite in a 2-hour orbit (i.e., about 8000-second period) and a value of $\varepsilon = 0.01$, we see that the attitude period, T_ω, is about 80 seconds. Thus, $R_A \sim 1/T_\omega = 0.0125 = O(\varepsilon)$.

For convenience, we consider a similarity transformation T to convert R_A to the diagonal (or Jordan) form. Let $\bar{u} = T\bar{v}$. Substituting for $\bar{u}$ in Eq. (11), we have

(12) $$\dot{\bar{v}} = \varepsilon\Lambda\bar{v} - \varepsilon Q(I^{-1}Q^{-1}\bar{v} \times)Q^{-1}\bar{v} + Q(\varepsilon\bar{M}_1 + \varepsilon^2\bar{M}_2 + \ldots)$$

where $Q = T^{-1}P_A I^{-1}$.

4.3 Multiple Scales Solution. In order to solve Eq. (12), we invoke the multiple scales technique. Accordingly, we extend the independent variable t into a multidimensional space as follows:

$$(13)\quad \begin{cases} t \to \{\vec{t}\} = \{\tau_0, \tau_1, \ldots\}; \quad \tau_i = \varepsilon^i t \text{ for } i=0,1,\ldots; \text{ and} \\ \bar{v}(t,\varepsilon) \to \bar{V}(\tau_0, \tau_1, \ldots, \varepsilon) \end{cases}$$

Therefore $d/dt \to (\partial/\partial\tau_0) + \varepsilon(\partial/\partial\tau_1) + \ldots$

Further, we expand $\bar{V}$ into an asymptotic expansion in ε

$$(14)\quad \bar{V}(\tau_0, \tau_1, \ldots, \varepsilon) = \bar{V}_0(\tau_0, \tau_1, \ldots) + \varepsilon\bar{V}_1(\tau_0, \tau_1, \ldots) + O(\varepsilon^2)$$

Using Eq. (13) and (14), the extended perturbation equations can be written to first order in ε as:

$$(15)\quad \frac{\partial \bar{V}_0}{\partial \tau_0} = 0$$

$$(16)\quad \frac{\partial \bar{V}_1}{\partial \tau_0} = -\frac{\partial \bar{V}_0}{\partial \tau_1} + \Lambda\bar{V}_0 - Q(I^{-1}Q^{-1}\bar{V}_0 \times)Q^{-1}\bar{V}_0 + QM_1$$

The partial differential equation (15) suggests that $\bar{V}_0$ is not a function of τ_0. The first-order solution $\bar{V}_1(\tau_0, \tau_1)$ can be written as

$$(17)\quad \bar{V}_1(\tau_0, \tau_1, \ldots) = \left[-\frac{\partial \bar{V}_0}{\partial \tau_1} + \Lambda\bar{V}_0\right]\tau_0 - \int_0^{\tau_0} Q(I^{-1}Q^{-1}\bar{V}_0 \times) \cdot Q^{-1}\bar{V}_0 \, d\tau_0 + \int_0^{\tau_0} Q\bar{M}_1 \, d\tau_0$$

In order to obtain an asymptotic solution, it is necessary to rewrite the nonlinear terms. The nonlinear term not containing the disturbance torque can be written as [11]:

$$(18)\quad Q(I^{-1}Q^{-1}\bar{V}_0 \times)Q^{-1}\bar{V}_0 = \sum_{i=1}^{3} \tilde{F}_i(\tau_0)\bar{V}_{0_i}\bar{V}_0$$

where $\tilde{F}_i(\tau_0 + T_\omega) = \tilde{F}_i(\tau_0)$, and V_{0_i}, $i = 1,2,3$ are the three components of the vector $\bar{V}_0$. We expand $\tilde{F}_1$, $\tilde{F}_2$, $\tilde{F}_3$ into Fourier series, and write

$$(19) \qquad Q(I^{-1}Q^{-1}\bar{V}_0 \times)Q^{-1}\bar{V}_0 \approx \sum_{i=1}^{3}\sum_{j=0}^{n}\left[E_{ij}\sin\left[\frac{2\pi j\tau_0}{T_\omega}\right] + F_{ij}\cos\left[\frac{2\pi j\tau_0}{T_\omega}\right]\right]V_{0_i}\bar{V}_0$$

where E_{ij} and F_{ij} are constant matrices.

The next step is to express the disturbance term $Q\bar{M}_1$ in a separable form to facilitate the integration in Eq. (17). For both the gravity-gradient and geomagnetic torques it can be shown that the QM_1 terms can be written in a separable form [11]. We can write

$$(20) \qquad Q\bar{M}_1 = G_1(\tau_1,\tau_2, \ldots) + Q_p(\tau_0)G_2(\tau_1,\tau_2, \ldots)$$

where the matrices G_1 and G_2 are functions of the slow variables τ_1, τ_2,... only, and $Q_p(\tau_0)$ is a matrix that is a function of the fast time scale τ_0 only. Therefore, Eq. (17) becomes:

$$(21) \qquad \bar{V}_1(\tau_0,\tau_1, \ldots) = \left[-\frac{\partial\bar{V}_0}{\partial\tau_1} + \Lambda\bar{V}_0 - \sum_{i=1}^{3} F_{i0}V_{0_i}\bar{V}_0 + G_1(\tau_1, \ldots)\right]\tau_0$$

$$- \int_0^{\tau_0} \sum_{i=1}^{3}\sum_{j=1}^{n}\left[E_{ij}\sin\left[\frac{2\pi j\tau_0}{T_\omega}\right] + F_{ij}\cos\left[\frac{2\pi j\tau_0}{T_\omega}\right]\right]V_{0_i}\bar{V}_0\, d\tau_0$$

$$+ \int_0^{\tau_0} Q_p(\tau_0)G_2(\tau_1, \ldots)\, d\tau_0$$

Now in order to achieve a uniformly valid asymptotic solution, we will impose the uniformity condition and require that the ratio of the norms of the vectors $\bar{V}_1$ and $\bar{V}_0$ (defined suitably) be bounded uniformly for all τ_0. Thus, to eliminate secularity in τ_0, we require that

$$\frac{d\bar{V}_0}{d\tau_1} = \Lambda\bar{V}_0 - \sum_{i=1}^{3} F_{i0} V_{0_i} \bar{V}_0 + G_1(\tau_1,\tau_2, \ldots) \tag{22}$$

with initial condition $\bar{V}_0(0) = 0$. Therefore, the solution $\bar{V}_1(\tau_0, \tau_1 \ldots)$ can now be written as

$$\begin{aligned} \bar{V}_1(\tau_0,\tau_1, \ldots) \approx & \int_0^{\tau_0} \sum_{i=1}^{3} \sum_{j=0}^{n} \left[E_{ij} \sin\left[\frac{2\pi js}{T_\omega}\right] + F_{ij} \cos\left[\frac{2\pi js}{T_\omega}\right]\right] V_{0i}\bar{V}_0 ds \\ & + \left[\int_0^{\tau_0} \theta_p(s)\, ds\right] G_2(\tau_1,\tau_2, \ldots) \end{aligned} \tag{23}$$

Equations (22) and (23) yield the first-order multiple scales asymptotic representation of $\bar{V}$. Retracing the transformations, the asymptotic solution to first order in ε of Euler's equation can therefore be written as:

$$\bar{\omega}(t) = \bar{\omega}_N(t) + \varepsilon P_A^{-1} T(\bar{V}_0 + \varepsilon\bar{V}_1) = \bar{\omega}_N + \varepsilon\overline{\delta\omega} \tag{24}$$

This solution gives us an alternative way of calculating the angular velocity vector $\bar{\omega}(t)$ instead of by direct integration of Euler's equation. By this new approach we obtain the fast varying oscillatory part of the dynamics analytically from Eq. (23). The slow dynamics are obtained by integrating Eq. (22) in the slow variable $\tau_1(t)$. This allows the use of a large integration time step, resulting in a saving of computer time.

4.2. Solution for Euler Symmetric Parameters. Euler parameters are kinematically related to the angular velocities in a manner that can be expressed by a linear differential equation:

$$\dot{\overline{\beta}} = \psi\{\overline{\omega}(t)\}\beta \tag{25}$$

where $\overline{\beta}$ is a 4-vector and $\psi(\overline{\omega})$ is a skew-symmetric matrix of angular velocities. Substitution of the asymptotic solution for $\overline{\omega}(t)$ will result in the equation:

$$\dot{\overline{\beta}} = \psi\{\overline{\omega}_N + \varepsilon\delta\overline{\omega}(t)\}\overline{\beta} \tag{26}$$

This equation is again solved asymptotically by the multiple scales technique. The extended perturbation equations are, order by order with $\overline{\beta} = \overline{\beta}_0(\tau_0,\tau_1, \ldots) + \varepsilon\overline{\beta}_1(\tau_0,\tau_1, \ldots) + \ldots$ and $\tau_i = \varepsilon^i t$:

$$\frac{\partial\overline{\beta}_0}{\partial\tau_0} = \psi[\overline{\omega}_N(\tau_0,\tau_1, \ldots)]\overline{\beta}_0 \tag{27}$$

$$\frac{\partial\overline{\beta}_1}{\partial\tau_0} = \psi(\overline{\omega}_N)\overline{\beta}_1 - \frac{\partial\overline{\beta}_0}{\partial\tau_1} + \psi[\delta\overline{\omega}]\,\overline{\beta}_0 \tag{28}$$

We solve them as follows: Let $\Phi_\beta(t, t_0)$ be the transition matrix for $\psi(\overline{\omega}_N)$, i.e.,

$$\frac{d\Phi_\beta}{dt} = \psi[\overline{\omega}_N(t)]\Phi_\beta \tag{29}$$

$\Phi_\beta(t)$ can be determined analytically and explicitly as shown by Morton, et al. [17]. The solution of the zeroth order equation (27) can be written as:

$$\overline{\beta}_0(\tau_0,\tau_1, \ldots) = \Phi_\beta(\tau_0,0)\overline{\beta}_{ON}(\tau_1,\tau_2, \ldots); \quad \overline{\beta}_{ON}(0) = \overline{\beta}(0) \tag{30}$$

$\overline{\beta}_{ON}(\tau_1,\tau_2,\ldots)$ is yet to be determined. The first-order equation can be integrated as:

$$\overline{\beta}_1(\tau_0,\tau_1, \ldots) = \Phi_\beta(\tau_0,0)\left[-\frac{\partial\overline{\beta}_{ON}}{\partial\tau_1}\tau_0 + \int_0^{\tau_0}\Phi_\beta^{-1}(s,0) \cdot \psi[\delta\overline{\omega}(\tau_0,\tau_1, \ldots)]\Phi_\beta(s,0)\overline{\beta}_{ON}\,ds\right] \tag{31}$$

where $\delta\overline{\omega}(\tau_0,\tau_1,\ldots)$ is given by Eq. (24) and $\Phi_\beta(\tau_0,0)$ is a function of τ_0 only. We can regroup terms as follows [11].

(32) $$\Phi_\beta^{-1}\psi(\delta\bar{\omega})\Phi_\beta = R_1(\tau_1) + P_B(\tau_0)R_2(\tau_1)$$

Equation (31) can now be written as:

(33) $$\bar{\beta}_1 = \Phi(\tau_0,0)\left[-\frac{\partial\bar{\beta}_{ON}}{\partial\tau_1} + R_1(\tau_1)\bar{\beta}_{ON}(\tau_1)\right]\tau_0 + \left[\int_0^{\tau_0} P_B(s)\,ds\right]R_2(\tau_1)\bar{\beta}_{ON}(\tau_1)$$

In order to solve this equation, we again appeal to the uniformity criterion and require that the uniformity ratio $\|\beta_1\|/\|\bar{\beta}_0\|$ be bounded in τ_0. Therefore it is sufficient that:

(34) $$\frac{\partial\bar{\beta}_{ON}}{\partial\tau_1} = -R_1(\tau_1)\bar{\beta}_{ON}$$

Now we have:

(35) $$\bar{\beta}_1 = \Phi_\beta(\tau_0,0)\left[\int_0^{\tau_0} P_B(s)\,ds\right]R_2(\tau_1)\bar{\beta}_{ON}$$

From these, the asymptotic solution to first order for the Euler symmetric parameters is given by

(36) $$\bar{\beta}(\tau_0,\tau_1,\ldots) = \Phi_\beta(\tau_0,0)\bar{\beta}_{ON}(\tau_1) + \varepsilon\bar{\beta}_1(\tau_0,\tau_1) + \ldots$$

$\bar{\beta}_1(\tau_0)$ gives the unbiased oscillatory motion.

The above theory is applicable to rigid-body satellites under the influence of gravity-gradient and geomagnetic torques.

5. Application. We now apply the theory developed to the problem of predicting the attitude motion of a rigid-body asymmetric satellite in an elliptic earth orbit. We will study the effect of the geogravity-gradient torques. The satellite parameters are given in Table 2.

TABLE 2

Satellite Parameters

MOMENTS OF INERTIA	ORBIT PARAMETERS	INITIAL CONDITIONS
I_x = 39.4 slug-ft^2	Eccentricity = 0.16	ω_x = 0.0246 rad/s
I_y = 33.3 slug-ft^2	Inclination = 0	ω_y = 0 rad/s
I_z = 10.3 slug-ft^2	Orbital Period = 10,000 s	ω_z = 0 rad/s
		$\beta_0 = \beta_3$ = 0.7071
		$\beta_1 = \beta_2$ = 0

In this case, the small parameter ε is

$$\varepsilon = \omega_{orbit}/\omega_{attitude} \approx 0.03$$

With the above initial conditions, the satellite attitude dynamics are first directly integrated using a fourth-order Runge-Kutta method with a small integration time step size of 10 seconds for a total interval of 8000 seconds. This result is considered to be very accurate and it is referred to as the reference case from here on. The other simulations are then compared to this reference case for checking their accuracy. A number of runs were tried, both by the asymptotic approach and by direct integration, with different step sizes. The errors in each case - i.e., the differences between each simulation and the reference case - are plotted against time and are given by Figures 1 through 10. Figures 1 through 8 describe the motion due to gravity-gradient torque.

Figure 9 is a plot of the maximum numerical errors as functions of the step size. From this plot, we see that with direct integration the step size ΔT should be no greater than 25 seconds. On the other hand, for the asymptotic approximation, the step size can be as large as 500 seconds, although the first-order asymptotic approximation errors approach zero as $\varepsilon \to 0$ but not as $t \to 0$. Figure 10 is a plot of the required computer time in terms of the step size ΔT. We note that for extrapolating a single step, the asymptotic approach requires about double the computer time using direct simulation. However, since the former allows use of a large time

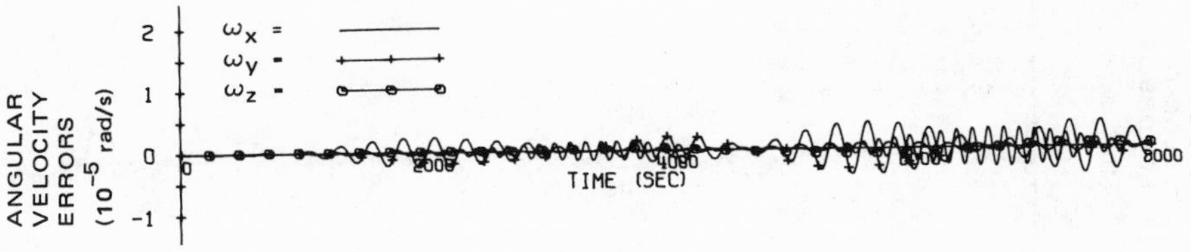

Figure 1: Simulation errors in rigid-body case. Angular velocity errors; MTS asymptotic solution, DT = 10 s; gravity-gradient torque.

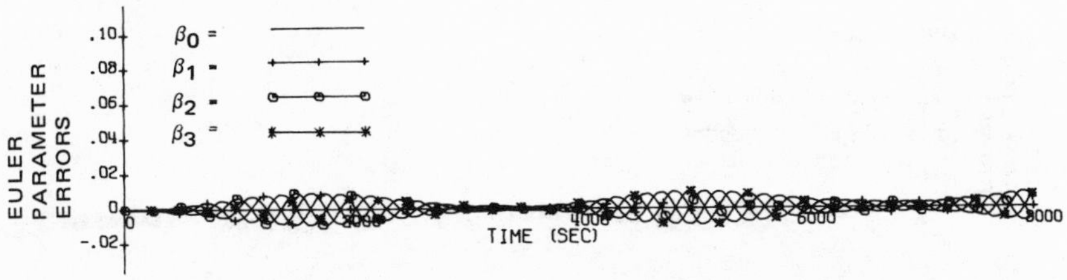

Figure 2: Simulation errors in rigid-body case. Euler parameter errors; MTS asymptotic solution, DT = 10 s; gravity-gradient torque.

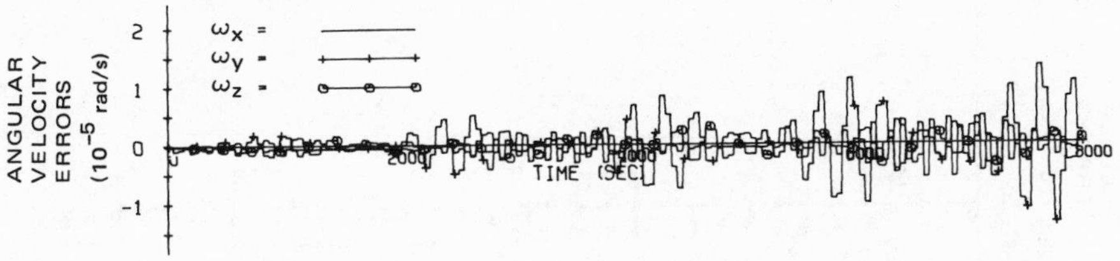

Figure 3: Simulation errors in rigid-body case. Angular velocity errors; direct integration, DT = 50 s; gravity-gradient torque.

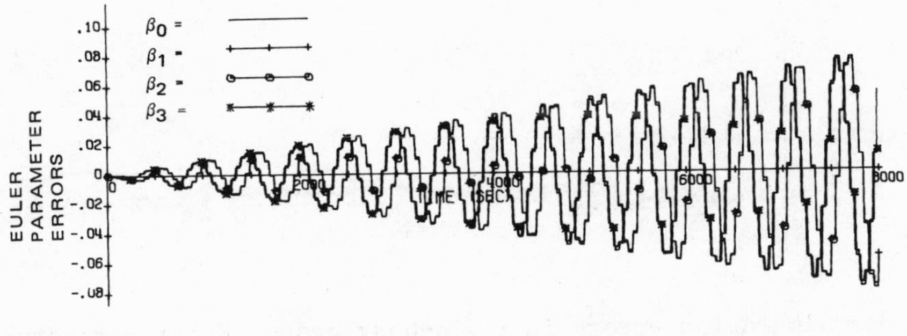

Figure 4: Simulation errors in rigid-body case. Euler parameter errors; direct integration, DT = 50 s; gravity-gradient torque.

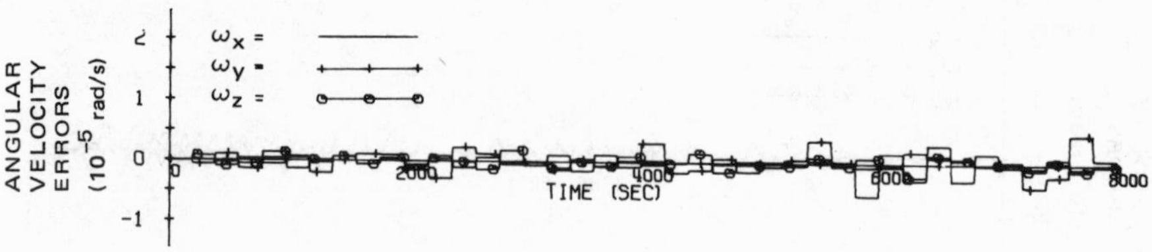

Figure 5: Simulation errors in rigid-body case. Angular velocity errors; MTS asymptotic solution, DT = 200 s, gravity-gradient torque.

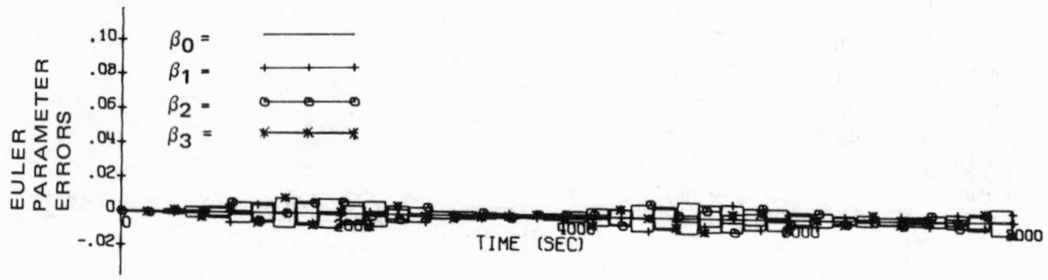

Figure 6: Simulation errors in rigid-body case. Euler parameter errors; MTS asymptotic solution, DT = 200 s; gravity-gradient torque.

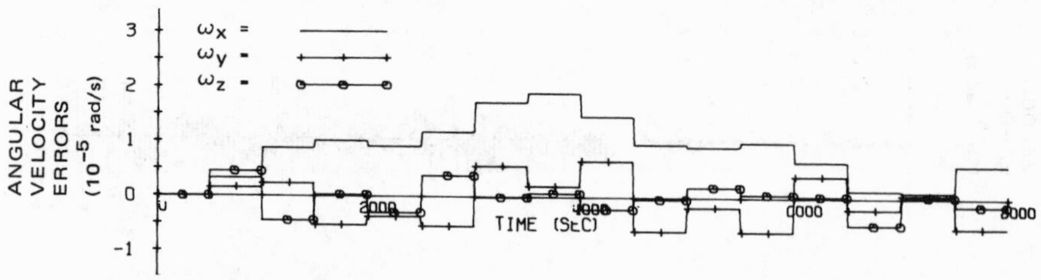

Figure 7: Simulation errors in rigid-body case. Angular velocity errors; MTS asymptotic solution, DT = 500 s; gravity-gradient torque.

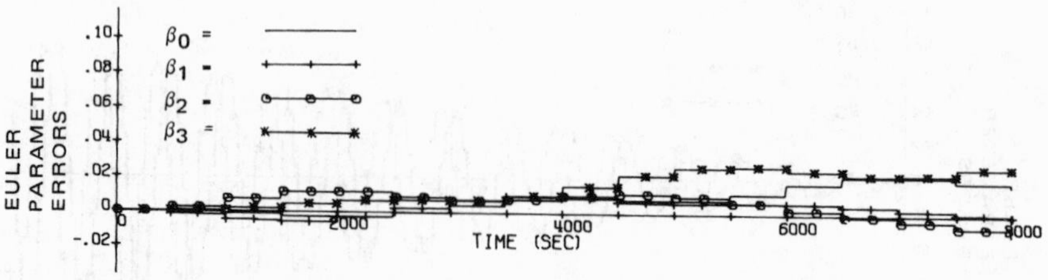

Figure 8: Simulation errors in rigid-body case. Euler parameter errors; MTS asymptotic solution, DT = 500 s; gravity-gradient torque.

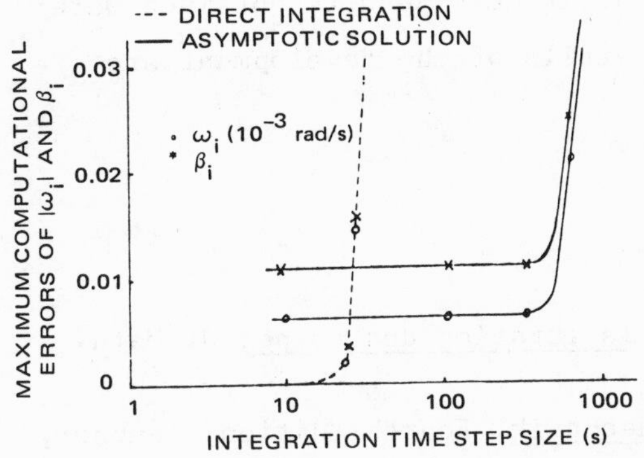

Figure 9. Maximum simulation errors.

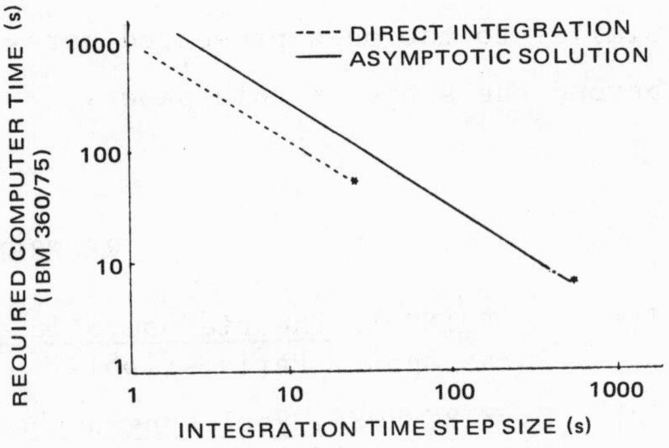

Figure 10. Computer time.

step, overall, this new approach will have a significant numerical advantage over direct simulation. In our particular case, the saving is of order 10.

6. <u>Summary and Conclusions</u>. A general method has been developed for fast prediction of the rotational motion of a rigid body orbiting the earth. The torques considered are due to geogravity-gradient and geomagnetic effects. An asymptotic solution is developed by the multiple scales approach such that the digital computer implementation of the approximation results in a significant saving of computer time vis-a-vis direct simulation. Furthermore, because our approach is general in that it is applicable to any orbit, initial conditions or mass distribution, it is hoped that the technique will provide a valuable tool in satellite engineering.

The approach was demonstrated by an application. The slow secular and the fast oscillatory effects of the gravity gradient and geomagnetic torques were separated and calculated individually. However, because of space limitations, only the figures for the gravity gradient case are given. Together, these two behaviors give a complete description, while separately, each describes a different aspect of the phenomenon.

The theory developed is also applicable to the motion of a class of dual-spin satellites - having one or more flywheels mounted onboard for control or stabilization of their attitude. Results obtained were similar to the ones presented here. Details of the development are beyond the scope of this paper.

REFERENCES

[1] L. POINSOT, Theorie nouvelle de la rotation des corps, J. Math. Pures Appl., Paris, (1851).

[2] G. KIRCHHOFF, Vorlesungen uber Mechanik, Fourth Edition, Teubner, Leipzig, 1897.

[3] F. KLEIN and A. SOMERFELD, Theorie des Kreises, Druck und Verlag von Teubner, (1898), pp. 475-484.

[4] N.N. BOGOLIUBOV and Y.A. MITROPOLSKY, Asymptotic Methods in the Theory of Nonlinear Oscillations, Gordon and Breach, New York, 1961.

[5] F.L. CHERNOUS'KO, The motion of a satellite about the mass center under gravity torques, Prikladnaya Matematika i Mekhanika, 27(1963), p. 474.

[6] R.I. HOLLAND and H.J. SPERLING, A first order theory for the rotational motion of a triaxial rigid body orbiting an oblate primary, J. Astron., 74(1969), p. 490.

[7] D.L. HITZL and J.V. BREAKWELL, Resonant and non-resonant gravity-gradient perturbations of a tumbling triaxial satellite, Celest. Mech., 3(1971), p. 346.

[8] J.E. COCHRAN, Effects of gravity-gradient torque on the rotational motion of a triaxial satellite in a precessing elliptic orbit, Celest. Mech., 6(1972), p. 127.

[9] R.J. PRINGLE, Satellite vibration-rotation motions studied via canonical transformations, Celest. Mech., 7(1973), p. 495.

[10] J. KEVORKIAN, The two variable expansion procedure for the approximate solution of certain nonlinear differential equations, Lectures in Appl. Math., Vol. 7, Space Math, Part 3, (J.B. Rosser, Ed.), Am. Math. Soc., (1966).

[11] Y.-C. TAO and R.V. RAMNATH, Satellite Attitude Prediction by Multiple Time Scales Method, The Charles Stark Draper Laboratory, Inc., Cambridge, Mass., Report R-930, December 1975. Also Ph.D. Thesis of Y.-C. Tao, Massachusetts Institute of Technology, 1975.

[12] Y.-C. TAO and R.V. RAMNATH, On the attitude motion of an orbiting rigid body under the influence of gravity gradient torque, presented at the AIAA/AAS Astrodynamics Conference, Palo Alto, Calif., August 1978.

[13] G. SANDRI, The foundations of nonequilibrium statistical mechanics, Ann. Phy. (N.Y.), 24(1963), pp. 332-380.

[14] R.V. RAMNATH and G. SANDRI, A generalized multiple scales approach to a class of linear differential equations, J. Math. Anal. and Appl., 28 (November 1969).

[15] A. KLIMAS, R.V. RAMNATH, and G. SANDRI, On the compatibility problem in the uniformization of asymptotic expansions, J. Math. Anal. and Appl., 32 (December 1970), pp. 481-504.

[16] R.W. BROCKETT, Finite Dimensional Linear Systems, Wiley and Sons, Inc., New York, 1970.

[17] H. MORTON, J. JUNKINS, and J. BLANTON, Analytical solutions for Euler parameters, presented at the AIAA/AAS Astrodynamic Conference, Vail, Colorado, July 1973.

[18] E.J. CHERNOSKY and E. MAPLE, Handbook of Geophysics, Macmillan, Chap. 10, New York, 1960.

[19] M. SHIGEHARA, Geomagnetic attitude control of an axisymmetric spinning satellite, J. Spacecraft, 9(1972).

Bifurcation Analysis of Aircraft High Angle-of-Attack Flight Dynamics

Raman K. Mehra* and James V. Carroll*

Abstract. Fighter aircraft operating in high angle-of-attack flight regimes invariably exhibit nonlinear phenomena, precluding effective global analysis by the usual techniques. In this paper, a new approach is presented, involving the application of bifurcation analysis and catastrophe theory methodology (BACTM) to specific jump, hysteresis, and limit cycle phenomena such as stall, departure, post-stall, spin entry, flat and steep spin, spin recovery, spin reversal, and wing rock. The aircraft considered in this report consist of a variable sweep fighter and a modern swept wing fighter; both are modeled using tabular aerodynamic data which are functions of several variables. A key to effective numerical implementation of the BACTM approach lies in the use of continuation methods, and it is shown how these are developed to provide a complete representation of the aircraft equilibrium and bifurcation surfaces in the state-control space consisting of velocity, roll rate, pitch rate, yaw rate, angles of attack and sideslip, pitch and roll angles, and aerosurface control deflections. Also presented is a particularly useful extension of continuation methods to the detection of stable attracting orbits (limit cycles). The use of BACTM for understanding high angle-of-attack phenomena, especially spin-related behavior, is discussed.

1. Introduction. Trends in fighter aircraft design in recent years have resulted in configurations noted for their high speed and performance capability. The cost of achieving this capability has been a drastic, often fatal loss of positive control of the aircraft as the pilot operates at or near the extremes of the flight envelope. This is especially true for aircraft motion at high angles-of-attack (α), where large deviations both in the state and control variables limit the application of the usual linearized analysis techniques. There is a conspicuous lack of techniques for analyzing global stability and large

*Scientific Systems, Inc., 186 Alewife Brook Parkway, Cambridge, Mass., 02138. This work was supported by Contract N00014-76-C-0780 from the Office of Naval Research.

maneuver response of aircraft. While certain phenomena (e.g., roll-coupling) have been analyzed in an isolated manner, there exists a clear need for a systematic approach to analyze global aircraft behavior at high α.

A suitable methodology for global stability analysis would, in addition to providing quantitative, global stability information, also contribute towards safer piloting procedures, control system design, stability augmentation, and aircraft model structure determination and design. The method should also be able to predict and explain such high-α phenomena as discontinuous motion (jumps), limit cycle behavior, and hysteresis effects. There is a particular need for a more complete understanding of limit cycle phenomena, because it is a critical part of high-α motion for all of the aircraft we have investigated.

The BACTM approach owes its success to several factors, including: (i) recent advances in experimental facilities, particularly new wind tunnel techniques [1,2]; (ii) the development of relatively inexpensive and efficient simulation facilities; and (iii) the development of the underlying theory used in BACTM, including results from topology, differential geometry, structural stability, bifurcation analysis, and catastrophe theory [3,4], as well as advanced mathematical techniques for stability analysis [5], and parameter identification [6], which can aid in the task of developing adequate high-α models.

The advances reported in this paper resulted from recognizing that discontinuous and limit cycle phenomena at high α could be analyzed using the new theoretical results. Independent work by others, such as Schy and Hannah [7], have supported this observation and our results have not only confirmed our original conjecture, but have also revealed new dynamic phenomena at high α. A major step in the realization of BACTM as an effective tool for high-α stability analysis has been the implementation of continuation methods for the solution of the aircraft equations. The continuation technique used here is derived from the work of several researchers [8,9,10]. It has led to successful application of the theory to real-world systems of complex structure and high order, by translating the BACTM equations into a form amenable to solution on digital machines.

The next section discusses high α dynamics and presents the simulation equations for the two aircraft models investigated here: a variable sweep fighter (aircraft F) and a swept-wing fighter (F-4). Section 3 discusses briefly the underlying principles of BACTM, and then describes how parametric continuation techniques work. Section 4 discusses specific results for the F-4 and aircraft F systems, and Section 5 summarizes the main results and presents our conclusions. Notation used in this article is defined in Appendix A.

2. High α Dynamics of Aircraft. Traditional aircraft stability analysis and control system synthesis has depended heavily on techniques involving linearization. This approach is very advantageous where applicable; however, as aircraft became faster and more maneuverable, it has become more difficult to justify linear methods. This is especially true in the case of high α motions. As an example, the stubby, thin wings of most modern fighter aircraft greatly reduce drag, but the consequently smaller value of axial inertia enhances roll coupling and autorotational effects. These undesirable responses are the consequence of the greater role that certain nonlinear terms achieve in the equations. Such motions are often counterintuitive as well, inducing the pilot to control actions which worsen the situation.

Equations of motion. The basic aircraft equations of motion for BACTM analysis are given by the general form

$$\dot{x} = f(x,\delta) \tag{2.1}$$

where f is the mapping $f : R^n \times R^3 \to R^n$ and $x \in R^n$, $\delta \in R^3$. For a rigid body aircraft with constant mass in a constant-gravity environment, (2.1) expands to the following [11]:

$$\dot{p} = -\left(\frac{I_z - I_y}{I_x}\right) qr + \bar{q}\,\frac{Sb}{I_x}\,C_\ell \tag{2.2}$$

$$\dot{q} = \left(\frac{I_z - I_x}{I_y}\right) pr + \bar{q}\,\frac{S\bar{c}}{I_y}\,C_m \tag{2.3}$$

$$\dot{r} = -\left(\frac{I_y - I_x}{I_z}\right) pq + \bar{q}\,\frac{Sb}{I_z}\,C_n \tag{2.4}$$

(2.5) $$\dot{\alpha} = q + [(\frac{\bar{q}S}{mV} C_z + \frac{g}{V} \cos\theta\cos\phi - p \sin\beta)\cos\alpha - (\frac{\bar{q}S}{mV} C_x - \frac{g}{V} \sin\theta + r \sin\beta)\sin\alpha]\sec\beta$$

(2.6) $$\dot{\beta} = (\frac{\bar{q}S}{mV} C_y + \frac{g}{V} \cos\theta\sin\phi)\cos\beta - [(\frac{\bar{q}S}{mV} C_x - \frac{g}{V} \sin\theta)\sin\beta + r]\cos\alpha - [(\frac{\bar{q}S}{mV} C_z + \frac{g}{V} \cos\theta\cos\phi)\sin\beta - p]\sin\alpha$$

(2.7)* $$(\frac{\dot{V}}{V}) = (\frac{\bar{q}S}{mV} C_x - \frac{g}{V} \sin\theta)\cos\alpha\cos\beta + (\frac{\bar{q}S}{mV} C_y + \frac{g}{V} \cos\theta\sin\phi)\sin\beta + (\frac{\bar{q}S}{mV} C_z + \frac{g}{V} \cos\theta\cos\phi)\sin\alpha\cos\beta$$

In this set, the body axis coordinatization is assumed to be in the body principal axes, so that all cross products of inertia are zero. Also, the origin of this axis system is the vehicle center of mass, so that no gravity torques exist in (p,q,r). Eqs. (2.2)-(2.7) represent a set of nonlinear, autonomous ordinary differential equations describing rotational and translational accelerations on the vehicle. Where gravity plays an important role (e.g., spin motion), it is necessary to add two kinematic equations:

(2.8) $$\dot{\theta} = q \cos\phi - r \sin\phi$$

(2.9) $$\dot{\phi} = p + (q \sin\phi + r \cos\phi)\tan\theta$$

See Appendix A for notation.

The elements of x, the state variable, are $(p,q,r,\alpha,\beta,V,\theta,\phi)$, so that the global dynamic system is eighth order. The advantage of BACTM and the continuation methods is that no further simplification is required, whereas other techniques often require several levels of simplification. In addition, more general equations, e.g., nonautonomous equations, are readily handled by our methodology.

Aerodynamic data. The aerosurface controls appear linearly in the (non-dimensional) aerodynamic coefficients, C_x, C_y, C_z, C_ℓ, C_m, C_n. These are aileron (δa), elevator (δe), and rudder (δr), so that δ is the set $(\delta a,\delta e,\delta r)$ from which we choose the continuation parameter. This parameter could also be b, S, I_y, etc., or any of the aero coefficients

*The non-dimensional force and moment terms C_x, C_y, C_z, C_ℓ, C_m and C_n are defined in Eqs. (2.10)-(2.15).

defined in the following expansions:

$$(2.10)\qquad C_x = C_{x_0} + C_{x_{\delta e}} + C_{x_{RB}} + C_{x_q}\left(\frac{\bar{c}}{2V}\right)q_0$$

$$(2.11)\qquad C_y = C_{y_0} + C_{y_{\delta a}}\delta a + C_{y_{\delta e}}\delta e + C_{y_{\delta r}}\delta r + C_{y_{RB}} + \left(\frac{b}{2V}\right)(p_0 C_{y_p} + r_0 C_{y_r})$$

$$(2.12)\qquad C_z = C_{z_0} + C_{z_{\delta e}}\delta e + C_{z_{RB}} + C_{z_q}\left(\frac{\bar{c}}{2V}\right)q_0$$

$$(2.13)\qquad C_\ell = C_{\ell_0} + C_{\ell_{\delta a}}\delta a + C_{\ell_{\delta e}}\delta e + C_{\ell_{\delta r}}\delta r + C_{\ell_{RB}} + \left(\frac{b}{2V}\right)(p_0 C_{\ell_p} + r_0 C_{\ell_r})$$

$$(2.14)\qquad C_m = C_{m_0} + C_{m_{\delta e}}\delta e + C_{m_{RB}} + C_{m_q}\left(\frac{\bar{c}}{2V}\right)q_0$$

$$(2.15)\qquad C_n = C_{n_0} + C_{n_{\delta a}}\delta a + C_{n_{\delta e}}\delta e + C_{n_{\delta r}}\delta r + C_{n_{RB}} + \left(\frac{b}{2V}\right)(p_0 C_{n_p} + r_0 C_{n_r})$$

Thus, the full aero model consists of tabular data for 33 coefficients. These data consist of static (S), forced oscillation (FO), and rotary balance (RB) derivatives. Because of the smoothness requirements on f, we are using cubic and bicubic splines to represent the α- and β-dependence of the aero data, which were taken under conditions of incompressible flow.

Aircraft models. Aircraft F is a variable sweep fighter whose aero data has been altered somewhat in order to study certain spin motions. This model has been extremely useful in developing the BACTM algorithms and in providing initial spin analysis results. We discuss spin behavior and its characteristics in more detail below.

The selection of the F-4 was motivated by the availability of its high α aero and flight test data base, and by the desire to investigate an actual aircraft. Also, the F-4 has a history of being excessively difficult to recover from certain types of spin motion [12], and so it is of practical importance to seek better understanding of its high α behavior.

Spin. Entry into spin is usually a consequence of post-stall departure, in which extreme control settings induce high α conditions and excessive yawing. Spin itself is a divergent motion featured by very high, persistent, values of α (up to 85°), and yaw rate (often between 100 and 300 deg. per sec.). Typically, the center of mass follows a tight helical path whose axis is vertical, and in which altitude decreases at a nearly steady rate. Except for α and r, the major spin indicators, the other variables do not attain extreme values in a developed spin.

During spin, the aerosurface controls lose almost all of their control effectiveness, due to the radical difference between spin and trim airflow patterns, and this makes recovery difficult. Control actions for recovery are so extreme that they can induce spin reversal, which is merely spin with r having the opposite sign.

Other high α motion. The evocative names found in the flight test literature include wing rock, nose slice, pre-stall buffeting, stall, and post stall departure, in addition to spin. Many of these phenomena are oscillatory in nature, adding emphasis to the application of limit cycle analysis and Hopf and global bifurcation theory to high-α dynamics. Wing rock is a condition of high frequency roll rate oscillation. It again represents high α longitudinal-lateral coupling effects. This condition is the result of a Hopf bifurcation (explained below), as δe decreases to its minimum value. Nose slice is a similar phenomenon, but contains more yaw rate in the motion than does wing rock. It is also a condition which is prominent near stall values of α. Stall and post stall departures are transient motions, and usually represent a transition phase between trim equilibrium and high α attracting limit cycles such as steep spin.

In general, departure refers to a situation in which the pilot loses positive control of the aircraft. This usually occurs at and shortly after stall. Fig. 1a shows a time history for the F-4 in which high α behavior was induced by control deflections. It represents data obtained from a .13 scale model radio-controlled drop flight test. The model is dropped from a helicopter at about 5000 feet. The post-stall motion (stall α for the F-4 is about 14°) shows a rather mild transition phase, before the onset of spin at about $t = 60$ sec. Note the change in sign of r (spin reversal) in response to the δr change at $t = 80$.

The above flight conditions are qualitative. Our BACTM analysis has

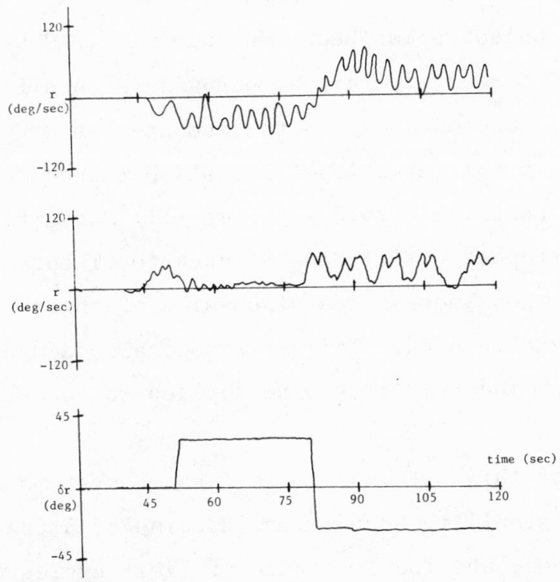

Figure 1: F-4 Correlation Study. (a) Flight test model yaw rate vs. time. (b) Simulation yaw rate vs. time. (c) Control input to both runs; $\delta e = -21°$, $\delta a = 0°$, $t < 50.$ and $t > 80.$; $\delta a = 30°$, $50 \le t \le 80$.

verified their presence quantitatively, as well as other types of high α motion, some of which are combinations of the above.

Correlation studies. A subsidiary problem to the ones discussed in this paper, yet one which is nonetheless very important, is that of obtaining adequate correlation between flight test results and the simulation results based on wind tunnel data. This correlation has been achieved with only limited success in high α regions for the F-4 [13]. At this time, F-4 results generated from wind tunnel data do not correlate well with scale model flight results, as the comparison in Fig. 1 indicates. In Fig. 1b, the same control inputs recorded in the flight test were used to generate the simulation trajectory. It is possible to expand the scope of BACTM to deal with this issue.

3. Continuation Methods. The continuation methodology which we will discuss in this section is the bridge by which the abstract mathematical results from topology, bifurcation analysis, and catastrophe theory are joined with numerical computation to generate global stability results for the complicated aircraft dynamic system at high α. Continuation methods effectively broaden the scope of application of the theory.

Before discussing continuation, we present a brief overview of the Bifurcation Analysis and Catastrophe Theory Methodology (BACTM) approach.

BACTM allows one to study the global behavior of nonlinear systems in an (n+m) dimensional space where n is the number of state variables and m is the number of control variables. The BACTM methodology is based on four theorems: (i) Center Manifold Theorem; (ii) Classification Theorem of Elementary Catastrophes; (iii) Hopf Bifurcation Theorem; and (iv) Global Implicit Function Theorem. A discussion of these theorems and a bibliography are found in [14]. This reference also demonstrates that Thom's Classification Theorem [4] may be applied to constant coefficient aircraft models.

The application of this methodology to dynamic systems enables global characterization of stability boundaries, domains of attraction, bifurcation and jump surfaces, and the existence of limit cycles. This characterization requires the generation of the system equilibrium and bifurcation surfaces, and only in simple cases is it possible to do this analytically. It is for this reason that we turn to continuation methods, as well as for reasons of efficiency, accuracy, and ability to deal readily with singularities.

Continuation methods deal with the problem of solving

$$g(x,\lambda) = 0 \tag{3.1}$$

for the state variables x as a function of the parameter λ. Our applications of continuation methods at this time consist of computing equilibrium surfaces, bifurcation surfaces (defined later), and limit cycle solutions (also defined and explained below). In each of these cases, x is an element of a finite dimensional space, and λ is a real parameter, $\lambda \in R^1$. The theory is actually more general, and x may be an element of an (infinite dimensional) Banach space B, with $\lambda \in R^m$. The case $x \in B$ occurs in trajectory optimization problems [15].

By solving (3.1) for the trajectory $x(\lambda)$ it is possible to determine x at some λ_1, say, based on an initial solution $x(\lambda_0)$. To generate a surface, we seek not only solutions at λ_0 and λ_1 (for the aircraft problem $\lambda \in \delta$), but λ at all points in between, and these are provided as a byproduct of the continuation method. The method, utilizing results from topology as well as differential calculus, is well suited to solving difficult nonlinear systems, as it is very adaptable to modern digital computers.

Our approach to solution by continuation is to combine the methods of Lahaye [9] and Davidenko [8]. The former begins with a solution at $\lambda=\lambda_0$, using $x(\lambda_{i-1})$ as the initial guess in a Newton-Raphson iteration at λ_i. Davidenko differentiated (3.1) with respect to λ to obtain

$$(3.2)\qquad G(x,\lambda)\,\frac{\partial x}{\partial\lambda} + \frac{\partial g}{\partial\lambda}(x,\lambda) = 0$$

which is usually (exceptions are noted below) integrated readily from λ_0 to any λ_1. In this equation, $G \overset{\Delta}{=} (\partial g/\partial x)$. The algorithm uses Davidenko's method for prediction and Lahaye's method as a corrector step. At each solution point, the eigenvalues of G are checked to provide local stability information. Of course, (3.2) cannot be solved for $x(\lambda)$ where G is singular.

G is singular at a <u>limit point</u> or at <u>bifurcation</u> points. A limit point occurs where the solution curve begins to "fold back" on itself; that is, λ is not monotonic; a <u>simple bifurcation</u> occurs where two curves intersect, and a <u>general bifurcation</u> where more than two curves intersect. See Fig. 2. Some authors [16] classify a limit point as a bifurcation point also, since the stable attractor is annihilated by an unstable attractor. Other types of bifurcations are <u>global bifurcations</u> (stable attracting orbit, or limit cycle, being annihilated by an unstable one), <u>Hopf bifurcations</u> (a stable attractor, or branch, changing into a limit cycle) and the <u>singular perturbation bifurcation</u> (in which $x(\lambda) \to \infty$ as $\lambda \to \lambda_1$; this class of singularity is not encountered in our problem).

It is now convenient to rewrite (3.1) as $g(z) = 0$, where z is the (n+1)-vector (x,λ). This emphasizes the fact that there is no distinction in the methods that follow between λ and the x_i. Since $g \in R^n$ and $z \in R^{n+1}$, the solution points generate a trajectory in R^{n+1}. At a limit point, the rank of

$$(3.3)\qquad \Gamma \overset{\Delta}{=} \left[\frac{\partial g}{\partial z}\right] = \left[G \,\middle|\, \frac{\partial g}{\partial\lambda}\right]$$

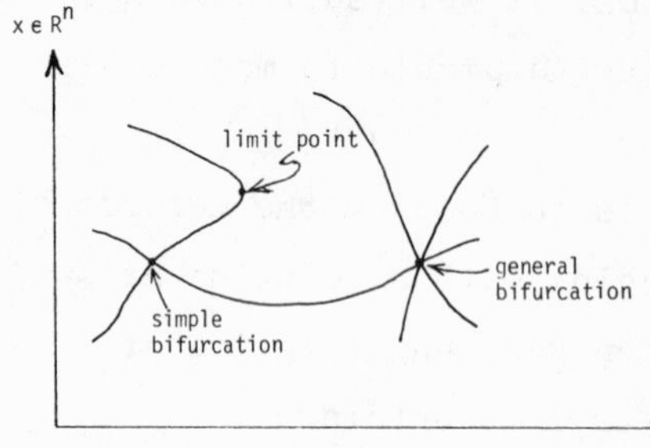

Figure 2: Limit and Bifurcation Points

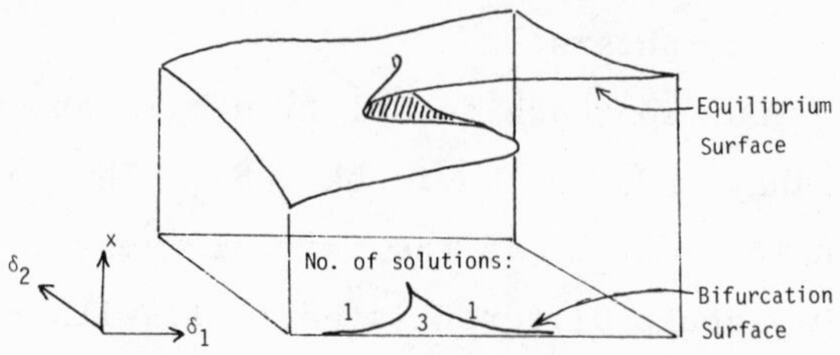

Figure 3: Equilibrium and Bifurcation Surfaces; Cusp Catastrophe

an $n \times (n+1)$ matrix, is n, the same as when G is nonsingular (hence, limit points are known as <u>weak singularities</u>). We exploit this by defining the arclength parameter s, and reformulating our problem to one of solving for those points $z(s) \in R^{n+1}$ which satisfy

$$(3.4) \qquad g(z(s)) = 0$$

Davidenko's equation (3.2) is then modified to reflect the new independent variable, s:

$$(3.5) \qquad Gx' + \frac{\partial g}{\partial \lambda} \lambda' = \Gamma x' = 0$$

where $(\;)' \triangleq d(\;)/ds$.

Keller [10] shows that the Frechet derivative of the system (3.4) and (3.5), which was expanded by introducing s, has similar features at regular and limit points; and that, therefore, continuation around limit points is feasible. Heuristically, replacing λ, which ceases to be monotonic at a limit point, with an arclength parameter s, which is inherently monotonic, restores regularity. Keller suggests several non-Euclidean arclength expressions ("normalizations"), but we have found success using the Euclidean one:

$$(3.6) \qquad N(z,s) = {z'_1}^2 + \ldots + {z'_{n+1}}^2 - 1 = 0$$

The augmented Frechet derivative is $\partial(g,N)/\partial z$.

Kubicek [17] has introduced a numerical algorithm for continuing past limit points, based on Gaussian elimination with controlled pivoting; it is this algorithm which we have incorporated in BACTM. Since Γ has rank n at a limit point, there must be at least one nonsingular Γ_k, where Γ_k is an $n \times n$ submatrix of Γ formed by eliminating the kth column of Γ, for $k \in (1,\ldots,n+1)$. (Note that $G \equiv \Gamma_{n+1}$.) Eq. (3.5) then may be rearranged

$$(3.7) \qquad \Gamma_k \{z_i'\}_{i \neq k} + \frac{\partial g}{\partial z_k} z_k' = 0$$

Eq. (3.7) is then solved for the n z_i', $i \neq k$, in terms of z_k', and the latter is found using (3.6). Sign ambiguity is resolved by choosing one of two possible directions for proceeding along the solution arc passing through λ_0. Column k represents the "singular" column at a limit point, and the "most singular" column at regular points.

Having obtained z', it may be used in either the predictor or corrector algorithms. In the corrector (Newton), it is solved for as Δz ($\Delta z_k \equiv 0$). In the predictor, z' is used in Adams-Bashforth variable order integration. Pivoting may be controlled by selective scaling of the columns of Γ.

At a bifurcation point, the method of Kubicek breaks down. This is because the rank of Γ is (n-1) for a simple bifurcation point, or less (general bifurcation point). Keller [10] has developed an algorithm for detecting bifurcation points, and for continuing along all branches emanating from them. It can be shown using Taylor expansions that, to second order at a simple bifurcation point, each intersecting branch is locally tangent to the subspace spanned by the eigenvectors associated with the zero eigenvalues of the $(n+1) \times (n+1)$ matrix $\Gamma^T\Gamma$. For a simple bifurcation point, there are two such eigenvectors, and the plane they span is normal to range space of Γ (i.e., (3.5) holds). Thus in this case, initial points to all branches emanating from the simple bifurcation point z* may be found by seeking zeros of

$$(3.8) \qquad g(z^* + \Delta z) = 0$$

where Δz has some small magnitude ε, and lies in the above plane. One finds the points, then, by varying the direction of Δz in this plane. The quantity ε should be large enough to avoid the numerical problems associated with z*, yet small enough so that the Taylor approximations are reasonable.

Applications of the continuation method. Here, we specify problems to which the BACTM continuation algorithm has been applied to the analysis of high α dynamics of aircraft. Results are discussed in Sec. 4.

Equilibrium surfaces. This is the system for which g in (3.4) becomes

$$g(x(s),\lambda(s)) = f(x(s),\lambda(s)) = \dot{x} = 0 \tag{3.9}$$

where x are the aircraft state variables, λ is one of the aerosurface controls (the other two are fixed while each equilibrium locus is generated). See Fig. 3.

Bifurcation surfaces. This is our designation for those manifolds resulting from the projection of the limit points of the system (3.9) onto the control subspace R^m. This condition adds to (3.9) the additional requirement

$$\Delta \triangleq \det(G) = 0 \tag{3.10}$$

Thus, g in (3.4) becomes

$$g(z(s)) = (f(z(s)),\Delta(z(s))) \tag{3.11}$$

Hence $g \in R^{n+1}$ here, so that $z \in R^{n+2}$. We therefore select two of the aerosurface controls to augment x, holding the third fixed. Every solution of (3.11), then, generates a curve in two-dimensional control space (Fig. 3).

Limit cycle continuation. A limit cycle is a nonlinear oscillation satisfying the condition

$$x(T;x_0) - x_0 = 0 \tag{3.12}$$

for some fixed $T \in (0,T_{max})$, $T_{max} < \infty$, where

$$x(t;x_0) \triangleq x_0 + \int_{t_0}^{t} f(x,\delta,\sigma)\, d\sigma \tag{3.13}$$

Our goal is to find x_0 and $T(x_0)$ for the system (2.1) such that (3.12) holds. We utilize results from Chua and Lin [18], who provide a method for finding one (x_0,T_0) pair; however, we have expanded the scope of their method by incorporating it into the framework of continuation methods, so that we can generate a continuum of limit cycle points

$x(T(s),\lambda(s))$, as the continuation parameter λ is changed. This extension represents a significant advance over previously known methods of limit cycle detection and analysis. By invoking the robustness and globality of the continuation approach, it is possible to start where a known limit cycle exists, at λ_0, and continue λ to regions where solutions are difficult to obtain.

To apply the continuation method, g in (3.4) becomes

$$g(z(s)) = x(T(z(s));x_0(z(s)) - x_0(z(s)) = 0 \tag{3.14}$$

Now $g \in R^n$ and $z_{n+1} = \lambda$, one of the aerosurface controls, so that one of the n+1 elements (x_0, T) must be dealt with. Utilizing the quadrature (3.13) and magnitude comparisons on the n elements of $f(x(T^{(0)}, x_0^{(0)}), \delta)$, where $(T^{(0)}, x_0^{(0)})$ are initial guesses for the limit cycle state, an x_{0_k} is found which remains fixed during the corrector steps which lead to the solution of (3.12). At each corrector step, $g(z(s))$ must be evaluated, and this requires determining $x(T;x_0)$ using (3.13). Γ is evaluated using Leibnitz' rule.

Other applications of continuation. If a bifurcation point z* is encountered in solving for equilibrium surfaces, it is possible to use continuation methods either to isolate this point more accurately or to generate a curve $z^*(\lambda(s))$, if the point is not isolated. For a simple bifurcation point, the system would be set up based on the rank $\Gamma = (n-1)$ condition, in addition to $f(x,\lambda) = 0$.

4. Application of BACTM to the Aircraft High α Problem. Fig. 4 shows a composite of four equilibrium branches which indicate how BACTM-generated equilibrium surfaces are used to understand and predict high α behavior, in this case spin and spin recovery. It shows curves of equilibrium yaw rate (r) vs. rudder (δr) for trim elevator ($\delta e = 0°$) and two values of aileron: $\delta a = 15°$ (Branch 1) and 0° (remaining branches). The letters defining the curves indicate the local stability results as follows: S is a stable point (all eigenvalues of the linear part of f have negative real parts), U is a simple unstable point (one positive root), L is an unstable complex pair, and H (lower part of Branch 1) is an LU combination. The S-segments are known as stable attractors, and motions beginning near them end up on them. All non-S segments are locally

divergent; however, globally the motion may grow to a limit cycle.

Jump phenomena occur at limit points. This is the manner in which one may escape an undesirable stable equilibrium such as the flat spin segment of Branch 1. Rudder is increased from its spin value (-25°), changing the equilibrium from point A, to a dynamic condition influenced by Branch 2. As rudder goes positive, the attracting closed orbit on Branch 2 is annihilated (global bifurcation) by an unstable one, and the motion becomes influenced by Branch 3's stable attractor. If δr is held at a high positive value, the motion will ultimately be attracted to Branch 4 (spin reversal); the desired result, return to trim at point B, is achieved by holding δr for a short time at 30°, and then returning to trim as $|r|$ gets small. Fig. 5 verifies that this strategy produces recovery. Returning to Fig. 4, the vertical bars represent projections of the limit cycle motions onto the r-δr subspace, taken at 2° intervals of <u>increasing</u> δr. The pattern for decreasing δr is much different; thus, hysteresis is a factor in studying this motion.

The F-4 model produces similar equilibrium branches, with one noteworthy difference. Its equivalent of Branch 1 is a flat branch over $\delta r \in \pm 30°$, changing from S to L as δr becomes positive. That is, we have not found it possible to effect a jump from spin equilibria to the less stable oscillatory spin branches and ultimately to trim, by using aerosurface controls. Our experience using BACTM thus corroborates actual F-4 flight test experience [12], in which it was found that flat spin recovery could be achieved only by using wing tip rockets or parachutes. Given that the BACTM curves are global and inexpensively obtained, our methodology offers much potential for reducing costs in preliminary design configuration studies.

The limit cycle continuation algorithm is not fully operational with the F-4; however, its numerical effectiveness has been demonstrated on this model. We have visually investigated a "limit cycle" motion based on oscillatory spin conditions obtained from Fig. 5. With an initial input of $x_0 = (p_0, q_0, r_0, \alpha_0, \beta_0) = (79.7, 14.6, 83.4, 58.7, -8.6)$ and $T^{(0)} = 2.5$ sec, we obtained after three Newton steps $x_0 = (70.33, 18.66, 73.50, 54.49, -7.4)$ and $T^{(3)} = 2.487$ to an error of less than one part in 10000. As a check, $x(T^{(3)}; x_0)$ was evaluated using (3.13), and was found to be (70.36,

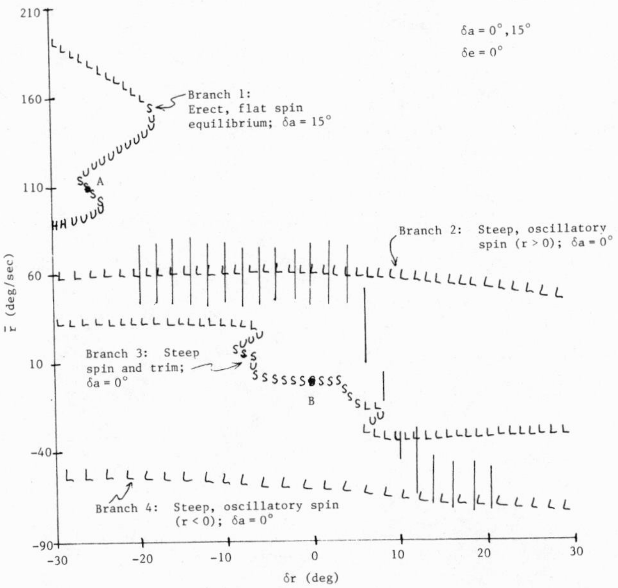

Figure 4: Aircraft F Equilibrium and Limit Cycle Trajectories for Spin Recovery; $\delta a = 0$, $\delta e = 0$; $V_0 = 450$ fps, $h = 30000$ ft; $\bar{r}$ vs. δr, δr increasing

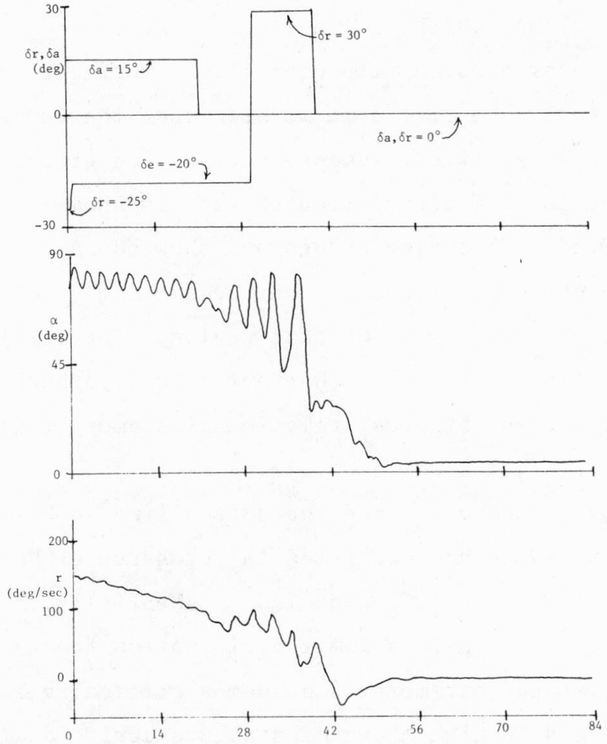

Figure 5: Aircraft F Spin Recovery; $\delta e = 0$, $h_0 = 30000$ ft, $V_0 = 450$ fps. Initial conditions are Point A of Figure 4.

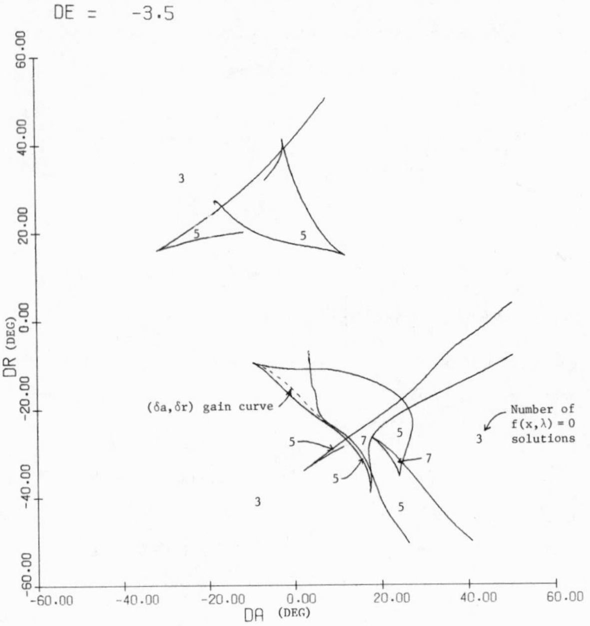

Figure 6: Bifurcation Surface, Showing Possible Gain Schedule for (δa, δr). The number of equilibrium solutions (not all stable equilibria) is shown for some regions.

18.63,73.50,54.49,-7.23), which is excellent agreement.

Bifurcation surfaces combine with equilibrium surfaces to provide a powerful analysis tool. In Fig. 3 it is seen that the number of equilibrium solutions changes (i.e., jumps occur) as one crosses a boundary. In Fig. 6 a part of the F-4 bifurcation surface is shown in the δa-δr plane, with $\delta e = -3.5°$. This figure does not show boundaries for Hopf bifurcations. The numbers in some of the regions indicate the number of equilibria (stable and unstable) for that region. The utility of surfaces such as this lies in the fact that autopilot synthesis can be greatly aided. From these figures, relationships among (δa,δe,δr) may be devised which avoid jumps while utilizing as fully as possible the envelope defined by the curves. See the dotted line in Fig. 6.

Finally, BACTM analysis has confirmed the presence of Hopf bifurcation phenomena in high α aircraft dynamics. Essentially, a Hopf bifurcation marks the bifurcation to a limit cycle motion from a stable equilibrium condition, as the parameter λ passes a critical value. The most common example of this in high α dynamics occurs during a pitch up maneuver, in which equilibrium α increases, under the influence of decreasing

δe, from trim values (about 3°) through its stall value (about 12° to 14° at subsonic speeds). As α increases, the motion changes from stable to pre-stall buffeting to wing rock, completing the Hopf bifurcation. Quite often, the S-to-L transition on an equilibrium surface indicates a Hopf bifurcation.

5. Conclusions. Based on analysis of the nonlinear models F-4 and aircraft F, we can conclude that: (i) a large number of unstable, jump, and limit cycle phenomena occurring at high angles-of-attack, can be analyzed in a unified fashion by using BACTM. (ii) BACTM provides a global representation of the equilibrium and bifurcation surfaces for nonlinear dynamic systems. The qualitative dynamics of the system for different initial conditions, controls, and system parameters can be obtained easily from the equilibrium and bifurcation surfaces. Also, the control and parameter values for which jumps and limit cycles appear are obtained directly from the equilibrium and bifurcation surfaces.

(iii) BACTM can generate equilibrium and bifurcation surfaces for aircraft models with tabular aerodynamic data. Based on these surfaces, control strategies for spin recovery have been developed, and these agree well with actual flight recovery procedures. In addition, the equilibrium surfaces have provided a basis for study of other high α phenomena such as stall, wing rock, and post-stall oscillatory motions.

(iv) We have demonstrated how BACTM may be used to synthesize a command augmentation system for the F-4, using its bifurcation surfaces along with the criterion of maximizing the range of control values which avoid jump or limit cycle behavior.

(v) BACTM can be expanded to include quantitative detection and analysis of limit cycle motions, and the generation of limit cycle solution points by means of continuation methods.

(vi) F-4 analysis has provided qualitative confirmation of behavior which has been determined beforehand using more exhaustive approaches. When correlation between the models based on wind tunnel data and flight test results is improved, BACTM will provide useful quantitative results on actual F-4 performance, as well as represent a basis for analyzing the effect of configuration or parameter changes made to the F-4.

It is our contention that the methodology presented here, and the manner in which it has been extended to complicated, nonlinear, high-order systems through the use of numerical techniques based on parametric continuation, adds to the analyst's repertoire a significant new tool. BACTM provides not only quantitative global stability information, but also the capability to predict and explain such non-intuitive phenomena as jumps, limit cycles, and hysteresis, and guidelines for stabilizing an unstable nonlinear system. In particular, for aircraft, the location and characteristics of the flight envelope can be determined more effectively, thereby reducing artificial constraints on the aircraft's control effectiveness.

REFERENCES

[1] J. R. CHAMBERS, "Status of Model Testing Techniques," Presented at the AFFDL Stall/Post-Stall/Spin Symp., Dec. 1971.

[2] J. CAMPBELL, "A Technique Utilizing Free Flying Radio-Controlled Models to Study Incipient- and Developed-Spin Characteristics of Airplanes," AGARD Report No. 76, 1959.

[3] F. TAKENS, (a) "Introduction to Global Analysis," and (b) "Notes on Forced Oscillations," in Communications 2 and 3, Math. Inst. Rijksuniversiteit, Utrecht.

[4] R. THOM, Structural Stability and Morphogenesis, Addison Wesley, Reading, Mass., 1974.

[5] R. F. STENGEL, J. H. TAYLOR, J. R. BROUSSARD and P. W. BERRY, "High Angle-of-Attack Stability and Control," Report ONR-CR-215-237-1, Office of Naval Research, 1976.

[6] W. E. HALL, N. K. GUPTA and J. S. TYLER JR., "Model Structure Determination and Parameter Identification for Nonlinear Aerodynamic Flight Regimes," in Methods for Aircraft State and Parameter Identification, AGARD-CP-172, 1974.

[7] A. A. SCHY and M. E. HANNAH, "Prediction of Jump Phenomena in Roll-Coupled Maneuvers of Airplanes," AIAA Third Atmospheric Flight Mechanics Conf., Arlington, Texas, 1976.

[8] D. DAVIDENKO, "On a New Method of Numerically Integrating a System of Nonlinear Equations," Dokl. Akad. Nauk. SSSR, 88(1953), pp. 601-604 (in Russian).

[9] E. LAHAYE, "Une Méthode de Résolution d'une Catégorie d'Équations Transcendantes," C. R. Acad. Sci. Paris, 198(1934), pp. 1840-1842.

[10] H. B. KELLER, "Numerical Solution of Bifurcation and Nonlinear Eigenvalue Problems," in Applications of Bifurcation Theory (ed. P. H. Rabinowitz), Academic Press, New York, 1977.

[11] W. M. ADAMS, "Analytic Prediction of Airplane Equilibrium Spin Characteristics," NASA TN D-6926, 1972.

[12] E. L. RUTAN, C. E. MCELROY and J. R. GENTRY, "Stall/Near Stall Investigation of the F-4E Aircraft," Technical Report No. 70-20, Air Force Flight Test Center, Edwards AFB, Calif., 1970.

[13] E. L. ANGLIN, "Aerodynamic Characteristics of Fighter Configurations During Spin Entries and Developed Spins," AIAA J. Aircraft, 15(1978), No. 11.

[14] R. K. MEHRA, W. C. KESSEL and J. V. CARROLL, "Global Stability and Control Analysis of Aircraft at High Angles-of-Attack," Report ONR-CR215-248-1, 1977.

[15] R. K. MEHRA, R. B. WASHBURN, S. SAJAN and J. V. CARROLL, "A Study of the Application of Singular Perturbation Theory," NASA CR-3167, 1979.

[16] R. ABRAHAM and J. E. MARSDEN, Foundations of Mechanics, Benjamin/Cummings, Reading, Mass., 1978.

[17] M. KUBICEK, "Algorithm 502, 'Dependence of Solution of Nonlinear Systems on a Parameter'," ACM-TOMS, 2(1976), pp. 98-107.

[18] L. O. CHUA and P. M. LIN, Computer-Aided Analysis of Electronic Circuits, Prentice-Hall, Englewood Cliffs, N.J., 1975.

APPENDIX A

Notation

b	wing span
B	Banach space
$\bar{c}$	wing chord
C_ℓ, C_m, C_n	rotational aerodynamic coefficients
C_x, C_y, C_z	translational aerodynamic coefficients

Note: derivatives, e.g., $\partial C_n/\partial p$ and $\partial C_x/\partial \delta e$ are abbreviated C_{n_p} and $C_{x_{\delta e}}$, respectively.

f	dynamic function
g	gravity; continuation function
G	Jacobian matrix, $[\partial g/\partial x]$
I_x, I_y, I_z	vehicle principal moments of inertia
ℓ, m, n	indices for rotational quantities with respect to vehicle roll, pitch, and yaw axes, respectively
L	limit cycle point

m	mass; dimension of control space
N	arclength normalization relation
p,q,r	vehicle angular velocity components, coordinatized in body axes--roll, pitch, and yaw rates, respectively
$\bar{q}$	dynamic pressure
R, R^1, R^n	space of real numbers; one, one, and n-dimensional
s	arclength parameter
S	aircraft reference area; stable equilibrium point
T	period of motion
U	refers to simple unstable equilibrium point
V	vehicle center of mass velocity
x	state variable
α, β	angles of attack and sideslip
Γ	augmented Jacobian matrix, $[\partial g/\partial z]$
Γ_k	square submatrix of Γ
δ	vector of aerosurface controls
$\delta a, \delta e, \delta r$	aileron, elevator, rudder, respectively
Δ	determinant of the Jacobian matrix G; differential operator
ϕ	roll angle
λ	continuation parameter

Nonlinear Aeroelasticity

Earl H. Dowell*

Abstract. Nonlinear aeroelasticity has many examples of complex motion of physical systems. For one class of problems, aeroelasticity of plates and shells, the equations of motion are well established. Here results obtained by numerical time integration may be compared to those obtained by qualitative theories of dynamics. For another class of problems, bluff body oscillators, the equations of motion themselves are still in an emerging stage. Here a qualitative theory may be useful in establishing a generic model. To use qualitative methods effectively, the system needs to be characterized by a relatively small number of degrees of freedom. Coordinate and equation reduction methods are being developed which may bring other classes of problems within the effective range of qualitative theories.

Introduction. Nonlinear aeroelasticity[1-4] is a rich source of static and dynamic instabilities and associated limit cycle motions. Fundamentally it combines the classical fields of fluid and solid mechanics. Usually, as the name implies, it does so within the context of the sub-fields of aerodynamics combined with elasticity.

The aerodynamic forces almost invariably are non-conservative. When acting upon a resonant elastic structure whose motion modifies these forces in a feedback sense, they lead to a complex and fascinating variety of dynamical behavior.

A further word should be said about the determination of the aerodynamic forces and the consequent implications for modeling aeroelastic phenomena. Even in the simplest, general theoretical model of (linearized) small perturbation, (inviscid, irrotational) potential flow, the aerodynamic forces <u>due to the structural motion</u> are related to the motion through a <u>convolution integral</u>. Hence the ultimate structural

*Department of Mechanical and Aerospace Engineering, Princeton University, Princeton, New Jersey, 08544.

equations of motion are integro-differential equations. Moreover the kernel of the convolution integral is frequently only known numerically. Hence these equations are often attacked numerically to obtain time histories of motion. The method of harmonic balance is also occasionally used. Of course, the great bulk of the theoretical literature is devoted to completely linear models[1,5-8] which may be solved by Fourier methods in the frequency domain.

In the present paper two aspects of the subject are addressed. Firstly, in Part I, a simplified representation of the nonlinear aeroelasticity of plates and shells is considered. The principal focus here is on the evolution in parameter space of chaotic motion from simple deterministic motion and its possible implications for the methods and results of differential dynamics as typified by Holmes' work[9,10]. For this class of problems, qualitative methods of the sort studied by Holmes may suggest effective ways of interpreting the results of numerical studies and, in turn, find the latter a useful check on their range of validity. There is a comprehensive theoretical and experimental literature [2] which the reader interested in the physical background of this problem may wish to consult.

In Part II a physical, intuitive approach is used to construct a theoretical nonlinear aeroelastic model based upon experimental information. This is done in the context of bluff body oscillators where no aerodynamic forces-structural motion relationship is presently known from first principles. It is suggested that the qualitative methods of differential dynamics may be able to place the construction of such theoretical models on a more rigorous and systematic basis.

Finally two methods are noted here for reducing the number of equations of motion. In one, component mode synthesis has been extended to nonlinear systems[11]. This method is particularly effective when the nonlinearity is confined to a few components of a multi-component system. In the other, for systems with distributed nonlinearities, the distinction between linear and nonlinear coordinates has been shown[12] to be of value. In this approach those modal coordinates are identified whose motion is sufficiently small that products of the coordinates and their time derivatives may be neglected. By conventional techniques these linear coordinates are eliminated in favor of the nonlinear ones. Typically the latter are few(er) in number. Both approaches appear useful in reducing the number of mathematical degrees of freedom so that there may be some hope of applying qualitative theories of dynamics.

The author would like to acknowledge several helpful discussions with colleagues at the Conference and, especially, a stimulating exchange with Philip Holmes.

Part I: Flutter of a Buckled Plate as an Example of Strange Attractor Motion. It has been known for some time that a plate under a compressive in-plane load with a fluid flow over its (upper) surface may undergo complex motions[2]. See SKETCH OF PLATE GEOMETRY below.

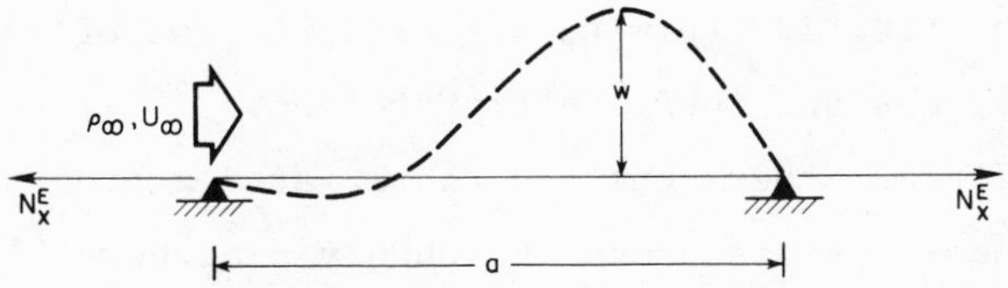

SKETCH OF PLATE GEOMETRY

Note: Plate deflection, w, is shown to a greatly exaggerated scale for clarity. w is the order of a plate thickness.

For no fluid flow, but with a sufficiently large compressive load, the plate will buckle into a statically deformed shape. By contrast, for no compressive load, but a fluid flow of sufficiently large velocity, the plate will flutter with a periodic, nearly harmonic motion. However, with both compressive load and a fluid flow, chaotic motion may occur which is thought to be a result of a strange attractor[9,10].

The problem is considered in its simplest formulation, i.e., a one-dimensional structural model (albeit nonlinear) and piston theory aerodynamics (appropriate to high supersonic Mach numbers). The governing partial differential equation is reduced to a system of ordinary differential equations by using a modal expansion and Galerkin's method. For this relatively simple model, Holmes[9,10] has obtained a number of interesting results using the methods of differential dynamics (for a two mode expansion). Here results are obtained (for two and four mode expansions) using numerical integration in time. Motivated by Holmes' results, the resulting data are examined in the phase plane and also Poincare plots. By systematically changing the compressive load and fluid flow parameters, the evolution of chaotic motion from simple, deterministic motion is studied.

It is worthy of emphasis that at lower Mach numbers the aerodynamic model needed becomes more elaborate and leads to integro-differential equations. These have been successfully solved by numerical integration techniques[2]. However, it is unclear to what extent the methods of differential dynamics may be used for such equations. Hence, for the present class of problems, these latter methods appear primarily to be of value in explaining the qualitative character of the motion and in suggesting effective formats for presentation of the results of numerical solutions and their interpretation.

Equations of Motion. Here only an abbreviated account is given as the underlying theory is discussed thoroughly elsewhere[2,13]. The governing partial differential equation is

$$(1) \quad D \frac{\partial^4 w}{\partial x^4} - (N_x + N_x^E) \frac{\partial^2 w}{\partial x^2} + m \frac{\partial^2 w}{\partial t^2} + \frac{\rho_\infty U_\infty^2}{M} \left[\frac{\partial w}{\partial x} + \frac{1}{U_\infty} \frac{\partial w}{\partial t}\right] = \Delta p$$

where

- w - plate transverse deflection
- x - streamwise spatial coordinate
- t - time
- D - $Eh^3/12(1 - \nu^2)$, plate bending stiffness
- N_x - $(Eh/2a) \int_0^a (\frac{\partial w}{\partial x})^2 \, dx$
- N_x^E - externally applied in-plane load
- E - modulus of elasticity
- ν - Poisson's ratio
- a - plate length
- h - plate thickness
- m - mass/per unit length
- ρ_∞ - flow density

U_∞ - flow velocity

M - flow Mach number

Δp - static pressure difference across the plate

A set of ordinary differential equations is obtained by using Galerkin's method with the modal expansion,

(2) $$w = \Sigma\, a_n(t) \sin \frac{n\pi x}{a}$$

The result is (in non-dimensional notation)

(3) $$A_n(n\pi)^4/2 + 6(1-\nu^2)\,[\sum_r A_r^2(r\pi)^2/2]\,A_n(n\pi)^2/2 + R_x A_n(n\pi)^2/2 + A_n''/2$$

$$+ \lambda\{\sum_m[nm/(n^2-m^2)]\,[1-(-1)^{n+m}]\,A_m + (\mu/M\lambda)^{1/2}\,A_n'\}$$

$$= P\,[1-(-1)^n]/(n\pi) \quad n = 1,2,\ldots,\infty$$

where

$A_n \equiv a_n/h$ and later $W \equiv w/h$

$\lambda \equiv \rho_\infty U_\infty{}^3\, a^3/MD$

$\mu \equiv \rho_\infty\, a/m$

$R_x \equiv N_x^E\, a^2/D$

$P \equiv \Delta p\, a^4/Dh$

$\tau \equiv t\,(D/ma^4)^{1/2}$

$' \equiv \partial(\ )/\partial\tau$

These equations may be numerically integrated to obtain time histories of motion. These are used to construct phase plane plots and Poincaré plots. Systematic numerical studies are discussed in the following section. The choice of parameters is guided by the earlier results of Dowell[2,13] and Holmes[9,10]. Among the parameters studied are

λ - a non-dimensional flow velocity parameter

R_x - a non-dimensional in-plane load parameter

Initial conditions

P - a non-dimensional static pressure differential

All results were obtained using a four mode expansion except for a few two mode calculations done for comparison.

To make the nature of the mathematical model as transparent as possible, consider a two mode representation from (3), where various numerical constants are omitted for clarity. One has

$$(4)\quad a_1'' + a_1 (1 + R_x) - \lambda a_2 + \lambda^{1/2} \zeta_1 a_1' + (a_1^2 + a_2^2) a_1 = P$$

$$a_2'' + a_2 (4 + R_x) + \lambda a_1 + \lambda^{1/2} \zeta_2 a_2' + (a_1^2 + a_2^2) a_2 = 0$$

The skew-symmetric terms involving λ are responsible for the dynamic instability, flutter, while the R_x terms (when $R_x < 0$) can cause a static instability, buckling. The form of flutter modeled by (4) is called coupled mode flutter and is so named because a minimum of two modes is required to produce it. If one examines a root locus of (4) for infinitesimal perturbations about the trivial equilibrium, $a_1 = a_2 = 0$, then the two eigenvalues, which were the system natural frequencies at $\lambda \equiv 0$, approach each other as λ increases and after near coincidence one of them passes into the unstable half plane of the root locus. Hence this form of flutter is also sometimes called merging frequency flutter[1,2]. Only this type of plate flutter has been treated by the qualitative theory of differential equations in the literature[9,10].

By contrast another form of flutter is called single mode flutter[1,2] and it arises because the aerodynamic forces create negative damping. Analogous to (4), a single equation for a_1 displays the essential features of this type of flutter.

$$(5)\quad a_1'' + \int_{-\infty}^{\tau} K_1[(\tau-\sigma)\lambda^{1/2}]\, a_1'(\sigma)\, d\sigma + a_1 (1 + R_x) + a_1^3 = P$$

The convolution integral is a mathematical representation of the physical fact that at time τ, the aerodynamic forces depend in general on the entire past history of the plate motion. This effect is important for Mach numbers, M, near unity, but is unimportant at large M where (3) and (4) apply.

A simple explanation of the nonlinear limit cycle behavior of (5) for $R_x = 0$, $P = 0$ may be obtained by using the method of harmonic balance.

Assume

$$a_1 = \bar{a}_1 \cos \Omega \tau \tag{6}$$

Substitute (6) into (5) and integrate over one period of motion (as the steady state is approached, $\tau \to \infty$). Then one obtains

$$K_1^* [\Omega/\lambda^{1/2}] = 0 \tag{7}$$

and

$$\Omega^2 = 1 + \bar{a}_1^{\,2} \tag{8}$$

where nonessential constants have been dropped. K_1^* is the (real part of the) Fourier transform of K_1 and physically is a damping coefficient. It is shown in the sketch below

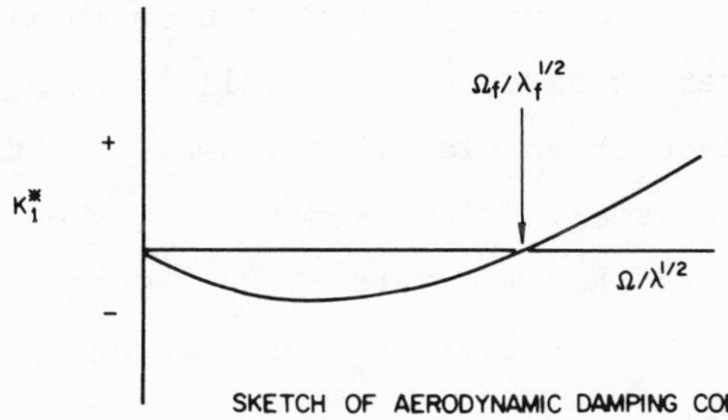

SKETCH OF AERODYNAMIC DAMPING COEFFICIENT

The infinitesimal stability is obtained when $\bar{a}_1 \equiv 0$. From (8) the frequency of the flutter oscillation is $\Omega_f = 1$ (i.e., the single mode natural frequency) while from (7) the flutter velocity parameter, λ_f, is found from the condition that $K_1^* = 0$ at $\lambda = \lambda_f$; see the above sketch.

Now consider what happens when $\lambda > \lambda_f$, $\bar{a}_1 > 0$. One still requires that $K_1^* = 0$, thus

$$\Omega/\lambda^{1/2} = \Omega_f/\lambda_f^{\,1/2} \tag{9}$$

and

$$\Omega^2 = 1 + \bar{a}_1^{\,2} \tag{10}$$

Solving (9) and (10), one has

$$\Omega^2 = \lambda/\lambda_f \tag{11}$$

and

(12) $$\bar{a}_1 = [\lambda/\lambda_f - 1]^{1/2}$$

(11) gives the limit cycle frequency and (12) the amplitude for $\lambda > \lambda_f$.

When $R_x < 0$ one anticipates that chaotic motion may occur[2], but we do not pursue that issue here.

Numerical Studies. For all results reported below, $\mu/M = 0.01$, $x/a = 0.3$ and $P = 0$ unless otherwise noted.

From previous numerical simulation studies[2], it is known that the parameters λ and R_x govern the type of motion which may occur. The sketch below displays a map in λ, R_x space which identifies the various types of motion which may occur. Only $R_x < 0$ is considered, since this proves to be the interesting case. For small λ and R_x, the steady state motion is a flat, undeformed plate. For small λ, but moderate R_x, the plate buckles. For small R_x, but moderate λ, the plate flutters with simple harmonic motion. For moderate λ, R_x a more complicated limit cycle motion occurs and for large R_x and moderate to large λ, chaotic motions ensues.

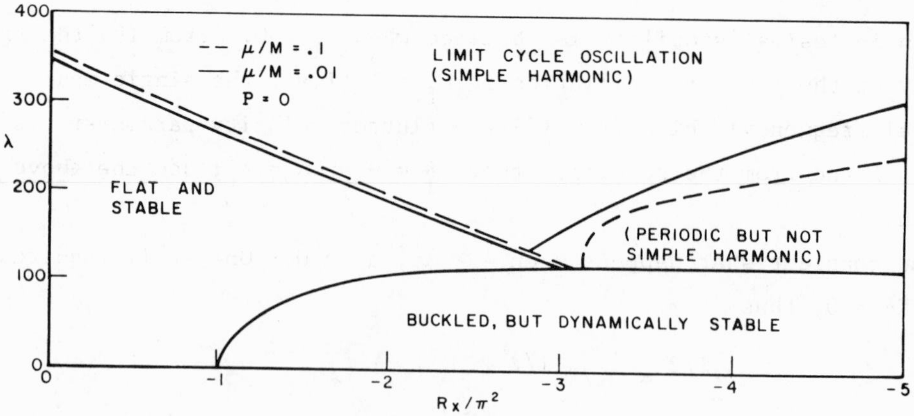

SKETCH OF STABILITY REGIONS

The results will be presented in the phase plane. Anticipating these, the distinctive types of motion which may occur are sketched as follows. The buckled plate corresponds to two (nonlinear static equilibria) points in the phase plane. Simple harmonic motion flutter is an ellipse. The more complicated limit cycle motion comprises a smaller orbit about each buckled state and a larger orbit which evolves from the flutter motion. The chaos is beyond the author's ability to sketch simply.

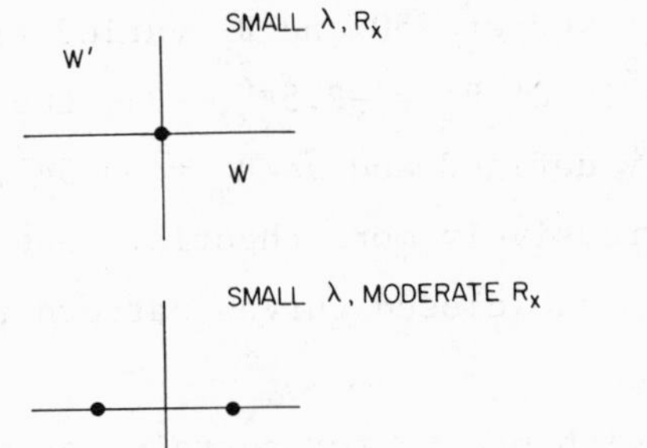

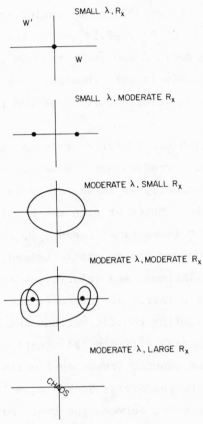

SKETCH OF REPRESENTATIVE PHASE PLANE ORBIT

To investigate the possibility of chaotic motion, two trend studies were made. First R_x was held fixed at $-4\pi^2$ and λ was varied from 300, 250, 200, 175, 150, 130, 115, 100. See Figures 1-4. At λ = 300 and 250 the phase plane plot, W' vs. W, of the limit cycle shows an ellipse typical of pure flutter motion. For λ = 200, the first deviations from the ellipse are evident, although the phase plane plot remains a single closed curve. However at λ = 175, 150, 130, 115 three closed curves are in evidence. The largest of these derives from the pure flutter motion, while the two smaller ones are associated with buckling or divergence. At λ = 175 the three closed curves are particularly clearly defined and as $\lambda = 150, 130, 115$ the motion becomes progressively more chaotic. At $\lambda = 100$ the phase plane plot is simply a point, representative of a pure buckling or divergence instability.

Secondly λ was held fixed at 150 and R_x varied from $-2.5\pi^2$, $-3\pi^2$, $-3.5\pi^2$, $-4\pi^2$, $-5\pi^2$, $-6\pi^2$. At $R_x = -2.5\pi^2$, $-3\pi^2$ the three closed curves are particularly clearly defined and as $R_x = -3.5\pi^2$, $-4\pi^2$. $-5\pi^2$, $-6\pi^2$, the motion becomes progressively more chaotic. See Figures 5 and 6. Nevertheless the basic three closed curves pattern remains albeit in a more obscure form.

The chaotic motion which occurs for certain parameter combinations of λ and R_x might well be termed random. However, it is clear that such motion evolves continuously in parameter space from motion which is decidedly deterministic. Moreover even taken solely as motion at a point in parameter space (some fixed λ and R_x), it is manifestly bounded in the phase plane and can be characterized, for example, by minima and maxima of displacement and velocity, W and W'.

Quite aside from its own intrinsic interest, the present results may serve as a paradigm for similar chaotic motions which result from limit cycles associated with partial differential equations. Perhaps the best known of these is the chaotic (turbulent) motion which is the limit cycle associated with the Navier-Stokes equations. There is at least one distinction, however, between the two. For the present problem there are two parameters, λ and R_x, each of which governs a distinctive instability, flutter and buckling (divergence). It is the interaction of these two essentially deterministic motions for certain combinations of λ and R_x that leads to chaotic motion. On the other hand only a single parameter, the Reynolds number $\equiv$ Re, appears in the Navier-Stokes equations (assuming incompressible flow). At this stage, one can only speculate that Re may play a dual role (as suggested by Lin's interpretation of <u>linear</u> hydrodynamic stability theory[5]) and/or that more than one instability mode is governed by Re.

<u>Effect of Initial Conditions</u>: For the base case of $\lambda = 150$ and $R_x = -4\pi^2$, various initial disturbance amplitudes were studied. In all

cases the initial deflection of the plate was assumed to be a pure first natural mode, i.e., $W(x,t = 0) = A_1(t = 0) \sin \pi x/a$, with various $A_1(t = 0)$ chosen. In addition to the base case value of $A_1(t = 0) = 0.1$, values of 0.2, 0.5, 1.0 and 2.0 were chosen. The results were essentially the same, showing no sensitivity to the initial disturbance amplitude chosen. It should be emphasized that this result is not universal since Ventres and Dowell[6] have already shown that for plates of finite (large) length/width ratio initial conditions can play an important role.

Effect of Static Pressure Differential: Another parameter of interest is a static pressure differential across the plate. This, of course, destroys the symmetry of the geometry by giving the plate deflection a preferred direction. Two trend studies were conducted. First of all, the simpler case of zero-in plane load, $R_x = 0$, was studied at $\lambda = 400$ for various pressure differentials, P. At P = 0 the plate flutters in a limit cycle oscillation whose phase plane trajectory is an ellipse centered about the origin, $W = W' = 0$. As P is increased from 0 to 100, 200, 250 the ellipse decreases in size and its center moves to the right. At P = 300 the ellipse has collapsed completely to a point; the static pressure differential having stiffened the plate sufficiently so that flutter is completely suppressed. As expected the motion does not exhibit any tendency to chaos for $R_x = 0$.

Next the base case of $\lambda = 150$, $R_x = -4\pi^2$ was reconsidered for various amounts of static pressure differential ranging from R = 0, 12.5, 25, 37.5, 50, 56.25, 62.5, 68.75, 75, 100, 200. For $P \geq 68.75$, the static loading was sufficient to suppress all flutter motion. For small P, say 0, 12.5, 25, 37.5,the motion was qualitatively similar and exhibited a somewhat chaotic appearance. However, it will be recalled that for $R_x = -3\pi^2$ the distinct three closed curves were present and, with that knowledge, these three close curves may still be seen, somewhat dimly to be sure, at $R_x = -4\pi^2$. The most remarkable result occurs, however, at P = 50, 56.25. See Figure 7. Here the motion has a very distinct, non-chaotic character with two closed orbits in evidence. That is, the static pressure differential has suppressed the chaotic character of the motion and one of the three orbits which appear for smaller P. For P = 62.5 the motion returns to a chaotic state with no readily discernible pattern and for $P \geq 68.75$, the limit cycle degenerates to a point and all motion ceases.

Poincaré Plots: To gain further insight into the nature of the motion, an alternative format for presentation of the data was considered, i.e., a Poincaré plot. For our purposes such a plot is one where the continuous phase plane diagram is sampled at discrete time intervals. Of course, since our time histories are generated by a digital computer and are thus necessarily discrete, all of the previous phase plane diagrams were, strictly speaking, Poincaré plots. However previously the time interval used was short compared to any time intervals characteristic of the motion, so that the diagrams were effectively continuous. The effect of systematically increasing this time interval was examined for two example cases, $\lambda = 150$ and $R_x = -3\pi^2$ or $-4\pi^2$. The former, $R_x = -3\pi^2$, is a periodic motion and as the sampling time interval becomes equal to the period of the motion, the Poincaré plot reduces to a single point. The latter, $R_x = -4\pi^2$, is a chaotic motion where no such reduction is expected, but other interesting insights may be obtained. A unit time interval was selected to be $\Delta\tau = 0.02$.

Results were obtained for $R_x = -3\pi^2$ and time intervals of 1, 2, 5, 10, 20, 34, and 40 units. For a small number of units the complete character of the phase plane diagram is evident; for 34 units essentially a single point is obtained indicating that this is the period of the periodic motion.

Similar results were obtained for $R_x = -4\pi^2$. Because the motion is now chaotic the character of the motion becomes very difficult to ascertain as the sampling time interval is increased. This is emphasized when the sampling points in the phase plane are unconnected. No period is found in which the phase plane diagram is reduced to a single point.

An alternative Poincaré plot may be constructed by specifying one of the dynamical variables and taking slices in variable space of the remaining coordinates. For example, one could specify $A_1' = 0$ and examine, at those times when this is true, the values of A_1, A_2, A_2', A_3, A_3',...

However for even four modes, the remaining variable space has dimension 7=4x2-1. Hence it would be difficult to display and interpret such results.

To reduce the dimension of this Poincaré plot, one might proceed as follows (which is suggested by physical considerations). Consider a particular physical displacement, W, and the corresponding velocity, W', for each such W which appears in the phase plane. (Of course, W is a linear combination of the A_n and W' of the A'_n. Hence they are suitable lower dimensional representations of the individual coordinates, A_n and A'_n.) Furthermore the range of W', call it $\Delta W'$, (for each such W) is a measure of the system chaos. A strictly periodic system would have a range of zero, i.e., for each such W, there is a single W'. Conversely if all W' are possible between W'_{min} and W'_{max} the system motion might be termed completely chaotic. The larger $R \equiv \Delta W/(W'_{max} - W'_{min})$, the more chaotic the motion. For multiple dynamic equilibria paths this definition may need to be refined to give a meaningful characterization of the motion. Also other characterizations may be useful, e.g., spatial averages of W and W'.

Effect of Number of Modes Retained: Two mode calculations were made which gave qualitatively similar results to those using four modes. Based upon the present results and earlier studies[1,4], it is expected that four mode calculations will give quantitatively accurate results for the parameter combinations studied here.

Conclusions. Chaotic motions occur for certain combinations of in-plane compressive load and fluid flow velocity. These evolve continuously from simpler motions with changes in these two parameters. Phase plane plots effectively display the results; Poincaré plots are less useful.

Results from the qualitative theory of nonlinear differential equations (differential dynamics) were helpful in motivating the present study and interpreting the results obtained.

Part II: Nonlinear Oscillator Models in Bluff Body Aeroelasticity. Such models[16-20] have been developed to provide a phenomenological description of observed motion of bluff (non-streamlined) elastic bodies in a streaming fluid. See SKETCH OF BLUFF BODY GEOMETRY IN A FLUID FLOW below.

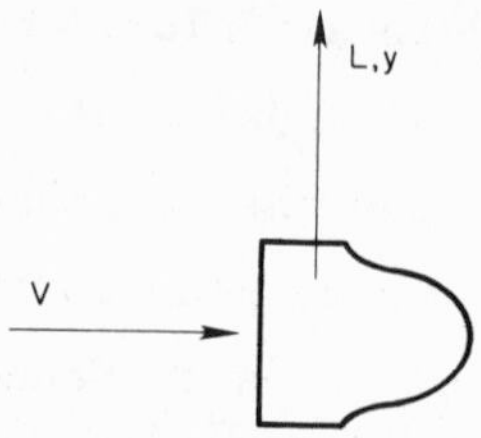

SKETCH OF BLUFF BODY GEOMETRY IN A FLUID FLOW

The dynamics of the elastic body are described by the equations of solid mechanics. These reduce, in their simplest form, to an oscillator with mass, stiffness and damping characteristics driven by a (fluid) force. By analogy, the dynamics of the fluid, as characterized by the fluid force on the body, are assumed to be described by a fluid oscillator. The properties of the fluid oscillator (forms of the terms in the differential equation for the fluid force and their coefficients) are deduced by requiring the coupled solid - fluid oscillator system to have solutions similar to those observed in practice.

In the present approach a more fundamental (but still phenomonological) approach is taken by deducing the fluid oscillator properties from fluid mechanical considerations (experimental and theoretical) alone. It should be noted that Iwan and Blevins[20] made a similar effort based upon largely the theoretical field equations of fluid mechanics. The desire, of course, is to obtain the equations of the solid oscillator and the fluid oscillator by separate, independent means and to use them together to predict the behavior of the combined fluid-elastic system. The present approach is based upon a combination of experimental and theoretical fluid mechanical considerations.

For informative critiques of nonlinear oscillator models, References 21 and 22 are recommended. A more extensive discussion of the present approach is contained in Reference 23.

Model Development. For simplicity, a single-degree-of-freedom system consisting of a bluff cylindrical structure transverse to an oncoming two-dimensional stream will be considered. Extensions to multidegree of freedoms and three-dimensional flow should be possible[24].

The structural equation of motion is well known

(1) $$m\,[\ddot{y} + 2\zeta_o\omega_o\dot{y} + \omega_o^2 y] = q\,D\,C_L$$

where

m - mass of structure

y - transverse deflection of the structure

ζ_o - critical damping ratio of in vacuo structural mode

ω_o - natural structural mode frequency

q = $1/2\ \rho V^2$, flow dynamic pressure

ρ - flow density

V - flow velocity

D - reference dimension of aerodynamic cross-section

C_L - aerodynamic lift coefficient

The crux of the matter is what oscillator equation may be used to determine C_L? This oscillator model will be determined by requiring that

(1) at high frequencies the fluid oscillator model give the expected virtual mass relationship between C_L and y, i.e.,

(2) $$C_L = -B_1\,\rho\,D^2\,\ddot{y} \quad \text{as } \omega \to \infty$$

where B_1 is a non-dimensional aerodynamic coefficient determined from (potential flow) theory and/or experiment (in a still fluid),

(2) at low frequencies the model give the expected quasi-steady relationship between C_L and y, i.e.,

(3) $$C_L = f\,(\dot{y}/V) \quad \text{as } \omega \to 0$$

For example, if one were to use the Parkinson galloping model[21,22] to evaluate the function, $f(\dot{y}/V)$, one might ask that

(4) $$C_L = A_1\,\dot{y}/V - A_3\,(\dot{y}/V)^3 + \ldots \quad \text{as } \omega \to 0$$

(3) for small C_L (and no structural motion $y \equiv 0$) the natural frequency of the fluid oscillation be determined by the Strouhal wake frequency, e.g.,

(5) $$\ddot{C}_L + \omega_s^2 C_L = 0 \quad \text{as } C_L, y \to 0$$

where ω_s is the dimensional fluid frequency, and $\omega_s = \frac{V}{D} k_s$, where k_s is a Strouhal number which depends upon Reynolds number,

(4) for a stationary cylinder the root mean square value of C_L be that observed in experiment. If one uses (5) with no additional terms involving C_L, this would require that the initial condition on C_L be

(6) $$C_L (t = 0) = 2^{1/2} C_{L_{rms}}$$

Alternatively one might add a nonlinear term to equation (5), which permits a limit cycle behavior whose amplitude is $C_{L_{rms}}$ (and which is independent of initial conditions). This latter procedure presumably is closer to the physics of the fluid, whose oscillations are those of a fluid limit cycle (turbulence) due to an unstable laminar flow. Previous authors[16-20] have suggested adding a nonlinear term to equation (5) which makes it into a van der Pol oscillator and we shall follow their lead here. The specification of the nonlinear term provides a basis for flexibility, but also arbitrariness, in the model. The next step beyond a van der Pol oscillator might be to introduce a strange attractor[9] nonlinearity which would produce a chaotic limit cycle motion. This would appear to model even more closely the fluid physics.

Derivation of Model Equation. There are several ways one may derive the fluid oscillator equation. The following is representative and reasonably straightforward.

From equation (2), one may deduce that

(7) $$\ddot{C}_L = - B_1 \frac{D}{V^2} \ddddot{y} \quad \text{as } \omega \to \infty$$

From equations (7) and (5), one infers that

(8) $$\ddot{C}_L + \omega_s^2 C_L = - B_1 \frac{D}{V^2} \ddddot{y}$$

and to satisfy equation (3) one modifies equation (8) to become

(9) $$\ddot{C}_L + \omega_s^2 C_L = -B_1 \frac{D}{V^2}\ddddot{y} + \omega_s^2 f\left(\frac{\dot{y}}{V}\right)$$

or, using equation (4) rather than equation (3),

(10) $$\ddot{C}_L + \omega_s^2 C_L = -B_1 \frac{D}{V^2}\ddddot{y} + \omega_s^2\left[A_1 \frac{\dot{y}}{V} - A_3\left(\frac{\dot{y}}{V}\right)^3 + A_5\left(\frac{\dot{y}}{V}\right)^5 - A_7\left(\frac{\dot{y}}{V}\right)^7\right]$$

Finally the nonlinear term may be added to equation (10), which for $y \equiv 0$ gives a limit cycle oscillation of magnitude $C_{L_{rms}} = C_{L_o}$. Here a van der Pol oscillator is used.

(11) $$\ddot{C}_L - \varepsilon\left[1 - 4\left(\frac{C_L}{C_{L_o}}\right)^2\right]\omega_s \dot{C}_L + \omega_s^2 C_L = -B_1 \frac{D}{V^2}\ddddot{y}$$

$$+ \omega_s^2\left[A_1 \frac{\dot{y}}{V} - A_3\left(\frac{\dot{y}}{V}\right)^3 + A_5\left(\frac{\dot{y}}{V}\right)^5 - A_7\left(\frac{\dot{y}}{V}\right)^7\right]$$

where ε is to be determined from subsequent fluid mechanical information. It is natural to introduce a non-dimensional time, $\tau \equiv \frac{V}{D}$ and a non-dimensional displacement, $x \equiv \frac{y}{D}$. Then equation (11) becomes

(12) $$C_L'' - \varepsilon\left[1 - 4\left(\frac{C_L}{C_{L_o}}\right)^2\right] k_s C_L' + k_s^2 C_L = -B_1 x''''$$

$$+ k_s^2\left[A_1 x' - A_3 (x')^3 + A_5 (x')^5 - A_7 (x')^7\right]$$

and equation (1) becomes

(13) $$x'' + 2\zeta_o k_o x' + k_o^2 x = \mu C_L$$

where $\mu \equiv \rho D^2/2m$ is a fluid/structural mass ratio. Thus, the problem is reduced to one with ten parameters which we divide into three classes.

I. A_1, A_3, A_5, A_7, B_1

II. $k_s, C_{L_{rms}}$

III. μ, k_o, ζ_o

Class I parameters are assumed to be universal, slowly varying functions of Reynolds number for a prescribed body cross-sectional shape; k_s and

$C_{L_{rms}}$ are assumed to be weak functions of Reynolds number and are often assigned typical values, $k_s \sim 1$, $C_{L_{rms}} \sim 0.25$. μ, k_o and ζ_o are usually varied to examine their inpact on the coupled fluid-oscillator motion[23].

Self-Consistency Checks on the Model and Other Remarks. The above derivation of the model is, perhaps, more properly termed a construction. Various analytical and numerical consistency checks have been proposed[23].

Check 1: One requires that $C_L \to - B_1 x''$ as $k \to \infty$. This requires that $x \to 0$ at a certain rate as $k \to \infty$.

Check 2: One requires that $C_L \to A_1 x' + A_3(x')^3 + \ldots$ as $k \to 0$.

Check 3: When $k \sim k_s \sim 1$, there is no formal guarantee that the present model is accurate. The high and low frequency terms have been combined in a patching process, rather than by a matching procedure[25].

Check 4: Neglecting all nonlinear terms a linearized stability analysis may be performed. One may show that the linearized system is always unstable. The question is then, is it reasonable for the coupled solid-fluid oscillator system to be always unstable in a linearized sense? This question would appear as much philosophical as technical. It is perhaps interesting to note that for $A_1 < 0$, the linearized system is stable at large k_o, i.e., small $V/\omega_o D$.

Check 5: As $x \to 0$, the model is well behaved.

Check 6: As $x \to \infty$, the model predicts that C_L will be dominated by its forcing due to structural motion and its own self-induced (turbulent) C_L will be small by comparison, i.e., $C_L >> C_{L_{rms}}$. This is plausible, but by no means confirmed by experimental data. A key question is whether there is significant interaction between the lift generated by turbulence and that due to motion. If so, the present model would require as a minimum the addition of terms of the form of products of C_L and x and/or their time rates of change.

Numerical Studies Including Comparisons with the Skop-Griffin Model. Skop-Griffin[18,19] have devised a more elaborate lift oscillator model which has the following equation of motion, cf. (11),

(14) $$\ddot{C}_L - [C_{L_o}^2 - C_L^2 - (\frac{\dot{C}_L}{\omega_s})^2](\omega_s G \dot{C}_L - \omega_s^2 H C_L) + \omega_s^2 C_L = \omega_s^2 F \frac{\dot{y}}{D}$$

where it is determined empirically that, to obtain good agreement with observed motion of elasticity supported cylinders in a fluid flow, F, G, H should be given by

(15) $$\log_{10} G = 0.25 - 0.21 S_G$$

$$\log_{10} hS_G^2 = - 0.83 + 0.98 S_G$$

$$H = \zeta_o h$$

and

$$F = 4 \frac{GS_G}{h}$$

where

$$S_G \equiv \zeta_o/\mu$$

It should be noted that Skop-Griffin use a mass ratio, μ, which is slightly different from the present one, i.e.,

(16) $$\mu_{Dowell} = \mu_{Skop-Griffin} k_s^2$$

For $k_s \sim 1$, the difference should be of no qualitative importance.

For $S_G = 0.5$ (say $\zeta_o = 0.005$ and $\mu = 0.01$), one finds that

$$G = 1.396$$

$$H = 0.009$$

$$F = 1.527$$

Non-dimensionalizing equation (14), one obtains

(17) $$C_L'' - [C_{L_o}^2 - C_L^2 - (\frac{C_L'}{k_s})^2](k_s G C_L' - k_s^2 H C_L) + k_s^2 C_L = k_s F x'$$

The term involving H is numerically small and calculations neglecting it show little change. Comparing equation (17) to equation (12), we see the equations are similar. The nonlinear term is different, however, and further study is needed to assess this difference. The right hand sides are essentially the same with a difference in coefficient.

$$(17) \quad k_s F x'$$

$$(12) \quad k_s^2 A_1 x'$$

if one ignores the virtual inertia terms ($B_1 \equiv 0$) and higher order galloping nonlinearities ($A_3 = 0$, etc.). Numerically

$$k_s F \sim 1.9$$

$$k_s^2 A_1 \sim 4.9$$

for $A_1 = 3.11$ (a typical value of a square cross-section cylinder[22]).

Without belaboring the point, it may be mentioned that the empirical relations (eq. 16) are incorrect from a fundamental physical point of view. Consider, for example, the special case of no structural motion, $x \equiv 0$. Clearly G and H cannot depend upon the damping and mass of the **structure** as embodied in ζ and μ. A similar argument may be made for F for prescribed structural motion. Nevertheless it is reassuring that numerically the present model and the Skop-Griffin model are similar

Determination of ε in the Present Model: ε (the fluid damping coefficient) would be determined from an experiment (or fluid mechanical theory) which measures C_L for a prescribed x. In the absence of such experimental data, we will compare the results of two calculations, one using the present model and the other using the Skop-Griffin model. ε will be determined by requiring the two results to agree. Specifically a sinusoidal structural motion will be specified, $x = x_o \sin k_s \tau$, with amplitude, x_o, and frequency, ω_s. In the steady state we shall consider $C_{L_{max}}$ vs. x_o. The results are shown in Figure 8. $\varepsilon = 0.2$ gives results from the present model which are in close agreement with Skop-Griffin.

It would be highly desirable to perform an experiment to compare with the "predictions" of Figure 8. Currie, et. al.[16,17] have also obtained similar results.

Coupled Fluid-Structure Oscillations: For fixed $\mu = 0.01$ and various ζ_o, the structural and fluid response was determined as a function of $k_o^{-1} = V/\omega_o D$. To allow a more meaningful comparison with the results of Skop-Griffin, in the present model B_1, A_3, A_5, A_7 were all set to zero.

The structural response (essentially a sinusoidal oscillation with frequency ω_o) is characterized by its maximum amplitude, x_{max}, and is plotted in Figure 9. The values $(V/\omega_o D)_G$ are the galloping velocities that would occur if one used the simple fluid model[21,22],

(12) $$C_L = \frac{F}{k_s} x'$$

or

(17) $$C_L = A_1 x'$$

As can be seen the present model predicts a plateau for x_{max} at high $V/\omega_o D$ while the Skop-Griffin model predicts a peak at intermediate $V/\omega_o D$ which experiences a modest decrease at higher $V/\omega_o D$. The latter behavior agrees better with experiment, but the differences among all the results are not dramatic by simply examining Figure 9.

A more significant difference between the Skop-Griffin model and the present results is obtained for the dominant frequency of the lift forces, ω_L[23]. In the present model the frequency at small $V/\omega_o D$ follows the Strouhal relation, $\omega_L = \omega_s = V/k_s D$; however for $V/\omega_o D > k_s^{-1} = 0.8$, the lift force has a dominant frequency at $\omega = \omega_o$, i.e., "lock-in" persists up to $V/\omega_o D = 2$. By contrast in the Skop-Griffin model lock-in occurs for $V/\omega_o D$ $0.6 \rightarrow 1.5$, but otherwise the Strouhal relation is followed. It should be recalled here that the <u>structure</u> oscillates basically at $\omega = \omega_o$ for all $V/\omega_o D$. Also it should be noted that in the Skop-Griffin model significant modulation of the lift force occurs. This is much less so in the present model.

In an effort to further understand the reasons for the differences between the Skop-Griffin model and the present one, in the latter A_1 was arbitrarily changed to 1.21 to force agreement between the right hand sides of equations (12) and (17). ε was then recomputed to give agreement between the two models for C_L vs x_o, cf. Figure 8; $\varepsilon = 0.08$ for $A_1 = 1.21$. The coupled structural-fluid response results for the present model were then recomputed as well. Now the results for the two models are in very close agreement, cf. Figure 9. Moreover this is also true for the results for lift force dominant frequency and magnitude[23].

The significance of this result is as follows: The details of the nonlinear term are not crucial, cf. equations (12) and (17). However, clearly the magnitudes of A_1 and ε are important. As explained earlier, these should be evaluated from a fluid mechanical experiment. In particular they are not functions of structural damping or mass, but will depend on geometrical shape.

As a final numerical evaluation of the present nonlinear fluid oscillator model, various terms on the right hand side were studied; (i) the normal case, A_1 only, (ii) a higher order case, A_1 and A_3 retained and (iii) the uncoupled case, $A_1 = A_3 = 0$. The results are shown in Figure 10. (ii) would allow one to reproduce the results of Parkinson for limit cycle oscillations due to galloping. However for the parameters studied _in this numerical example_, the retention of A_3 leads to little difference from the nominal case.

The uncoupled case, $A_1 = A_3 = 0$, neglects the effect of structural motion on the fluid lift force. Hence the structural response is that due to a prescribed force, C_L. Note however that the structural motion does not show a resonant peak near $V/\omega_o D = k_s^{-1}$, but rather shows a plateau behavior in Figure 10. Apparently this is because C_L is a periodic, but non-harmonic function of time (it is after all a limit cycle solution to van der Pol's equation) and therefore has a richer frequency content than a simple sinusoidal force.

Concluding Remarks. A rational, self-consistent nonlinear oscillator model has been suggested which takes into account the known (theoretical and experimental) behavior of the _fluid_. Results from such a model have been compared to those obtained from the empirical Skop-Griffin model. The results are broadly similar. It was pointed out that from a fundamental point of view the published method for determining the coefficients in the Skop-Griffin model is physically inconsistent. Means for determing the coefficients of the nonlinear fluid oscillator model by fluid mechanics experiments (with zero and prescribed structural motion) have been suggested.

Much of the impetus for using nonlinear fluid oscillator models comes from the careful experimental work of Bishop and Hassan[26]. They were fully conscious that their experimental data for lift (and drag) on an oscillating cylinder (with prescribed motion) could be interpreted as describing the behavior of a nonlinear oscillator. Previous authors[16-22] have suggested that the van der Pol oscillator will give results qualitatively similar to the experimental data of Reference 26. Although not attempted here, it would be of some interest to make quantitative comparisons with the data of Bishop and Hassan using the present fluid oscillator model. The experimental data are not sufficient to make an unambiguous comparison, nor is the present theoretical model capable of predicting all the features of the experimental results, e.g., oscillating drag. However some further insight might be obtained from such comparisons.

Finally, in the longer term, one may hope that the qualitative theory of differential equations (e.g., see Holmes[9] and the paper of Guckenheimer in this volume) may lend insight and provide more rigorous methods for establishing generic models for this class of dynamical systems.

REFERENCES

[1] E. H. Dowell, H. C. Curtiss, Jr., R. H. Scanlan and F. Sisto, A Modern Course in Aeroelasticity, Sitjhoff-Noordhoff, Leyden, The Netherlands, 1979.

[2] E. H. Dowell, Aeroelasticity of Plates and Shells, Noordhoff, Leyden, The Netherlands, 1975.

[3] E. Breitbach, Effects of Structural Nonlinearities on Aircraft Vibration and Flutter, AGARD Report R-665, 1977.

[4] E. Simiu and R. H. Scanlan, Wind Effects on Structures, John Wiley and Sons, New York, 1978.

[5] R. L. Bisplinghoff and H. Ashley, Principles of Aeroelasticity, John Wiley and Sons, New York, 1962.

[6] R. L. Bisplinghoff, H. Ashley and R. L. Halfman, Aeroelasticity, Addison-Wesley Publishing Co., Cambridge, Mass., 1955.

[7] Y. C. Fung, An Introduction to the Theory of Aeroelasticity, John Wiley and Sons, New York, 1955.

[8] R. H. Scanlan and R. Rosenbaum, Aircraft Vibration and Flutter, MacMillan Company, New York, 1951.

[9] P. J. Holmes, Bifurcations of Divergence and Flutter in Flow-Induced Oscillations: A Finite Dimensional Analysis, J. Sound and Vibrations, 53(1977), pp. 471-503.

[10] P. J. Holmes and J. Marsden, Bifurcations to Divergence and Flutter in Flow-Induced Oscillations: An Infinite Dimensional Analysis, Automatica, 14(1978), pp. 367-384.

[11] E. H. Dowell, Component Mode Analysis of Nonlinear and Nonconservative Systems, Accepted for publication in the J. Applied Mechanics.

[12] E. H. Dowell, On Linear and Nonlinear Coordinates, in preparation.

[13] E. H. Dowell, Nonlinear Oscillations of a Fluttering Plate I., AIAA Journal, 4(1966), pp. 1267-1275. Also Part II, AIAA Journal, 5(1967), pp. 1856-1862.

[14] C. C. Lin, The Theory of Hydrodynamic Stability, Cambridge University Press, 1955.

[15] C. S. Ventres and E. H. Dowell, Comparison of Theory and Experiment for Nonlinear Flutter of Loaded Plates, AIAA Journal, 8(1970, pp. 2022-2030.

[16] R. T. Hartlen and I. G. Currie, Lift Oscillator Model of Vortex-Induced Vibration, Proc. ASCE, EM 5(1970) pp. 577-591.

[17] I. G. Currie, R. T. Hartlen and W. W. Martin, The Response of Circular Cylinders to Vortex Shedding, IUTAM-IAHR Symposium on Flow-Induced Vibrations (1974), Edited by E. Naudascher, pp. 128-142.

[18] O. M. Griffin, R. A. Skop and G. H. Loopman, The Vortex-Excited Resonant Vibrations of Circular Cylinders, J. Sound and Vibration, 31(1973) pp. 235-249.

[19] R. A. Skop and O. M. Griffin, A Model for the Vortex-Excited Response of Bluff Cylinders, J. Sound Vibration 27(1973) pp. 225-233.

[20] W. D. Iwan and R. D. Blevins, A Model for the Vortex-Induced Oscillation of Structures, J. Applied Mechanics, ASME, 41(1974) pp. 581-585.

[21] E. Simiu, and R. H. Scanlan, Wind Effects on Structures, John Wiley and Sons, New York (1978), pp. 194-212.

[22] G. V. Parkinson, Mathematical Models of Flow-Induced Vibrations of Bluff Bodies, IUTAM-IAHR Symposium on Flow-Induced Vibrations (1974), Edited by E. Naudascher, pp. 128-142.

[23] E. H. Dowell, Nonlinear Oscillator Models in Bluff Body Aeroelasticity, submitted for publication.

[24] R. A. Skop and O. M. Griffin, On a Theory for the Vortex-Excited Oscillations of Flexible Cylindrical Structures, J. Sound Vibration, 41(1975, pp. 263-274.

[25] M. Van Dyke, Perturbation Methods in Fluid Mechanics, The Parabolic Press, Stanford, California (1975).

[26] R. E. D. Bishop and A. Y. Hassan, The Lift and Drag Forces on an Oscillating Cylinder, Proc. Roy. Soc. A, 277(1964) pp. 32-75.

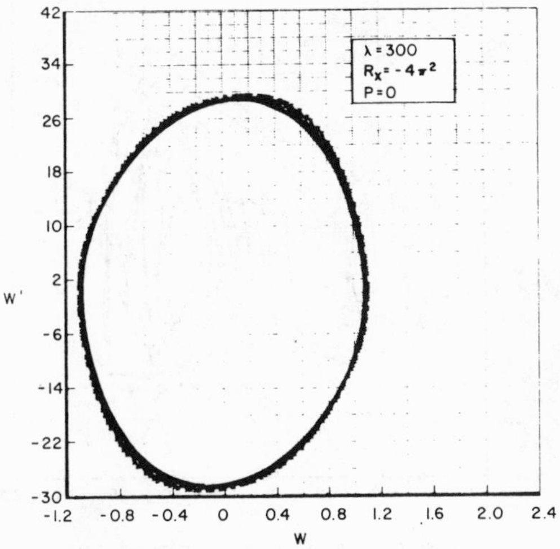

PHASE PLANE PLOT: EFFECT OF FLOW VELOCITY

FIGURE 1

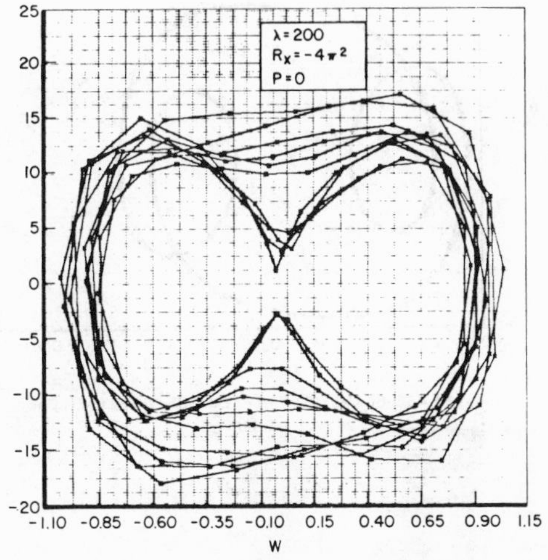

PHASE PLANE PLOT: EFFECT OF FLOW VELOCITY

FIGURE 2

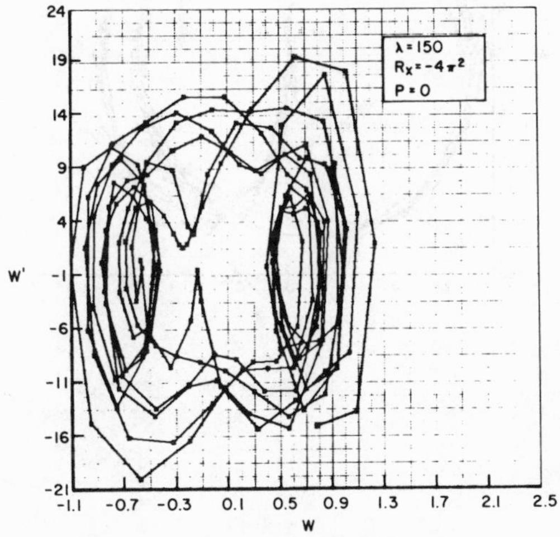

PHASE PLANE PLOT: EFFECT OF FLOW VELOCITY

FIGURE 3

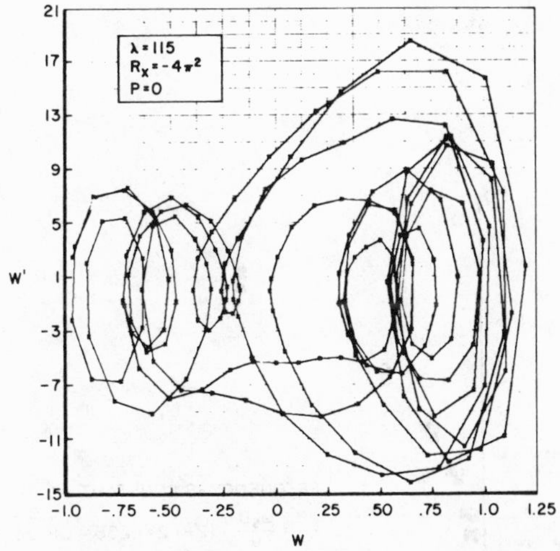

PHASE PLANE PLOT: EFFECT OF FLOW VELOCITY

FIGURE 4

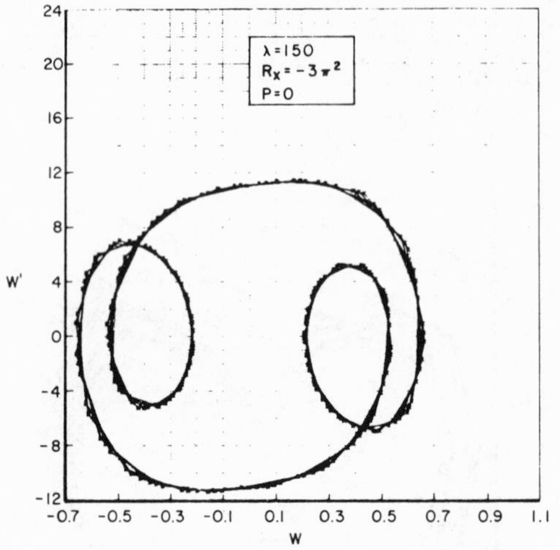

PHASE PLANE PLOT : EFFECT OF IN-PLANE LOAD
FIGURE 5

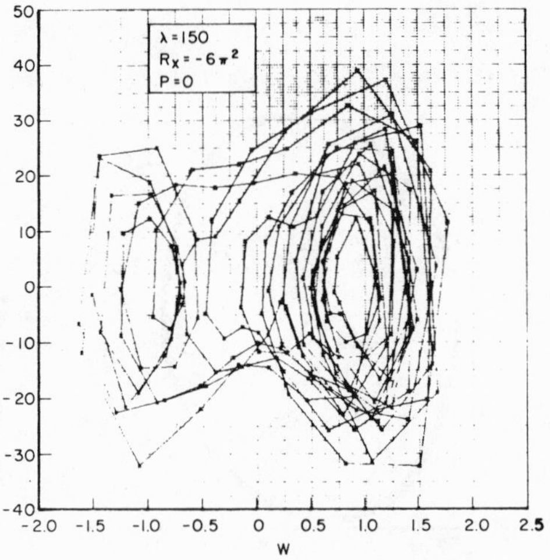

PHASE PLANE PLOT : EFFECT OF IN-PLANE LOAD
FIGURE 6

PHASE PLANE PLOT : EFFECT OF STATIC PRESSURE DIFFERENTIAL

FIGURE 7 ⟶

MAXIMUM LIFT VS. STRUCTURAL AMPLITUDE
⟵ FIGURE 8

FREQUENCY IS CONSTANT AT

$$\frac{\omega D}{V} = \frac{\omega_S D}{V} = (2\pi).2 = 1.256 \text{ FOR BOTH MODELS}$$

AND FOR ALL X_0.

SKOP & GRIFFIN   DOWELL

● $S_G = .5$ ○ $\epsilon = .2$ ■ $\epsilon = .08$
                       $A_1 = 3.11$ $A_1 = 1.21$

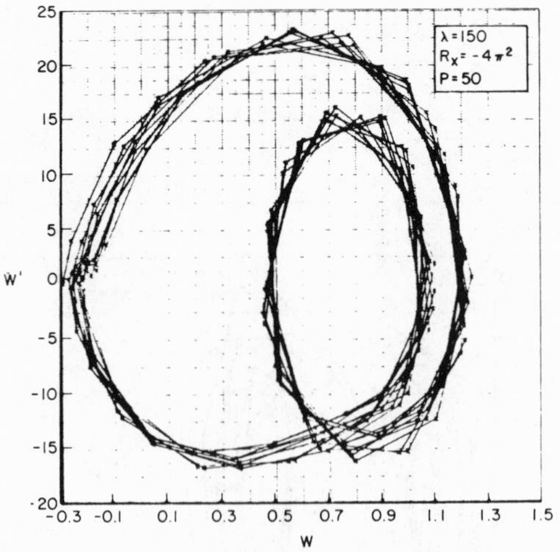

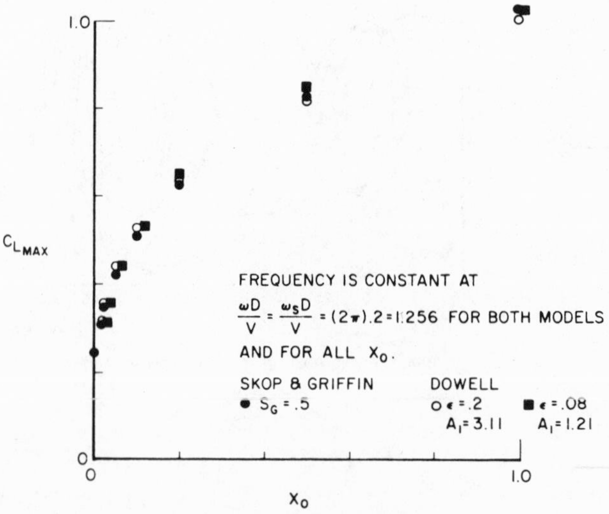

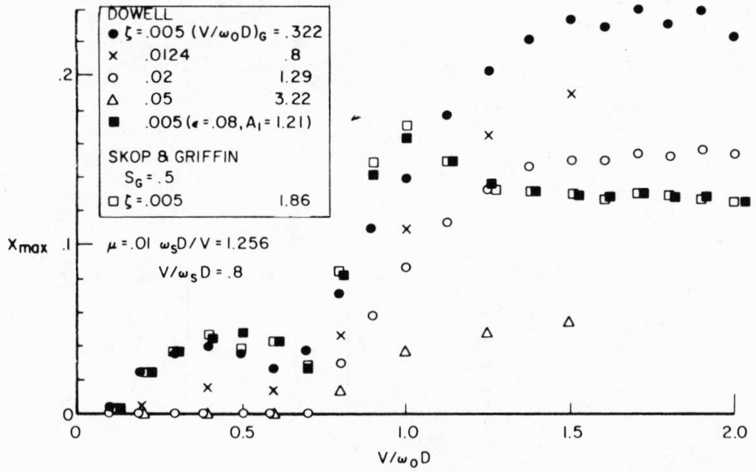

MAXIMUM STRUCTURAL DEFLECTION VS. FLOW VELOCITY

FIGURE 9

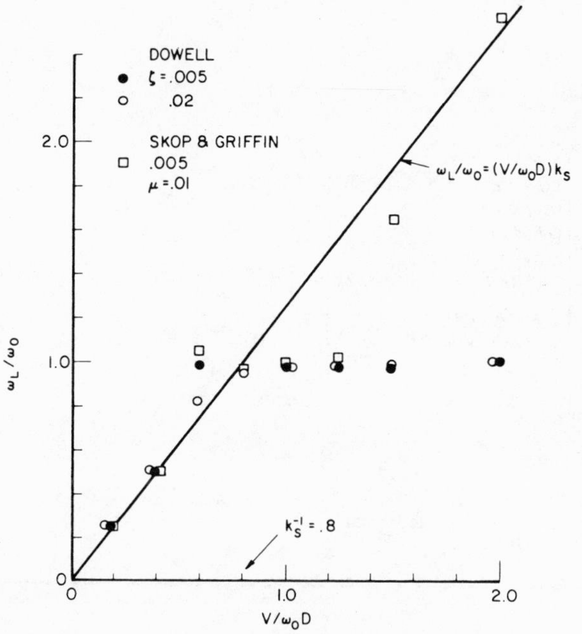

DOMINANT FREQUENCY OF LIFT FORCE VS. FLOW VELOCITY

FIGURE 10

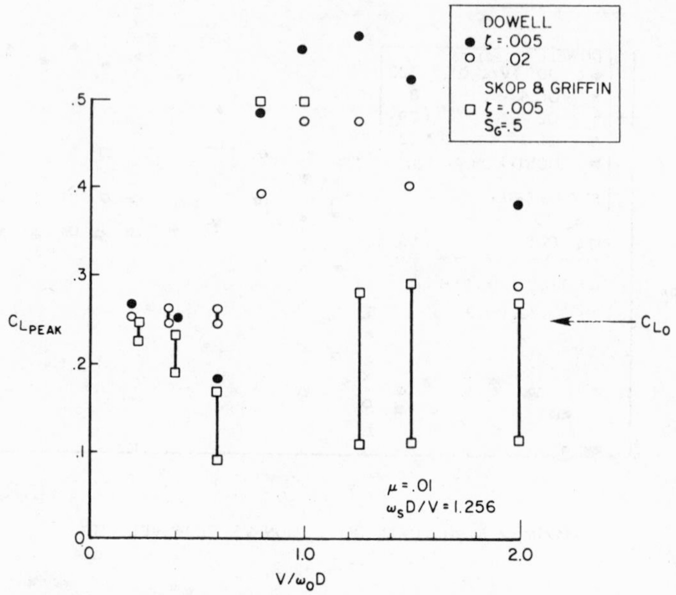

LIFT FORCE PEAKS VS. FLOW VELOCITY

FIGURE 11

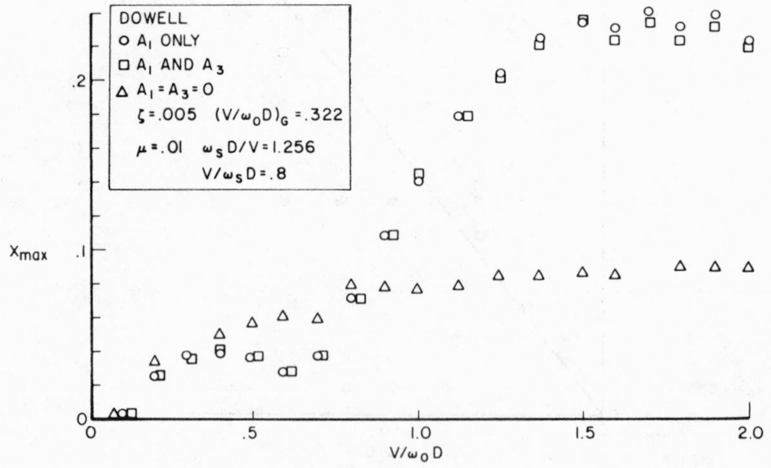

MAXIMUM STRUCTURAL DEFLECTION VS. FLOW VELOCITY

FIGURE 12

Nonlinear Behavior in Rail Vehicle Dynamics

Neil K. Cooperrider*

Abstract. Significant practical problems are associated with the nonlinear dynamics of rail vehicles. Nonlinearities in the vehicle suspension and wheel-rail interconnections lead to characteristic nonlinear behavior such as limit cycles and jump resonance. Because realistic problems involve high order systems with multiple nonlinearities, approximation and numerical techniques are used in their solution.

The important rail vehicle nonlinearities, typical behavior and solution techniques are described in this paper. Areas for future study are outlined.

Introduction. A significant number of practical problems associated with the operation of rail vehicles can be attributed to individual rail vehicle dynamic behavior. Instability and associated limit cycle motions cause severe wear and riding comfort difficulties and may lead to derailment. Poor vehicle guidance and response in curves also leads to wear and derailment problems. Less severe, but economically important, questions concern riding quality, noise generation, and component maintenance requirements. Nonlinear dynamics is unavoidable in dealing with many of these problems. The majority of the North American railcar fleet, for example, utilizes nonlinear suspension components with dry friction damping mechanisms out of cost considerations. In addition, one of the fundamental rail vehicle guidance mechanisms, the wheel flange, is inherently nonlinear. The safety questions mentioned above involve extreme motions where nonlinear mechanisms are also likely to be most important.

*Professor, Mechanical Engineering Department, Arizona State University, Tempe, Arizona 85281

The objectives of this paper are to describe the nonlinear problems found in rail vehicle dynamics, to show some illustrations of nonlinear response phenomena encountered in rail vehicle motions, and to discuss current solution techniques and their inadequacies.

Rail Vehicle Description. The basic element of the rail vehicle guidance and support system is the wheelset. In conventional rail vehicles, the wheelset consists of two steel wheels mounted on a solid axle without any possibility of relative rotation between the wheels and the axle, as shown schematically in Figure 1. The wheel

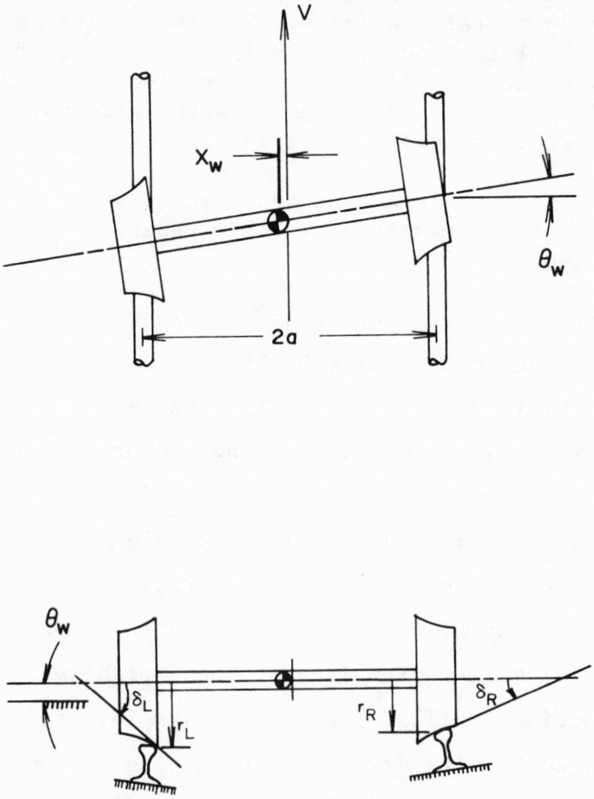

Figure 1. Wheelset Schematic

cross-sectional profile is not cylindrical, and may have a variety of shapes ranging from a conical taper to a nearly circular profile. The lateral behavior of a single wheelset may be described in most cases by two degrees of freedom, lateral and yaw motions, as shown in Figure 1.

The wheelset may be incorporated in various ways in a complete rail vehicle. The simplest rail vehicles utilize only two wheelsets connected to the body of the vehicle by a suspension system. Vehicles of this configuration are widely used on the European continent in freight service. The majority of conventional vehicles incorporate two or more wheelsets into an undercarriage, termed the "truck" in North American usage, and the "bogie" in European usage. A wide variety of mechanisms or suspensions are employed to connect the wheelsets to the bogie.

Rail Vehicle Nonlinearities. In rail vehicles nonlinear characteristics occur in the laws governing the interaction forces between the vehicle system bodies. Every connection from the wheel-rail interface to the bogie-car body suspension may involve significant nonlinearities.

In the discussion below attention has been focused on the dominant nonlinearities found in present vehicles. Space does not permit description of nonlinear effects that may be found in future vehicles, such as nonlinear control systems for improved guidance and suspension.

Wheel-Rail Geometry. An analysis of rail vehicle dynamic behavior in the lateral plane requires a description of the constrained motion of the wheelset on the rails. In the simplest model, the following variables are functions of the wheelset lateral position relative to the rails, $\Delta x_{w/r}$:

r_L, r_R -- left and right wheel rolling radii

δ_L, δ_R -- left and right wheel-rail contact angle

ϕ_w -- wheelset roll angle

These constrained variables are illustrated in Figure 1.

Although these constraint functions may be expressed analytically for simple wheel and rail geometries, realistic wheel and rail configurations require numerical computation of the constraint functions [1]. A typical numerical solution is shown in Figure 2. Here, as is often true, the constraint functions are nearly linear until the

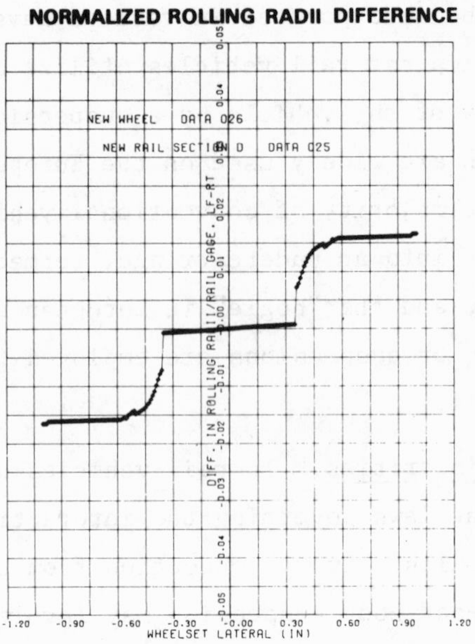

Figure 2. Numerical Wheel-Rail Geometric Constraint Function [1]

wheel/rail contact point shifts to the wheel flange. In many cases, including that shown here, the contact point shifts abruptly leading to discontinuities in the constraint functions.

The wheel-rail geometry may be linearized for wheel set-roadbed relative displacements less than flange clearance, i.e.

$$\frac{r_L - r_R}{2a} \simeq \lambda \frac{\Delta x_{w/r}}{a}$$

$$\frac{\delta_L - \delta_R}{2} \simeq \Delta \frac{\Delta x_{w/r}}{a} \qquad (1)$$

$$\phi_w \simeq \Gamma \frac{\Delta x_{w/r}}{a}$$

where: a -- semi-rail gauge

For larger wheel set-roadbed displacements each constrained variable must be treated separately. Because the contact angles reach values as large as 70°, well beyond the limit of small angle trignometric function assumptions, nonlinear trignometric functions also enter the equations of motion.

Wheel-Rail Contact Forces

The tangential forces between the wheel and rail play a major role in rail vehicle dynamics. These forces arise out of the difference in strain rates of the two bodies in the contact region, which, when normalized by the forward velocity of the wheel, is termed the creepage. The creepage is a function of the wheelset independent and constrained variables, and may be expressed in the following components for the left wheel of a wheelset:

$$\text{Lateral:}\quad \xi_{XL} = \left[\frac{1}{V}(\dot{x}_w + r_L\dot{\phi}_w) - \theta_w\right]\cos(\delta_L+\phi_w) + \frac{a\dot{\phi}_w}{V}\sin(\delta_L+\phi_w) \qquad (2)$$

$$\text{Longitudinal:}\quad \xi_{ZL} = \left(1 - \frac{r_L}{r_o}\right) - \frac{a\dot{\theta}_w}{V} \qquad (3)$$

$$\text{Spin:}\quad \xi_{\theta L} = \frac{\dot{\theta}_w}{V}\cos(\delta_L+\phi_w) - \frac{1}{r_o}\sin\delta_L \qquad (4)$$

The general form of the creep force-creepage functions are inherently nonlinear [2], as seen in Figure 3, due to the limitation on the force imposed by the Coulomb friction law. Closed form analytical expressions for the general nonlinear creep force law do not exist. Solutions for this problem are found by iterative numerical procedures [3,4], that can be employed in tabular or approximate form in vehicle dynamics analysis [5,6].

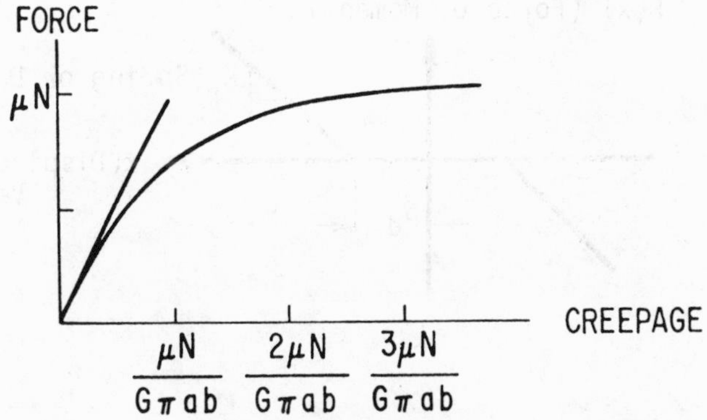

Figure 3. Nonlinear Creep Force Law [2]

A linear relationship of the following form is valid for small values of creepages [3]:

Lateral Force: $T_x = -f_{11}\xi_x - f_{12}\xi_\theta$ (5)

Longitudinal Force: $T_z = -f_{33}\xi_z$ (6)

Moment: $M_\theta = f_{12}\xi_x - f_{22}\xi_\theta$ (7)

Although the linear relationship is valid for rail vehicles in normal operation on straight track, analysis of curving or extreme situations that may lead to derailment requires the full nonlinear creep laws.

Suspension Forces

The force-displacement and force-velocity characteristics of the components connecting the vehicle members are in nearly all practical cases nonlinear, although they often can be linearized over certain operating ranges. Nearly all the nonlinearities found in rail car suspensions may be described by combinations of the following three nonlinear elements:

1. Deadband spring or damper:

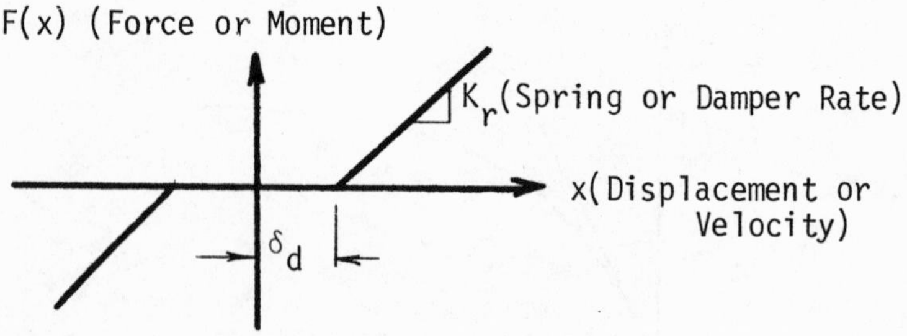

2. Hardening/Softening Spring or Damper

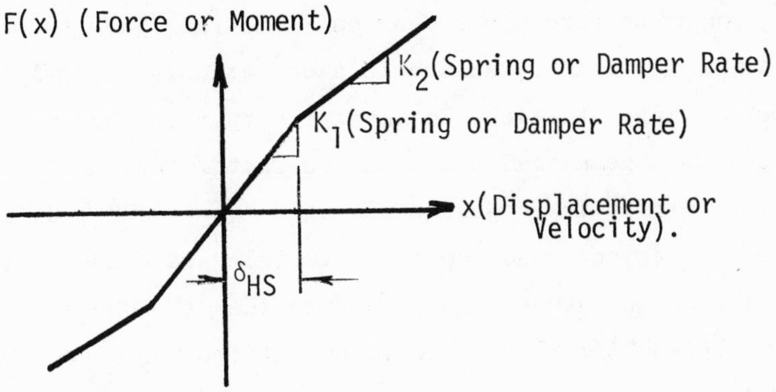

3. Coulomb Friction:

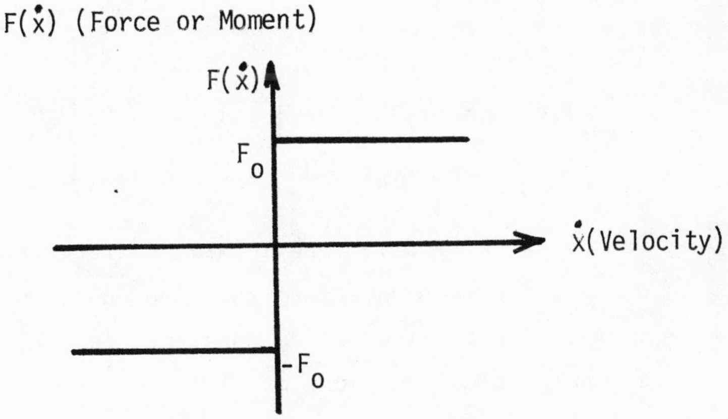

Although the hardening/softening element shown above is an approximation, in many cases, to continuous functions, it is frequently easier computationally to use the hardening/softening approximation, and the results are often insensitive to approximations of this sort.

Nonlinear Rail Vehicle Behavior

Many examples of uniquely nonlinear behavior such as multiple equilibrium states, jump resonance, and limit cycle motion occur in rail vehicle dynamics. In most cases, this behavior is distinctly undesirable. The range of nonlinear behavior that can occur is best illustrated by example.

Equilibrium Solutions (Rail vehicle curving behavior)

The ability of a rail vehicle to traverse curves without excessive or dangerous forces between wheel and rail is extremely important for safe and efficient rail vehicle operation. To deal realistically with the curving situation, one must incorporate models for nonlinear wheel-rail geometry, nonlinear contact forces and nonlinear suspension relationships in the equations of motion. The equilibrium solution of these equations under constant vehicle speed and constant roadbed radius of curvature conditions provides the wheelset tracking error and the wheel/rail contact force information needed to evaluate performance.

The equilibrium equations for a railway vehicle at constant velocity in a curve of constant curvature have the following form when a linear creepforce law is assumed [6],

$$A(\underline{\lambda},\underline{k})\ x = B(\underline{\lambda},\underline{k}) \begin{bmatrix} 1/R \\ \\ \theta_d \end{bmatrix} + E \begin{bmatrix} Ho \\ Mo \\ No \end{bmatrix} \qquad (8)$$

where: $\underline{\lambda}$ -- vector of "effective" wheel-rail coefficients (i.e. λ, Δ, Γ, and δ_o, for each wheelset) that are functions of $\underline{x}$

$\underline{k}$ -- vector of effective spring constants for suspension elements, also functions of $\underline{x}$.

$\left[\frac{1}{R}, \theta_d\right]^T$ -- inputs determined by the curve geometry and the vehicle speed

$\begin{bmatrix} Ho \\ Mo \\ No \end{bmatrix}$ -- external forces and torques from adjacent cars

E -- influence matrix for external forces and movements

A typical solution for these equations, obtained by iterative means [5] depicts the lateral wheel-rail forces as functions of cant deficiency in the manner shown in Figure 4. Note the distinct differences between linear and nonlinear solutions for this situation.

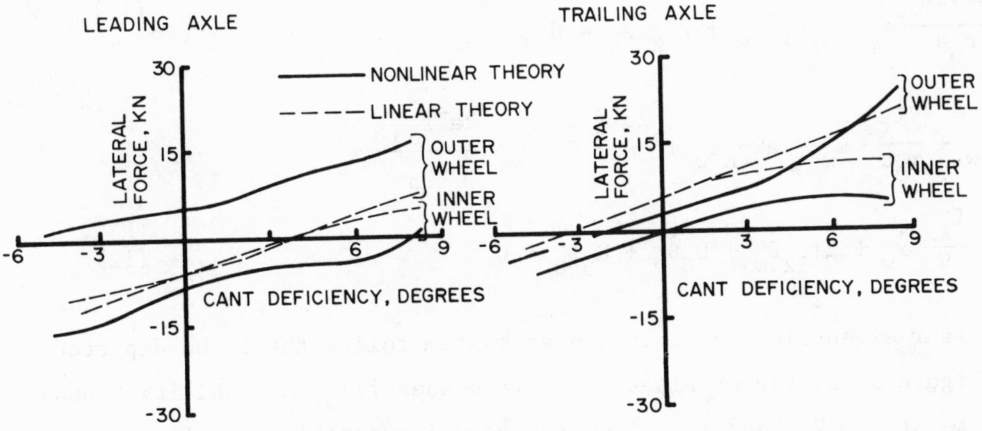

Figure 4. Nonlinear Curving (Equilibrium) Behavior [5]

Stability

Perhaps the most interesting aspect of rail vehicle behavior concerns the stability of rail vehicle motions. The stability problem can be illustrated by the lateral behavior of a single wheelset described by the following coupled, nonlinear second order equations of motion:

$$m\ddot{x}_w - F_{XL} \cos(\delta_L + \phi_w) - F_{XR} \cos(\delta_R - \phi_w) + W\Delta = F_{SUSP} \quad (9)$$

$$I_w \ddot{\Theta}_w - a(F_{ZR} - F_{ZL}) - M_L \cos(\delta_L + \phi_w) - M_R \cos(\delta_R - \phi_w) = M_{SUSP} \quad (10)$$

where: ϕ_w, δ_L, δ_R Δ - nonlinear wheel-rail geometric constraint functions of x_w.

F_{XL}, F_{XR}, F_{ZL}, F_{ZR}, M_L, M_R - nonlinear contact force and moment functions

F_{SUSP}, M_{SUSP} - suspension force and moment functions

When the creep forces and the suspension forces can be represented by linear characteristics, these equations take the following form:

$$m\ddot{x}_w + \frac{\Delta W}{a} x_w + 2\frac{f_{11}}{V}\dot{x}_w - 2f_{11}\theta_w + \left[\frac{2f_{12}}{V} - \frac{VI_X}{ar_o}\right]\delta_o \dot{\theta}_w$$

$$- \frac{2f_{12}\Delta}{r_o a} x_w + 2K_X x_w + 2D_X \dot{x}_w = 0 \tag{11}$$

$$I_y\ddot{\theta}_w + \frac{VI_X\Gamma}{r_o a}\dot{x}_w - aW\delta_o\theta_w + \frac{2a^2 f_{33}}{V}\dot{\theta}_w + \frac{2a^2 f_{33}\lambda}{ar_o} x_w$$

$$- 2\frac{f_{12}}{V}\dot{x}_w + 2f_{12}\theta_w + D_\theta\dot{\theta}_w + K_\theta\dot{\theta}_w = 0 \tag{12}$$

The four eigenvalues of this linear system follow the locus depicted in Figure 5, as the wheelset speed increases [7]. A stability boundary exists at $V = V_c$, and this linear system is unstable at all speeds beyond V_c. Such critical speeds occur well within the operating speed range of many rail vehicles.

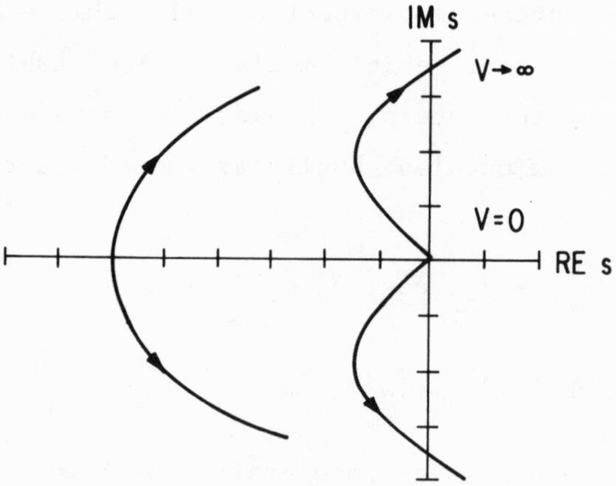

Figure 5. Wheelset Eigenvalue Locus with Speed [7]

When nonlinear wheel-rail contact and suspension forces are important, the stability situation becomes considerably more complex. For example, if we retain linear suspension characteristics, but account for the nonlinear wheel and rail geometry, then the linear analysis still provides stability information for motion about the origin. At speeds above V_c however, a stable limit cycle exists, due to the large stabilizing forces generated by the wheel flanges [8,9]. This behavior can be seen in the plot of limit cycle wheelset lateral motion amplitude vs. wheelset speed of Figure 6. Operation under limit cycle conditions occurs frequently in rail vehicle operation.

The stability situation is further complicated by nonlinear suspension characteristics. Coulomb friction, for example, causes the stability boundary to depend on the amplitude of the motion as well

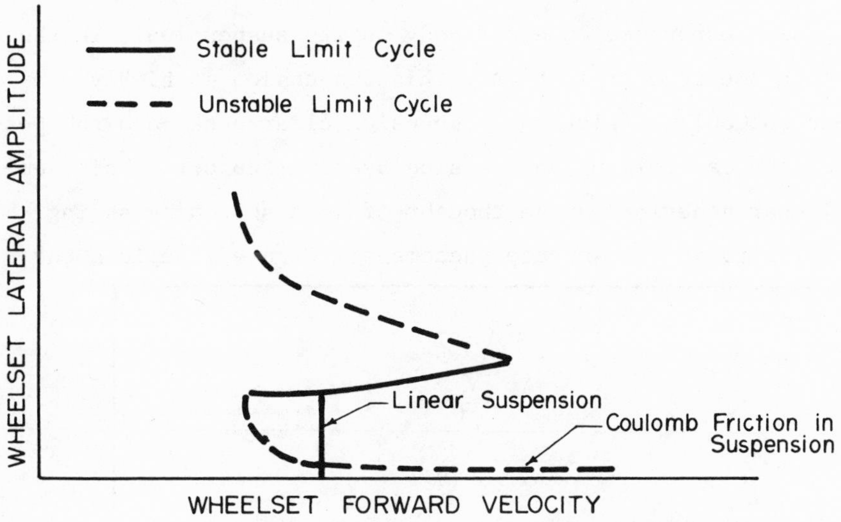

Figure 6. Wheelset Limit Cycle Behavior [26]

as on system parameters such as vehicle speed. Also illustrated in Figure 6 are the stability boundaries (unstable limit cycles) and limit cycle amplitudes for the single wheelset with Coulomb friction in the suspension.

The single wheelset example discussed here is a highly simplified representation of the rail car behavior that is only useful for qualitative analysis. Realistic, quantitative analysis requires a minimum of 14 state variables and 8 nonlinearities.

Forced Harmonic Response

Harmonic forced response occurs when a railcar travels over jointed rails with equal joint spacing. The input disturbances caused by the roadbed irregularities are periodic at the rail joint spatial frequency and its harmonics. A practical case of great importance is the freight car "rock and roll" phenomena, where the rocking motions of the car may be large enough to lift the wheels from the rails.

The rock and roll phenomenon occurs when the railcar travels at a speed where the frequency of the rail joint irregularities (cross-level or rolling displacements, in this case) coincide with the resonant roll response of the car body on its suspension. In the typical North American freight car, this suspension is highly nonlinear, due to Coulomb friction, suspension clearances, and the possibility of the car rolling onto a side bearing support. This combined nonlinear behavior may be thought of as a softening spring, and gives rise to jump resonance phenomena. Figure 7 depicts the

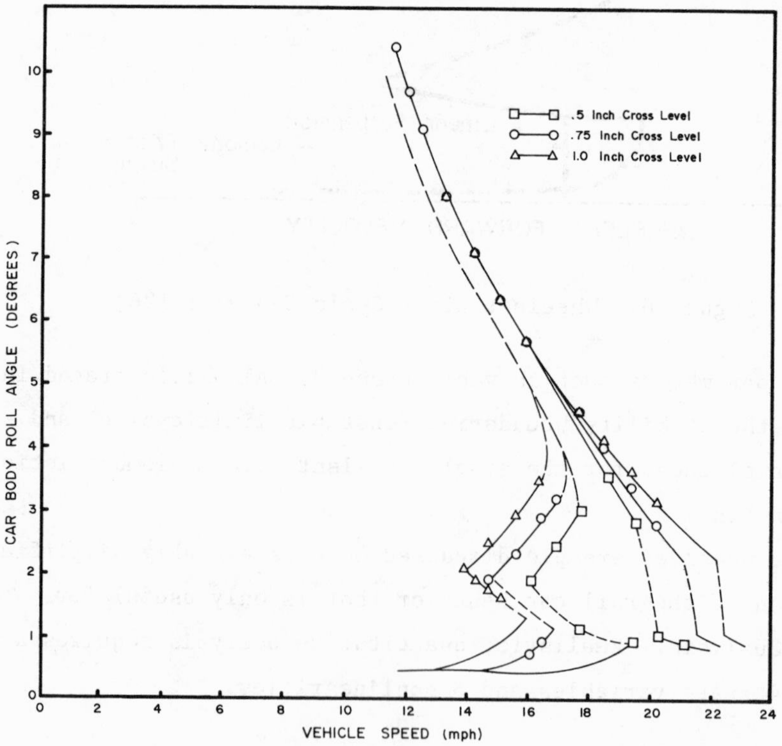

Figure 7. Freight Car Roll Response with Jump Resonances [10]

roll response of such a vehicle as a function of the vehicle speed [10]. Jump resonance in the speed range 14-20mph may be seen here.

Forced Random Response

Equally important, but less spectacular is the forced response of nonlinear rail vehicles to random variations in the roadbed. The random response of the vehicle is utilized in assessing ride comfort, evaluating the load environment seen by the vehicle equipment, and determining the adequacy of clearances and safety limits in the vehicle configuration.

The random guideway irregularity is normally specified in terms of its power spectral density (PSD). The form and parameters of rail roadbed PSD's have been well documented through extensive measurements on European and American railroads [11].

Solutions to the nonlinear random response problem have been obtained by numerical integration [12], analog integration [13], and equivalent linearization [14] methods. A typical result for a single wheelset, Figure 8, depicts the lateral to vertical wheel-rail force ratio as a function of the R.M.S. roadbed irregularity. A linear analysis would, of course, yield a linear relationship here. The rapid increase in L/V ratio seen in this figure can be attributed to the nonlinear effect of the wheel flange.

Solution Techniques

Solution of practical rail vehicle dynamics problems poses the most difficult sort of nonlinear analysis problem, due to high system order and strongly nonlinear, often numerical, multiple nonlinearities found in realistic mathematical models. In most cases, the vehicle equations of motion can be expressed in standard second order form,

$$M\underline{\ddot{x}} + C\underline{\dot{x}} + K\underline{x} = \sum_i \underline{F}_i\,(\underline{x},\, \underline{\dot{x}},\, \underline{u}_R,\, \underline{\dot{u}}_R) + B_1\underline{u}_R + B_2\underline{\dot{u}}_R \qquad (9)$$

where: M,C,K -- mass, damping and stiffness matrices

$\underline{F}_i\,(\underline{x},\underline{\dot{x}},u_R,\dot{u}_R)$ -- vector of nonlinear functions

$\underline{x}$ - vector of system degrees of freedom

$\underline{u}_R$ -- vector of roadbed input variables

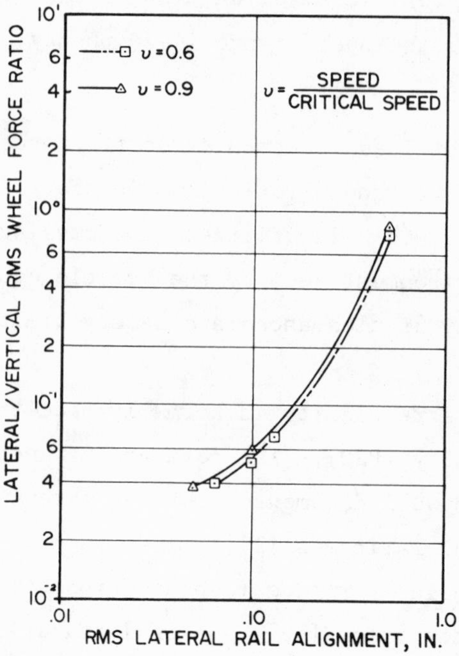

Figure 8. Wheelset Random Response [12]

In practical cases, a minimum of 9 degrees of freedom are required. The components of the F vector normally contain a minimum of 8 distinct nonlinear functions. These nonlinear functions are often strongly nonlinear, such as Coulomb friction, and may only exist in numerical tables, precluding analytic expressions for the system Jacobian. General approaches to this high order, multiple nonlinearity problem are still evolving.

Equilibrium Solutions

The equilibrium solution of nonlinear equations of motion is of great practical interest in the case of rail vehicle curving behavior. Such problems involve a minimum of nine algebraic equations containing multiple nonlinearities.

Solution of the equilibrium problem without the nonlinear contact forces has been obtained by a functional iteration approach that entails successive solution of equivalent linear problems [6]. This procedure involves determination of effective "linear relationships for the nonlinear functions at each iteration step, and subsequent solution of resulting linear system of equations. Convergence difficulties are often encountered with discontinuous functions such as Coulomb friction.

The general equilibrium problem that includes the creep force nonlinearities has not yet been tackled. The step wise linear procedure used to date will not be suitable in this case, and a more general search technique will be required. It is quite possible that the viewpoint and techniques of bifurcation theory, described in accompanying papers, may find fruitful application to this problem.

Linearization and Piecewise Linearization

Linearization is naturally the most widely used technique for treating the nonlinear rail vehicle dynamics problem. It cannot be used, however, when the nonlinear characteristics are discontinuous or strongly nonlinear in the range of interest, as are Coulomb friction, certain wheel-rail constraints, and the creep force relationship. These difficulties are occasionally handled in a linear manner by employing equivalent linear representations. This approach requires assumptions concerning the nature of the inputs to each nonlinearity (frequency and amplitudes, for example) that must be closely reviewed after obtaining a solution. If this process is carried through several iterations it approaches, in essence, the equivalent linearization procedure.

Although rail vehicle nonlinearities are often piecewise nonlinear, this characteristic is rarely utilized in dynamic analyses. This can be attributed to the fact that an extremely large number of piecewise linear states may exist in a high order, multiple nonlinearity system, leading to excessive computational expenses.

Numerical Integration

The most straight forward numerical approach to solution of high order nonlinear equations of motion is numerical integration. However, this approach is extremely demanding of computer and human resources. Solution of the stability problem to obtain a chart such as shown in Figure 6 might require as many as 50 individual simulations at varying initial conditions and velocities. Nevertheless, numerous studies have used this approach to study stability [15,16,17], harmonic forced response [18,19], random forced response [12,20], and rail vehicle curving [5,6,21].

Direct numerical integration appears to be the only approach to studying the dynamics of rail vehicle curving. This curving problem poses particular difficulties in numerical integration due to the nat-ture of the wheel geometry. When the wheel contact point moves from treat to flange contact, the system equations change from a relatively "soft" system to a very "stiff" system. As a result, a very small numerical integration step size is required while in flange contact, although the high frequency dynamics are not of particular interest. Efficient numerical integration methods for integration of such stiff, nonlinear systems may provide a solution to these difficulties.

Analog and Hybrid Techniques

Analog and hybrid analog/digital equipment provide a means to avoid some of the numerical integration problems. Stability [8] and forced response [13,22] of rail vehicles have been studied with such equip-ment. Although the set up for such simulations may require consider-able ingenuity and effort, hybrid results have been extremely useful for comparison with both experimental [23] and approximate theoretical [9] solutions.

Equivalent Linearization

Perhaps the most promising approach for efficient solution of non-linear rail car dynamics problems lies in the equivalent lineariza-tion approaches. Early studies to apply the Krylov and Bogoluibov method to simple rail vehicle analyses [24,25] demonstrated its ef-fectiveness. Similarly, statistical linearization proved useful for simple vehicles [20]. More recent work has demonstrated the feasibil-ity of applying these methods to more complex vehicles with multiple nonlinearities [26,27], although this work has also uncovered some inherent difficulties.

The equivalent linearization process involves replacing the non-linear element with a linear element whose gain is a function of the properties of the nonlinearity. The process may be interpreted [28] in terms of finding the gain that minimizes the mean square error between the nonlinear element and its linear equivalent. For a sinusoidal input, the gain may be complex and depend on the input amplitude and frequency. When the input is random, the equivalent gain depends on the probability density function and variance of the input signal. For example, the equivalent sinusoidal input and Gaussian input describing functions for the $\frac{r_1 - r_R}{2a}$ wheel-rail geometric function are shown in Figure 9.

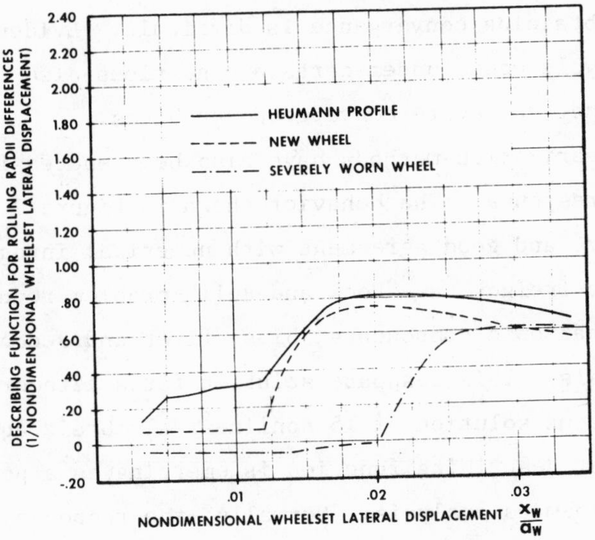

Figure 9. Typical Rolling Radii Difference Sinusoidal Input Describing Functions [26]

System stability is investigated by transforming the nonlinear problem to an equivalent linear problem of the form

$$\dot{\underline{x}} = \underline{N}(\underline{a},\omega)\ \underline{x} \tag{10}$$

where $\underline{N}$ is a sinusoidal input describing function matrix, $\underline{a}$ a vector of amplitudes input to the nonlinear elements, and ω the frequency. Limit cycle existence requires that one of the eigenvalues of the describing function matrix have a zero real part, while the input amplitude vector, $\underline{a}$, satisfies the eigenvector conditions associated with that eigenvalue. For high order systems with multiple nonlinearities, the solution requires iterative numerical methods. Several iterative algorithms have been tried [26].

The limit cycle behavior similar to that depicted in Figure 6 was found by this approach and checked with hybrid computer integration results. Nearly perfect agreement was found. Experience to date indicates that equivalent linearization provides accurate limit cycle results for rail vehicle dynamics problems, when convergence of the iterative method is obtained. However, when the describing function derivatives are large, as occurs with Coulomb friction at small amplitudes and with the wheel-rail constraint functions near the jump to flange contact, obtaining convergence is difficult. Evidence of multiple, limit cycle modes under certain conditions also exists, a situation that leads to severe convergence problems.

Equivalent linearization methods have also been successful in solving for harmonic response. The behavior shown in Figure 7 was obtained in this way, and good agreement with numerical integration results found. This freight car "Rock and Roll" problem required inclusion of three harmonic components, bias, first and second harmonics, in the input signals. This response solution for a 12th order system required simultaneous solution of 15 nonlinear algebraic equations.

The random input describing function is emerging as a powerful tool in rail vehicle dynamics analysis. Normally, the response power spectral density is the objective of such an analysis. Two iterative approaches have been employed in rail vehicle problems. In one approach, equivalent linear second order system equations are formed by assuming R.M.S. values for inputs to nonlinearities. PSD's for the response variables are computed using standard frequency domain techniques [27]. R.M.S. values are found by integrating the PSD's, comparing with initial assumptions and updating. This approach converges well with weak nonlinearities but often does not converge when the describing function derivitives are large.

The second method differs in that a time domain solution of the steady state covariance propagation equation is utilized to find the output R.M.S. values [29]. For mechanical systems, the frequency domain approach appears to be faster computationally, because one can use the system equations in second order form, while the time domain solution requires first order state equations [27].

Conclusions

The inherently nonlinear nature of certain important rail vehicle dynamics problems, and the status of solutions for these problems have been described in this paper. Realistic rail vehicle dynamics problems characteristically involve high order systems, a minimum of 18 states, and multiple, discontinuous nonlinear functions that may only be described in numerical or algorithmic terms. A number of difficulties exist in solving these dynamics problems.

Equilibrium solutions of the vehicle dynamic equations of motion are of particular interest in studying rail vehicle curving. Bifurcation analysis similar to that discussed in other papers at this conference may be fruitful in treating this problem. The existence of multiple solutions and the convergence difficulties caused by "hard" nonlinearities require further research.

The direct, numerical integration, approach to the nonlinear dynamics problem must deal with the computational expense caused by the high system order and the "stiff" character of the system in certain ranges of the nonlinearities. Coulomb friction and flange contact, for example, lead to stiff equations of motion. In addition, the possibilities of jump resonance and multiple limit cycles require simulations for a wide range of initial conditions and forcing functions.

The high system order and multiple nonlinearities has led to investigation of approximate solution techniques such as the describing function methods to solve for vehicle response. Such methods appear reasonably accurate and efficient when they converge, but encounter difficulties when discontinuous nonlinearities are present. The possibility of multiple limit cycle modes also presents numerical problems. Areas for development include methods for handling multiple input/multiple output nonlinearities such as the creepage-creep force function, and reliable techniques for efficiently solving the nonlinear algebraic (or integral) equations that arise with such methods.

It appears that future efforts on nonlinear rail vehicle dynamics will focus on simplification of system models, characterization of nonlinear response behavior, and improvement of approximation techniques.

REFERENCES

[1] COOPERRIDER, N.K., LAW, E.H., HULL, R., KADALA, P., and TUTEN, J.M., "Analytical and Experimental Determination of Nonlinear Wheel/Rail Geometric Constraints," Federal Railroad Administration FRA OR&D-76-244 (PB25290), Dec. 1975.

[2] VERMEULEN, P.J. and JOHNSON, K.L., "Contact of Nonspherical Elastic Bodies Transmitting Tangential Forces," Transactions of the ASME, June 1964, pp. 338-340.

[3] KALKER, J.J., "On the Rolling Contact of Two Elastic Bodies in the Presence of Dry Friction," Doctoral Dissertation, Technische Hogeschool, Delft, The Netherlands, 1967.

[4] KALKER, J.J., "Survey of Wheel-Rail Rolling Contact Theory," Vehicle Systems Dynamics, Vol. 8, No. 4, Sept. 1979, pp. 317-358.

[5] ELKINS, J. A., and GOSTLING, R. J., "A General Quasi-Static Curving Theory for Railway Vehicles," The Dynamics of Vehicles on Roads and on Tracks, Slibar, A. and H. Springer, eds., Swets and Zeitlinger, 1978, pp. 388-406.

[6] LAW, E. H. and N. K. COOPERRIDER, "Nonlinear Dynamics and Steady State Curving of Rail Vehicles," presented at 1978 ASME Winter Annual Meeting and submitted for publication.

[7] COOPERRIDER, N. K., "The Lateral Stability of Conventional Railway Trucks," Proc. 1st Int. Conf. on Veh. Mech., 1968, pp.37-67.

[8] COOPERRIDER, N. K., HEDRICK, J. K., LAW, E. H., and MALSTROM, C. "The Application of Quasi-linearization Techniques to the Prediction of Nonlinear Railway Vehicle Response," Vehicle System Dynamics, Vol. 4, No. 2-3, July 1975, pp. 141-148.

[9] HANNEBRICK, D. N., LEE, H. S. H., WEINSTOCK, H. and HEDRICK, J.K., "Influence of Axle Load, Track Gauge, and Wheel Profile on Rail-Vehicle Hunting," Transactions of the ASME, Journal of Engineering for Industry, Series B, Vol. 99, No. 1, February 1977, pp. 186-195.

[10] HEDRICK, J.K., "Nonlinear System Response: Quasi-Linearization Methods," Nonlinear System Analysis and Synthesis, ASME, New York, 1978, pp. 97-124.

[11] CORBIN, J.C., and KAUFMAN, W.M., "Classifying Track by Power Spectral Density," Mechanics of Transportation Systems, AMD-15, ASME, Dec. 1975, pp. 1-20.

[12] LAW, E. H., "Nonlinear Wheelset Dynamic Response to Random Rail Irregularities," Journal of Engineering for Industry, Transactions of the ASME, Series B, Vol. 96, No. 2, November 1974.

[13] ARAI, S. and K. YOKOSE, "Simulation of Lateral Motion of 2 Axle Railway Vehicle Running," Proc. IUTAM Symposium on Dynamics of Vehicles on Roads and on Tracks, Delft, The Netherlands, 1975, pp. 326-344.

[14] HEDRICK, J. K. and I. A. CASTELAZO, "Statistical Linearization of the Nonlinear Rail Vehicle Wheelset," 6th IUTAM Symposium on Dynamics of Vehicles on Roads and Tracks, Berlin, 1979.

[15] COOPERRIDER, N. K., "The Hunting Behavior of Conventional Railway Trucks," Journal of Engineering for Industry, Trans. ASME, Series B, Vol. 94, May 1972, pp. 752-762.

[16] GILCHRIST, A. O., HOBBS, A.E.W., KING, B. L., and WASHBY, V., "The Riding of Two Particular Designs of Four Wheeled Railway Vehicles," Interaction Between Vehicle and Track, Proc. Institution of Mechanical Engineers, Vol. 180, Part 3F, 1966, p. 99-113.

[17] SAUVAGE, G., "Nonlinear Model for the Study of Rail Vehicle Dynamics," Proc. IUTAM Symposium on Dynamics of Vehicles on Roads and Tracks, Delft, August 1975, pp. 326-344.

[18] WEIBE, D.,"The Effect of Lateral Instability of High Center of Gravity Cars," Trans. ASME, Vol. 90-B, No. 4, November 1968, pp. 462.

[19] PLATIN, B. E., BEAMAN, J. J., HEDRICK, J. K. and WORMLEY, D. N., Computational Methods to Predict Railcar Response to Track Cross-Level Variations, Federal Railroad Administration, Report No. FRA-OR&D-76-293, September 1976.

[20] STASSEN, H. G., "Random Lateral Motions of Railway Vehicles," Doctoral dissertation, Technische-Hogeschool, Delft. The Netherlands, 1967.

[21] WILLIS, T., and SMITH, K. R., "A Mathematical Simulation of the Curve Entry and Curve Negotiation Dynamics of Flexible Two Axle Railway Trucks," ASME Paper 76-WA/RT-14.

[22] OSBON, W. O. and T. H. PUTNAM, "Engineering Design Study of Active Ride Stabilizer for the Dept. of Transportation's High Speed Test Cars," Westinghouse Electric Research Laboratories, Final Report, Contract No. 3-0267, June 1969.

[23] LAW, E. H., COOPERRIDER, N. K. and R. H. FRIES, "Freight Car Dynamics: Field Test and Comparisons with Theory," Federal Railroad Admin. Report, in preparation.

[24] DEPATER, A. D., "The Approximate Determination of the Hunting Movement of a Railway Vehicle by Aid of the Method of Krylov and Bogoliubov," Applied Scientific Research. Section A, Vol. 10, 1961, pp. 205-228.

[25] LAW, E. H., and BRAND, R. S., "Analysis of the Nonlinear Dynamics of a Railway Vehicle Wheelset," Journal of Dynamics, Measurement, and Control, Trans. ASME, Series G, Vol. 95, No. 1, March 1973, pp. 28-35.

[26] HEDRICK, J. K., COOPERRIDER, N. K., and LAW, E. H., "The Application of Quasi-Linearization Techniques to Rail Vehicle Dynamic Analysis," Federal Railroad Administration Report No. FRA/ORD-78/56, November 1978.

[27] HEDRICK, J. K. and A. V. ARSLAN, "Nonlinear Analysis of Rail Vehicle Forced Lateral Response and Stability," Trans. ASME J. Dynamic Systems, Measurements and Control, Trans. ASME, Series G, Vol. 101, No. 3, Sept. 1979.

[28] GELB, A., and VANDERVELDE, W. E., Multiple-Input Describing Functions and Nonlinear System Design, McGraw-Hill, New York, 1968.

[29] BRYSON, A. E. and Y. C. HO, Applied Optimal Control, Ginn and Co., Waltham, Mass. 1969.

Accuracy of Statistical Linearization

J. J. Beaman*

Abstract. One of the most general approximate methods for the analysis of nonlinear stochastic systems without resorting to simulation is the method of statistical linearization. The accuracy of this method is frequently based on the "filter hypothesis," but results of an accuracy study shown in this paper reveal that this hypothesis can often be in error for closed-loop systems; instead the relative magnitude of the input into the nonlinearity seems to be a safer accuracy guide. It is also shown that the variances predicted by statistical linearization are a lower bound for the actual variances in a class of Hamiltonian-like systems.

1. Introduction. The most general approximate method for the analysis of nonlinear dynamic stochastic systems without resorting to digital simulation is the method of statistical linearization which was initially developed independently by Booton [1] and Kazakov [2]. Besides the use of the method as an analytical tool, it has also been proposed as a synthesis tool (Douce and King [3], Sawaragi, Sugai, and Sunahara [4], Hedrick [5], and Beaman [6]). If the statistical linearization method gives a good representation of the nonlinear system then these synthesis techniques are quite general and useful.

The method can be described by considering the general nonlinear dynamical system represented as

$$d\underline{x} = \underline{f}(\underline{x},t)dt + d\underline{b} \tag{1}$$

where $\underline{x}$ and $\underline{f}$, are n-vectors, $\underline{b}$ a n-vector Brownian motion process with $\langle d\bar{b}\, d\underline{b} \rangle^{\dagger} = Qdt$ and (1) can be formally interpreted as an Itô

*Assistant Professor, Department of Mechanical Engineering, The University of Texas at Austin, Austin, Texas.

†< > denotes expectation.

stochastic differential equation (Jazwinski [7]). The basic problem in linearization is to find an equivalent linear system which in some sense represents (1). One way this can be done is by approximating the nonlinearity as

$$\underline{f}(\underline{x},t) \simeq \underline{a} + N(\underline{x} - \underline{m}) \tag{2}$$

where $\underline{m}$ is the expected value of $\underline{x}$, $\underline{a}$ is an n-vector, and N is an n x n matrix. The coefficients $\underline{a}$ and N are chosen such that (Gelb [8], Sunahara [9], Phaneuf [10])

$$\begin{aligned} \underline{a} &= \langle \underline{f} \rangle \\ N &= \langle \underline{f}(\overline{x} - \overline{m}) \rangle P^{-1} \end{aligned} \tag{3}$$

*

where P is the covariance matrix, $\langle (\underline{x} - \underline{m})(\overline{x} - \overline{m}) \rangle$.

Although statistical linearization can be applied formally to a large number of different systems, it suffers from the fact that the method is an approximate one in which estimates of accuracy are not easy to come by. Experimental evidence regarding the accuracy of the method (Smith [11]) indicates that the approximation is acceptable in a wide variety of situations. It can be shown (Phaneuf [10] and Beaman [6]) that the corresponding statistically linearized system

$$d\underline{x} = [\underline{a} + N(\underline{x} - \underline{m})]dt + d\underline{b} \tag{4}$$

will have an identical mean, $\underline{m}(t)$, and covariance, P(t), to that of the nonlinear system (1) provided the exact probability density function by which to evaluate the expectations is known. Unfortunately, the probability density function is usually not known *á priori*, and one must be assumed. The density function is usually taken to be Gaussian. The reason for this assumption is based on the so-called "filter hypothesis" which appeals to central limit theorems for validity. When the Gaussian assumption is incorrect to a large degree, the statistical linearization technique can have considerable error.

2. *Current Perspective*. The accuracy of statistical linearization, or the degree to which the nonlinear system is represented by

---

*$\underline{v}$ column vector; $\overline{v}$ row vector.

the statistically linearized one, is currently an open issue. General discussion of the accuracy of the method can be found in Atherton [12] in which he states:

> Attempts have been made by various authors to find methods for estimating the accuracy of quasilinear analysis of nonlinear systems with random inputs but no truly satisfactory solution has resulted.

As regards the accuracy of the method, he concludes that it ". . . is in some way dependent on the amount of distortion produced by the nonlinearity and the effectiveness of the low pass plant in filtering the signal and producing a near Gaussian input to the nonlinearity." These comments are indicative of the general consensus of many authors in the field of statistical linearization. As can be seen by these statements, the justification for the accuracy of statistical linearization has been primarily intuitive in nature, although some quantitative accuracy studies can be found in Sawaragi [13], [4], Hugnh [14], and Smith [11]. Included in the work of Smith is a rule of thumb, based primarily on experimental evidence, to access the accuracy of statistical linearization. Also, theoretical estimates of accuracy which can be applied in certain cases are examined in Kolovskii [15] and Holtzman [16].

2.1 Critique of the Filter Hypothesis.

> Everybody believes in the law of errors, the experimentalists because they think it is a mathematical theorem, the mathematicians because they think it is an experimental fact.
>
> Lippman (quoted in Poincaré [17])

As stated earlier, if the probability density function* is known exactly, the exact mean and covariance propagation will be predicted by statistical linearization. Since this function is usually not known, a Gaussian distribution is typically assumed, and, as mentioned previously, a prime reason for this assumption is the "filter hypothesis." This hypothesis can be visualized by noting the following single-loop system (Figure 1) forced by a Gaussian random process. The "filter hypothesis" would imply that if G(s) is a stable low-pass linear system

*All probability functions refer to 1st order distributions.

then the density function of y would be close to Gaussian even if the density function of u is not. This is based on central limit theorems and open-loop experimental results, primarily from the communications and filtering fields. Examples can be found in Middleton [18], Sawaragi [4], and Pervozvanskii [19] for open-loop systems.

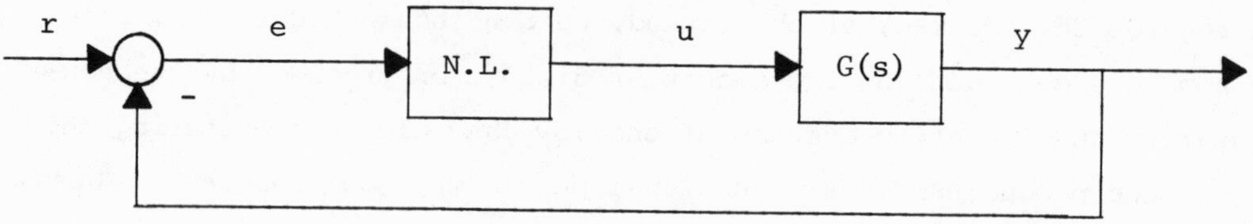

FIGURE 1. System for Description of Filter Hypothesis

These examples are convincing testaments to the "filter hypotheses" in the open-loop case, and one would expect that the hypothesis would extend naturally to closed-loop systems. One such study of a closed loop system can be found in Sawaragi [13] and is depicted in Figure 2.

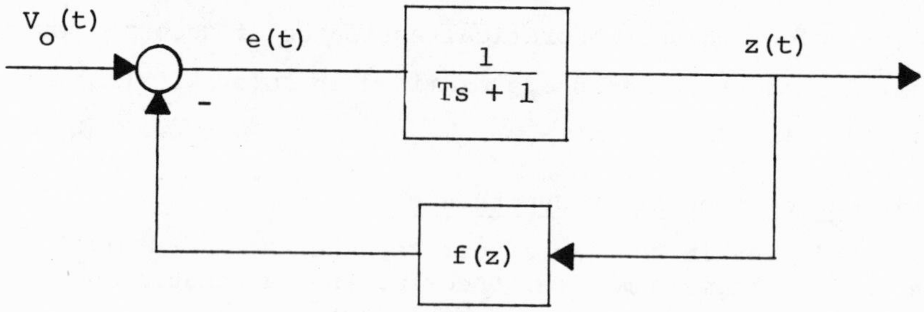

FIGURE 2. System Studied by Sawaragi

For $f(z) = z^3$ and $V_o(t)$ a Gaussian white noise process of intensity q, Sawaragi shows that as the linear plant time constant increases (more low-pass) then the density function of z approaches Gaussian. Unfortunately, this is <u>not</u> a general result, not even for the system of Figure 2. This can be readily seen by returning to this system with a general nonlinearity f(z). The system can be represented as

$$T\dot{z} + z + f(z) = V_o(t) \tag{5a}$$

or as

(5b) $$T\,dz + (z + f(z))dt = db$$

where formally $V_o = \frac{db}{dt}$ and $\langle db^2 \rangle = q dt$. If we now time scale (5b) with $\tau = t/T$, (5b) can be rewritten as

(6) $$dz + (z + f(z))d\tau = Tdb = d\beta$$

where $\langle d\beta^2 \rangle = \langle T^2\, db^2 \rangle = \frac{q}{T}\, d\tau$. Therefore, the only effect of T on the output z is to scale the input white noise intensity, q. As T increases, the effective intensity decreases, or as the system becomes more low-pass the effect of the noise on the system diminishes. The results of Sawaragi can now be inferred directly. For small amplitudes, $f(z) = \gamma z^3$ is small when compared with the linear term z, and the system will like a linear one, i.e. Gaussian distribution of z. This is the case for T large which is the result of Sawaragi.

Now suppose f(z) is a nonlinearity which is not negligible relative to a linear term for small amplitudes such as f(z) = sgn(z)*. In this case, one would reach a conclusion in opposition to the filter hypothesis, i.e. that of increasing T makes the system more non-Gaussian. For this simple system, the Fokker-Planck-Kolmogorov equation (Fuller [20]) can be solved for the exact steady-state density p(z). This density function is plotted in Figure 3 for two values of $q^* = q/T$,

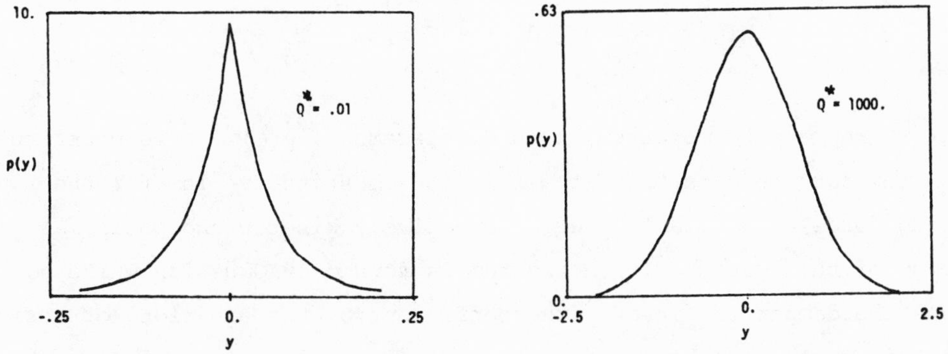

FIGURE 3. Plot of Density Function vs. Normalized Dependent Variable of $T\dot{z} + z + \text{sgn}(z) = V_o$ for Two Values of $q^* = q/T$

and as can be seen as q* decreases (more low-pass) the density function becomes more non-Gaussian.

*sgn(z) is the signum function.

This counterexample to the filter hypothesis for a closed-loop system is in no way unique. There are many functions, $f(z)$, which are not negligible or linear for small z. This includes all functions with singularities of the first derivative at the origin such as $f(z) = \text{sgn}(z)|z|^r \; r < 1$. Other counterexamples for different systems can be found in Beaman [6].

3. Accuracy of Statistical Linearization for Hamiltonian-like Systems. As was discussed in the last section, the "filter hypothesis" does not in general give an adequate assessment of accuracy. Therefore, in this section, a study is presented of a class of systems for which the exact steady-state solution to Fokker-Planck-Kolmogorov equation can be found in closed form. This class of systems, termed Hamiltonian systems, are representable as

$$\dot{\underline{x}} = \frac{\partial h}{\partial \underline{\xi}} \quad , \qquad \dot{\underline{\xi}} = -\frac{\partial h}{\partial \underline{x}} - B\frac{\partial h}{\partial \underline{\xi}} + \underline{w} \tag{7}$$

where B is a damping coefficient matrix, $\underline{w}$ is a vector white noise process with intensity matrix Q, and h is a scalar function.

Included in this class of systems is the general nonlinear mass-spring-damper system for which an accuracy study is presented in this paper. The Hamiltonian $h(\underline{\xi},\underline{x})$ for this system can be written as

$$h(\underline{\xi},\underline{x}) = \int_0^{\xi} v(\xi)d\xi + \int_0^{\xi} f(x)dx = T_{\xi} + T_x$$

where ξ represents momentum, x is displacement, $v(\xi)$ is the constitution law for the mass velocity, f is the constitutive law for the spring force, and T_ξ and T_x are respectively kinetic and potential energy of the system. If the system is forced by Gaussian white noise w and the damper is linear with coefficient b, the equations of motion for this system can be represented as

$$\dot{x} = \frac{\partial h}{\partial \xi} = v(\xi) \quad ; \quad \dot{\xi} = -\frac{\partial h}{\partial x} - b\frac{\partial h}{\partial \xi} + W = -f(x) - bv(\xi) + W \tag{8}$$

The steady-state probability distribution is (Fuller [20])

$$p(x,\xi) = C \exp\left[-\frac{2b}{q}\left(\int_0^x f(x)\,dx + \int_0^{\xi} v(\xi)\,d\xi\right)\right] \tag{9}$$

where q is the intensity of the white noise and C is a normalizing constant chosen such that

$$\int_{-\infty}^{+\infty} \int p(x,\xi)\,dx\,d\xi = 1 \quad .$$

If a Gaussian distribution is assumed, the N matrix can be found as

$$N = \begin{bmatrix} 0 & N_I(\sigma_\xi) \\ -N_k(\sigma_x) & -b\,N_I(\sigma_\xi) \end{bmatrix}$$

where

$$\sigma_\xi = \sqrt{\langle(\xi - \langle\xi\rangle)^2\rangle} \quad , \qquad \sigma_x = \sqrt{\langle(x - \langle x\rangle)^2\rangle}$$

$$N_I(\sigma_\xi) = \frac{1}{\sqrt{2\pi}\,\sigma_\xi^3} \int_{-\infty}^{\infty} (\langle\xi - \langle\xi\rangle) f \exp\left(-(\xi - \langle\xi\rangle)^2/2\sigma_\xi^2\right) d\xi$$

$$K_k(\sigma_x) = \frac{1}{\sqrt{2\pi}\,\sigma_x^3} \int_{-\infty}^{\infty} (x - \langle x\rangle) v \exp\left(-(x - \langle x\rangle)^2/2\sigma_x^2\right) dx$$

For the purposes of the study, assume odd nonlinearities such that the steady-state means $\langle x\rangle = \langle\xi\rangle = \langle f\rangle = \langle v\rangle = 0$. The steady-state covariance of (4) then yields

$$N_I\,\sigma_\xi^2 = q/2b \qquad ; \qquad N_k\,\sigma_x^2 = q/2b \tag{10}$$

The two nonlinear algebraic equations can be solved for the statistical linearization estimates of the variances $\tilde{\mu}_x = \sigma_x^2$ and $\tilde{\mu}_\xi = \sigma_\xi^2$.

Four different nonlinearities were then considered to ascertain the accuracies of the estimates $\tilde{\mu}_x$ and $\tilde{\mu}_\xi$. These nonlinearities are depicted in Figure 4. Noting from equation (9) that the variables ξ and

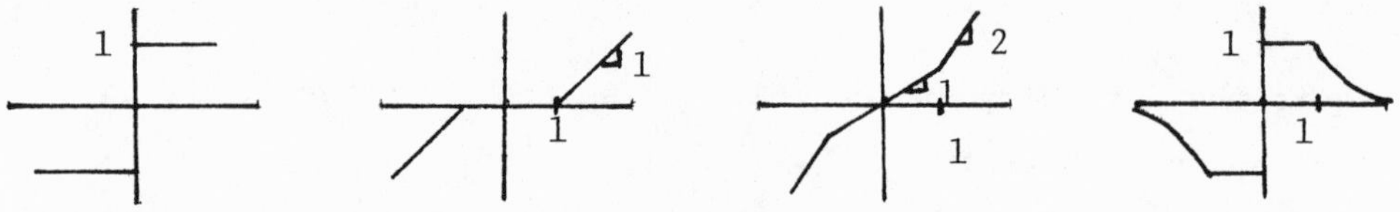

Signum Threshold Gain Change Signum with 1/x

FIGURE 4. Nonlinearities Used in Accuracy Study

x are statistically independent in steady-state, it is only necessary to study one of the variables, say x. The results of this study are presented in Figures 5 through 8. In these figures, $e = (\mu_2 - \tilde{\mu}_2)/\mu_2$ is the relative error and $\lambda_4 = \langle x^4\rangle/\langle x^2\rangle - 3$ is the kurtosis, a measure of the non-Gaussian nature of x. The results for the signum function are presented in Figure 5. Note that λ_4 and e, are both constant.

The results of the threshold nonlinearity are presented in Figure 6. For large q, the nonlinearity appears linear and λ_4 and e go to zero, but as q gets small, e goes to one. This means, that for small q, the error approaches 100%. In order to understand this phenomenon, consider the following first order stochastic differential equation

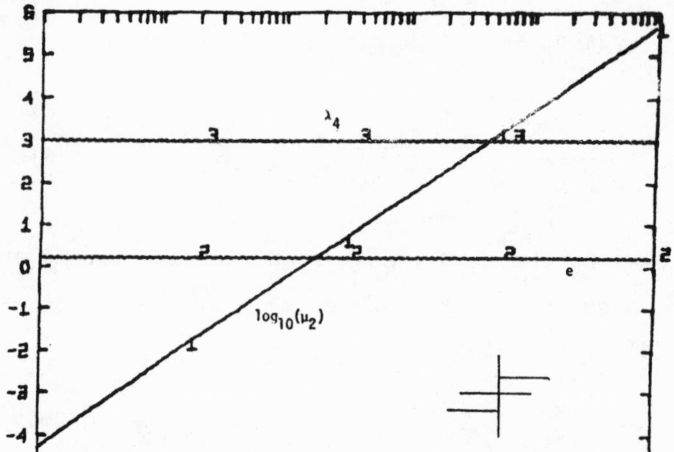

FIGURE 5. Plot of Kurtosis, λ_4 and Relative Error of Gaussian Statistical Linearization, e, vs. Noise Intensity q for f(x) Signum Function.

$dx = - f(x)dt + d\beta$, where $f(x)$ is the threshold element. For large inputs, this equation would be well approximated by the linear equation, $dx = - xdt + d\beta$, and $\mu_2 = q/2$. Now consider small inputs, for which one would be tempted to approximate the original differential equation with the linear differential equation, $dx = d\beta$. The solution for the variance for this equation is, $\mu_2 = qt/2$, which is unbounded as $t \to \infty$. Therefore, for small q one would not expect the values of x to go to zero, but rather lie approximately between the two points

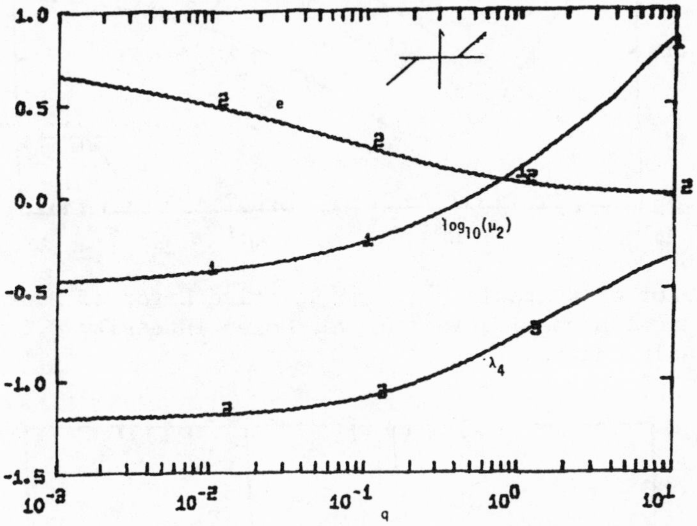

FIGURE 6. Plot of Kurtosis, λ_4, and Relative Error of Gaussian Statistical Linearization, e, vs. Noise Intensity q for f(x) a Threshold.

$x = \pm 1$ (where the gain increases). The variance of this distribution approaches a finite value while that of Gaussian statistical linearization approaches zero. This accounts for the 100% error. A similar nonlinearity to the threshold is the gain change, but, in this case, for both large and small q the error goes to zero. This is to be expected since for both large and small inputs the nonlinearity appears linear and its gain gives a stable system.

It is also possible to obtain 100% error for large input as shown in Figure 8. For sufficiently large $q (q \geq 2/3)$ the variance μ_2 is not finite. This also occurs for $\tilde{\mu}_2$, but not until $q \geq 2$. Therefore, for values of $2/3 \leq q < 2$ statistical linearization has 100% error. The reason for this result is similar to that of the threshold result, except now for large q instead of small q the nonlinearity approaches zero effective gain which gives an unstable system.

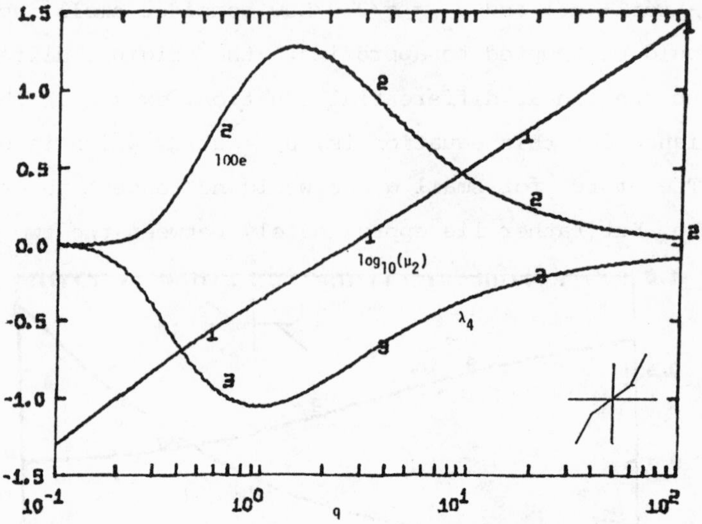

FIGURE 7. Plot of Kurtosis, λ_4, and Relative Error of Gaussian Statistical Linearization, e, vs. Noise Intensity q for f(x) a Gain Change.

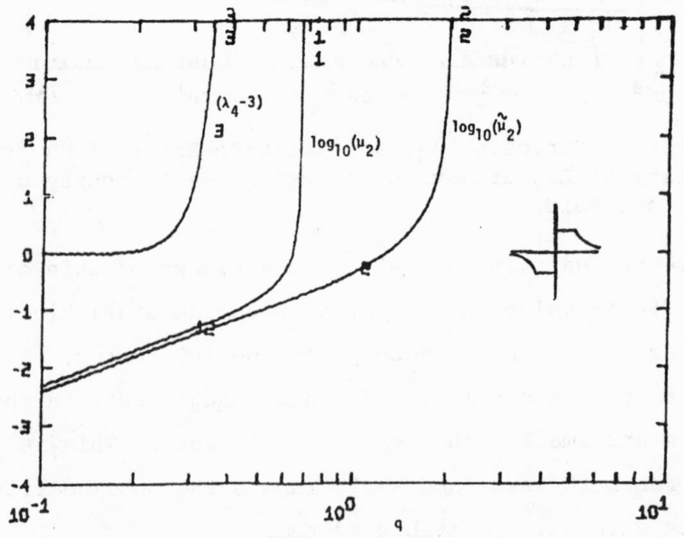

FIGURE 8. Plot of Kurtosis, λ_4, Gaussian Statistical Linearization Variance $\tilde{\mu}_2$, and the Exact Variance μ_2 vs. Noise Intensity q.

3.1 <u>Gaussian Statistical Linearization as a Lower Bound</u>. There is one extremely interesting feature of the previous results which has not yet been discussed. For every nonlinearity, e is always greater than zero which implies $\tilde{\mu}_2 < \mu_2$ in every case. This feature was also noted by Budgor [21] for the Duffing oscillator forced by white noise. It would be of practical value to know if this is a general result. In this section, a theorem is given which partially supplies this result.

For Hamiltonian systems described by equation (7), the probability density function for a variable is often of the form of

$$(11) \qquad p(x) = C\left[\exp - \int_0^x f(t)dt\right] \quad ; \quad C = 1/\int_{-\infty}^{\infty} \exp\left[- \int_0^x f(t)dt\right] dx$$

<u>Theorem</u>: (Lower Bound) Let x be a random variable with density function in the form of (11), let f(x) be an odd function with f(0) = 0. In addition, let n > 0 be the Gaussian statistical linearized gain and define g as, g(x) = f(x) - nx. If g(x) changes sign at most two times for x > 0, then $\tilde{\mu}_2 < \mu_2$.

The proof of this theorem can be found in Beaman [6]. It is the author's feeling that further analysis would extend this result to a much wider class of functions and systems, but this is left as an open question.

Further discussion of explicit error bounds can be found in Kolovskii [15], Holtzman [16], and Beaman [6].

4. <u>Conclusion</u>. This has been concerned with the accuracy of statistical linearization. In particular, it was found that the "filter hypothesis" can readily be in error for closed-loop nonlinear systems. As a alternative, the relative magnitude of the inputs in the nonlinearities seems to be a safer intuitive guide to the accuracy of the method, and that the most severe inaccuracies occur when the effective gain of the nonlinearity approaches a value which drives the system unstable. Also, it was found that the non-Gaussian nature of the input, characterized by the kurtosis, is also a good indicator of accuracy. (However, for functions which are not odd, one would expect the skewness to be just as, or more, important than kurtosis.) It was also

shown that statistical linearization provides a lower bound for the vaιiances of a class of Hamiltonian-like system.

REFERENCES

[1] R. C. BOOTON, Nonlinear control systems with statistical inputs, DACL Report No. 61, Cambridge, Mass: MIT, March 1, 1952.

[2] I.E. KAZAKOV, An approximate method for the statistical investigations of nonlinear systems, Tr. VVIA iva. Prof. N. E. Zhukovskogo, 394(1954).

[3] J. L. DOUCE and R. E. KING, The effect of an additional nonlinearity on the performance of torque-limited control systems subjected to random inputs, Proc. IEE, 107C(1960), pp. 190-197.

[4] Y. SAWARAGI, Y. N. SUGAI, and Y. SUNAHARA, Statistical studies on nonlinear control systems, Nippon Press, Osaki, Japan, 1962.

[5] J. K. HEDRICK, The use of statistical describing functions with linear quadratic controller design, presented at 1976 JACC, Purdue Univ., Indiana, July 1976.

[6] J. J. BEAMAN, Statistical linearization for the analysis and control of nonlinear stochastic systems, ScD Thesis, MIT, Dec. 1978.

[7] A. H. JAZWINSKI, Stochastic processes and filtering theory, Academic Press, 1970.

[8] A. Gelb, Applied optimal estimation, MIT Press, 1974.

[9] Y. SUNAHARA, An approximation method of state estimation for nonlinear dynamical systems, JACC, Univ. of Colorado, 1969.

[10] R. J. PHANEUF, Approximate nonlinear estimation, Ph.D. Thesis, MIT, May 1968.

[11] H. W. SMITH, Approximate analysis of randomly excited nonlinear controls, MIT Press, Research Monograph No. 34, 1964.

[12] D. P. ATHERTON, Nonlinear control engineering, Van Nostrand Reinhold, 1975.

[13] Y. SAWARAGI and N. SUGAI, Accuracy considerations of the equivalent linearization technique for the analysis of a nonlinear control system with a Gaussian random input, Mem. Fac. Eng., Kyoto, 23(1961), No. 1, pp. 300-319.

[14] M. T. HUGNH, N. MOREAU, and J. G. PACQUET, Comparative study of the methods of analysis of nonlinear systems subject to random inputs, Ann. Tekcommuns., 27(1972), No. 11-12, pp. 518-525.

[15] M. Z. KOLOVSKII, Estimating the accuracy of solutions obtained by the method of statistical linearization, Autom. Rem. Cont., 22(1966), No. 10, pp. 43-53.

[16] J. M. HOLTZMAN, Analysis of statistical linearization of non-linear control systems, SIAM Jour. Control, 6(1968), No. 2, pp. 235-243.

[17] H. POINCARÉ, Calcul des probabilités, 2nd Edition, Paris, 1912.

[18] D. MIDDLETON, An introduction to statistical communication theory, McGraw-Hill, 1960.

[19] A. A. PERVOZVANSKII, Random processes and filtering theory, Academic Press, 1920.

[20] A. T. FULLER, Analysis of nonlinear stochastic systems by means of the Fokker-Planck equation, Int. Jour. Control, 9(1969), No. 6.

[21] A. B. BUDGOR, K. LINDENBERG, and K. E. SHULER, Studies in non-linear stochastic processes II, the duffing oscillator revisited, J. Statistical Phy., 15(1976), pp. 355-374.

SECTION THREE
CHEMICAL ENGINEERING

Bifurcations of a Model Diffusion-Reaction System

C. R. Kennedy* and R. Aris**

Abstract. After some general observations on kinetic, compartmental and distributed systems a model suggested by Rössler is explored. Bifurcations to unsymmetric steady and uniform oscillatory states from the uniform steady states are demonstrated and the former are shown to break down into oscillations of increasing complexity.

Introduction. The bifurcation of solutions to equations representing the interplay of diffusion and reaction has received a great deal of attention in recent years both from a theoretical (1, 2, 3, 4) and from an experimental (5, 6) point of view, while, since the discovery of chaos (7, 8) many have wondered whether the strange flickerings and turbulent behavior of quite ordinary systems may not be ascribed to some such cause. Rössler has demonstrated how several types of system might be patient of strange attractors (9, 10) and we wish to pursue one of his suggestions here. Marek has considered a somewhat similar system (11) while a bimolecular reaction has engaged Othmer and Kádas (12).

The Systems Considered. The vector $\underset{\sim}{x}$ may be thought of as a vector of compositions in R^n with components $x^1, x^2, \ldots, x^n$. It may exist in three contexts which are worth comparing: the kinetic, where it is

*Mobil Research and Development Corporation, Paulsboro, NJ 08066.
**Department of Chemical Engineering & Materials Science, University of Minnesota. The senior author is indebted to the Mathematics Research Center of the University of Wisconsin for a visiting appointment which allowed the writing up of this work. This part of the doctoral work of the junior author was not supported by any agency, but by his own industry as a teaching assistant.

a function only of time, $\underset{\sim}{x}(t)$; the compartmental, where it also depends on an enumerative index, $\underset{\sim}{x}_j(t)$, $j = 1,\ldots,N$; and the distributed, where (for simplicity) it depends on one position variable, z say, as $\underset{\sim}{x}(z,t)$. The three habitats will be distinguished by their equations:

(K) Kinetic system: $\dot{\underset{\sim}{x}} = \underset{\sim}{f}(\underset{\sim}{x})$

(C) Compartmental system: $\dot{\underset{\sim}{x}}_j = \underset{\sim}{D}(\underset{\sim}{x}_{j+1} - 2\underset{\sim}{x}_j + \underset{\sim}{x}_{j-1}) + \underset{\sim}{f}(\underset{\sim}{x}_j)$

$\underset{\sim}{x}_0 = \underset{\sim}{x}_1,\ \underset{\sim}{x}_{N+1} = \underset{\sim}{x}_N$

(D) Distributed system: $\dot{\underset{\sim}{x}} = \underset{\sim}{D}\underset{\sim}{x}_{zz} + \underset{\sim}{f}(\underset{\sim}{x}),\ \underset{\sim}{x}_z = 0,\ z = 0, L.$

We observe that all three systems are closed except in so far as the function $\underset{\sim}{f}$ may contain terms which imply the removal or supply of any species. The kinetic system is the single compartment; the distributed is the limit of an infinite number of compartments if $\underset{\sim}{D}$ is made proportional to (N/L). Of course there are more subtle ways of approximating the distributed system by a discrete one and we shall use a Galerkin method in calculations on (D).

The equilibrium or steady state of (K) is $\underset{\sim}{x}_e$ given by

(1) $$\underset{\sim}{f}(\underset{\sim}{x}_e) = 0$$

and if $\underset{\sim}{x} = \underset{\sim}{x}_e + \underset{\sim}{\phi}$ the linearization about this state is

(2) $$\dot{\underset{\sim}{\phi}} = \underset{\sim}{J}\underset{\sim}{\phi}$$

where $\underset{\sim}{J}$ is the Jacobian matrix $\nabla\underset{\sim}{f}(\underset{\sim}{x}_e)$. Local stability is then determined by the eigenvalues of $\underset{\sim}{J}$. These solutions are also steady states of (C) and (D) with $\underset{\sim}{x}_j = \underset{\sim}{x}_e$, $j = 1,\ldots,N$, $\underset{\sim}{x}(z,t) = \underset{\sim}{x}_e$, $0 \le z \le L$, and will be called uniform steady states. If κ_j, $j = 1, 2,\ldots,N$ are eigenvalues of the N x N matrix

$$\begin{bmatrix} 1 & -1 & \cdot & \cdot & \cdot & \cdot \\ -1 & 2 & -1 & & \cdot & \cdot \\ \cdot & -1 & 2 & -1 & \cdot & \cdot \\ \cdot & \cdot & -1 & \cdot & \cdot & \cdot \\ \cdot & \cdot & \cdot & \cdot & 2 & -1 \\ \cdot & \cdot & \cdot & \cdot & -1 & 1 \end{bmatrix}$$

then those of the linearization of the uniform state of (C) are given by

(3) $$|\lambda \underset{\sim}{I} + \kappa_j \underset{\sim}{D} - \underset{\sim}{J}| = 0, \; j = 1,2,\ldots,N$$

Since the κ_j are distinct positive numbers with $\kappa_1 = 0$, the first equation gives the eigenvalues of (K) and the remaining (N - 1)n eigenvalues depend on $\underset{\sim}{D}$. Thus the stability of (K) is a necessary, but not sufficient, condition for the stability of (C). Moreover it is differences of diffusivity that destabilize (C) when (K) is stable, for if $\underset{\sim}{D} = D\underset{\sim}{I}$ the nN roots of (3) are

$$\lambda_i - \kappa_j D, \; i = 1,\ldots,n, \; j = 1,\ldots,N,$$

where λ_i are the eigenvalues of $\underset{\sim}{J}$. Similarly the linearization about the uniform steady state of (D) is

(4) $$\dot{\underset{\sim}{\phi}} = \underset{\sim}{D}\underset{\sim}{\phi}_{zz} + \underset{\sim}{J}\underset{\sim}{\phi}$$

and a perturbation of the form $\underset{\sim}{c}e^{\lambda t}\cos(n\pi z/L)$ is stable if

(5) $$|\lambda \underset{\sim}{I} + (n\pi/L)^2 \underset{\sim}{D} - \underset{\sim}{J}| = 0$$

has roots with negative real parts for all n. It follows that the stability of the uniform steady state of (D) bears the same relation to (K) as does (C).

(C) and (D) also have non-uniform steady states unrelated to those of (K). They appear in symmetric pairs, for if x_{ej} satisfy

(6) $$0 = \underset{\sim}{D}(\underset{\sim}{x}_{e,j+1} - 2\underset{\sim}{x}_{e,j} + \underset{\sim}{x}_{e,j-1}) + \underset{\sim}{f}(\underset{\sim}{x}_{e,j})$$

then so do $\underset{\sim}{x}^*_{e,j} = \underset{\sim}{x}_{e,N+1-j}$, and if $\underset{\sim}{x}_e(z)$ satisfies

(7) $$0 = \underset{\sim}{D}(\underset{\sim}{x}_e)_{zz} + \underset{\sim}{f}(\underset{\sim}{x}_e)$$

so does $\underset{\sim}{x}^*_e(z) = \underset{\sim}{x}_e(L - z)$.

If (K) has periodic solutions $\underset{\sim}{x}(t) = \underset{\sim}{p}(t) = \underset{\sim}{p}(t + T)$, then (C) and (D) will also have uniform periodic solutions independent of j and z respectively. Similar remarks apply to the linearized stability which must of course be determined by the Floquet multipliers.

Rössler's Model. For the model system proposed by Rössler (10) we take $\underset{\sim}{x} = (u,v)$ and then (K) is

$$\dot{u} = u + 1 - uv/(\kappa + u) \tag{8}$$

$$\dot{v} = \nu(\sigma u - v) \tag{9}$$

and has three parameters κ, ν, and σ. On the first and last of these the steady state depends and is unique and "physical" (i.e. $u_e > 0$, $v_e > 0$) when $\sigma > 1$

$$v_e = \sigma u_e = \frac{\sigma}{2(\sigma - 1)} \{(1 + \kappa) + \sqrt{(1 + \kappa)^2 + 4\kappa(\sigma - 1)}\} \tag{10}$$

It can be most easily found as the intersection of the line $v = \sigma u$ with the hyperbola $v = u + (1 + \kappa) + \kappa u^{-1}$ as shown in Fig. 1.

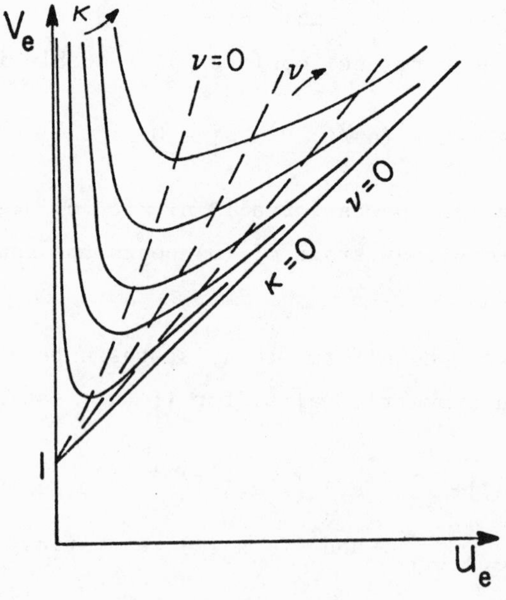

Figure 1 The u, v-plane for the system (K) showing the loci of steady-states for various κ. These curves must be intersected by a line of slope σ through the origin. The broken lines show the loci of critical stability for various values of ν.

Of the conditions for stability one (that the produce of the eigenvalues be positive) is automatically satisfied and the other is satisfied when

$$(11) \qquad \nu > g(\kappa,\sigma) = (u_e^2 - \kappa)/u_e(\kappa + u_e)$$

where u_e is given by (10). The dotted lines in Fig. 1 are contours of ν and in Fig. 2 the surface g is shown. If the parametric point lies below the surface shown in Fig. 2 the unique steady state is unstable. Moreover as we cross this surface the eigenvalues are

$$(12) \qquad \pm i\omega = \pm i \sqrt{(u_e^2 - \kappa)(2\kappa + u_e + \kappa u_e)/u_e(u_e + \kappa)}$$

Applying the criterion for stability of the bifurcating limit cycle shows that it is stable for $\nu < g(\kappa,\sigma)$ and in the neighbourhood of that surface. Thus as ν decreases the steady state loses its stability and sheds a stable limit cycle. Fig. 3 shows this for $\sigma = 2.5$, $\kappa = 0.144$ for which $u_e = 0.87$, $v_e = 2.18$ and $g = 0.7$; limit cycles for $\nu = 0.6$, 0.5 and 0.3 are shown. These become larger and more triangular as $\nu \to 0$.

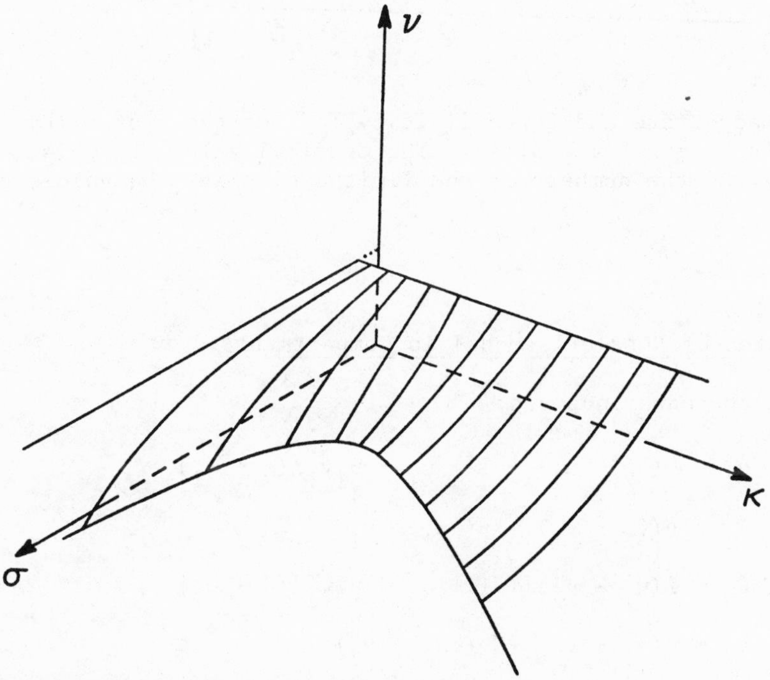

Figure 2 Surface in κ,ν,σ-space below which (K) is unstable.

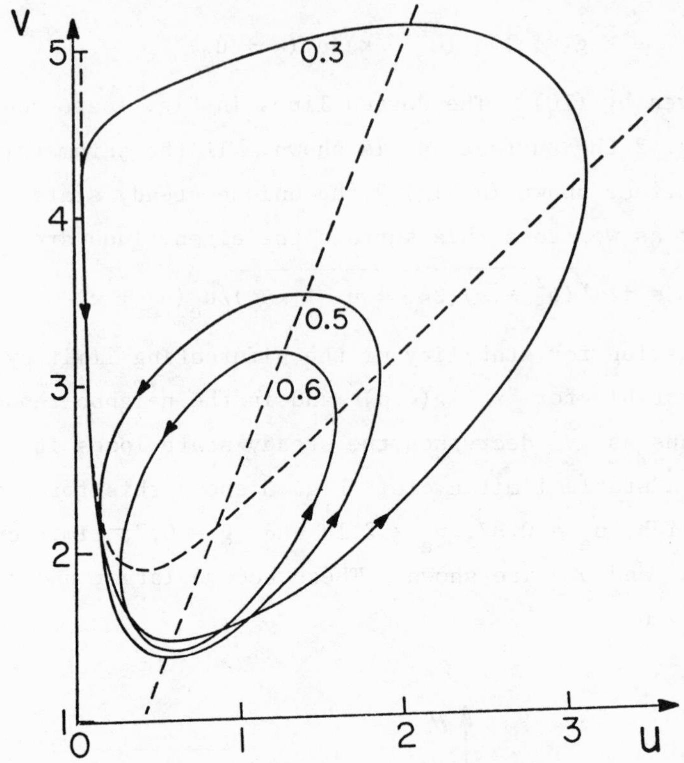

Figure 3 Steady-state and three limit cycles for the system (K) with $\sigma = 2.5$ $\kappa = 0.144$. The critical value of ν is 0.7 and the numbers on the limit cycles are the values of ν.

Steady States of Rössler's Model in Compartments.

For $N = 2$ the equations are:

(13) $$\dot{u}_1 = \tilde{\omega}(u_2 - u_1) + 1 + u_1 - u_1 v_1/(\kappa + u_1)$$

(14) $$\dot{v}_1 = \mu(v_2 - v_1) + \nu(\sigma u_1 - v_1)$$

(15) $$\dot{u}_2 = \tilde{\omega}(u_1 - u_2) + 1 + u_2 - u_2 v_2/(\kappa + u_2)$$

(16) $$\dot{v}_2 = \mu(v_1 - v_2) + \nu(\sigma u_2 - v_2)$$

For the moment we will follow Rössler in considering only the exchange of the second chemical species and so put $\tilde{\omega} = 0$. Since (14) and (16) are linear they can be solved for the steady-state v_{1e} and v_{2e} in terms of u_{1e} and u_{2e}

(17) $$v_{je} = \frac{\sigma(\mu + \nu)}{\nu + 2\mu} u_{je} + \frac{\sigma\mu}{\nu + 2\mu} u_{ie}, \quad i \neq j;\ i,j = 1,2$$

Substituting into (13) and (15) then gives

(18) $$u_{2e} = F(u_{1e}), u_{1e} = F(u_{2e})$$

where

$$F(u) = \frac{\alpha + 2}{\sigma} \{ \frac{\kappa}{u} + [1 - \sigma\frac{\alpha + 1}{\alpha + 2}]u + 1 + \kappa \},\ \alpha = \frac{\nu}{\mu} .$$

The equation $u_{je} = F(F(u_{je}))$ breaks down into two quadratics, the first of which is the one already encountered giving the uniform state $u_{je} = u_e$, $v_{je} = v_e$ where u_e and v_e are given by (10). The second is

(20) $$u_e^2 + \frac{(1 + \kappa)(\alpha + 2)}{(\alpha + 2) - \sigma(\alpha + 1)} u_e - \frac{\kappa(\alpha + 2)}{\sigma\alpha - (\alpha + 2)} = 0$$

When the two roots of this quadratic (u_+ and u_-, say) are both positive we have a symmetric pair of non-uniform solutions

(21) $$\begin{aligned} u_{1e} &= u_+,\ v_{1e} = f(u_+),\ u_{2e} = u_-,\ v_{2e} = f(u_-) \\ u_{1e} &= u_-,\ v_{1e} = f(u_-),\ u_{2e} = u_+,\ v_{2e} = f(u_+) \end{aligned}$$

where

(22) $$f(u) = (1 + u)(\kappa + u)/u$$

It is not hard to see that such solutions can only obtain if

(23) $$\frac{\alpha + 2}{\alpha + 1} < \sigma < \frac{\alpha + 2}{\alpha} .$$

The stability of the solutions is governed by the eigenvalue of the matrix

$$(24) \qquad \begin{bmatrix} g_1 & -h_1 & \cdot & \cdot \\ \nu\sigma & -(\nu+\mu) & \cdot & \mu \\ \cdot & \cdot & g_2 & -h_2 \\ \cdot & \mu & \nu\sigma & -(\nu+\mu) \end{bmatrix}$$

where

$$(25) \qquad g_j = 1 - \sigma v_{je}/(\kappa + u_{je})^2, \; h_j = u_{je}/(\kappa + u_{je}) .$$

For the uniform steady state, the characteristic equation has two quadratic factors:

$$(26) \qquad \{\lambda^2 + (\nu - g)\lambda + \nu(\sigma h - g)\}\{\lambda^2 + (\nu - g + 2\mu)\lambda + \nu(\sigma h - g) - 2\mu g\} = 0.$$

The first of these is the same as for (K) and $\sigma h - g > 0$, $\sigma > 1$. Since $\mu > 0$ we now have the conditions

$$(27) \qquad \nu > \mathrm{Max}\{g, 2\mu g/(\sigma h - g)\}$$

for stability.

Again we expect interesting behavior as ν decreases and the bifurcation picture will be governed by the order in which the coefficients in the two quadratic factors change sign. There are two possibilities which can best be illustrated by Fig. 4 which is drawn for the same fixed values of κ and σ as Fig. 3. The exchange parameter μ is also held constant so that decreasing ν is represented by a parametric point travelling horizontally to the left. The uniform steady state is stable to the right of A B C and if the parametric point cross AB the uniform state loses its stability to the uniform periodic state in which the two compartments behave exactly as in (K). The line OBC is $\nu = 2\mu g/(\sigma h - g)$ so that, if the parametric point cross BC, the last coefficient in (26) changes sign first and the system loses stability by one real eigenvalue becoming positive (rather

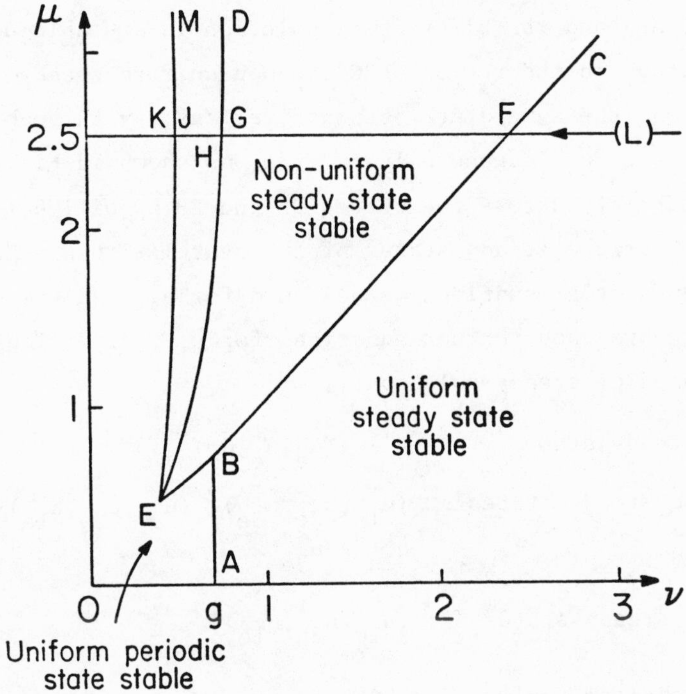

Figure 4 Bifurcation diagram in the μ,ν-plane for $N = 2$, $\kappa = 0.144$, $\sigma = 2.5$.

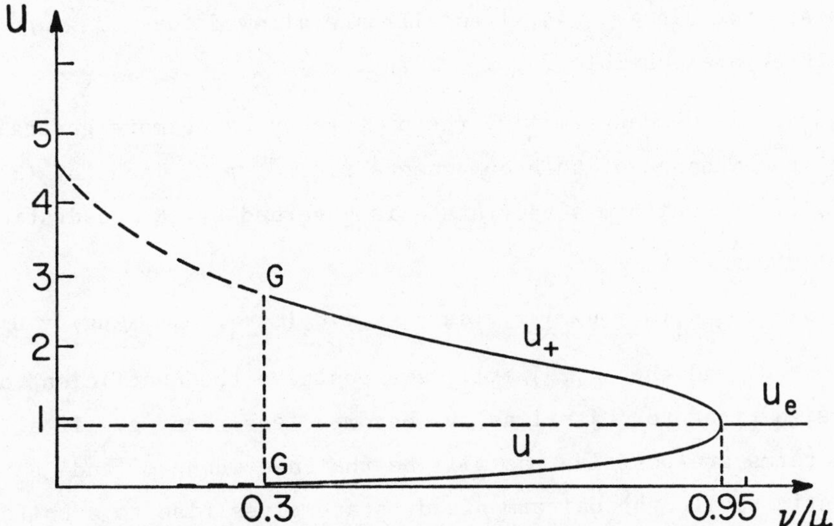

Figure 5 Uniform and non-uniform steady states as functions of ν/μ.

than a pair of complex conjugates passing across the imaginary axis, as in the crossing of AB) and the bifurcating solution is a stable non-uniform steady state. In the region DEBC the non-uniform steady states are stable and the same state obtains over any ray through the origin, since u_+ and u_- depend only on ν/μ, as shown in Fig. 5. They lose their stability across the curves DE and EB as will be discussed after a look at the steady states of the system with $N > 2$.

For $N = 3$, a quadratic equation can be found for u_{2e} (at the expense of considerable algebra) and further equations for u_{1e}, u_{3e}. They give rise to three classes of steady solution:

I. uniform steady state (u_e, u_e, u_e)

II. asymmetric steady states (u_{1e}, u_{2e}, u_{3e}), (u_{3e}, u_{2e}, u_{1e}) $u_{1e} \neq u_{3e}$

III. symmetric steady states (u_{1e}, u_{2e}, u_{1e}),

with similar dispositions for v.

We shall not give the formulae but show in Fig. 6 the solutions that make all the u_{je} positive. For this set of parameters only Class I can obtain for $\nu/\mu > 1.44$; I and III are allowed for $.57 < \nu/\mu < 1.45$ and all three are possible for $\nu < .57\mu$.

For large N we can simplify the picture by being more general and allowing the exchange of both components i.e. $\tilde{\omega} \neq 0$. Then, by (3), the stability of the uniform steady state is governed by N quadratics, $j = 1,\ldots,N$,

$$\lambda^2 + \{\nu - g + \kappa_j(\tilde{\omega} + \mu)\}\lambda + \nu(\sigma h - g) + \kappa_j(\tilde{\omega}\mu\kappa_j + \tilde{\omega}\nu - \mu g) = 0$$

Since $\kappa_1 = 0$ and the κ_j, $j > 1$, are positive the coefficient of λ changes sign first for $j = 1$ as ν becomes less than g. If all the constant terms are positive, as will be the case when $\tilde{\omega}$ and μ are sufficiently small, the uniform steady state gives rise to a uniform periodic one. Across the line

$$\mu = \nu \operatorname*{Min^+}_{j>1}\left\{\frac{\sigma h - g + \tilde{\omega}\kappa_j^2}{\kappa_j(g - \tilde{\omega}\kappa_j)}\right\} = \nu \operatorname*{Min^+}_{j>1} M_j \tag{28}$$

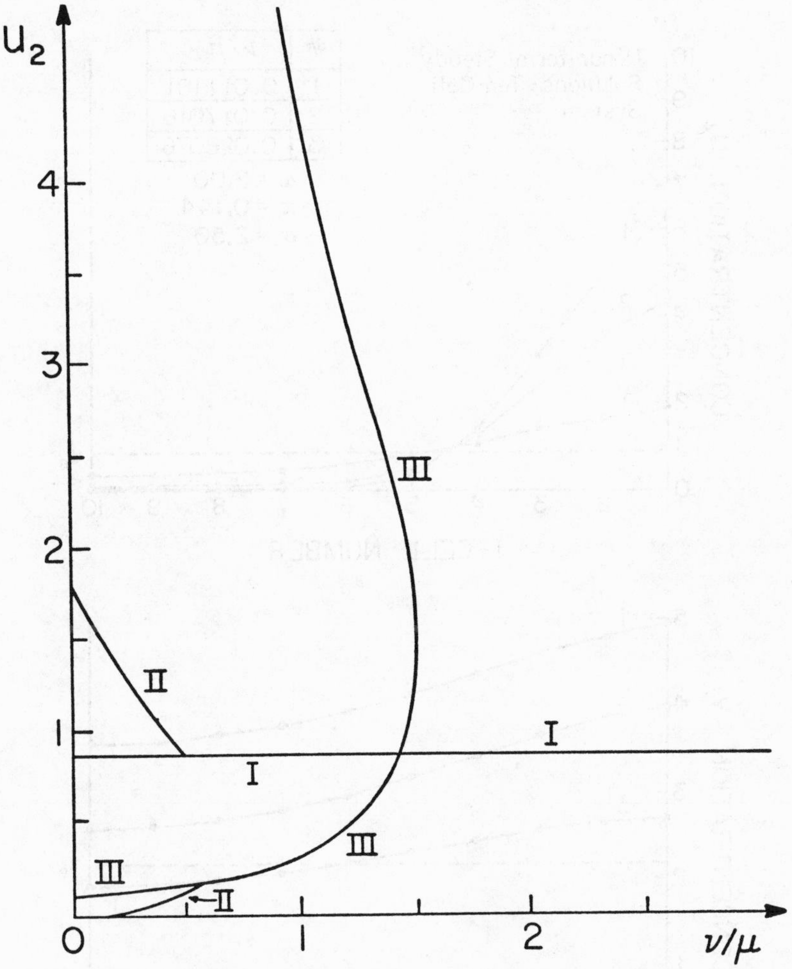

FIgure 6 Uniform and non-uniform steady states for $N = 3$. I denotes the uniform steady state, II denotes the axisymmetric steady state, III the symmetric one.

there is a bifurcation to a non-uniform steady state. (Min^{+} is the least positive value.) Since we are primarily interested in the dynamic behavior we will not try to explore this complex picture, but choose $\tilde{\omega}$ so as to simplify it. We note that $j = 2$ may not be the minimizing value of j in (28). For example with $N = 10$ and $\tilde{\omega} = 2$ only one of the expressions in (28) is positive and we have a diagram like that in Fig. 4. The non-uniform steady states in the DEBC region are the same along rays of constant ν/μ. Three examples are given in Fig. 7.

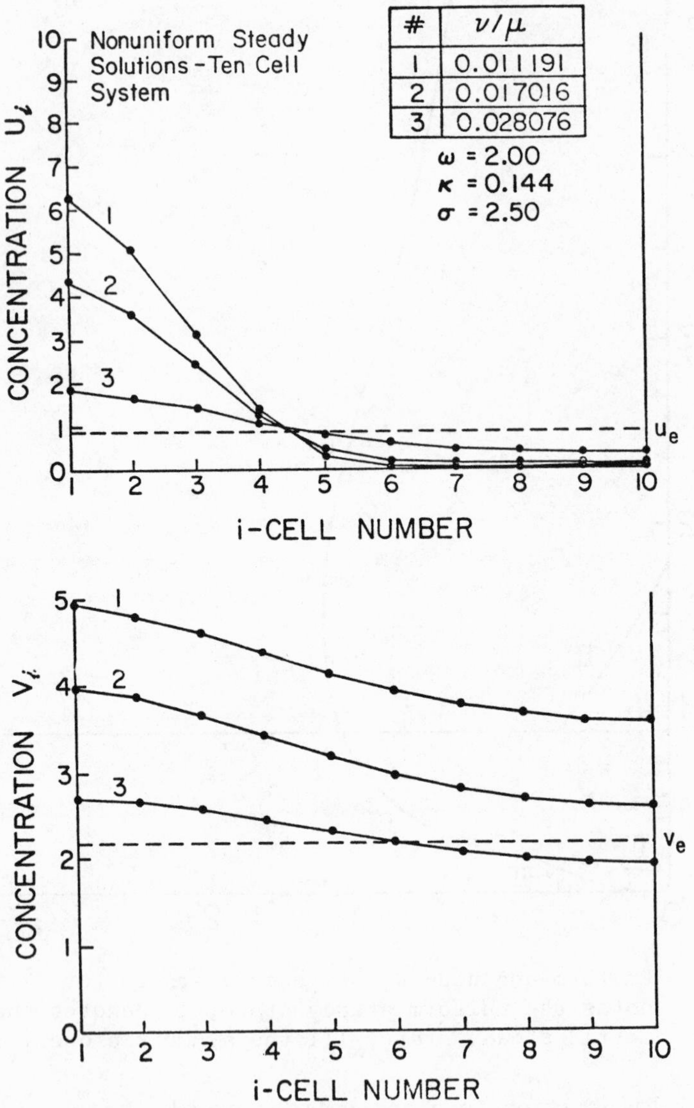

Figure 7 Non-uniform steady states for $N = 10$ and three values of the ratio ν/μ.

Dynamics of Rössler in Compartments. Rössler gave several examples of the dynamics with $N = 2$ and we shall follow his lead by systematically varying ν for $N = 2$ before increasing N. In particular we keep $\kappa = 0.144$, $\sigma = \mu = 2.5$ and $\tilde{\omega} = 0$, so that we move to the left on the line (L) of Fig. 4. To the right of F there is only the uniform state $u_1 = u_2 = 0.87$, $v_1 = v_2 = 2.18$. Between F and G the non-uniform steady states can be followed in Fig. 5, but at G this becomes unstable by a Hopf bifurcation, ν being 0.755. The kind of oscillation that arises is shown in Fig. 8 for which $\nu = .75$ (point H in Fig. 4). At $\nu = .696$ (point J of Fig. 4), this oscillation loses its stability as a Floquet multiplier passes out of the

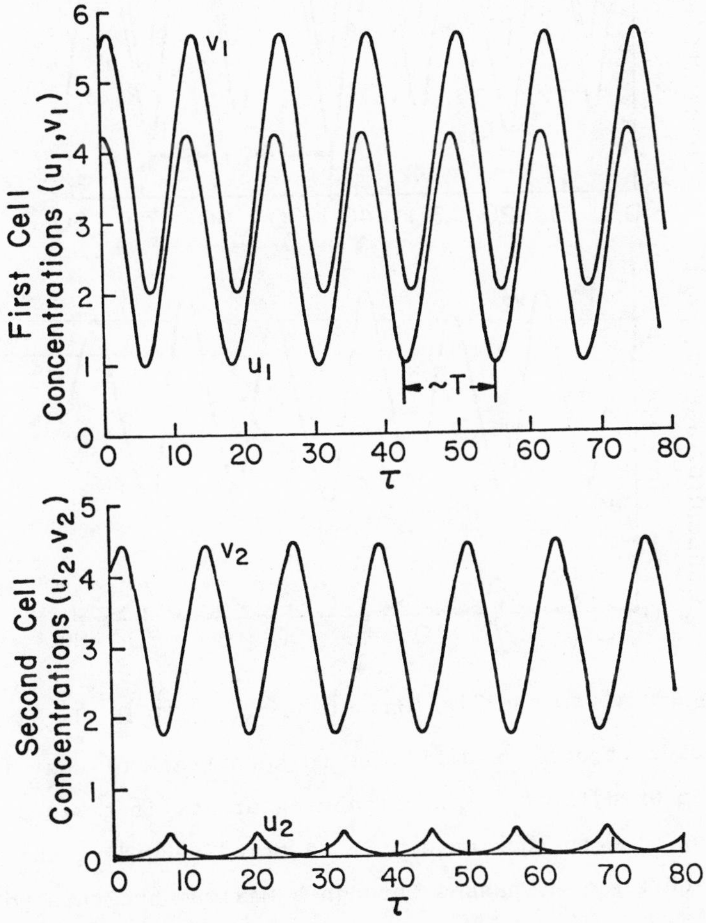

Figure 8 Non-uniform oscillations of period ~T in the two cells.

unit circle through -1, and a stable periodic solution of twice the period arises; this solution, for $\nu = 0.690$, is shown in Fig. 9. By tracking the critical Floquet multiplier as in Fig. 10, we find a bifurcation which again doubles the period to about four times the original one. The stability of this quickly breaks down and the period again doubles.

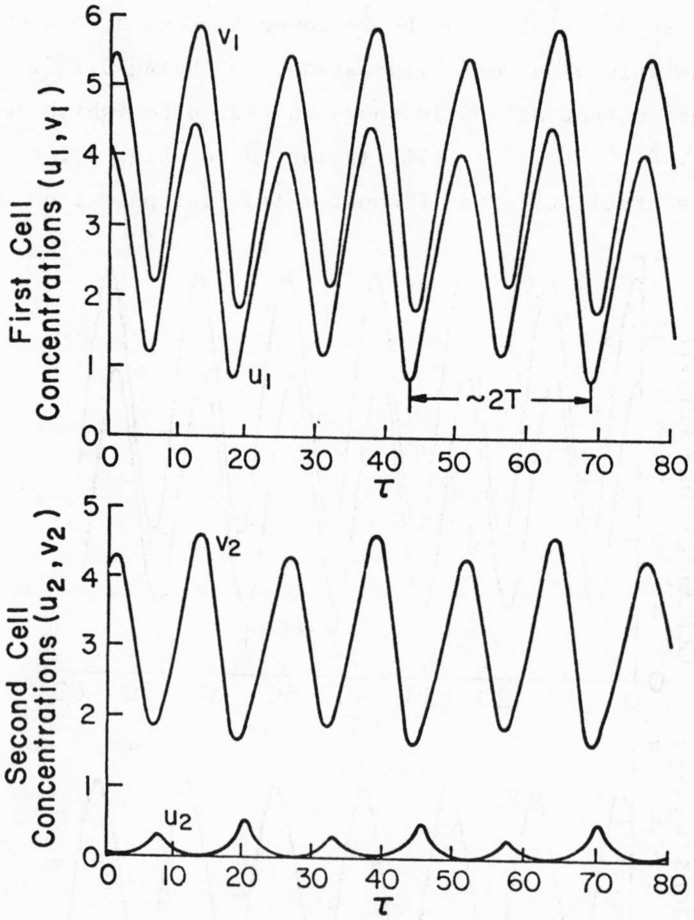

Figure 9 Non-uniform oscillations of period ~2T in the two cells.

At this point it becomes difficult to keep track of what is happening for, even when there is an attracting orbit, it takes an unconscionably long time to lock onto it. We tried recording the values of everything each time u_1 passed through a maximum and this poor-man's-Poincaré-map gives clear support for the picture just drawn. There was less clear evidence of period 3 oscillations but much indication

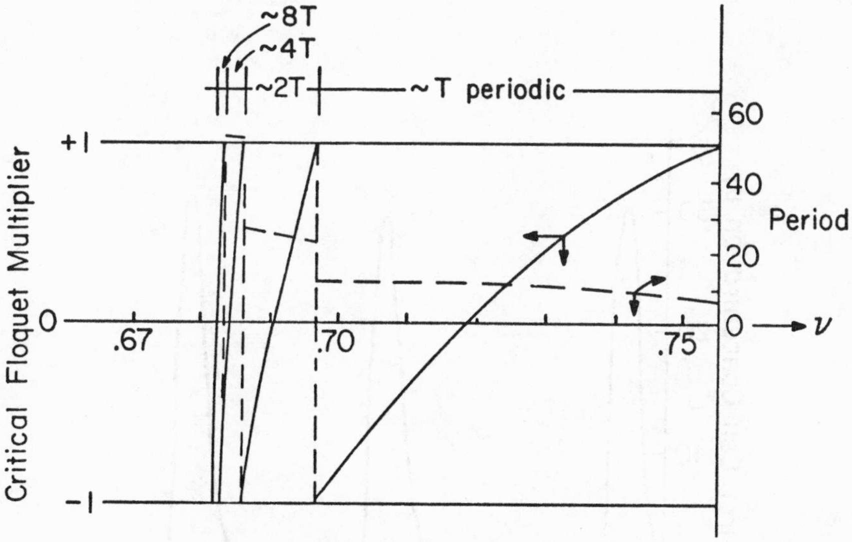

Figure 10 Critical Floquet multiplier and period as functions of ν.

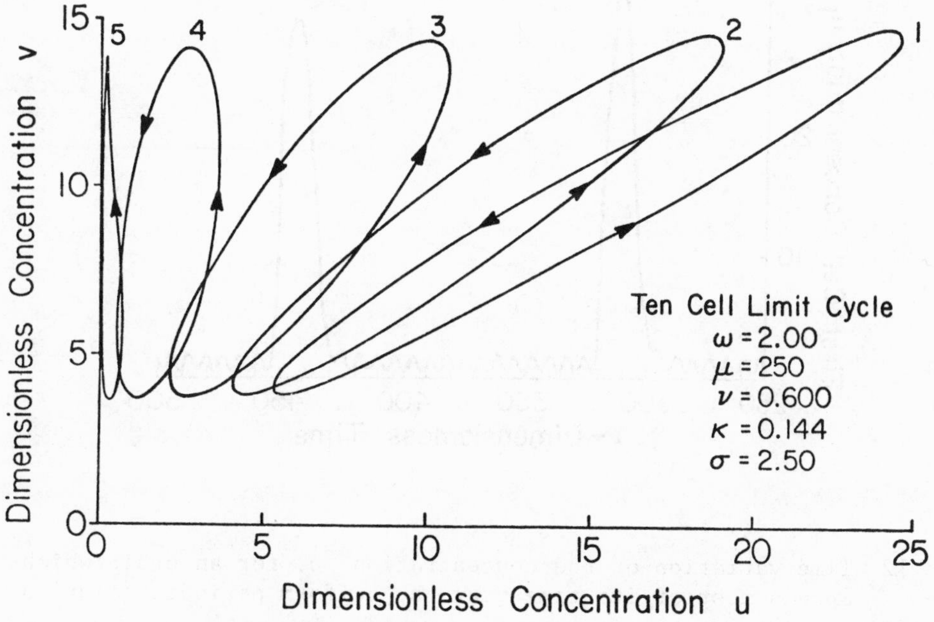

Figure 11 Non-uniform periodic solutions in five of the ten cells shown as phase-plane diagrams.

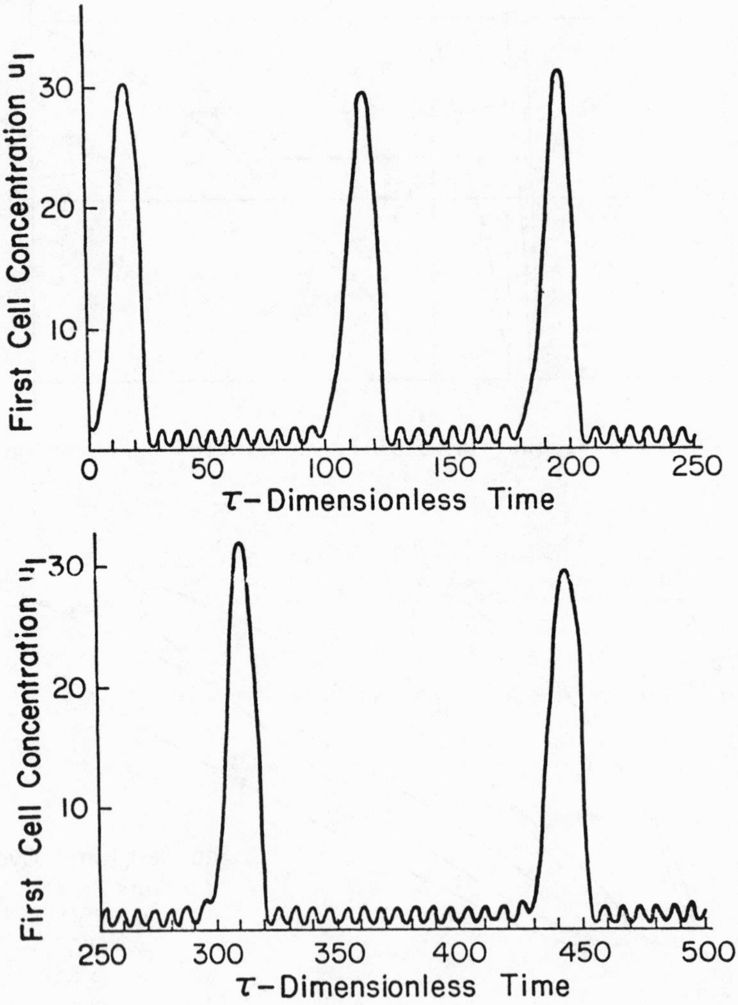

Figure 12 Time variation of the concentration u_1 for an orbit which spends most of its time near the uniform periodic solution.

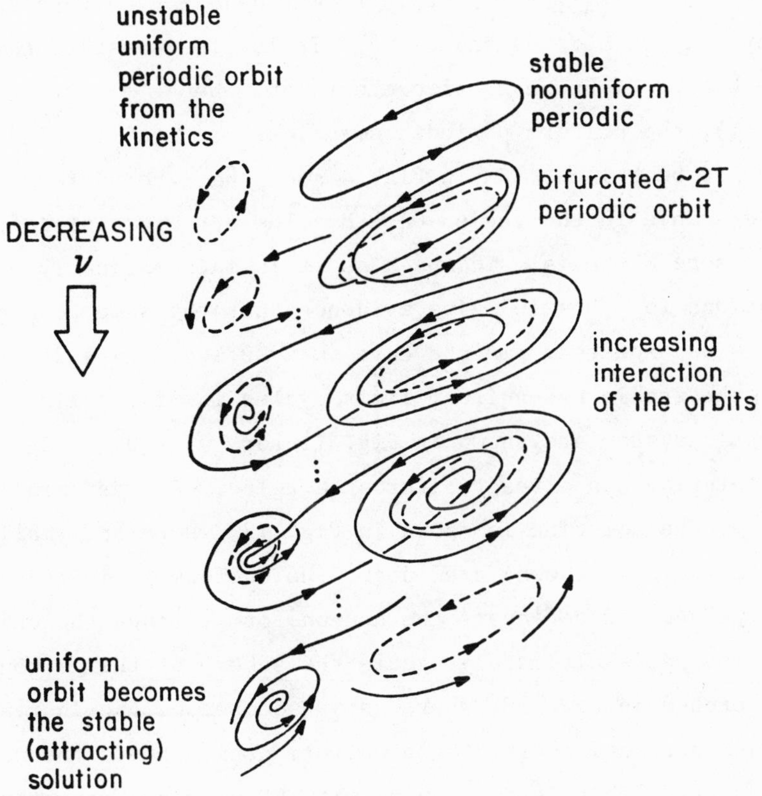

Figure 13 Schematic of how the attracting orbits evolve.

of irregular behavior, though whether this is chaos or a solution of long period cannot be determined empirically. It may be that as ν decreases we have a sequence of bifurcations such as is known (12) to obtain for $x_{n+1} = \lambda x_n(1 - x_n)$ or there may be a Hopf bifurcation of the Poincaré map to a torus and so on. It is also possible that the successive bifurcations cease altogether for, when $\nu < 0.515$ (point K in Fig. 4), the uniform periodic solutions are stable. There would thus appear to be a locus EKM in Fig. 4 such that the uniform periodic solution is stable in the region OABEKM. The structure of left hand part of the zone MKEGD is as complex as it is mathematically erogenous.

Explorations for $N = 10$ give evidence that the same general behavior is to be expected. In the case $\kappa = 0.144$, $\mu = 250$, $\tilde{\omega} = 2$, $\sigma = 2.5$ the stable non-uniform limit cycles in five cells of the same symmetric system are shown in Fig. 11 for $\nu = 0.6$. By $\nu = 0.57$ the solution has already bifurcated twice. For yet smaller values of ν the solution is shown in Fig. 12, where the smaller ripples (only u_1 is shown) are 'near' the uniform periodic solution and the large departures 'near' the non-uniform. Since the uniform periodic solution is ultimately stable the scheme of things would seem to be as sketched in Fig. 13, where as ν decreases the increasing stability and attractiveness of the uniform orbits tears the non-uniform orbit apart. This is in keeping with Rössler's observations (10). No explorations were made in the neighborhood of the point B in Fig. 4. This omission will be repaired by ongoing studies, but, as was pointed out at the conference, it is likely that the picture is much more complicated - and interesting - than is shown here.

Rössler Distributed. Here we will take $\xi = z/L$ and work with

(29) $$u_\tau = \tilde{\omega} u_{\xi\xi} + 1 + u - uv/(\kappa + u)$$

(30) $$v_\tau = \mu v_{\xi\xi} + \nu(\sigma u - v)$$

with

(31) $$u_\xi = v_\xi = 0 \quad \text{at} \quad \xi = 0,1$$

The uniform solution $u_e(\xi) = u_e$, $v_e(\xi) = v_e$ is still given by (10), while by (5), its stability is governed by

$$(32)\quad \lambda^2 + (\nu - g + (\tilde{\omega} + \mu)n^2\pi^2)\lambda + [\nu(\sigma h - g) + n^2\pi^2(n^2\pi^2\tilde{\omega}\mu + \omega\nu - \mu g)] = 0$$

If $\tilde{\omega}$ and μ are non-zero, these eigenvalues tend to $-n^2\pi^2\tilde{\omega}$ and $-n^2\pi^2\mu$ (i.e. to $-\infty$) as $n \to \infty$, but if $\tilde{\omega} = 0$ one tends to g and the other to $+\infty$. We shall avoid the case of $\tilde{\omega} = 0$ for this reason.

We again ask how the uniform solution bifurcates and again have a diagram as in Fig. 4 for which the line OBC is

$$(32)\qquad \mu = \left[\operatorname*{Min^+}_{n>0} \frac{\sigma h - g + n^2\pi^2\tilde{\omega}}{(g - \tilde{\omega}n^2\pi^2)n^2\pi^2}\right]\nu = \left[\operatorname*{Min^+}_{n>0} M(n)\right]\nu$$

Since $g < 1$, a necessary condition for instability by violation of (32) is $\tilde{\omega}\pi^2 < 1$. If μ is small enough, reduction of ν will take the solutions of the quadratic for $n = 0$ into the unstable half plane as a pair of complex conjugates. If μ is above the point B, a disturbance of wave number (n_0 say) that minimizes the right hand side of (31) begins to grow. This will be illustrated by some numerical examples, but we mention in passing that, using the usual type of analysis, a criterion to distinguish between sub- and super- criticality was obtained. It is too complicated a formula to permit a general conclusion, but it is suspected that as ν decreases the uniform steady-state gives up its stability to the non-uniform and that, whilst the uniform state is stable, stable non-uniform states do not exist when n_0 is odd but do exist when n_0 is even.

Three cases were calculated, with $\sigma = 2.5$, $\kappa = 0.144$ and $\tilde{\omega} =$ 0.01, 0.02 and 0.05. When $\tilde{\omega} = 0.01$, $0 < M(2) < M(1)$, $M(n) < 0$, $n>2$, so that the minimizing n_0 is 2 and we should expect a symmetric solution to arise first and perhaps give its stability to an unsymmetric one as ν decreases further since $M(1)$ is also positive. Fig. 14 is the result of transient calculations which converged to the stable non-uniform steady states. Though these are far from the transition it is the larger ν that gives the symmetric state $(n = 2)$. There is an analogy with the case of three compartments where, as ν decreases, the symmetric solutions (class III) are superseded by the

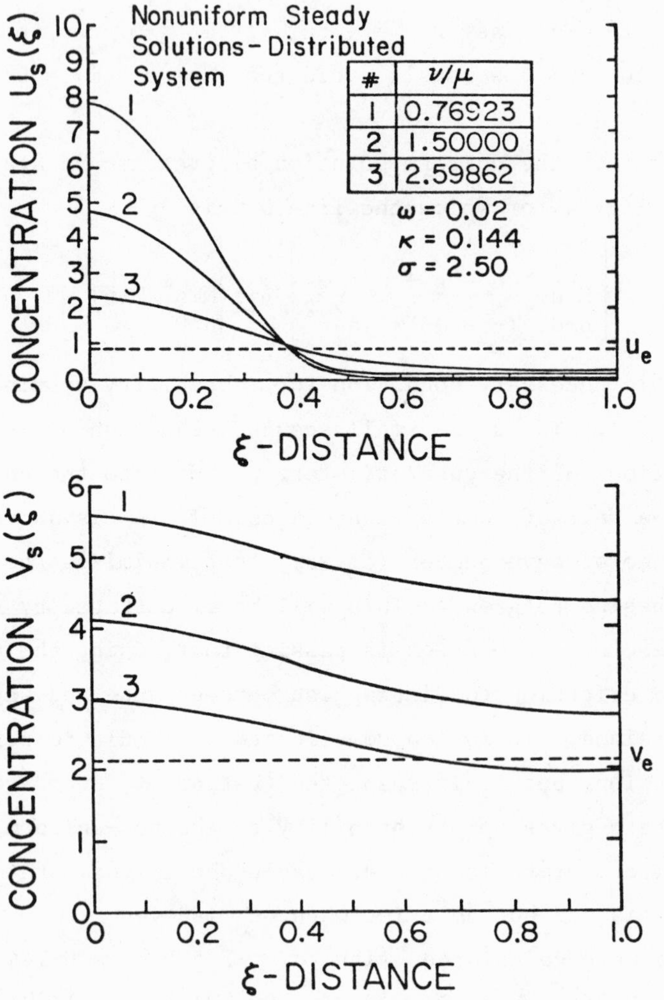

Figure 14 Symmetric and asymmetric steady-states for the distributed system with $\tilde{\omega}$ equal to 0.01.

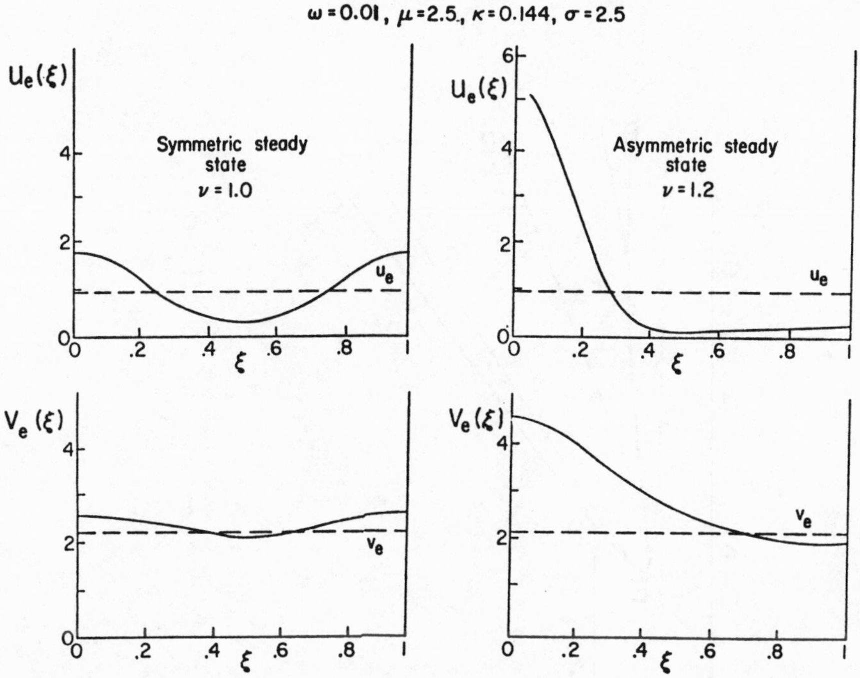

Figure 15 Unsymmetric steady-states of the distributed system for $\tilde{\omega} = 0.02$.

unsymmetric (class III).

For $\tilde{\omega} = 0.02$, $n_0 = 1$ and $M(n) < 0$ for $n > 1$. The non-uniform steady states depend only on the ratio μ/ν and three are shown in Fig. 15 that correspond to the rays in Fig. 16. The eigenvalues of the linearization about the non-uniform steady-state were calculated and it was found that stability was lost, by the real part of a pair of complex conjugate eigenvalues becoming positive, on the line BED. Since it was confirmed that the uniform periodic solution was stable for small enough ν, it seems likely that there is a region of complex behavior, KED, just as in the case $N = 2$. Calculations with $\tilde{\omega} = 0.05$ concur and the agreement between calculations for $N = 10$ and (D) give some confidence that this case of (C) gives a good indication of the dynamics of (D). Certainly the worst can be expected of (D) for which Pismen (14) has shown how chaos may arise.

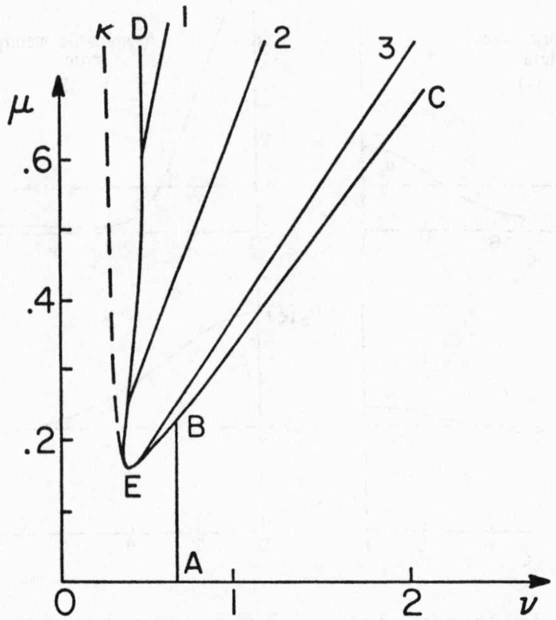

Figure 16 The presumed bifurcation diagram for the distributed system. The rays labeled 1, 2, and 3 are the loci of the solutions shown in Figure 15. The uniform solution is stable to the right of ABC, the uniform periodic solution to the left of KABA. The non-uniform steady solutions are stable within the arms of DEBC and a region of exotic behavior exists in KED.

REFERENCES

[1] O. GUREL and O. E. RÖSSLER, (Eds.), Bifurcation Theory and Applications in Scientific Disciplines, Ann. N. Y. Acad. Sci. 316 (1979), pp. 1-708. (The whole volume contains many papers of importance).

[2] P. C. FIFE, Mathematical Aspects of Reacting and Diffusing Systems, Springer-Verlag, New York, 1979.

[3] R. ARIS, The Mathematical Theory of Diffusion and Reaction in Permeable Catalysts, Clarendon Press, Oxford, 1975.

[4] J. J. TYSON, The Belousov-Zhabotinskii Reaction, Springer-Verlag, New York, 1979.

[5] Papers in Part VIII, pp. 623-684 of 1 above.

[6] Kinetics of Physico-chemical Oscillations, Proc. confce. Aachen. Sept. 19/22, 1979.

[7] E. LORENZ, Deterministic nonperiodic flows, J. Atmos. Sci., 20 (1963), pp. 130-141.

[8] T-Y. LI and J. A. YORKE, Period three implies chaos, Amer. Math. Monthly, 82 (1975) pp. 985-992.

[9] O. E. RÖSSLER, Chaotic behavior in simple reaction systems, Z. Naturforsch, 31a (1976), pp. 259-264.

[10] O. E. RÖSSLER, Chaos in abstract kinetics: two prototypes, Bull. Math. Biol., 39 (1977)pp. 275-289.

[11] M. MAREK, Personal Communication.

[12] Z. KÁDAS and H. G. OTHMER, Stable limit cycles in a two-component bimolecular reaction, J. Chem. Phys., 70 (1979), pp. 1845-1850.

[13] R. M. MAY, Simple mathematical models with very complicated dynamics, Nature 261 (1976), pp. 459-467.

[14] L. M. PISMEN, A source of chaos in distributed kinetic systems, Chem. Engrg. Sci., 34 (1979), pp. 129-131.

Bifurcation Phenomena in Stirred Tanks and Catalytic Reactors

W. H. Ray* and K. F. Jensen*

ABSTRACT. A review of the present state of understanding of continuous stirred tank reactors is presented along with certain new results on bifurcation phenomena for catalytic surfaces. A few examples are discussed which show repeated Hopf bifurcation, complex oscillations, and chaotic behaviour arising from a simple catalyst surface model.

1. Introduction. The existence of bifurcation phenomena in common types of chemical reactors has been known for more than 150 years (e.g.; [1 - 3]) and yet new experimental and mathematical results are being discovered and reported even today. A rather good history of the these developments may be found in recent review articles (e.g. [4 - 6]). Probably the simplest type of chemical reactor which exhibits interesting bifurcation behaviour is the continuous stirred tank reactor (CSTR). Equally interesting are

---

*Mathematics Research Center and Department of Chemical Engineering, University of Wisconsin-Madison. This work was supported in part by theUnited States Army under Contract Number DAAG29-75-C-0024, National Science Foundation and the donors of the Petroleum Research Fund administered by the American Chemical Society. The authors are indebted to Tunde Ogunnaike for contributing his artistic talents to Figure 1.

catalytic surfaces and catalytic reactors which are only slightly more complex systems but which provide a wide variety of bifurcation phenomena (e.g.; multiple steady states, both simple and complex non-linear oscillations, chaotic behaviour, standing and travelling waves, etc.) [4 - 7]. In this paper we plan to briefly discuss new results which have been reported for the CSTR and then move on to show some of the intriguing bifurcation phenomena which arise from a new model for catalytic surfaces.

2. The Continuous Stirred Tank Reactor. The continuous stirred tank reactor (CSTR) is a stirred pot (Figure 1) into which chemicals of a certain recipe flow continuously and are stirred and heated or cooled while undergoing chemical reaction. These reactors often occur in connected multiples which can have ever increasing static and dynamic complexity. The bifurcation behaviour of the CSTR has been studied for more than 25 years [6, 8-15] and yet qualitatively new results seem to continually appear. To begin the discussion let us consider the simplext case: a CSTR in which a single irreversible first order exothermic reactor is being carried out. The modelling equations take the form

(1) $$\frac{dx_1}{dt} = -x_1 + Da(1-x_1)\exp\{\frac{x_2}{1+x_2/\gamma}\}$$

(2) $$Le\,\frac{dx_2}{dt} = -x_2 + BDa(1-x_1)\exp\{\frac{x_2}{1+x_2/\gamma}\} - \beta(x_2-x_{2c})$$

Figure 1. The Three Sisters

where x_1 denotes the conversion of reactant and x_2 a dimensionless temperature. The system parameters are B, a dimensionless heat of reaction, γ a dimensionless activation energy, x_{2c} a dimensionless coolant temperature, Le a ratio of characteristic times for transport, β a dimensionless heat transfer coefficient, and Da a dimensionless ratio of reactor residence time to reaction time. If desired, these latter two parameters can be further parameterized in terms of the reactor residence time, τ,

$$Da = Da_0 \tau \tag{3}$$

$$\beta = \beta_0 \tau \tag{4}$$

where Da_0, β_0 are reference parameters.

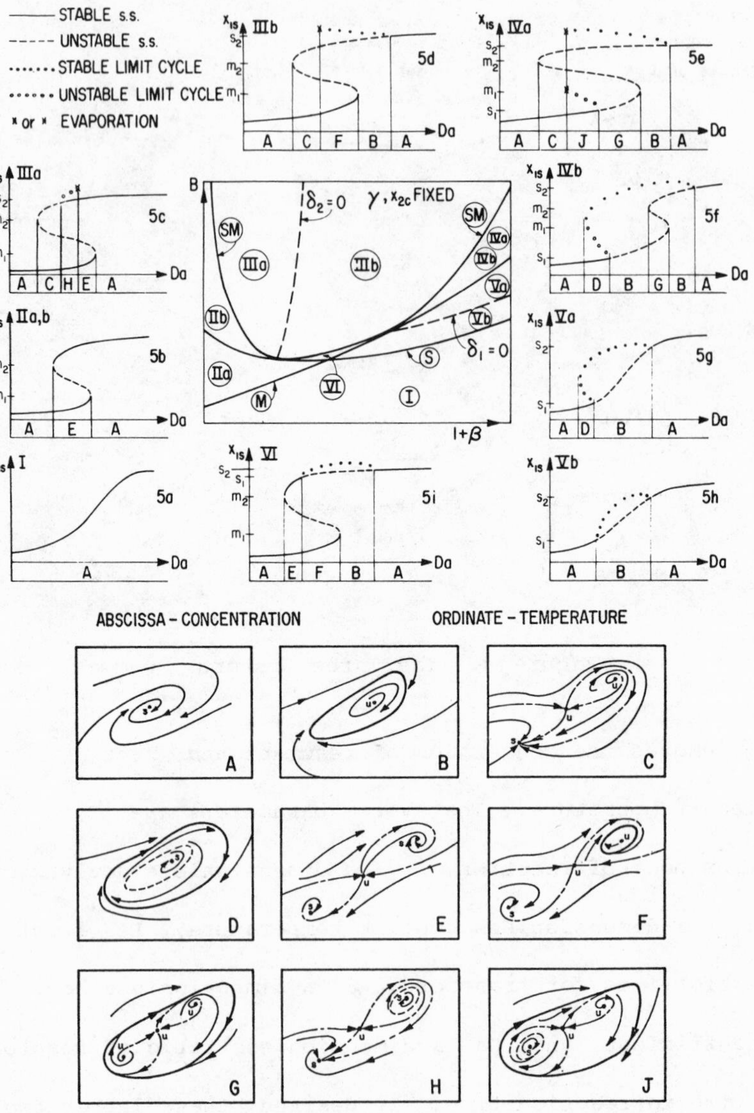

Figure 2. Bifurcation with respect to Da and resulting phase portraits.

The structure of the static and Hopf bifurcation behaviour for Equations (1,2) has been studied (e.g. cf [6, 12, 13]) and may be summarized in Figure 2. The formation of single limit cycles from saddle loops was conjectured in [13] as a means of explaining the disappearance of limit cycles of large amplitude and the numerical computations in [13] support this conjecture. Recently, some questions have been raised [16,17] about the possible bifurcation of pairs of limit cycles from degenerate saddle loops. Speculations that this could lead to a vast array of new phase portraits have

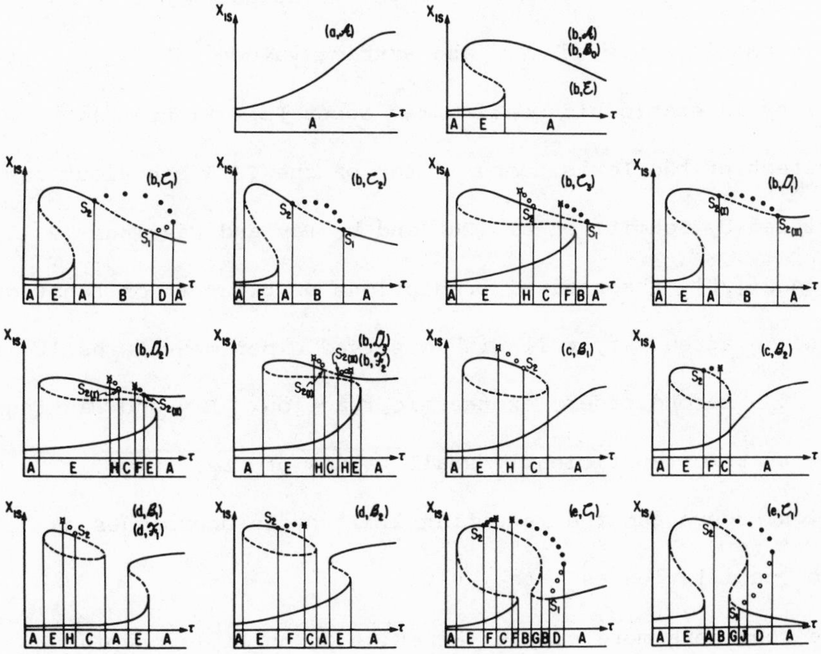

Figure 3. Bifurcation with respect to τ for $\gamma \to \infty$.

been made [17]; however, no computations have been reported which show these new phase portraits actually exist. Some attempts at such calculations have been unsuccessful [16].

When the reparameterization given by Equation (3,4) is made there is very complex static and Hopf bifurcation behaviour. For the case where $\gamma \to \infty$, the structure has been calculated (cf. [14]) and may be seen in Figure 3. Recent results by Golubitsky and Keyfitz [18] as well as by Huang and Varma [19] seem to indicate that for relatively small values of γ and extreme values of x_{2c}, even more variety in static bifurcation may occur (cf. Figure 4).

The effect of the Lewis number, Le on the CSTR behaviour has been analyzed by Schmitz et al [20] and by Ray and Hastings [21]. In reference [20], theoretical predictions of bifurcation behaviour were found to agree very well with observed experimental results for the case of a second order exothermic reaction. Ray and Hastings [21] show that for sufficiently small values of Le, Hopf bifurcation occurs and the resulting limit cycle approaches a relaxation oscillation as $Le \to 0$.

If one considers more complex kinetics or multiple coupled stirred tank reactors, the behaviour becomes even more interesting. For example, Halbe and Poore [22] have carried out the bifurcation analysis for the case of two consecutive reactions $A \to B \to C$ in a CSTR and have shown secondary static branching and a rich variety of Hopf bifurcation behaviour. Recently, Aris and his

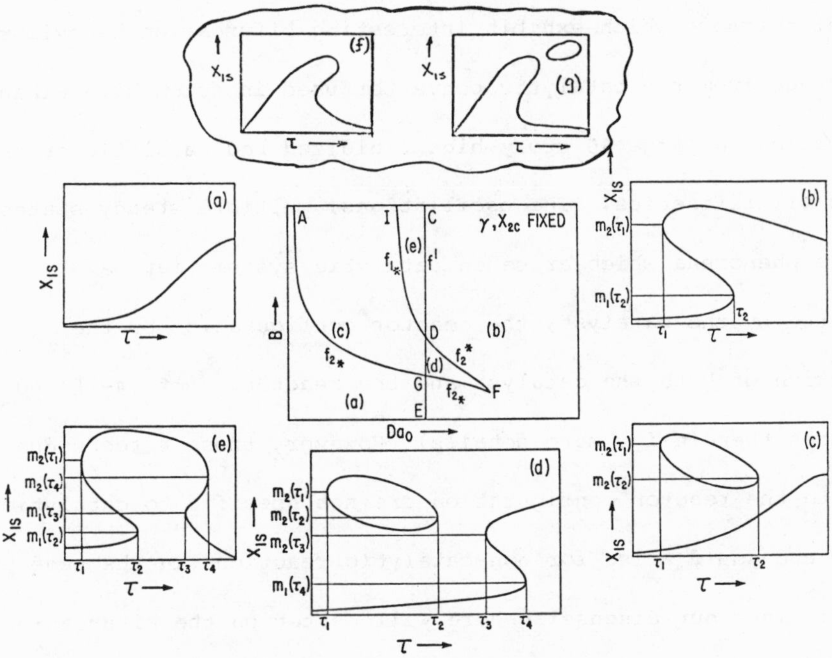

Figure 4. Static bifurcation with respect to τ for both finite and infinite γ .

students [26] have also studied the bifurcation behaviour of this problem. Marek [23-25] and Aris [26] have considered the behaviour of systems of coupled stirred tanks, and their results illustrate quite clearly that for multiple CSTRs the story of CSTR bifurcation behaviour is far from complete.

3. Catalytic Reactors. There are many types of catalytic chemical reactors which exhibit interesting bifurcation behaviour. These range from the catalytic converter used in automobile exhaust purefication to large 10 story high fluidized bed catalytic crackers used in oil refineries. The oscillations, multiple steady states, and wave phenomena which arise in catalytic systems may be instigated by the catalyst, the reactor configuration or the interaction of both the catalyst and the reactor. (cf [4-7] and references therein for more details) However, those effects due solely to the reactor configuration are not specific to catalytic systems and would arise for non-catalytic reactions in the same reactor. Thus our discussion here will center on the bifurcation behaviour of catalytic surfaces.

Although there have been many chemical explanations put forth for the wide variety of dynamic behaviour exhibited by catalytic surfaces (e.g.; [4-7]), these phenomena are so pervasive that a physical explanation would seem plausible. Recently, the authors developed such a physical model for unsupported [27,28] and supported [29] catalytic surfaces.

For unsupported catalysts such as catalytic wires, the model arose from the following experimental observations:

(i) Completely smooth wires do not readily ignite, but must be "activated" by heat treatment which roughens the surface.

(ii) Smooth wires do not oscillate and analysis of simple models confirms that such oscillations should not be possible.

(iii) Roughened wires oscillate for a wide range of catalysts and reactant gases, and these oscillations tend to be very complex.

(iv) Optical observations of the roughened oscillating surface show very high oscillating temperatures locally on the roughened surface.

When one considers all of this evidence together, a new view of the dynamics of a catalytic metal surface emerges. In particular, it appears that the roughened surface of a catalytic wire or gauze plays a key role in the dynamic behaviour of the reaction. Electron micrographs of platinum and platinum alloy catalytic wires such as shown in Figure 5 [31] clearly show the presence of rough, porous protrusions on the surface of the catalytic wire. To mathematically model this "fuzzy" wire, we shall choose the simple picture shown in Figure 6. The protrusions on the wire surface are modelled as cylinders of radius R_p and length L_p. These protrusions have a size distribution, $F(R_p)\ dR_p$ which defines the number of protrusions per unit area of bulk wire surface in each size range R_p to $R_p + dR_p$. Each size protrusion can oscillate with a

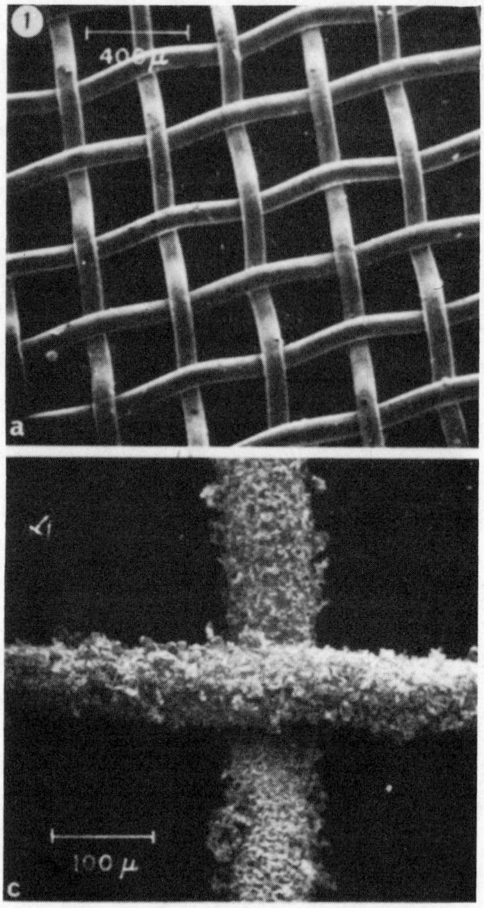

Figure 5. Scanning electron micrographs of platinum rhodium gauzes used for ammonia oxidation, above: smooth wires before use, below: rough wires after use. (Photograph kindly provided by Professor L. D. Schmidt.)

frequency dependent on R_p and the distribution of protrusions can give rise to complex oscillations on the wire. Thermal communication between each protrusion and the bulk wire occurs at the end of the protrusion over a circular cross section of area

πR_p^2. The fraction of the bulk wire surface covered by these protrusions is thus given by

(5) $$\varepsilon = \int_0^\infty F(R_p)\pi R_p^2 \, dR_p \qquad 0 < \varepsilon < 1$$

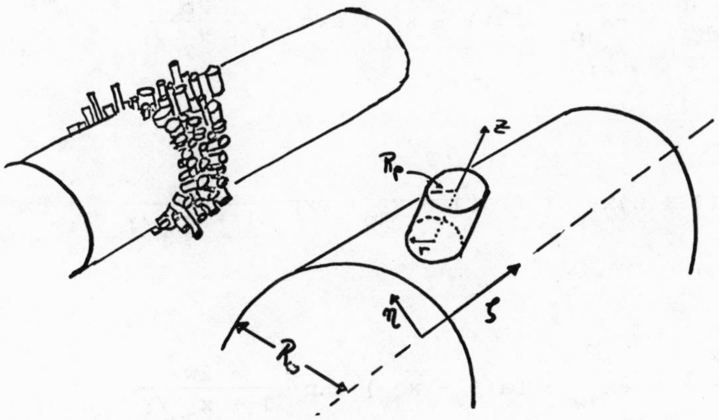

Figure 6. The fuzzy wire model.

and is assumed uniform over the entire wire surface. One may formulate heat and material balances for this composite system in a very general way [27,28]. However, a particularly simple model may be developed by assuming a first order irreversible reaction on the wire (in excess oxygen) with rate of mass transfer controlled by adsorption to the surface. If one assumes both the wire and protrusions to be uniform in temperature and concentration and the heat transfer between protrusions and wire to be modelled by Newton's law of cooling, then the modelling equations take the dimensionless form

$$\frac{dx_{1p}}{dt} = -x_{1p} + Da(1 - x_{1p}) \exp \frac{x_{2p}}{1 + x_{2p}/\gamma} \tag{6}$$

$$Le \frac{dx_{2p}}{dt} = -(1 + \beta)x_{2p} + BDa(1 - x_{1p}) \exp \frac{x_{2p}}{1 + x_{2p}/\gamma} + \beta x_{2w} \tag{7}$$

$$\frac{dx_{1w}}{dt} = -x_{1w} + Da(1 - x_{1w}) \exp \frac{x_{2w}}{1 + x_{2w}/\gamma} \tag{8}$$

$$\begin{aligned} Le_w \frac{dx_{2w}}{dt} = {} & -(1 - \varepsilon + \beta_w \varepsilon)x_{2w} + \beta_w \varepsilon \langle x_2 \rangle \\ & + (1 - \varepsilon)B_w Da(1 - x_{1w}) \exp \frac{x_{2w}}{1 + x_{2w}/\gamma} \end{aligned} \tag{9}$$

where x_{1p}, x_{2p} are the composition and temperature of a protrusion of size R_p and x_{1w}, x_{2w} are the corresponding values on the bulk wire. Here

(10) $$\langle x_2 \rangle = \frac{\int_0^\infty a_c F(R_p) x_{2p} \, dR_p}{\int_0^\infty a_c F(R_p) \, dR_p}$$

is the dimensionless area average protrusion temperature. For purposes of simulation it is useful to assume a discrete set of protrusion sizes, R_{p_i}, so that

(11) $$F(R_p) = \sum_{i=1}^{N} n_{p_i} \, \delta(R_p - R_{p_i})$$

where n_{p_i} is the number of protrusions/unit wire area of size R_{p_i}. Thus Equations (6 - 10) represent a set of $2N + 2$ model equations.

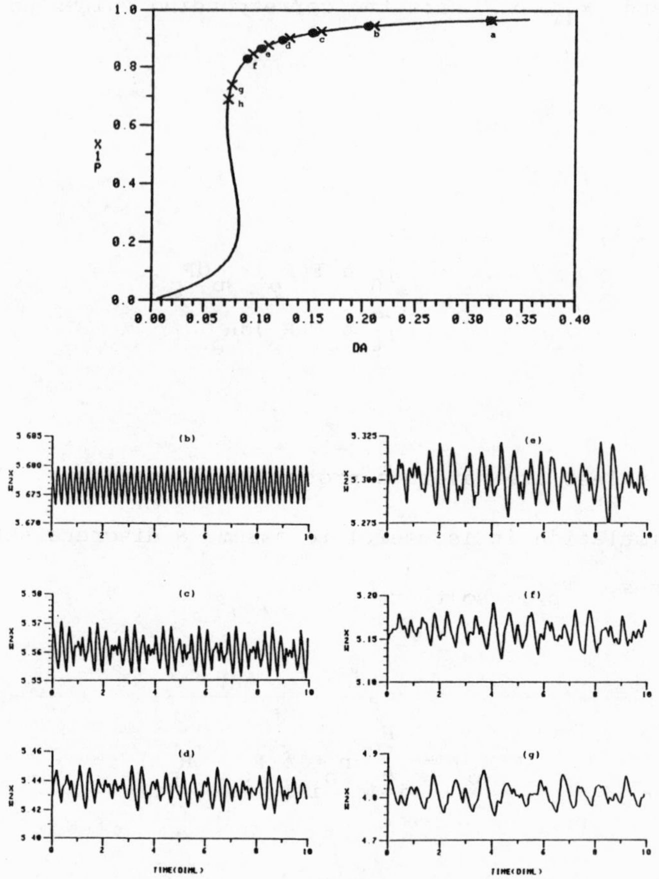

Figure 7. Hopf bifurcation points for the fuzzy wire model simulations, with varying catalyst activitiy, (.) bifurcation point, (x) simulation. The wire totally covered by six protrusion sizes evenly distributed, B = 6.0, Le = 0.20, 0.17, 0.14, 0.11, 0.08, 0.05, β = 1.0, γ =20.0, τ =2.0.

This model gives an amazing variety of ignition, extinction and oscillatory behaviour [28], but space only allows the presentation of a few results here. For example, in the case where there are 6 different protrusion sizes on the wire, there is secondary, tertiary, and higher order Hopf bifurcation as each size protrusion begins to oscillate with decreasing values of Da. The bifurcation points for the individual protrusions and the resulting complex dynamic behaviour of the entire wire is shown in Figure 7.

An even more interesting case involving only two protrusion sizes may be seen in Figure 8. Here as Da decreases the smaller protrusion bifurcates first and begins to oscillate as shown in Figure 8(d). Then as Da is decreased only slightly, the larger protrusion is "excited" and participates in a large complex oscillation in tandem with the smaller protrusion. As Da is decreased still further, the larger protrusion bifurcates and oscillations become less complex and more nearly regular. The fact that this rather complex oscillation arises for only two protrusion sizes on an inert bulk wire suggests that even simple interacting reaction systems can show a wide variety of dynamic behaviour.

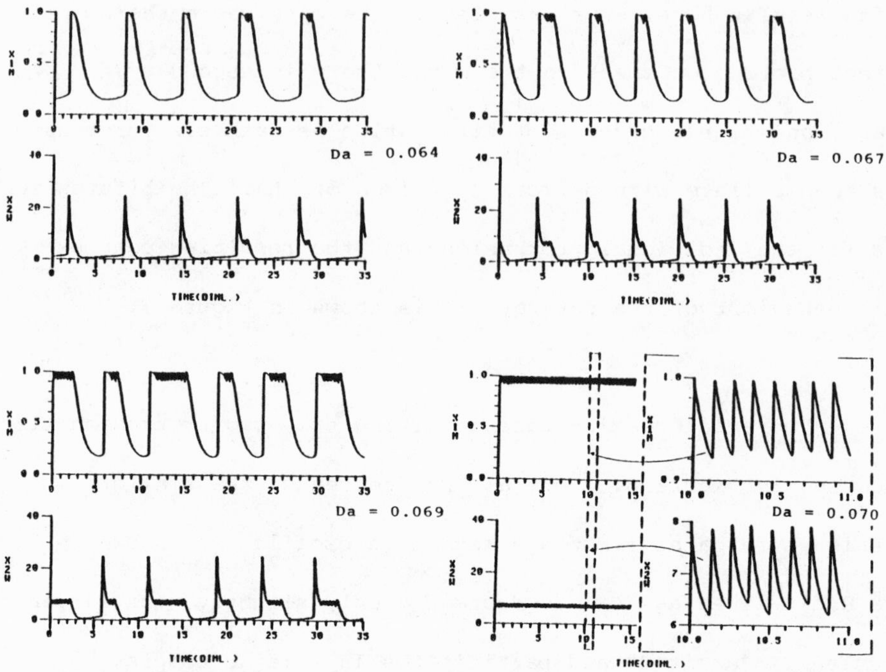

Figure 8. Bulk wire oscillations for varying catalytic activity. The wire is totally covered by equal areas of two protrusion sizes, B = 7.2, Le = 0.09, 0.005, β = 2.0, γ_0 = 30.0, τ = 0.053
(a) Da = 0.064, (b) Da = 0.067, (c) Da = 0.069, (d) Da = 0.070.

Two forms of oscillating behaviour are possible when the bulk wire is only partially covered by protrusions. In the first case the bulk wire serves to attenuate the protrusion generated oscillations [27], while in the second and more interesting case the wire also sustains oscillations. A particularly interesting case arises when both the protrusion and wire are about the same size, in this case the Hopf-bifurcation points for wire and protrusion equations can coincide. Figure 9 illustrates the time evolution of the state variables, x_{1p} and x_{2w} for a wire half covered by protrusions of a uniform size as Da is decreased away from such a multiple Hopf-bifurcation point. Initially the oscillations are complex, but as Da decreases further they become single peaked due to synchronization of wire and protrusion dynamics.

To illustrate the prediction of apparent chaotic behaviour from our model and a comparison with experiment, we provide in Figure 10 a simulation with four protrusion sizes and the parameters for the catalytic oxidation of butane over Pt. Note that the chaotic temperature dynamics predicted by the model are in rather good agreement with the experimental observations of Edwards et al [30].

4. Concluding Remarks. From the very few examples presented here it is clear that commonly encountered chemical reactors provide an abundance of mathematically interesting bifurcation phenomena. In the case where the number of chemical reactor equations rises above two, then complex oscillations, chaos, and other exotic behaviour

are predicted even from simple models. The fact that these dynamic systems are also very important from the practical viewpoint, means that they make very good candidates for further study both by engineers and mathematicians.

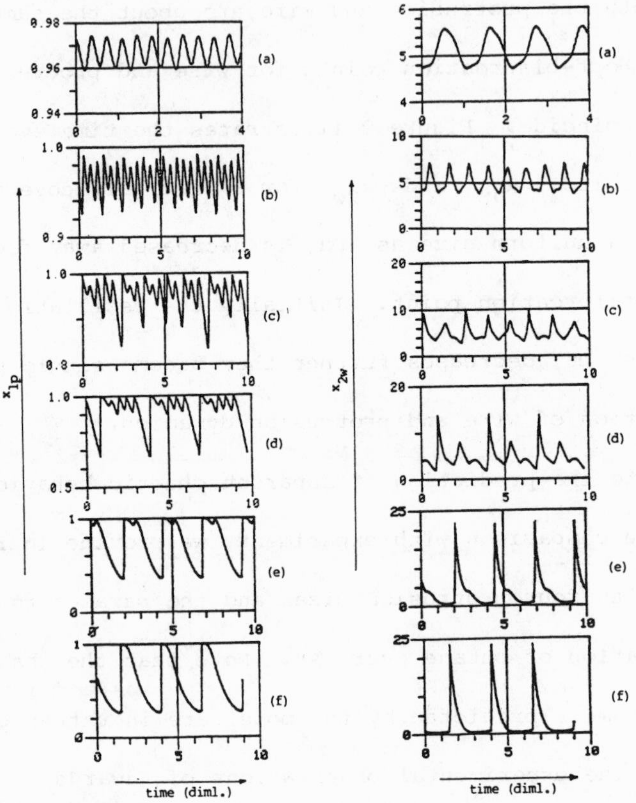

Figure 9. Oscillations in x_{1p} and x_{2w} for Da decreasing away from a multiple Hopf-bifurcation point Da = 0.1089, B = 8.32, $B = 4.99$, Le = 0.1, $Le_w = 0.1$, $\beta = 0.1$, $\beta_w = 0.3$, $\gamma = 20$, $\epsilon = 0.5$ (a) Da = 0.1083, (b) Da = 0.1029, (c) Da = 0.09808, (d) Da = 0.09696, (e) Da = 0.09179, (f) Da = 0.07637.

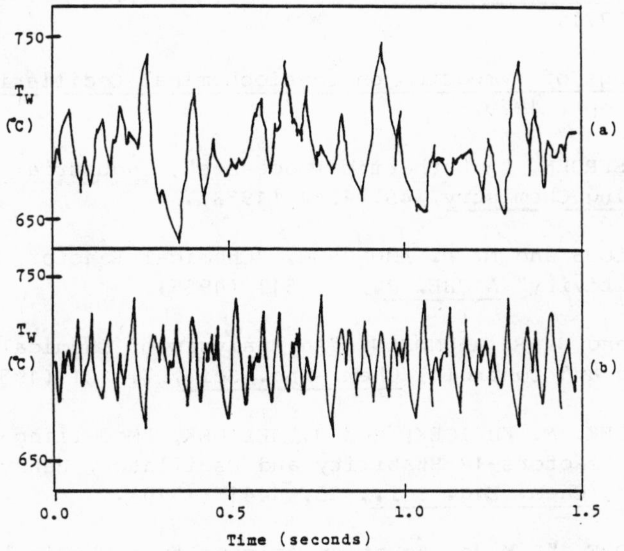

Figure 10. Comparison of experimentally observed and theoretically predicted oscillations in butane oxidation. (a) Experiments (Ref. [30]), (b) Fuzzy wire model prediction.

REFERENCES

[1] A. T. FECHNER, Schweigg J. Chem. Phys. 53, 141 (1828).

[2] F. G. LILJENROTH, "Starting and Stability Phenomena of Ammonia Oxidation and Similar Reactions" Chem. and Met. Eng. 19, 287 (1918).

[3] W. C. BRAY, "A Periodic Reaction in Homogeneous Solution and its Relation to Catalysis" J. Am. Chem. Soc. 43, 1262 (1921).

[4] M. G. SLINKO and M. M. SLINKO, "Self-Oscillations of Heterogeneous Catalytic Reaction Rates" Catal. Rev. Sci. Eng. 17, 119 (1978).

[5] M. SHEINTUCH and R. A. SCHMITZ, "Oscillations in Catalytic Reactions" Catal. Rev. Sci. Eng. 15, 107 (1977).

[6] W. H. RAY, "Bifurcation Phenomena in Chemically Reacting Systems" Applications of Bifurcation Theory, 285, Academic Press (1977).

[7] Proceedings of Symposium on Physiochemical Oscillations, Aachen, Sept. 1979.

[8] C. VAN HEERDEN, "Autothermic Processes", Industrial and Engineering Chemistry, 45, 1242 (1953).

[9] O. J. BILOUS and N. R. AMUNDSON, "Chemical Reactor Stability and Sensitivity" AICHE. J., 1, 513 (1955).

[10] R. ARIS and N. R. AMUNDSON, "An Analysis of Chemical Reactor Stability and Control", Chem. Eng. Sci., 7, 121 (1958).

[11] V. HLAVACEK, M. KUBICEK, and J. JELINEK, "Modelling of Chemical Reactors-18 Stability and Oscillatory Behaviour of the CSTR", Chem. Eng. Sci., 25, 1441 (1970).

[12] A. B. POORE, "A Model Equation Arising from Chemical Reactor Theory", Archive for Rational Mechanics and Analysis, 52, 358 (1973).

[13] A. UPPAL, W. H. RAY and A. B. POORE, "On the Dynamic Behaviour of Continuous Stirred Tank Reactors", Chem. Eng. Sci., 29, 967 (1974).

[14] A. UPPAL, W. H. RAY and A. B. POORE,"The Classification of the Dynamic Behaviour of Continuous Stirred Tank Reactors - Influence of Reactor Residence Time", Chem. Eng. Sci., 31, 205 (1976).

[15] A. B. POORE, "On the Theory and Application of the Hopf-Friedrichs Bifurcation Theory", Archive for Rat. Mech. and Anal., 60, 371 (1976).

[16] J. W. REYN and A. H. BLOKLAND, "A Note on the Dynamic Behaviour of CSTRs", Delft Progress Report 3, 198 (1978).

[17] D. A. VAGANOV, N. G. SAMOILENKO and V. G. ABRAMOV, "Periodic Regimes of Continuous Stirred Tank Reactors", Chem. Eng. Sci. 33, 1133 (1978), also published in Doklody Acad. Nauk SSSR (in Russian) 234, 640 (1977).

[18] M. GOLUBITSKY and B. L. KEYFITZ, "A Qualitative Study of the Steady-state Solutions for a CSTR", SIAM J. Math. Analysis (in press).

[19] D. T-J. HUANG and A. VARMA, "On the Reference Time in the Multiplicity Analysis for CSTR's", Chem. Eng. Sci. (in press).

[20] R. A. SCHMITZ, R. R. BAUTZ, W. H. RAY and A. UPPAL, "The Dynamic Behaviour of a CSTR: Some Comparisons of Theory and Experiment" AIChE J. 25, 289 (1979).

[21] W. H. RAY and S. P. HASTINGS, "The Influence of the Lewis Number on the Dynamics of Chemically Reacting Systems", Chem. Eng. Sci. (in press).

[22] D. C. HALBE and A. B. POORE, Private Communication (1975).

[23] M. MAREK and I. STUCHL, "Synchronization in Two Interacting Oscillatory States", Biophys. Chem. 3, 241 (1975).

[24] M. MAREK and J. JUDA, "Synchronization and Coupled Chemical Oscillators - Comparison of Experiments with Theory", Biophys. Chem. 4, (1976).

[25] M. MAREK, Private Communication (1979).

[26] R. ARIS, Private Communication (1979).

[27] K. F. JENSEN and W. H. RAY, "A New Approach to Modelling the Dynamics of Catalytic Surfaces", in ref. [7].

[28] K. F. JENSEN and W. H. RAY, "A Microscopic Model for Catalytic Surfaces - I. Catalytic Wires and Gauzes". Preprints of 1979 Annual AIChE meeting, San Francisco, Nov. 1979. (Also submitted for publication).

[29] K. F. JENSEN and W. H. RAY, "A New View of Ignition, Extinction, and Oscillations on Supported Catalyst Surfaces". Proceedings 6th Int. Symposium on Chemical Reaction Engineering, Nice, March 1980.

[30] W. M. EDWARDS, F. L. WORLEY and D. LUSS, "Temperature Fluctuation of Catalytic Wires and Gauzes", Chem. Eng. Sci. 28, 1479 (1973).

(31) L. D. SCHMIDT and D. LUSS, "Physical and Chemical Characterization of Platinum-Rhodium Gauze Catalyst", J. Catalysis 22, 269 (1971).

A Singularity Theory Approach to Qualitative Behavior of Complex Chemical Systems

Martin Golubitsky,* Barbara L. Keyfitz,** and David Schaeffer†

Abstract. The application of singularity theory to the study of steady-state bifurcation problems should help substantially in the analysis of certain types of chemical systems. Two specific examples are described here. The first is the classic model by a single reaction in a continuous flow stirred tank chemical reactor. For this problem all of the qualitatively different local bifurcation diagrams (seven in number) for the steady-state temperature as a function of the flow rate are found. The second problem is the analysis of a four reaction thermal-chainbranching model for the combustion of hydrogen in a closed container. The equilibrium equations are considered as a bifurcation problem by studying the steady-state temperature as the initial pressure is varied. For a particular range of the parameters in this model, 17 stable bifurcation diagrams are found. Those diagrams show, in particular, that the thermal-chainbranching model contains the complexity necessary to model the observed phenomena of an explosion peninsula.

Both the results were obtained using a local analysis suggested by singularity theory. In particular, a distinguished point, or organizing center, was found in the parameter space of each problem. It has the property that all the qualitatively different bifurcation diagrams of the problem appear in a neighborhood of the organizing center. This provides an immense simplification of these two problems, and should apply to other equilibrium problems involving many parameters, such as are typical of chemical systems.

* Department of Mathematics, Arizona State University, Tempe, AZ 85281.
Research supported in part by NSF Grant MCS-79-05799
** Department of Mathematics, Arizona State University, Tempe, AZ 85281.
Research supported in part by NSF Grant MCS-77-04164.
† Department of Mathematics, Duke University, Durham, NC 27706.
Research supported in part by NSF Grant MCS-79-02010 and US Army Contract DAA-29-78-G0127.

1. INTRODUCTION. In this paper, we outline an attempt to use qualitative mathematical methods to obtain information about chemical reactions with complex kinetics. For some problems involving chemical reactions the techniques of singularity theory afford the possibility of a compromise between modelling by a single overall reaction which does not have the correct qualitative behavior and taking into account the full details of elementary reactions. The latter process can be carried out only with difficulty even on a computer and may well result in answers that are too detailed to present a clear picture of the overall features of the reaction. So far, we have applied our analysis only to a sampling of problems, but the results suggest that further applications are possible. Details of the results presented here can be found in [2] and [3] ; background material is provided in [4] .

In this talk I will present a description of our results for two problems : one, the well-known model of a continuously stirred tank chemical reactor (CSTR) with a single one-step exothermic reaction, has been studied extensively, but singularity theory is able to organize the qualitative description of the steady-state solutions efficiently and even enabled us to discover some new results. The second problem is a four-equation model for a combined thermal and chainbranching reaction, specialized somewhat to describe the combustion of hydrogen. It has been suggested that this model contains enough complexity to explain the explosion peninsula in the $H_2 - O_2$ reaction ; we show that singularity theory is a good tool to explore the model, and that, in a certain sense, qualitative features that characterize the explosion peninsula are present.

These two problems have in common that they involve only analysis of the steady states of the system, not the full dynamics. In both cases, the steady-state problem reduces to a single algebraic equation, although this is not an essential feature of the method [4] . Both problemsinvolve many parameters, and this is why the analysis of even the steady states may be intractible. In both problems there is a distinguished parameter which makes it natural to express the problem as a bifurcation problem in that parameter. And in both problems what is required is to enumerate the qualitatively different bifurcation diagrams. The theory of imperfect bifurcation via singularity theory developed in [4] is expressly designed to give the precise qualitative information required, using relatively simple calculations more or less equivalent to computing low-order Taylor series expansions.

In the next two sections we outline the problems and our results.

2. THE CSTR PROBLEM. In this standard model problem, a reactant R is converted to product P via a first-order, exothermic volume-preserving reaction in a homogeneous mixture with inflow, outflow and cooling [5]. The equations for the evolution of product and for the temperature of the mixture are respectively

(1a) $$\frac{dx}{dt} = -\varepsilon x + D(1-x)A(y) = f_1(x,y)$$

(1b) $$\frac{dy}{dt} = -\varepsilon y + BD(1-x)A(y) - (y-\eta) = f_2(x,y)$$

where ε measures the flow-rate, D is the ratio of the chemical time based on inflow temperature to the cooling rate (Damköhler number), B is the exothermicity, η the coolant temperature, and A(y) is the temperature-dependent reaction rate normalized so that A = 1 at inflow temperature 0 ; usually A(y) is an Arrhenius term $e^{\frac{\gamma y}{1+y}}$, where γ is the activation energy.

The question we address concerns the multiplicity of steady-state solutions as the parameter ε is varied. Solving $f_1 = f_2 = 0$ for x leads to a single equation

(2) $G(y,\varepsilon;B,D,\eta) = \eta - (1+\varepsilon)y + B\varepsilon/(1+\varepsilon/DA(y)) = 0$ for the steady-state temperature y in terms of ε . The information required on multiplicity is simply the number of solutions y for each ε and whether they represent high-or-low-temperature steady states. Hence we are willing to replace G = 0 by another problem $H(x,\lambda) = 0$ where $\lambda = \lambda(\varepsilon), x = x(y,\varepsilon)$ with $\lambda' > 0$, $\frac{dx}{dy} > 0$, since any such change of variables preserves the qualitative features of the bifurcation dia-

grams of (2). In the language of singularity theory this is a contact equivalence, and if G has finite codimension (the only case we can consider), H can be taken to be a low-degree polynomial. The role of the parameter $\beta = (B,D,\eta)$ is as follows. For each β the changes of variables may depend on β - i.e. $G(y,\varepsilon,\beta)$ is contact equivalent to $H(x,\lambda,\alpha)$ if $\alpha = \alpha(\beta)$, $\lambda = \lambda(\varepsilon,\beta)$, $x = x(y,\varepsilon,\beta)$ are nonsingular coordinate changes as before. As α varies near some fixed value, say $\alpha = 0$, the functions $H(x,\lambda,\alpha)$ describe small perturbations of $H(x,\lambda,0)$. The set of such perturbations is called an unfolding of $H(x,\lambda,0)$. If, as α varies, all qualitively different perturbations of $H(x,\lambda,0)$ (under the notion of contact equivalence) are described, $H(x,\lambda,\alpha)$ is called a universal unfolding. Of course, at most parameter values the bifurcation diagram is stable under small perturbations. But there will always be distinguished values of β at which the bifurcation diagram changes type and there may be a value β_0 in the neighborhood of which all the qualitatively different diagrams will occur. Such a point we call an organizing center for the problem. When it exists we can take $\alpha(\beta_o)=0$ and enumerate the bifurcation diagrams simply by analyzing the universal unfolding of $H(x,\lambda,0)$ which, in the case of finite codimension, is again a polynomial. It must be emphasized that all these coordinate changes are local and all results therefore apply only in a neighborhood of the organizing center. Generally, the organizing center is characterized by the vanishing of a large number of derivatives of G with respect to y and ε (defining conditions) and H involves the first non-vanishing derivatives (nondegeneracy conditions). One cannot give a recipe for when an organizing center will exist in a given problem or what it will be ; however, previous investigations of this problem [6] and the number of parameters present together suggest as a candidate the winged cusp singularity $H(x,\lambda) = x^3 + \lambda^2 = 0$, which is defined by $H = H_x = H_\lambda = H_{xx} = H_{x\lambda} = 0$, $H_{xxx} \cdot H_{\lambda\lambda} > 0$. This is a codimension three singularity whose universal unfolding is

$$F(x,\lambda,\alpha_1,\alpha_2,\alpha_3) = x^3 + \lambda^2 + (\alpha_1 + \alpha_2\lambda)x + \alpha_3 = 0$$

The organizing center and the seven stable perturbed diagrams are shown in Figure 1 :

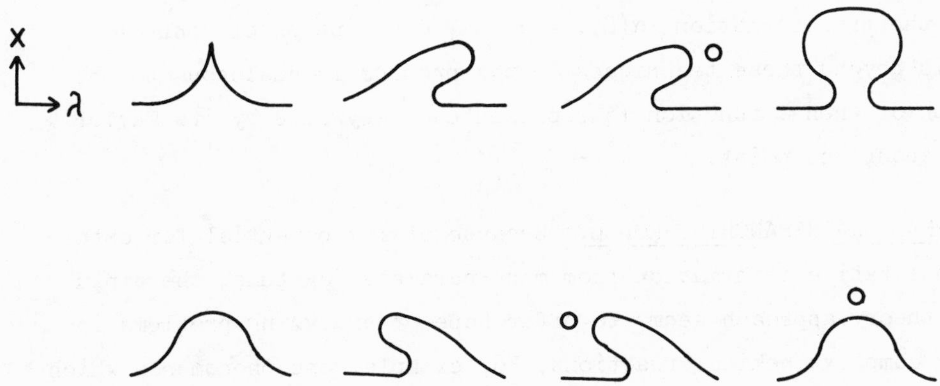

Figure 1.

For the function G given by equation (2) we have

THEOREM : For a class of functions A(y) containing a C^3-open neighborhood of Arrhenius terms with $\gamma > 8/3$, there is a unique point $P_0 = (y_0,\varepsilon_0;B_0,D_0,\eta_0)$ such that $G(y,\varepsilon;B_0,D_0,\eta_0)$ is contact equivalent to $H = x^3 + \lambda^2 = 0$ in a neighborhood of (y_0,ε_0). At (B_0,D_0,η_0), the parameters (B,D,η), form a set of universal unfolding parameters for the singularity. Furthermore, for any values of (B,D,η), only the bifurcation diagrams that appear in the unfolding of H appear locally.

The proof of the theorem involves solving the five defining equations for the singularity and verifying the nondegeneracy condition and the independence of (B,D,η) near (B_0,D_0,η_0). The last statement involves showing that no other, higher-order singularity appears in G. For an Arrhenius term , with $\gamma >> 1$, the coordinates of P_0 are

$$y_0 \sim \frac{1}{\gamma}, \varepsilon_0 \sim \frac{\gamma}{2}, \eta_0 \sim -\frac{1}{2}, D_0 \sim \frac{\gamma}{2e}, B_0 \sim \frac{4}{\gamma}.$$

Because $D_0 \to \infty$ with γ , the organizing center moves to the boundary of the domain as $\gamma \to \infty$ and this may explain why two of the seven stable diagrams were not found in [6] . Near P_0 the regions in (B,D,η) space corresponding to each diagram can be found by investigating the transformation $\alpha(\beta)$. A global description of control space is beyond these techniques -- the problem is analogous to the question of when a function is represented everywhere by its Taylor series about one point.

3. THERMAL CHAINBRANCHING MODEL. Because of its potential for obtaining qualitative information from many-parameter systems, the singularity theory approach seems to offer hope of analyzing problems involving complex chemical reactions. For example, one phenomenon which cannot be understood without taking the underlying chemical kinetics into account is the explosion peninsula in the hydrogen-oxygen reaction. The curve in figure 2 separates the $P - T_0$ plane into two regions: the reaction of hydrogen and oxygen in a closed container at initial pressure P immersed in a bath of temperature T_0 takes place explosively fast or at a moderate speed according as (P,T_0) lies above or below the curve.

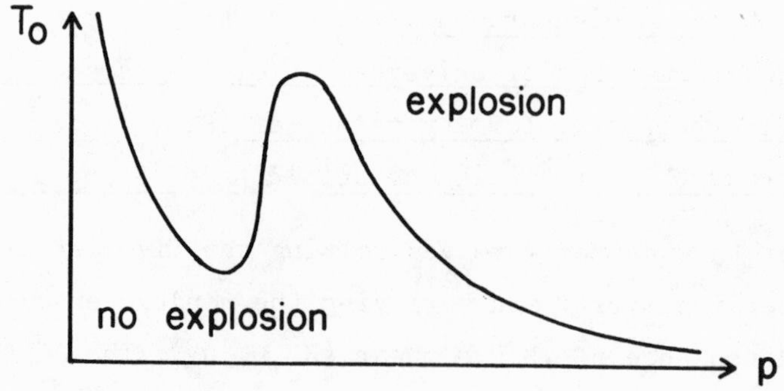

Figure 2.

Explanations for this fact that the explosion limit curve is not a monotonic function of P have centered on the existence of two types of reactions in the complete mechanism: some which are strongly exothermic and some which produce large quantities of unstable radicals which in turn must react further in a chainbranching sequence. Different temperature and pressure dependences of the rates of these reactions could cause an overall reaction which is explosive at a fixed T_0 and P to be inhibited when the pressure is either increased or decreased. Various elementary reactions have been proposed as the important ones [1] ; in [7] , Yang and Gray advanced a specific mechanism which fits into a four-reaction overall model combining thermal and chainbranching effects. Their scheme contains a single active intermediate, X, which is evolved via an initiation step, multiplies in a branching step, and is destroyed in two ways: by a pressure-sensitive gaseous recombination or a pressure-independent reaction at the wall. The notation is given in Table 1.

TABLE 1.

Step		Rate	Heat Release
initiation	$A \rightarrow X$	$k_i = Z_i \exp(-\gamma_i/T)P^{\alpha_i}$	h_i
branching	$X \rightarrow nX$	$k_b = Z_b \exp(-\gamma_b/T)P^{\alpha_b}$	h_b
wall destruction	$X \rightarrow B$	$k_w = Z_w \exp(-\gamma_w/T)$	h_w
gaseous recombination	$X \rightarrow B'$	$k_g = Z_g \exp(-\gamma_g/T)P^{\alpha_g}$	h_g

For purposes of illustration and in carrying out the later calculations, it is assumed that each step is a single reaction with mass-action and Arrhenius kinetics. In fact, this is not quite true, and even if there is a single rate-determining reaction for each step, there is no general agreement on values to assign to the Z's, γ's, α's or h's. However, our results on the system described here will hold for any more complicated system which includes the steps outlined above in its behavior.

This model leads to the following equations for the evolution of $x = [X]$ and T, the temperature of the reacting mixture :

(3a) $$\frac{dx}{dt} = k_i + ((n - 1)k_b - k_w - k_g)x$$

(3b) $$\frac{dT}{dt} = k_i h_i + (h_b k_b + h_w k_w + h_g k_g)x - \ell(T - T_0).$$

The second important assumption in this model is to ignore consumption of the reactants and to assume that the overall pressure P does not change during the reaction. Thus, since the initial concentrations may be taken proportional to P, we are assuming that the Z's and P are constant in Table 1.

In the steady-state theory considered here, we are interested only in the steady-state solutions of (3). Thus the assumption of constant P may be reworded ; it is assumed that the system (3) comes to equilibrium on a time-scale which is rapid compared to the rate of change of pressure of the system. This assumption has been validated for simpler ignition models [5] .

Upon eliminating x from the steady-state equations (3), we obtain a single equation G = 0 involving T, P and eighteen parameters. Nondimensionalizing reduces this number to nine. We assume in addition that the pressure dependences in Table 1 may be taken as $\alpha_i = \alpha_g = 2$, $\alpha_b = 1$. This is true for Yang and Gray's model [7] ; see also [1] . Finally, we make an ad hoc assumption to simplify the calculations : $h_i = -h_w = - h_g$; that is, the only net heat release is in the branching step. This is approximately justified by the numerical estimates for these parameters.

The reduction for the steady-state temperature is now

(4) $$G(T,v,a,T_0,Z,\gamma_2,\gamma_3) = F - a/(T - T_0) = 0$$

where $F = E_3 v^3 - 3E_2 v^2 + ZE_1 v$;

here $E_j = \exp(\gamma_j/T)$ $j = 1,2,3$;

$$\gamma_1 = \gamma_i + \gamma_b - \gamma_g = 1,$$

$$\gamma_2 = \gamma_i \ ; \ \gamma_3 = \gamma_i + \gamma_b - \gamma_w$$

$$v = \sqrt{\frac{B_b}{3B_w}} \cdot \frac{1}{p} \; ; \; Z = \frac{B_g}{B_w}\sqrt{\frac{3B_w}{B_b}} \; ; \; a = \frac{B_i(h_b - h_i)B_b}{\ell \, B_w} \; .$$

The pressure has been replaced by the specific volume, v, for convenience, and T and T_0 are scaled so that $\gamma_1 = 1$. For the $H_2 - O_2$ reaction γ_i is much larger than the other activation energies, so it is in fact reasonable to assume that γ_1, γ_2 and γ_3 are positive and approximately equal.

We now consider how questions involving the explosion peninsula of Figure 1 can be phrased as qualitative bifurcation problems for Equation 4. If we choose values of T_0 for which all three explosion limits exist , and hold the last five variables in (4) fixed, then the graph of G = 0 is a bifurcation diagram for the steady-state temperature, T, as a function of the bifurcation parameter, v. Physically, this corresponds to performing a series of experiments in which the pressure is varied quasistatically ; the explosion limit curve is in fact found by varying the pressure (or T_0) until the steady-state temperature jumps abruptly [1]. Thus a solution of (4) of the form sketched in Figure 3a or 3b would be consistent with the form of the limit curve in Figure 2.

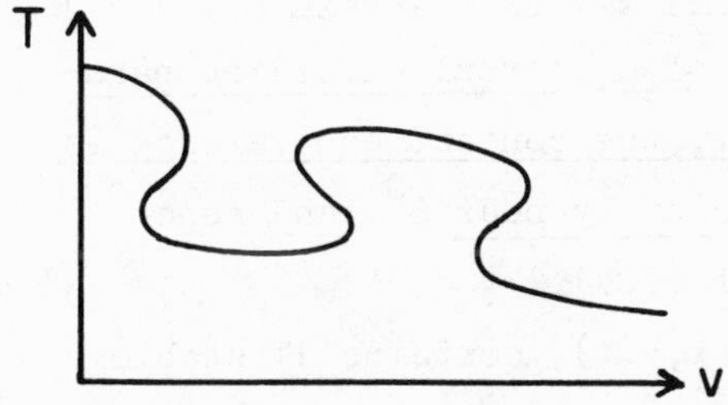

Figure 3a

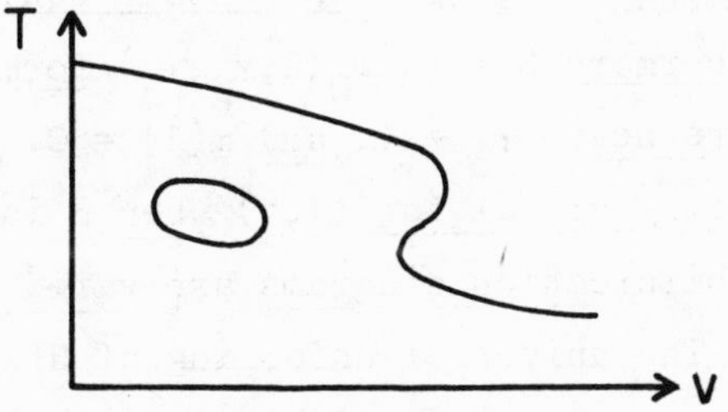

Figure 3b

Showing that bifurcation diagrams of the qualitative nature of 3a or 3b along with the simpler diagrams corresponding to a single explosion limit, are present among the solutions of (4) would thus indicate that the model from which (3) and (4) were derived does indeed have sufficient complexity to provide an explanation for the explosion peninsula. This we have been able to do by finding an organizing center for equation (4). Our organizing center is the one-root non-degenerate cubic singularity, the "London Underground",

$$H(x,\lambda) = x^3 - 3mx\lambda^2 + 2\lambda^3, \; m < 1 \tag{5}$$

whose defining conditions and nondegeneracy conditions are

$$H = H_x = H_\lambda = H_{xx} = H_{x\lambda} = H_{\lambda\lambda} = 0, \; H_{xxx} \neq 0 \text{ and}$$

$$m^3 = \frac{(B^2 - AC)^3}{(A^2D - 3ABC + 2B^3)^2} < 1 \text{ where } A = \frac{1}{6} H_{xxx}, \; B = \frac{1}{2} H_{xx\lambda},$$

$$C = \frac{1}{2} H_{x\lambda\lambda}, \; D = \frac{1}{6} H_{\lambda\lambda\lambda}.$$

This singularity has codimension five. We have

THEOREM. *For each γ_3 near 1, there is a unique point*

$$P^0(\gamma_3) = \frac{3-\gamma_3}{4}, \; \exp\left(\frac{2(1-\gamma_3)}{3-\gamma_3}\right), \; \frac{3-\gamma_3}{8} e^2, \; \frac{3-\gamma_3}{8}, \; 3, \; \frac{\gamma_3+1}{2}, \gamma_3)$$

$$= (T^0, v^0, \beta^0)$$

such that $G(T,v,\beta^0)$ is contact equivalent to (5) at P^0, with $m(\gamma_3) < 1$. Furthermore $\beta = (a, T_0, Z, \gamma_2, \gamma_3)$ form a set of universal unfolding parameters near $\gamma_3 = 1$, and $m(1) = 0$. Finally, the contact equivalence is actually global for (T,v) when β is sufficiently near β^0, and hence the bifurcation diagrams are valid for all (T,v).

The universal unfolding of $H(x,\lambda)$, $H(x,\lambda,\alpha)$, contains 17 stable diagrams. These are depicted in Figure 4.

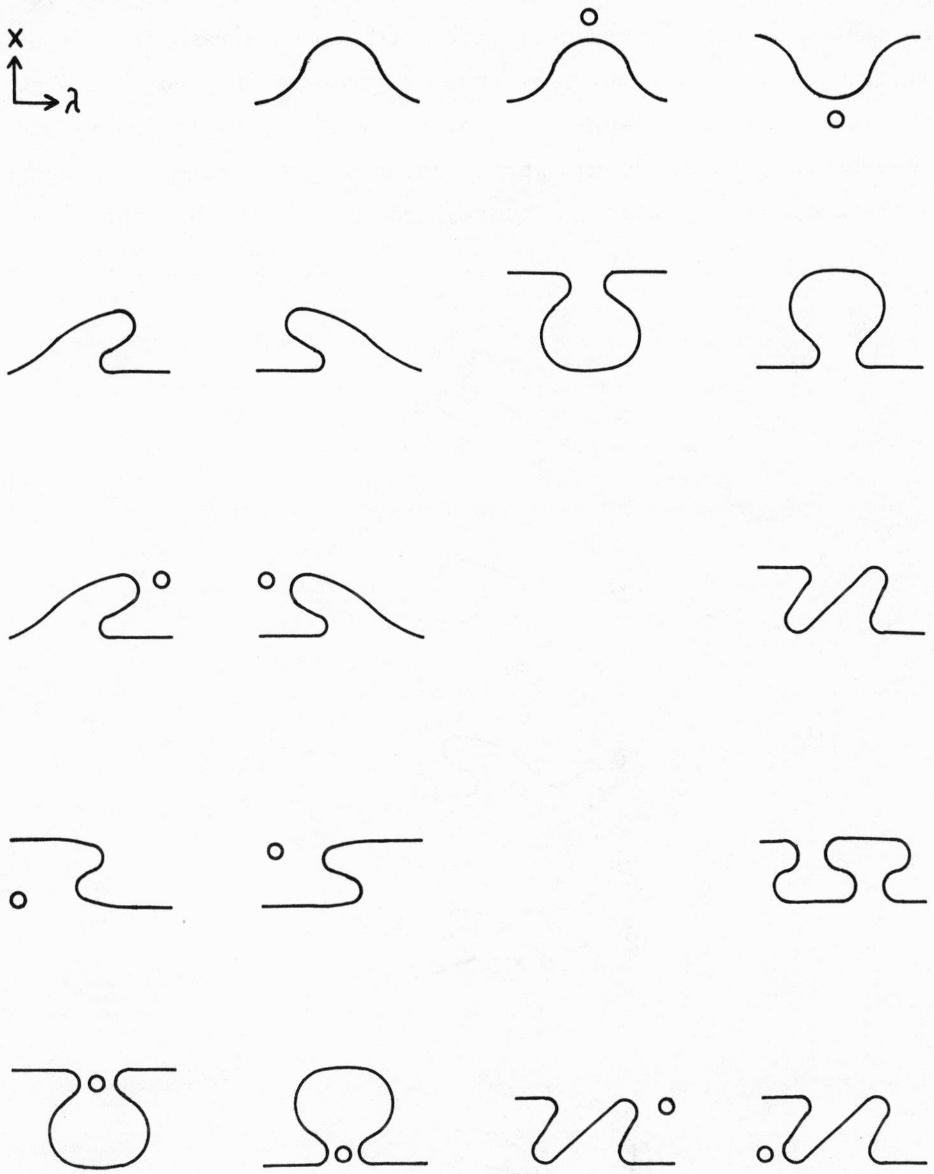

Figure 4.

To find the values of β corresponding to different diagrams would again involve inverting the mapping $\alpha(\beta)$. To show that the explosion peninsula, as interpreted in this steady-state model, actually occurs as T_0 varies in (4) with (a,Z,γ_2,γ_3) held constant, would involve verifying that a particular sequence of bifurcation diagrams from Figure 4 is traversed. A plausible sequence is pictured in Figure 5, showing the formation and disappearance of the multiple explosion limits as T_0 decreases (reading down). Verification of this sequence is again beyond the scope of singularity theory, and we have not attempted it yet.

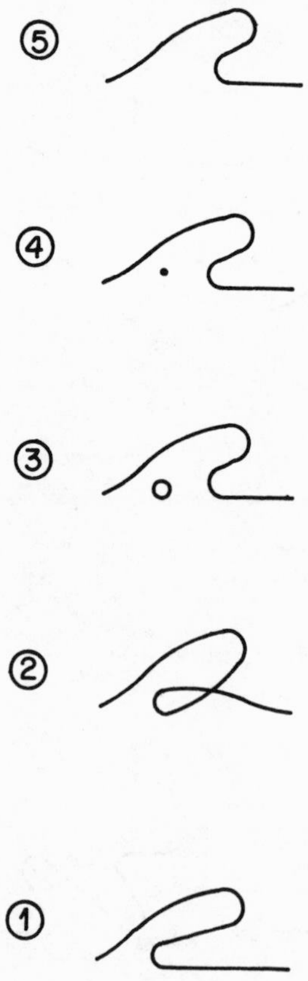

Figure 5.

We conclude with some comments on the parameter range explored here. Since

$\gamma_2 = \frac{\gamma_1+\gamma_3}{2}$ at the organizing center, our assumption $\gamma_3 \sim \gamma_1$ is equivalent to $\gamma_i >> \gamma_b, \gamma_g, \gamma_w$ for the activation energies of the four steps. In fact, $\gamma_b/\gamma_i \sim .2$ for one proposed mechanism [1], and $\gamma_g, \gamma_w \sim 0$, so this is reasonable. At the organizing center, $T_0 = \frac{1}{2} T = \frac{1}{4}$ when $\gamma_3 = 1$; since γ_i is in excess of 50,000 °K for this same proposed mechanism, we have ambient and steady-state temperature of 12,500° and 25,000° respectively. These are a different order of magnitude from experiment, and suggest that the parameter values are not too close to the organizing center -- although they may be sufficiently close (in the topology used here) for the results to be valid. The assumption on heats of reaction is not strictly true either; we can, however, add the extra heats of reaction to our parameter set and the theorem still predicts the complete qualitative behavior when they are small (in the topology). Much further work remains if a complete parameter exploration of the thermal-chainbranching model is to be carried out : possibly singularity theory will help in some of these steps. Perhaps more important, this analysis indicates the way in which singularity theory can contribute to the understanding of mathematical models for chemical reactions.

REFERENCES

[1] G. DIXON-LEWIS and D.J. WILLIAMS, "The Oxidation of Hydrogen and Carbon Monoxide", in C.H. BAMFORD and C.F.H. TIPPER, Eds. "Comprehensive Chemical Kinetics vol. 17 : Gas-Phase Combustion," Elsevier, Amsterdam, 1977.

[2] M. GOLUBITSKY and B.L. KEYFITZ, "A qualitative study of the steady state solutions for a continuous flow stirred tank chemical reactor". SIAM J. Math. Anal. To appear.

[3] M. GOLUBITSKY, B.L. KEYFITZ, and D. SCHAEFFER. "A Singularity Theory Analysis of the Thermal-Chainbranching Model for the Explosion Peninsula", submitted to Comm. Pure Appl. Math.

[4] M. GOLUBITSKY and D. SCHAEFFER, "A theory for imperfect bifurcation via singularity theory," Comm. Pure Appl. Math.. 32 (1979) pp. 21-98.

[5] P. GRAY and P.R. LEE, "Thermal Explosion Theory," in C.F.H. TIPPER, ed., "Oxid & Comb. Rev. 2," Elsevier, Amsterdam, 1967.

[6] A. UPPAL, W.J. RAY and A.B. POORE, "The classification of the dynamic behavior of continuous stirred tank reactors -- influence of reactor residence time," Chem. Eng. Sci., 31 (1976), pp. 205-214.

[7] C.H. YANG and B. F. GRAY, "The determination of explosion limits from a unified thermal and chain theory," 11th Symp. (Int.) Comb. (1967), pp. 1099-1106.

Waves in Distributed Chemical Systems: Experiments and Computations

Pavel Raschman,* Milan Kubicek,* and Milos Marek**

Abstract. Several experimental observations of the concentration waves arising due to reaction and transport interaction in the solution of bromate, malonic acid, sulfuric acid and ceric/cerous ions and/or ferroin are discussed. Measurements in systems with multiple electrodes have confirmed that several periodic wave-like regimes can be observed. An example of an apparently chaotic regime is also shown. The results of simulation of the system of two PDE´s describing one dimensional diffusion and reaction show qualitatively the same phenomena. The Brussellator model is used to illustrate the fact that a large number of different wave-like periodic regimes can coexist and to study subsequent development of more complex wave-like regimes during slow increase of the characteristic dimension of the system. The implications of the effects of external noise on the stability and/or observability of the higher wave number wave-like regimes is stressed.

1. Introduction. Concentration wave-like profiles in the systems with reaction and transport are well established phenomena extensively studied both experimentally and theoretically.[1-7] We shall not attempt to review here the available literature but shall concentrate on several characteristic experimental observations and try to juxtapose them with some results of mathematical modelling in a hope to point out some of the unsolved problems.

2. Some experimental observations. Winfree[8,9] has reviewed numerous experimental arrangements exhibiting concentration waves in the Zhabotinski system. Even if a large number of both experimental and theoreti-

---

[x]Dept. of Chemical Eng. Prague Institute of Chemical Technology, Prague, Czechoslovakia
[xx]on leave at Dept. of Chemical Eng., University of Wisconsin Madison.
The author is indebted to the Dept. of Chemical Engrg., University of Wisconsin for support.

cal papers were devoted to the subject, there are still principal disagreements between individual authors on the proper mechanism of generation of the target patterns generally and leading centers specifically. We have performed experiments with controlled generation of leading centers in the quiescent layer of reacting solution by local temperature inhomogeneity[10]. Leading centre acted as hard oscillator (finite temperature disturbance was necessary for initiation of the leading centre; once initiated, no further externally induced temperature heterogeneity was required for continuous function of the leading centre) and the frequency of the wave emitted at the leading centre was higher than frequency of oscillations in the well stirred cell at the same conditions (concentrations, temperature).

Smoes[11] has recently reported on her extensive observations of the characteristics of the concentration waves in an unstirred layer of solution with Zhabotinski reaction. Both oscillatory and nonoscillatory compositions of reacting solutions were studied. She has observed waves of varying characteristics with nonuniform velocity in different concentration regions. The majority of the observed leading centers had periods between 90 and 100 % of the bulk period (in the well stirred solution). The number of observed centers increased with increasing frequency of the bulk oscillations.

In the case of the leading center we can assume that concentration inhomogeneity within the volume with the characteristic dimension L was created. The course of concentrations of characteristic components can then be described by the set of reaction diffusion equations

$$\frac{\partial c_i}{\partial \tau} = D_i \nabla^2 c_i + \sum_{j=1}^{R} \gamma_{ij} \cdot f_j(c_1 \ldots c_R), \quad i=1,2,..,R \qquad (1)$$

with the boundary conditions defining the flux at the external surface of the leading center volume

$$D_i \nabla c_i = \mu (c_{is} - c_i) \qquad (2)$$

Characteristic dimension of the leading center will change slowly with time (ε small)

$$\frac{\partial L}{\partial \tau} = \varepsilon f_L(\tau) \qquad (3)$$

The above set of equations (1) - (3) will describe the formation of the leading center if solutions with periodically varying concentrations (or fluxes) at the boundary exist.

More detailed characterization of the wave-like regimes can be obtained in experimental arrangements shown schematically in Fig. 1a - tubular reactor with an annular cross section and in Fig. 1b - the reactor of the ring shape. Pt wire electrodes are positioned along the reactors and the time course of the redox potential corresponding to concentration changes within the reaction mixture is followed. Redox potential at individual electrodes is checked with equimolar solution of Ce^{4+}/Ce^{3+} ions before and after measurements. Reactors are filled with the reacting solution of the given composition, temperature is kept constant and the time course of the redox potential on individual electrodes is recorded, some time after perturbation of concentration or temperature was applied. Various time dependent and stationary regimes are observed.

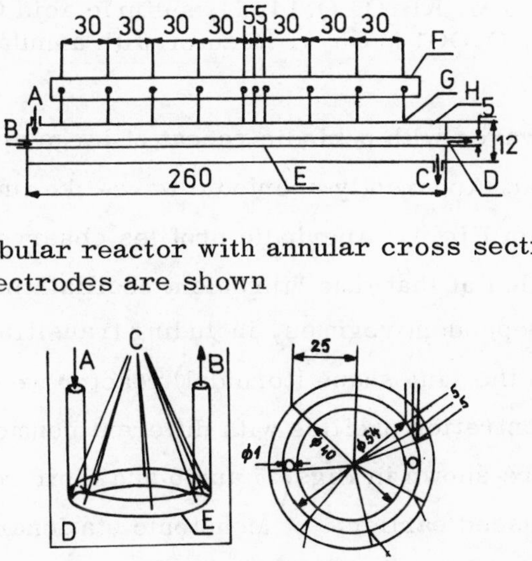

Figure 1a) Tubular reactor with annular cross section. Positions of Pt electrodes are shown

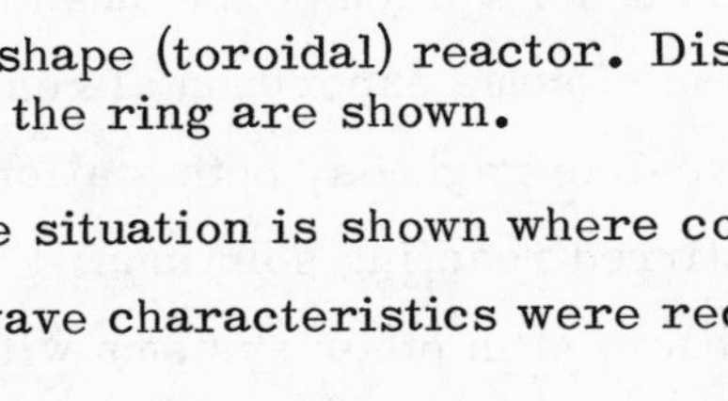

Figure 1b) Ring shape (toroidal) reactor. Distances between the electrodes along the ring are shown.

In Fig. 2 the situation is shown where concentration waves with approximately plane-wave characteristics were recorded.

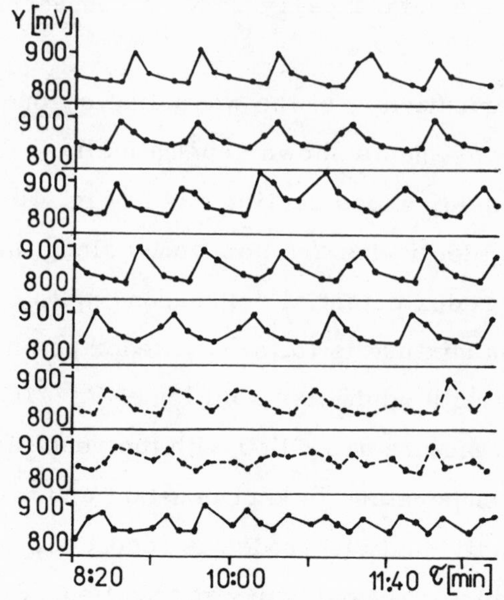

Figure 2) Concentration waves with approximately plane-wave characteristics. Temperature 25 °C. Concentrations of reaction components: malonic acid 0.3 M, $KBrO_3$ 0.14. M, sulfuric acid 0.8 N, ferroin 0.0125 M, 0.001 M Ce^{4+}. Reactor with annular cross section.

In Fig. 3 concentration waves with periodic repeat of large and small amplitude pulses are presented. Apparently aperiodic wave-like time course of concentrations is shown in Fig.4. Aperiodic profiles observed in tubular reactor in 1973 were called at that time "turbulent oscillations".(12)
A variety of other time dependent regimes, including transitions between them, were observed. In the ring shape (toroidal) reactor we can, morever, observe stationary concentration profiles with different nonmonotonic characteristics. Two of them are shown in Figs. 5 and 6 and their relation to modelling studies was discussed earlier.(14) Monotonic stationary concentration profiles were also observed under various experimental conditions. We can conclude the short exposition of some experimental results with the statement, that rich variety of wave-like regimes, both stationary and time dependent, can be observed in unstirred reacting solutions. Similar wave-like and stationary patterns were observed in other systems with organic reactants (cf. paper by Körös in(6)) and in enzymatic systems (cf. review by

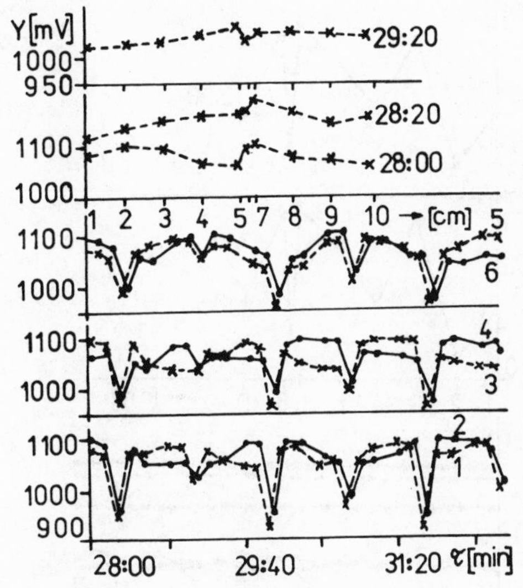

Figure 3) Concentration waves with more complex characteristics. Temperature 25 °C. Concentrations of reaction components: malonic acid 0.05 M, $KBrO_3$-0.05 M, 0.001 M Ce^{4+}, 3N sulfuric acid. Upper figure-longitudinal profiles; lower figure-time dependence of the redox potential at individual electrodes. Reactor with annular cross section.

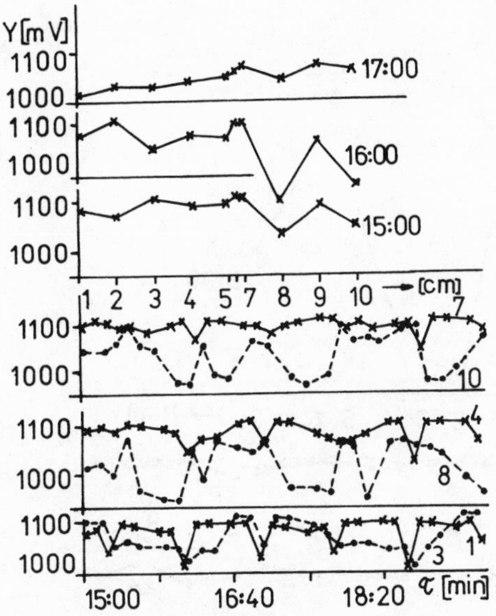

Figure 4) Apparently aperiodic concentration profiles. Temperature 25 °C. Concentrations of reaction components: malonic acid 0.032 M, $KBrO_3$ 0.032 M, Ce^{4+} 0.001 M, sulfuric acid-3N; upper figure; longitudinal profiles; lower figure-time dependence of the potential at individual electrodes. Reactor with annular cross section.

Hess in[6]). We believe that there currently exists an urgent need for theoretical studies of realistic model reaction diffusion systems, where for the chosen system the existence, experimentally verifiable characteristics and transitions between various regimes in dependence on the change of parameters could be quantitatively described. However, as mostly only qualitatively different predictions of various models can be tested by comparison with experiments, the knowledge of the behavior of the solutions over the entire range of the value of the studied parameter (and not just a behavior for small or large parameter values) is required.

We have attempted to obtain such dependences for the reaction-diffusion

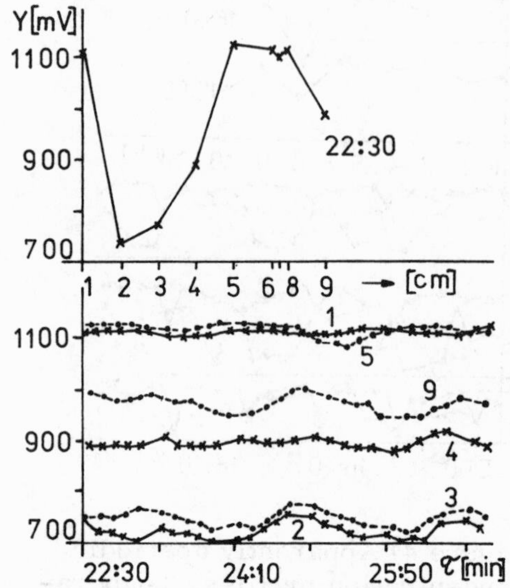

Figure 5) Nonmonotonic stationary concentration profile in the ring shape reactor. Temperature: 25 °C; concentration of reaction components: malonic acid 0.032 M, $KBrO_3$ – –0.032 M, Ce^{4+} –0.001 M, sulfuric acid – 3N; upper figure–longitudinal profile; lower figure–time course of redox potentials on individual electrodes.

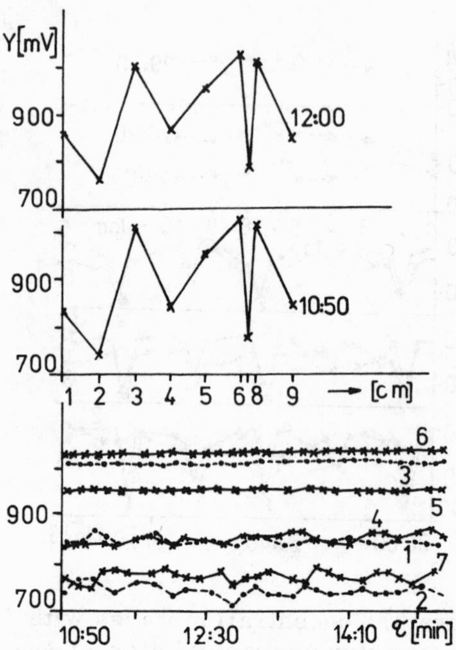

Figure 6) Nonmonotonic stationary concentration profile in the ring shape (toroidal) reactor. Temperature: 25 °C; concentration of reaction components: malonid acid–0.032 M, $KBrO_3$–0.01 M, Ce^{4+}–0.001 M, sulfuric acid 3 N; upper figure–longitudinal profiles; lower figure–time course of redox potential on individual electrodes.

systems numerically, using continuation techniques, in the case of the nonmonotonic (spatially periodic) stationary solutions.[15-19] In the following we would like to discuss some results of two simulation studies of the time dependent and stationary solutions of the reaction diffusion equations.

3. Some computed examples. The reaction diffusion equations (1) describing transport by Fickian diffusion and reaction of two characteristic reaction components x and y can be written in the form

$$\frac{\partial x}{\partial \tau} = \frac{D_x}{L^2} \frac{\partial^2 x}{\partial z^2} + f(x, y) \tag{4}$$

$$\frac{\partial y}{\partial \tau} = \frac{D_y}{L^2}\frac{\partial^2 y}{\partial z^2} + g(x,y) \tag{5}$$

$$z \in \langle 0,1 \rangle, \quad \tau = 0: \quad x = x_0(z), \; y = y_0(z) \tag{6}$$

Here f (x,y), g (x,y) describe reaction kinetic functions.

For the Belousov-Zhabotinski reaction a number of models have been suggested.[1,4,5,6] The model currently considered as closest to the prevailing concepts of reaction mechanism was proposed by Noyes and coworkers and its application to the description of the plane wave-like phenomena is discussed by Reusser and Field.[20] Fife[21] has recently used singular perturbation analysis of the sharp wave fronts for general oscillatory or excitable kinetics. His model of the leading center requires three variables in the kinetic equations. Howard and Kopell[22] have recently developed a theory of target patterns for a λ-ω systems, a special type of two component reaction diffusion systems. We shall illustrate the possibility of the modelling of the experimentally observed phenomena by several results of simulation of Eqs. (4), (5) using a semi-empirical kinetic model proposed by Zhabotinski.[1] In this model

$$f(x,y) = y(1-x) - \delta x \tag{7}$$

$$g(x,y) = y\{1 - x[1+\alpha+(y-\alpha)^2]\}/\varepsilon + 1 \tag{8}$$

α, $\delta = \gamma\beta$ and ε are parameters dependent on the initial concentration of reactants. Let us consider zero flux (Neumann) boundary conditions

$$z = 0,1: \quad \frac{\partial x}{\partial z} = \frac{\partial y}{\partial z} = 0 \tag{9}$$

and fixed (Dirichlet) boundary conditions

$$z = 0,1: \quad \begin{aligned} x(0,\tau) &= x(1,\tau) = \bar{x} \\ y(0,\tau) &= y(1,\tau) = \bar{y} \end{aligned} \tag{10}$$

to Eqs. (4), (5). Here $\bar{x}$, $\bar{y}$ are roots of kinetic equations (7), (8) (i.e. $f(\bar{x},\bar{y}) = 0$, $g(\bar{x},\bar{y}) = 0$) and also trivial solutions of (4) and (5).

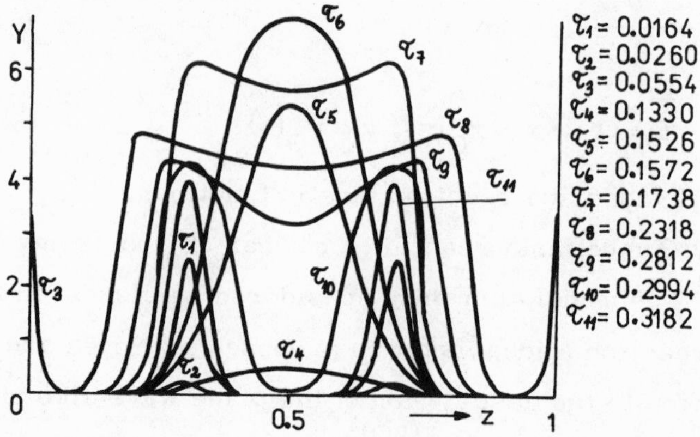

Figure 7) Transient spatial concentration profiles of the component y over one period. Zhabotinski model, Eqs. (7), (8). $\alpha = 4.0$, $\beta = 0.002$, $\gamma = 0.03$, $\varepsilon = 0.005$, $D_x/L^2 = 0.25$, $D_y/L^2 = 0.0625$. Boundary conditions (10), $\bar{x} = 0.1951$, $\bar{y} = 3.63591$; $\tau_1 - \tau_n$ - characteristic times within one period.

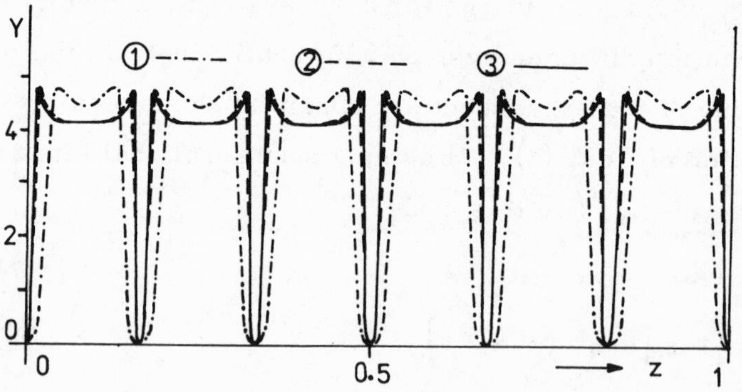

Figure 8) Stationary concentration profiles of the component y-Zhabotinski model, Eqs. (7),(8) . $\alpha = 4.0$, $\beta = 0.0018$, $\gamma = 0.03$, $\varepsilon = 0.005$, $D_x/D_y = \rho = 4$.
① $D_x/L^2 = 1.928 \times 10^{-2}$, ② $D_x/L^2 = 4.82 \times 10^{-3}$,
③ $D_x/L^2 = 2.41 \times 10^{-3}$, boundary conditions (9).

The parameters $\alpha, \delta \sim O(1)$ and $\varepsilon \sim O(10^{-3})$ and thus two widely different time scales appear in the equations . After discretization the system (4) and (5) is stiff and certain degree of care has to be exercised in choosing numerical techniques. Two difference methods of the second order approximation with respect to both variables were used (boundary value technique and the Crank-Nicolson scheme) and the results were compared.[23] In Fig. 7 periodic spatial concentration profiles of the component y calculated for boundary conditions (10) are shown. We can observe that center of the system, "leading center" sends out periodically concentration waves with a period equal to $\tau_{11} = 0.3182$. Stable stationary concentration profiles can also be predicted by this kinetic model. In Fig. 8, three such profiles are shown for different values of the diffusion coefficients. Simultaneous existence of the stationary and periodic wave-like regimes in a version of the Noyes model was recently discussed by Tomita et al.[24] Both Noyes and Zhabotinski kinetic equation models describe relaxational oscillations in the well mixed system. Corresponding differential equations are stiff and even if this suggests convenient choice of the small parameter to use in application of asymptotic expansion techniques, it precludes efficient numerical analysis over the entire region of values of parameters. We are therefore using a model system - well known Brussellator - for development of the techniques of analysis, which would enable us to obtain complete bifurcation diagrams and dependences of the solution on characteristic parameters. Analysis of the stationary states along these lines was published elsewhere[17] and here we would like to discuss some wave-like periodic concentration profiles which arise if the pair of eigenvalues of the system (4), (5) linearized around trivial solution crosses imaginary axis (Hopf bifurcation).

4. Brussellator model-waves. Kinetic relations are in the form

$$f(x,y) = A - (B+1)x + x^2 y \tag{11}$$

$$g(x,y) = Bx - x^2 y \tag{12}$$

and the trivial stationary solution is $\bar{x} = A$, $\bar{y} = B/A$.

Let us denote

$$\begin{bmatrix} a_{11} & a_{12} \\ a_{21} & a_{22} \end{bmatrix} = \begin{bmatrix} \dfrac{\partial f(\bar{x},\bar{y})}{\partial x} & \dfrac{\partial f(\bar{x},\bar{y})}{\partial y} \\ \dfrac{\partial g(\bar{x},\bar{y})}{\partial x} & \dfrac{\partial g(\bar{x},\bar{y})}{\partial y} \end{bmatrix}$$

The linear stability analysis of (4), (5) with boundary conditions (9) leads to the characteristic equation for the eigenvalues λ_n

$$\lambda_n^2 + P_n(L)\lambda_n + Q_n(L) = 0, \quad n = 0,1,2,.. \tag{13}$$

For boundary conditions (10) is the characteristic equation the same but for $n = 1,2,..$

Characteristic dimension L is chosen as bifurcation parameter. In (13)

$$Q_n(L) = D_x D_y k_n^2 - (D_x a_{22} + D_y a_{11})k_n + a_{11}a_{22} - a_{12}a_{21}$$

$$P_n(L) = a_{11}a_{22} - (D_x + D_y)k_n$$

$$k_n = (n\pi/L)^2$$

A pair of purely imaginary eigenvalues will occur for $P_n(L) = 0$, i.e. at

$$L_n^{xx} = n\pi\sqrt{\frac{D_x + D_y}{a_{11} + a_{22}}} \tag{14}$$

Positive value of L_n^{xx} will exist for the models where simultaneously

$$\frac{a_{11} + a_{22}}{\sqrt{|a_{12}\ a_{21}|}} > 0 \qquad \frac{|a_{11} - \rho a_{22}|}{\sqrt{|a_{12}\ a_{21}|}} < \rho + 1$$

where $\rho = D_x/D_y$. Conditions can be fulfilled also for $\rho = 1$. Close to the bifurcation points (pair of eigenvalues cannot be multiple here) $L_n^{xx} = nL_1^{xx}$, the frequency of the bifurcating time periodic solutions can be approximated by $(Q_n)^{1/2} = (Q_1)^{1/2}$, i.e. will be the same for the solutions with the wavenumbers 1,2,3 ... bifurcating at L_1^{xx}, L_2^{xx}, L_3^{xx} ... The frequency of the bifurcating time periodic solutions close to the bifurcation points, (i.e. to the points where $P_n(L) = 0$) will be lower than the frequency of the solutions in the well mixed case if we shall consider $\rho = 1$. Here in the well mixed (lumped parameter) system $Q_0 = a_{11}a_{22} - a_{12}a_{21}$. As in the distributed system at the bifurcation point $P_n = a_{11}a_{22} - 2Dk_n = 0$ $(D = D_x = D_y)$ then

$Q_n = Q_O + Dk_n (Dk_n - (a_{11} + a_{22})) Q_O$. If $\rho \neq 1$ we can then have any relation between Q_n and Q_O.

The question of stability of the bifurcating solutions can be answered either by construction of solutions close to the bifurcation points (e.g. using asymptotic expansions) or directly by numerical solution of Eqs. (4), (5). Let us choose for the Brussellator model the values of parameters $A = 2$, $B = 5.45$. In the system without concentration gradients (lumped parameter system) limit cycle solution surrounding unique unstable steady state will exist, with the period of oscillations equal to $T_p = 3.84$. Let us choose further the values of diffusion coefficients in the distributed system $D_x = 0.008$, $\rho = D_x/D_y = 2$. Then we can calculate $L_1^{xx} = 0.51302$ from (14). For $\rho = 1$ we would have obtained a lower value of L_1^{xx}.

Behavior of various periodic solutions of Eqs. (4), (5) was studied numerically for different values of L. Difference approximations to Eqs. (4), (5) were constructed and profiles for various mesh sizes were computed until the difference between the solutions over the entire period was acceptable for finer and coarser mesh size.[25]

Let us consider the case of Dirichlet boundary conditions (10) first. For small values of L the homogeneous (trivial) solution is stable. When the characteristic dimension of the system is increased beyond L_1^{xx} periodic solution with a spatial profile similar to the eigenfunction for $n = 1$ appears; further increase of length brings subsequent appearance of more complicated solutions. The course of two stable periodic solutions - spatial profiles of the component x at chosen characteristic times are shown in Fig. 9a ($L = 2.25$, $T_p = 3.35$) and in Fig. 9b ($L = 4.5$, $T_p = 3.35$). In both situations would these time periodic spatial profiles fulfill requirements for the leading centers - their periods are lower than in the well mixed systems and in both cases concentration waves "travel" to the boundary (with varying velocity), and fluxes at the boundaries are periodic.

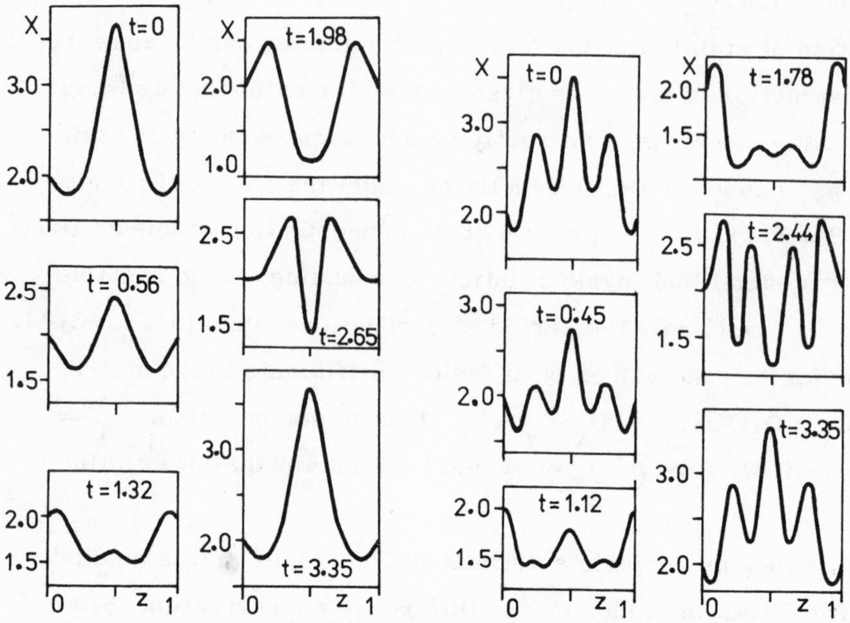

Figure 9) Time periodic spatial concentration profiles of the component x - Brussellator, Eqs. (11), (12), (10); $A = 2$, $B = 5.45$, $\rho = 2$, $D_x = 0.008$. a) $L = 2.25$, $T_p = 3.35$; b) $L = 4.5$, $T_p = 3.35$.

When considering the Neumann boundary conditions, Eq. (9), we have to take into account that the trivial solution in the distributed system here is unstable and thus the solutions bifurcating from this solution will be also unstable if no change of stability (e.g. secondary bifurcation) would occur. This apparently happens in our case.

Homogeneous (spatially uniform) periodic solution appears to be stable for $L \in (0,2)$; close to $L = L_1^{xx}$ stable inhomogeneous solution appears with the form similar to corresponding eigenfunction, $x(z) \sim A_x \cos n\pi z$, $n = 1$. With further increase of length, solutions with higher wavenumbers appear at L_2^{xx}, L_3^{xx} .... In Fig. 10a, b, c, d such solutions are shown- 10 a : n = 1, 10 b : n = 2, 10 c : n = 3 and 10 d : n = 4. With respect to the symmetry properties of the Eqs. (4), (5), the solutions for n = 2, 3,4 can be composed from the solution for n = 1 as the corresponding lengths are multiples of $L = 1.125$. Let us denote the profile in the Fig. 10 a

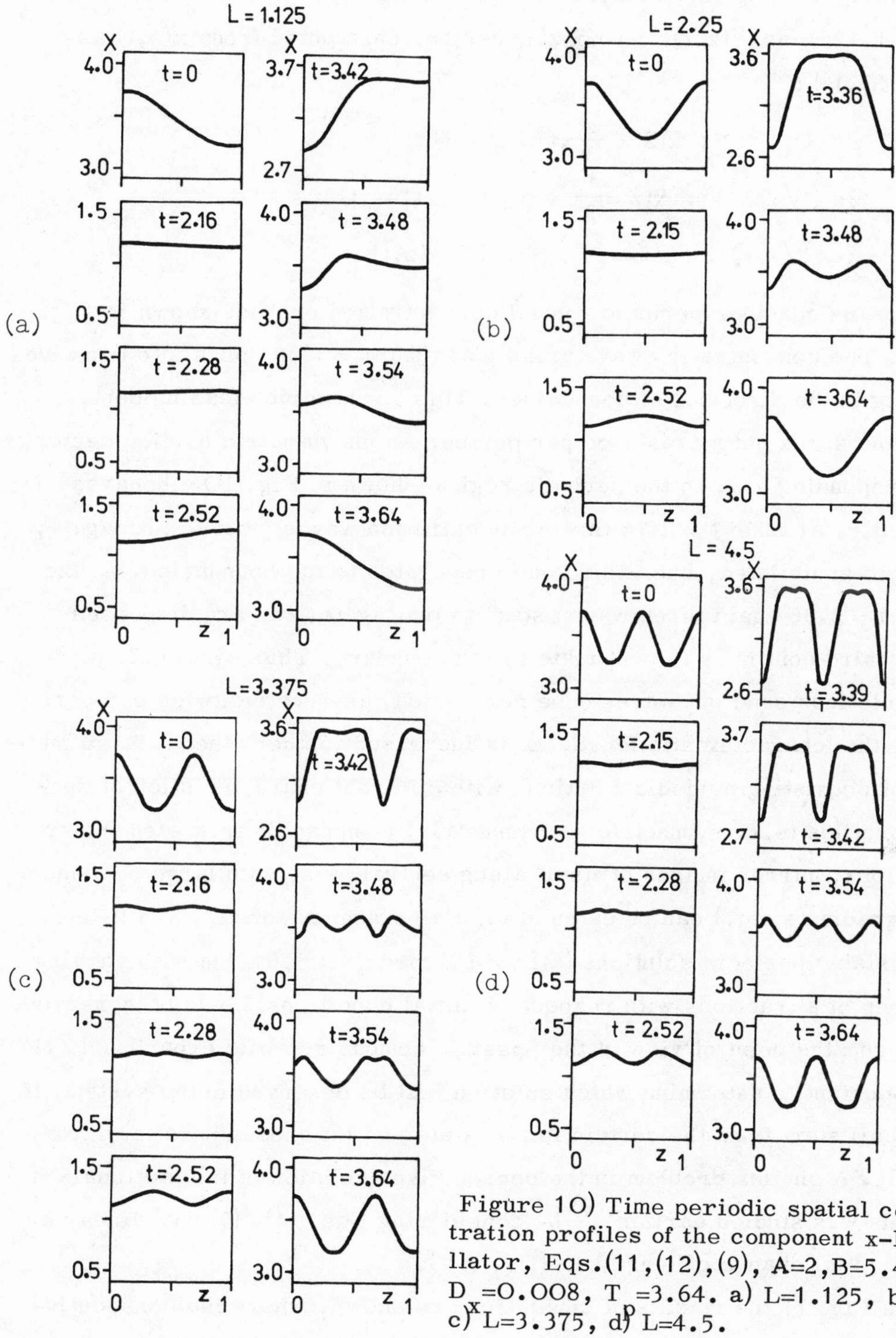

Figure 1O) Time periodic spatial concentration profiles of the component x-Brusselator, Eqs.(11,(12),(9), A=2, B=5.45, ρ=2, D_x=0.008, T_p=3.64. a) L=1.125, b) L=2.25, c) L=3.375, d) L=4.5.

as $x(z,t)$ and, for example, the profile in the Fig. 10 c ($L = 3.375 = 3 \times 1.125$) as $\tilde{x}(z,t)$. Then $\tilde{x}(z,t)$ can be constructed from $x(z,t)$ as ($z \in \langle 0,1 \rangle$):

$$\tilde{z} \in \langle 0, 1/3 \rangle: \quad \tilde{x}(\tilde{z} = \frac{z}{3}, t) = x(z,t)$$

$$\tilde{z} \in \langle 1/3, 2/3 \rangle: \quad \tilde{x}(\tilde{z} = \frac{1}{3} + \frac{z}{3}, t) = x(1-z,t)$$

$$\tilde{z} \in \langle 2/3, 1 \rangle: \quad \tilde{x}(\tilde{z} = \frac{2}{3} + \frac{z}{3}, t) = x(z,t)$$

Let us consider periodic spatial concentration profiles shown in Fig. 10 b. The concentration wave arising at time $t = 3.36$ splits into two waves moving in the direction of boundaries. This "symmetric" leading center thus sends out waves once per period. An unsymmetric leading center, corresponding e.g. to the periodic regime shown in Fig. 10 c behaves differently. At time $t = 2.16$ three concentration waves start to propagate; two are annihilated, but a third one propagates to the boundaries. At the time $t = 3.36$ again three waves start to propagate, two are dispersed but a third continues to propagate to the boundary. Thus symmetric periodic solutions send out waves once per period, asymmetric twice per period.

If the characteristic length L is increased further, then a large number of coexisting periodic solutions with different characteristics of spatial profiles (e.g. symmetric – asymmetric, composable from elementary solutions, mirror images of itself along central axis, with large concentration gradients etc.) can be calculated. For example, for $L = 4.5$ fifteen such stable periodic solutions (all with period $T = 3.64$ but with varying regions of attractions with respect to initial conditions) could be observed.

From the point of view of the possible comparison with experiment, it is important to determine which solution will be observed in the system, if we shall start from the certain initial state and let chosen parameter vary slowly. A similar problem in the consecutive evolution of the stationary solutions was studied earlier[19] by considering Eqs. (4), (5) with linear or exponential change of length L (Eq. (3)).

In Fig. 11 the results of simulation are shown, where the consecutive evolution of periodic wave-like regimes was studied by simulation of

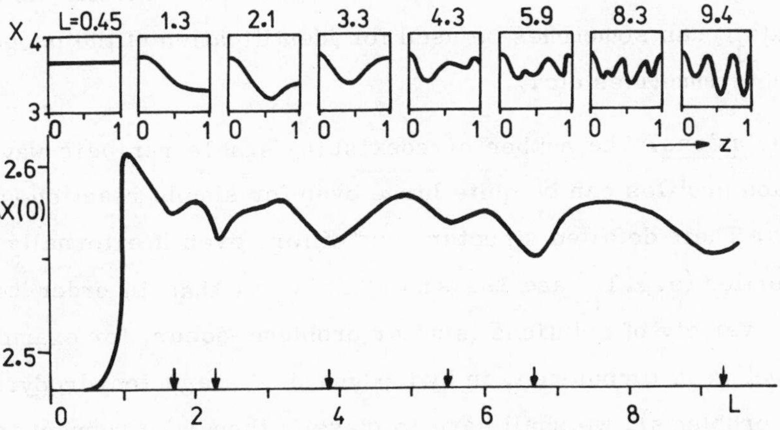

Figure 11) Evolution of stable time periodic regimes with increasing characteristic dimension. Numerical solutions of Eqs. (3), (4),(5). Brussellator, Eqs. (11), (12), (9). A = 2, B =5.45, ϱ = 2, D_x= = 0,008, L_O= 0.45, a = 0.0055. Period constant T_p=3.64.

Eqs. (3), (4), (5). The slow increase of the characteristic dimension was described as $L = L_O + at$, a = 0,0055 ($a = \Delta L/T_p$, ΔL = 0.02). The simulation started at L_O = 0.45, where homogeneous periodic solution is stable. Random perturbations of the magnitude 0.01 were applied for $L \in (0.45, 2.2)$, further on no perturbations were used. In the upper part of Fig. 11 the characteristic spatial profiles of the eight individual stable periodic regimes are shown. We can follow consecutive development of more complex wave-like solutions. The same value of the period (within computational error) was observed in the all calculated cases. The transitions between individual solutions probably occur at the limit points or bifurcation points, where the original solution either ceases to exist or loses its stability. Which one from the multiple existing solutions is actually realized is a problem deserving an attention of mathematicians.

In the lower part of Figure 11 the dependence $\Delta x(0) = \max_{T_p} [x(0,t)] - \min_{T_p} [x(0,t)]$ (maximum amplitude over the period at the boundary) on L is shown. Every minimum denoted by an arrow corresponds to the appearance of a new periodic regime. We can see that an inhomogeneous periodic solution is under the effect of random fluctuations realized at $L \doteq L_1^{xx} = 0.52$, even if the region of stability of the homogeneous limit cycle extends over the interval $L \in \langle 0, 2 \rangle$. As more extensive studies have shown, the ratios $\Delta x(z_i)/\Delta x(z_j)$ can sometimes be used for identification of the properties of solutions (symmetries etc.).

5. Conclusions. The number of coexisting stable periodic wave-like concentration profiles can be quite large even for simple reaction-diffusion models. Their detailed structure can differ, even if externally observable properties (e.g. T_p) are the same. It seems that in order to classify such a variety of solutions (similar problems occur, for example, in the studies of weak turbulence, in hydrodynamics, magnetohydrodynamics and plasma problems), we shall have to classify them with respect to different phase portraits. The classification should probably be made using simple model dissipative systems, similarly as it is done for conservative systems - see e.g. Arnold.[26]

In experimental systems we shall consider observability of periodic regimes with higher wave-numbers, and we have to consider the effects of noise. Disturbances (concentration fluctuations) with higher wave numbers will be damped fast by diffusion. However, we can expect resonance and possible entrainment and/or excitations of the naturally occuring periodic solutions by the fluctuations with the commensurate frequencies in the noise spectra in those cases where they will have sufficient amplitude. To estimate approximately the region of wave-numbers affected by noise, let us assume that we have a noise level of 0.03. Then, if we set characteristic wave-numbers of our periodic regime equal to r and if we consider components of noise with a characteristic wave number p, the resonance between them will occur if r/p will be sufficiently close to a rational fraction, i.e. if for example[26] holds $|r/p - m/n| < 1/n^2$ (m/n is rational). Let us choose $n = 6$, then $0.03 > 1/n^2$. Thus if 0.03 is considered as a characteristic amplitude of the followed component of noise, then periodic regimes with wave numbers higher than $n = 6$ could be affected by resonance with noise. Then we shall have to turn to the stochastic description of our system and calculate the distribution of probabilities of individual regimes, e.g. in the form of the times of the first passages out of the regions of attraction under the effects of small fluctuations.[27]

REFERENCES

(1) A.M.ZHABOTINSKI, Concentration Oscillations, Nauka, Moscow, 1974 (in Russian)

(2) R. ARIS, The Mathematical Theory of Diffusion and Reaction in Permeable Catalysts, Clarendon Press, Oxford, 1975.

(3) P.C. FIFE, Mathematical Aspects of Reacting and Diffusing Systems, Springer Verlag, New York, 1979

(4) J.J.TYSON, The Belousov-Zhabotinski Reaction, Springer Verlag, New York, 1976

(5) Synergetics, Far from Equilibirum, Proc. of the Conference Far from Equilibrium: Instabilities and Structures, Editors: A.Pacault and C.Vidal, Springer Verlag, New York, 1979

(6) Kinetics of Physicochemical Oscillations, Proc. Confer. Aachen, Sept. 19/11, 1979

(7) G.P.IVANICKII, V.I.KRINSKII, E.E.SELKOV, Mathematical Biophysics of the Cell, Nauka, Moscow, 1978 (in Russian)

(8) A.T.WINFREE, Stably rotation patterns of reaction and diffusion, in Theoretical Chemistry 4, pp 1-51, ed. M.Eyring and D.Henderson, Academic Press, New York, 1978

(9) A.T.WINFREE, The Geometry of Biological Time, Springer Verlag, New York, 1979

(10) M.MAREK, J.JUDA, Controlled generation of reaction-diffusion waves, Sci. Papers of the Prague Inst. Chem. Technol., K 13, pp 129-137, 1978

(11) M.-L.SMOES, Chemical waves in the oscillatory Zhabotinski systems; a transition from temporal to spatio-temporal organization. International Symposium on Synergetics-Bielefeld, Sept. 23-30, 1979 to appear in "Lecture Notes on Synergetics", Springer Verlag, New York, 1980.

(12) M.MAREK and P.SVOBODOVA, Experimental study of the effect of transport processes in an oscillatory reaction system Sci Papers of the Prague Institute of Chemical Technology, K 10 (1979)

(13) O.VAFEK, Experimental study of the oscillating reaction in the distributed parameter system, MSc Thesis, Prague Institute of Chemical Technology, 1975

(14) M.MAREK, Dissipative structures in chemical systems - Theory and experiment, paper in (5) above.

(15) M.KUBICEK and M.MAREK, Evaluation of limit and bifurcation points for algebraic equations and nonlinear boundary value problems, Appl. Math. and Comp. 5 (1979)

(16) M.KUBICEK and M.MAREK, Steady state spatial structures in dissipative systems: numerical algorithm and detailed analysis, J. Chem. Phys. 67, 1997-2006 (1977)

(17) M.KUBICEK, V.RYZLER and M.MAREK, Spatial structures in a reaction-diffusion system - detailed analysis of the Brussellator, Biophys. Chem. 8 , 235-246 (1978)

(18) M.KUBICEK, M.MAREK, P.HUSTAK and V.RYZLER, Bifurcation, multiplicity and stability in reaction-diffusion systems, Proc. of the Symp. Computers in Chem. Engng., High Tatras 5-9th, Oct. 1977; p. 903-939

(19) M.MAREK, M.KUBICEK, Patterns of spatial organization in growth, submitted to Z.f. Naturforschung

(20) E.J.REUSSER, R.J.FIELD, The transition from phase waves to trigger waves in a model of the Zhabotinski reaction. J. Am.Chem. Soc. 101 , 1063 (1979)

(21) P.C.FIFE, Wave Fronts and Target Patterns, International Symposium on Synergetics - Bielefeld, Sept. 23-30, 1979, to appear in "Lecture Notes on Synergetics", Springer Verlag, New York 1980.

(22) N.KOPELL, L.N.HOWARD, Target patterns and horseshoes from a perturbed central force problem : some temporally periodic solutions to reaction diffusion equations. Advanced seminar on dynamics and modelling of reactive systems, Madison , Oct. 22-24, 1979

(23) O.VAFEK, P.POSPISIL , M.MAREK, Transient and stationary spatial profiles in Belousov-Zhabotinski reaction , Sci. Papers of the Prague Inst. Chem. Technolog., Ser. K., 1979

(24) K.TOMITA, A.ITO, T.OHTA, Simplified model for Belousov-Zhabotinski reaction, J. Theor. Biol. 68, 459 (1977)

(25) P.RASCHMAN, Concentration waves in reaction-diffusion systems, MSc Thesis, Prague Institute of Chemical Technology, 1979

(26) V.I.ARNOLD, Additional Chapters in the Theory of Ordinary Differential equations. Nauka, Moscow, 1978 (in Russian)

(27) A.D. WENTZEL, M.I.FREIDLIN, Fluctuations in Dynamic Systems - Effects of Small Random Perturbations, Nauka, Moscow, 1979 (in Russian)

Computer-Aided Analysis of Nonlinear Problems in Transport Phenomena

Robert A. Brown,* L. E. Scriven,† and William J. Silliman††

Abstract. The nonlinear partial differential equations of mass, momentum, energy, species, and charge transport, especially in two and three dimensions, can be solved in terms of functions of limited differentiability — no more than the physics warrants — rather than the analytic functions of classical analysis. Particularly convenient for expanding density, pressure, velocity, temperature, concentration, and so on are basis sets consisting of low-order polynomials, each function being nonzero only within a subdomain of the transport region. Organizing the polynomials into so-called finite-element basis functions facilitates systematically generating and analyzing solutions by large, fast computers employing modern matrix techniques.

In the algorithms described here for equilibrium and steady-state problems the coefficients in the expansions are derived by the Galerkin weighted residual scheme and are calculated from the resulting sets of nonlinear algebraic equations by the Newton-Raphson method. As a parameter is varied, initial approximations are gotten from nearby solutions by continuation methods. The Newton-Raphson technique is greatly preferred because the Jacobian of the solution is a treasure trove, not only for continuation, but also for analyzing stability of solutions, for detecting bifurcation of solution families, and for computing asymptotic estimates of the effects on any solution, of small changes in parameters, boundary conditions, and boundary shape.

These matters are illustrated by analyses of the shape and stability of interfaces between immiscible fluids at equilibrium, and of meniscus shape in viscous free-surface flow of liquids.

*Department of Chemical Engineering, Massachusetts Institute of Technology, Cambridge, MA 02139.
†Department of Chemical Engineering & Materials Science, University of Minnesota, Minneapolis, MN 55455.
††Exxon Production Research Company, P.O. Box 2981, Houston, TX 77001.
This paper is based on research supported by the National Aeronautics and Space Administration's Fund for Independent Research under Grant NSG 7296 and by the National Science Foundation's Engineering and Applied Science Directorate under grant ENG 77-28315.

1. Introduction. The thermodynamic principles, the conservation laws, and the constitutive relations governing equilibrium, transport, and transformation in fields and phases and across interphase boundaries give rise to integro-partial differential equations, partial differential equations, and interface and boundary conditions that defy classical analysis except in special circumstances that begin with symmetries and usually end with functions of a single variable. Engineering applications require efficient methods for calculating the solutions of such problems and for determining the stability of those solutions to small disturbances, their relation to other solutions, and their sensitivities to small changes in parameters, boundary conditions, and boundary shape. Knowledge of the sensitivity, stability, and multiplicity of a solution (especially the existence of limit points and bifurcation points; see [24]) is in many applications as valuable as knowing the solution. These by-products of a complete analysis give increased meaning to mathematical modeling of complex processes and lead to systematic optimization of processes based on transport principles.

Such a complete analysis is rarely possible by traditional methods which seek solutions in terms of the eigenfunctions of special linear eigenproblems defined on separable coordinate domains. These eigenfunctions are each nonzero almost everywhere in the domain and hence all are involved in satisfying each boundary and interface condition. The limitations of traditional analysis make treatment of complex transport problems difficult or, in many cases, virtually impossible. Even when traditional techniques are successful, the results may only be valid for parts of the domain and also may be restricted to special ranges of parameters.

An alternative has emerged with the advent of high-speed computers and the evolution of efficient software. It is a mathematical generalization and systemization of an idea that goes back at least to the von Karman-Polhausen integral method for boundary-layer analysis, in which a flow is divided into two subdomains, and approximated by simple functions satisfying the momentum principle in an integral sense. The finite element bases invented by structural engineers [2,23,26] are organizations of low-order polynomials constructed to have a given amount of continuity and to allow ease in computer-aided manipulation. The weight of each basis function in the approximate solution is determined so that some weighted residual approximation of the original transport problem is satisfied [6]. That the finite element idea can be mathematically regularized was first recognized by Courant [5] and Prager and Synge [13].

The result of finite element analysis is a system of nonlinear algebraic equations for the basis function weights that mimics the behavior of the differential form of the transport problem. Using new techniques based on Newton's method for the solution of this equation set, it is now possible to perform the complete analysis proposed above. Several of the numerical techniques applicable to this analysis are described in Section 3. Other methods for computer-aided analysis of steady and dynamic problems appear elsewhere in this volume.

The feasibility of computer-aided analysis of complicated transport problems is demonstrated in Sections 4 and 5 with two problems of capillary hydrodynamics — the study of the dynamics of interfaces with appreciable surface tension. First, the shape and stability of a rigidly rotating liquid drop held captive between two parallel disks is considered. Here, the only forces acting on the system are conservative and an extremum principle is used to define equilibrium shapes and shape stability. Second, the flow of a Newtonian liquid issuing from a slit die is calculated. Here, no variational principle exists and a Galerkin weighted residual formulation is used. The sensitivity of the shape of the liquid sheet to the Capillary number, the ratio of normal viscous stress to capillary pressure, is predicted by computer-aided analysis.

The issues addressed here are general and the methods illustrated here by these examples are applicable to a wide range of transport problems.

2. Finite Element Representations. The computer-aided extension of the von Karman-Polhausen two-domain analysis is analysis on subdomains where within each segment or subdomain the solution is approximated as a simple, low-order polynomial or specially chosen function, with coefficients determined so the entire approximation has a specified level of continuity and so the differential equation and boundary conditions are satisfied. Finite-element bases as presented in standard texts [2,23,26] are reorganizations of these polynomials into sets that automatically satisfy the continuity conditions.

The bilinear, biquadratic, and reduced quadratic bases used in the examples that are presented here are built by tesselating the domain into quadrilateral elements and by laying out on these elements a grid of nodes consisting of the vertices of the quadrilaterals (for all three elements), the midpoints of each element boundary (reduced quadratic and biquadratic only), and the centroid of each element (biquadratic only). One basis function is associated with each node and is defined to be unity at that node and to be zero at all other nodes in the domain; see [23] for details.

Since each basis function is nonzero only in elements contiguous to the node with which the function is associated, the basis set has only minimal overlap compared to eigenfunction expansions and globally defined polynomials which, for general problems, overlap over the entire domain.

The solution, call it f(x,y), is represented by an expansion of unknown coefficients $\{\alpha_i\}$ and basis functions $\{\Phi^i(x,y)\}$ as

$$f(x,y) = \sum_{i=1}^{N} \alpha_i \Phi^i(x,y) \qquad (1)$$

where N is the total number of functions in the basis. When the transport problem has an extremum principle the coefficients are determined by minimizing the appropriate functional; when no such principle exists, a weighted-residual method is used to satisfy the equations and boundary conditions. Galerkin weighted residuals, which are formed by weighting the differential equation with each basis function and integrating the product over the domain, are applied here. The Galerkin method as well as an extremum principle mathematically require only limited differentiability of the basis functions which make up the approximate solution (1). For transport problems which result from the application of conservation principles the highest-order derivative usually appears in divergence form and integration-by-parts and application of the divergence theorem always can be used to reduce the amount of differentiability required of the basis.

Finite element representation of the fields in a steady-state non-linear transport problem produces a finite set of nonlinear algebraic equations or residuals for the weighting coefficient α_i of each basis function Φ^i; see equation (1). The dimension N of the set reflects the number of subdomains or elements into which the system has been divided and can in principle be increased without limit to refine the representation to whatever extent required. Moreover, the relative sizes of the subdomains can be controlled to advantage by constructing smaller elements and hence incorporating more basis functions in regions of the domain where the solution is varying rapidly.

Now while a finite element representation lacks the higher-order differentiability that representation in domain-spanning eigenfunctions would have, its status as an approximation to physical reality is scarcely any less. For it can be argued that local constitutive relations and boundary conditions drawn from that reality require existence of no more than a gradient ordinarily (a second gradient rarely). Whether the finite element solutions of a transport problem, even in the limit of vanishingly small elements, have exactly the same parametric sensitivity and pathology, i.e. multiplicity of solutions, limit points and bifurcation points, is a mathematical issue. Were there physically significant differences one would wonder first of all if the constitutive relations and boundary conditions, and perhaps even the system shape, were themselves adequately accurate descriptions of reality.

3. Computer-aided analysis of nonlinear equation set. The set of nonlinear algebraic equations or residuals that result from applica-

tion of the Galerkin/finite-element method to a transport problem can be written as

$$R_i\ (\underline{\alpha};\ \underline{p}) = 0 \qquad i=1,\ldots,N \qquad (2)$$

where $\underline{\alpha} = (\alpha_1, \alpha_2, \ldots, \alpha_N)^T$ is the vector of unknown coefficients and $\underline{p} = (p_1,\ p_2, \ldots, p_M)^T$ is the vector of parameters that define the problem. In general, parameters may enter the problem explicitly, such as physical properties or boundary input data, or implicitly, such as the shape of the boundary itself. If the Newton-Raphson method is used to solve (2) iteratively, the approximation at the k+1-st iteration $\underline{\alpha}^{(k+1)}$ is related to the approximation at the k-th iteration by

$$\underline{\alpha}^{(k+1)} = \underline{\alpha}^{(k)} - [\underline{\underline{J}}(\underline{\alpha}^{(k)})]^{-1}\underline{R}(\underline{\alpha}^{(k)}) \qquad (3)$$

where $\underline{\underline{J}}$ is the Jacobian matrix of the equation set (2), i.e. the matrix of the derivatives $J_{ij} \equiv \partial R_i/\partial\alpha_j$.

When, for a given value of the parameters $\{p_i^o\}$, the solution has converged within the desired tolerance, the Jacobian matrix can be used to examine the dependence of the solution on parameters, and changes within the solution itself. Our view of such a computer-aided algorithm is diagrammed in Figure 1 and elaborated on in the remainder of this Section.

Following the developments in [18, 20], if the Jacobian was inverted, the entries in the inverse would describe the variation of each local value α_i with respect to the residuals R_j on each subdomain. That is, $(\underline{\underline{J}}^{-1})_{ij} = (\partial\alpha_i/\partial R_j)_{R_k}$. Owing to the local character of the finite element basis functions, each residual R_j involves only the nodal values α_k at nodes in the same subdomain. Hence $\underline{\underline{J}}^{-1}$ describes the extent to which changes in each subdomain influence the nodal values everywhere. The inverse Jacobian $\underline{\underline{J}}^{-1}$ is analogous to a Green's function for the linear problem at the parameter values $\{p_i^o\}$; the elements of $\underline{\underline{J}}^{-1}$ give the domain of influence of the equations $\{R_j\}$.

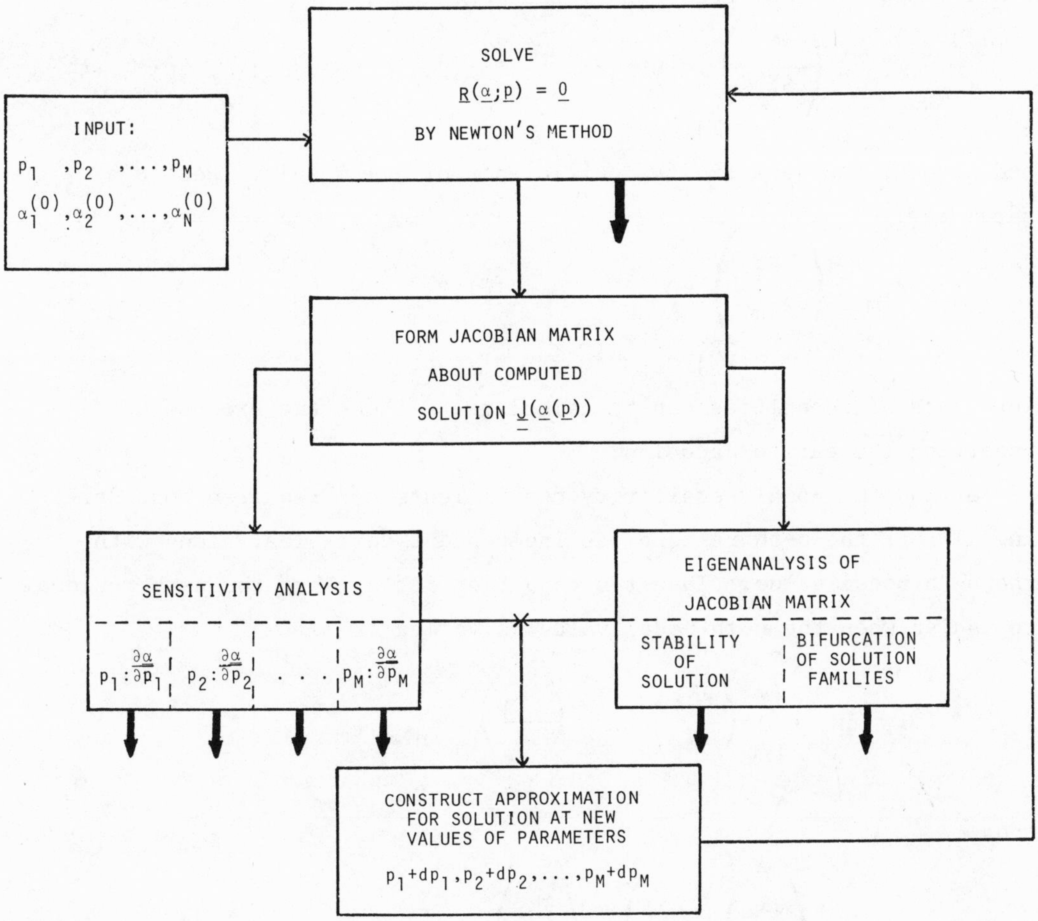

Figure 1. Flowsheet for computer-aided analysis.

If the boundary-value problem is elliptic, the region of influence is, in mathematical principle, the entire domain; but from a practical viewpoint some may be quite restricted while others are large. Knowledge of the influence of each equation is valuable to the analyst who wants to make the nodal mesh more nearly optimal, or to make simplifying approximations in regions that do not affect strongly the practically important areas of the solution.

For all these purposes there are two kinds of quantities that are more useful and easier to calculate than the inverse of the Jacobian. They are, first, the sensitivity coefficients of other nodal values to one particular nodal value when the equations are satisfied everywhere except on the subdomain of that one node,

$$c_{in} \equiv \left(\frac{\partial\alpha_i}{\partial\alpha_n}\right)_{R_j} = 0, \; j\neq n \qquad i=1,2,\ldots,n-1,\; n+1,\ldots,N; \tag{4}$$

and second, the sensitivity coefficients of all nodal values to a parameter:

$$d_{im} \equiv \left(\frac{\partial\alpha_i}{\partial p_m}\right)_{R_j} = 0 \quad , \qquad i=1,2,\ldots,N \quad . \tag{5}$$

Both sets of quantities can be calculated without the expense of inverting the entire Jacobian.

To find the nodal sensitivity coefficients c_{in} requires considering all but the n-th residual as independent variables, along with the n-th nodal value. Then the condition for all but the n-th residual to vanish when the n-th nodal value is varied is

$$0 = dR_j = \sum_{i\neq n}^{N} \left(\frac{\partial R_j}{\partial\alpha_i}\right)_{\alpha_k} d\alpha_i + \left(\frac{\partial R_j}{\partial\alpha_n}\right)_{\alpha_k} d\alpha_n \quad , \qquad j\neq n. \tag{6}$$

Thus,

$$0 = \sum_{i\neq n}^{N} \left(\frac{\partial R_j}{\partial\alpha_i}\right)_{\alpha_k} \left(\frac{\partial\alpha_i}{\partial\alpha_n}\right)_{R_j} + \left(\frac{\partial R_j}{\partial\alpha_n}\right)_{\alpha_k} \quad , \qquad j\neq n, \tag{7}$$

or, in matrix form,

$$\underline{0} = \underline{\underline{J}}_n \underline{c}_n + \underline{r}_n \quad , \quad \underline{c}_n = -\underline{\underline{J}}_n^{-1} \underline{r}_n \tag{8}$$

where $\underline{\underline{J}}_n$ is the Jacobian $(\partial R_j/\partial\alpha_i)$ with the n-th row and the n-th column deleted, $\underline{r}_n$ is the n-th column, and $\underline{c}_n$ is the vector with components c_{in} as defined by (4).

Similarly, to find the nodal sensitivity coefficients d_{im} requires considering all N residuals as independent variables along with the m-th parameter p_m. Then the condition for evolution of a solution as the parameter is varied is

$$0 = dR_j = \sum_{i=1}^{N} \left(\frac{\partial R_j}{\partial \alpha_i}\right)_{\alpha_k, p_k} d\alpha_i + \left(\frac{\partial R_j}{\partial p_m}\right)_{\alpha_k, p_q, q \neq m} dp_m \tag{9}$$

Thus,

$$0 = \sum_{i=1}^{N} \left(\frac{\partial R_j}{\partial \alpha_i}\right)_{\alpha_k} \left(\frac{\partial \alpha_i}{\partial p_m}\right)_{R_j} + \left(\frac{\partial R_j}{\partial p_m}\right)_{\alpha_k} , \tag{10}$$

or, in matrix form,

$$\underline{0} = \underline{\underline{J}}\,\underline{d}_m + \underline{s}_m \quad ; \quad \underline{d}_m = -\underline{\underline{J}}^{-1}\,\underline{s}_m \tag{11}$$

where $\underline{s}_m$ is the column vector, $(\partial R_j / \partial p_m)$, and $\underline{d}_m$ is the vector with components as defined by (5). Once found, the set of sensitivity coefficients $\underline{d}_m$ permits an extremely useful multiple-Taylor-series expansion of a family of solutions around any member of the family, i.e. a base solution $\underline{\alpha}_0(\underline{p}^0) \equiv \underline{\alpha}_0(p_1^0, p_2^0, \ldots p_m^0)$,

$$\underline{\alpha}(\underline{p}^0 + d\underline{p}) = \underline{\alpha}_0 + \sum_{q=1}^{M} \underline{d}_m \, d\underline{p} + \text{H.O.T.} \tag{12}$$

where $d\underline{p}$ denotes the m-dimensional vector of differential changes in the parameters. From (11), the first-order terms in the Taylor series (12) are calculated simply by solving a linear system of equations with the Jacobian of the base solution $\underline{\alpha}_0$. The Taylor series (12) is equivalent to the result of a traditional perturbation analysis where the linearized problem, for a small change in $\{p_i\}$ from a known state $\underline{\alpha}_0(\underline{p}^0)$, is solved. These conclusions drawn from (12) were first reached by Kubíček [11] and by Stewart and Sørensen [22].

As shown in Figure 1, the expansion (12) also forms the basis of continuation or embedding methods for generating the first approximation to the solution of (2) when the parameters are incrementally changed in a sequential solution strategy [3, 25]. Equation (12) leads to an essentially prediction-corrector scheme; the new solution is predicted by (12) and then corrected by subsequent Newton iterations. When the iteration is converged, the new Jacobian is ready to use in the next step.

Continuation methods have recently been extended to treat nonlinear systems where the Jacobian matrix may be singular as at bifurcation or limit points [10,12,17,21]. Once a bifurcation point of two solution families has been located, iteration techniques are also available for forcing the nonlinear equation solver to converge to a particular solution family [1,10,14].

For transport problems with an extremum principle eigenanalysis of the Jacobian matrix is an effective means of studying both stability and bifurcation of steady states. For these systems steady states are found as stationary points of an energy functional and are stable and persist in time only if the stationary points are at least local minima of the functional. The eigenvalues of the Jacobian when evaluated at a stationary point describe the curvature of the energy surface near the equilibrium state. When all eigenvalues are positive the stationary point is a local minimum.

Bifurcating states are conveniently detected by the occurrence of a neutrally stable state which is signalled by the vanishing of an eigenvalue. Since only eigenvalues near to zero are needed in bifurcation and stability analysis, iterative methods for calculating only the lowest several are preferred (see [15]).

4. Analysis of viscous free-surface flow. To illustrate the applicability of the Galerkin/finite-element method coupled with computer-aided analysis, parts of a study [19] of the viscous, two-dimensional flow of a liquid jet issuing from a slit die as shown in Figure 2 are summarized. When the effects of gravity are negligible, the liquid jet reaches a final thickness h_∞ within a short distance from the die, provided the Reynolds number $N_{Re} \equiv Vb/\upsilon$ is small, where V is a characteristic velocity of the jet, υ is the kinematic viscosity of the liquid and b is the half-width of the die. The swell ratio $\Delta \equiv h_\infty/b$ depends on the interplay of surface tension, viscous, and inertial forces acting throughout the flow and on the free-surface of the flow.

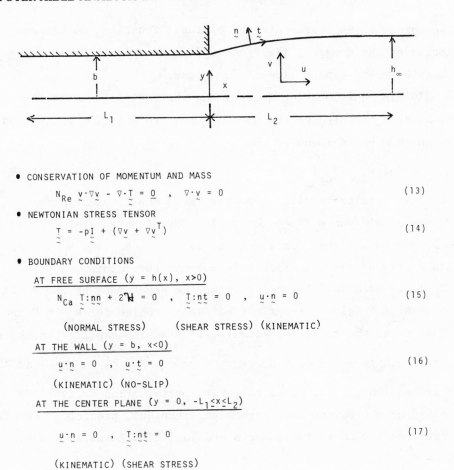

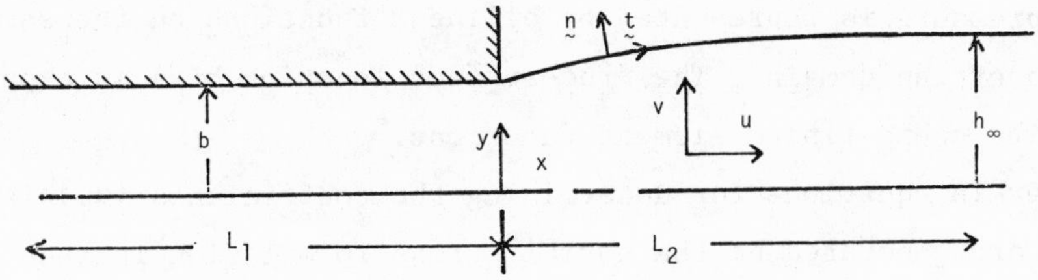

• CONSERVATION OF MOMENTUM AND MASS

$$N_{Re}\, \underset{\sim}{v}\cdot\nabla\underset{\sim}{v} - \nabla\cdot\underset{\approx}{T} = \underline{0} \quad , \quad \nabla\cdot\underset{\sim}{v} = 0 \qquad (13)$$

• NEWTONIAN STRESS TENSOR

$$\underset{\approx}{T} = -p\underset{\approx}{I} + (\nabla\underset{\sim}{v} + \nabla\underset{\sim}{v}^{T}) \qquad (14)$$

• BOUNDARY CONDITIONS

AT FREE SURFACE $(y = h(x),\ x>0)$

$$N_{Ca}\, \underset{\approx}{T}:\underset{\sim}{n}\underset{\sim}{n} + 2\mathcal{H} = 0 \quad , \quad \underset{\approx}{T}:\underset{\sim}{n}\underset{\sim}{t} = 0 \quad , \quad \underset{\sim}{u}\cdot\underset{\sim}{n} = 0 \qquad (15)$$

(NORMAL STRESS) (SHEAR STRESS) (KINEMATIC)

AT THE WALL $(y = b,\ x<0)$

$$\underset{\sim}{u}\cdot\underset{\sim}{n} = 0 \quad , \quad \underset{\sim}{u}\cdot\underset{\sim}{t} = 0 \qquad (16)$$

(KINEMATIC) (NO-SLIP)

AT THE CENTER PLANE $(y = 0,\ -L_1 \le x \le L_2)$

$$\underset{\sim}{u}\cdot\underset{\sim}{n} = 0 \quad , \quad \underset{\approx}{T}:\underset{\sim}{n}\underset{\sim}{t} = 0 \qquad (17)$$

(KINEMATIC) (SHEAR STRESS)

INFLOW AND OUTFLOW BOUNDARIES

$$\underset{\approx}{T}:\underset{\sim}{n}\underset{\sim}{n} = 0 \quad , \quad \underset{\sim}{u}\cdot\underset{\sim}{t} = 0 \quad \text{at} \quad x = -L_1 = L_2 \qquad (18)$$

Figure 2. Governing equations and boundary conditions for a liquid jet issuing from a slit die.

The velocity $\underline{v}(x,y)$ and the pressure $p(x,y)$ of the liquid and the location of the surface $h(x)$ of the jet are governed by the principles of conservation of linear-momentum and mass, which are given in dimensionless form in Figure 2 along with the interface and boundary conditions. A solution of the viscous, free-surface flow problem specified by equations (13-18) is sought by the Galerkin weighted residual method in terms of the mixed-order finite element representations recommended by Hood and Taylor [9]. Here x- and y-components of velocity are approximated by expansions of biquadratic basis functions and pressure is represented by bilinear functions on the same tesselation of the domain. The free-surface location h(x) is expanded in Hermite cubic finite-element functions.

The Galerkin equations for determining the coefficients $\{\alpha_i\}$ in these expansions are generated by the normal stress formulation introduced in [19] and can be represented as

$$\underline{R}\ (\underline{\alpha};\ N_{Re},\ N_{Ca}) = \underline{0} \tag{19}$$

where $N_{Ca} \equiv V\mu/\sigma$ is the capillary number (μ is the liquid viscosity, and σ is the liquid/gas surface tension). The equation set (19) is solved by Newton's method for specified values of N_{Re} and N_{Ca}. The Jacobian matrix of the converged solution and the column vector $(\partial \underline{R}_i/\partial N_{Ca})$ are used in (11) to determine the dependence of the swell ratio Δ on N_{Ca}. Solving the finite-element problem for $N_{Re} = 0$ and $N_{Ca} = 0$ leads to the following formula for Δ at low capillary numbers

$$\Delta \equiv h_\infty/b = 1 + 0.36\ N_{Ca}. \tag{20}$$

Richardson [16], by means of elaborate perturbation derivations that relied on the Weiner-Hopf expansion technique, predicted that the swell ratio at low capillary numbers and vanishing Reynolds number should be

$$\Delta \equiv h_\infty/b = 1 + 0.356\ N_{Ca} \tag{21}$$

The perturbation-like prediction of finite-element analysis agrees to within one percent to the result of more traditional analysis. However, the computer-aided analysis is more general; the limitation $N_{Ca} \ll 1$ inherent to Richardson's analysis is not required for calculations based on finite-element analysis. Figure 3 shows the difference between the swell ratio calculated by finite-element analysis and that predicted by (20) or (21) for four decades of N_{Ca}. These results quantity the range of validity of the asymptotic formulae (20-21). Clearly for N_{Ca} approaching unity neither formula is valid, but equations like (20) can be formed by computer-aided analysis in the neighborhood of <u>any</u> value of N_{Ca}.

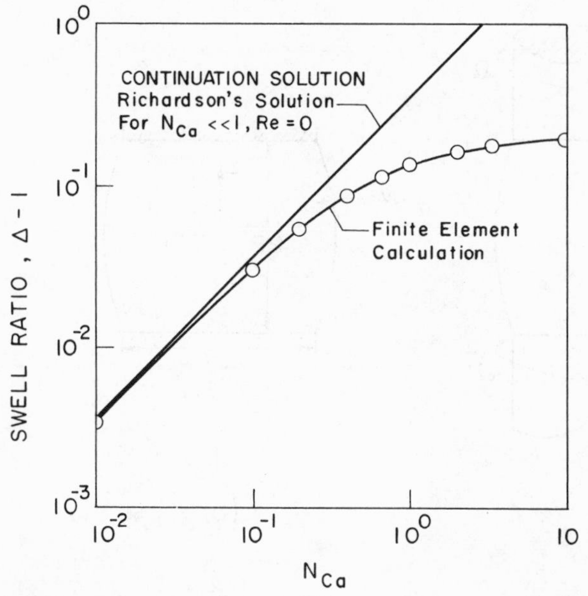

Figure 3. Comparison of asymptotic and finite-element calculations for the swell ratio Δ of a sheet jet as a function of Capillary number N_{Ca}.

5. <u>Shape and stability of captive rotating drops</u>. To illustrate the applicability of a finite-element based algorithm for calculating equilibrium states, their stability, and bifurcation points and limit points, parts of a study [4] of the axisymmetric and three-dimensional shapes of captive, rotating liquid drops as shown in Figure 4 are summarized. The drops are of Volume V and density ρ and are held by surface tension σ between parallel, coaxial, circular, solid surfaces of radius R a distance 2B apart. Both faces rotate at angular velocity Ω and when the surrounding fluid is tenuous, the captive drop rotates rigidly with a shape dictated by local capillary and centrifugal forces.

The shape of the drop is represented in cylindrical polar coordinates as $\tilde{r} = \tilde{f}(\theta,\tilde{z})$ and is presumed to have a plane of reflective symmetry that passes through the axis of rotation; this limits the domain of the drop to $-B\le\tilde{z}\le B$ and $0\le\theta\le\pi$. The drop also is presumed to hold fast to the circular edges of the solid faces.

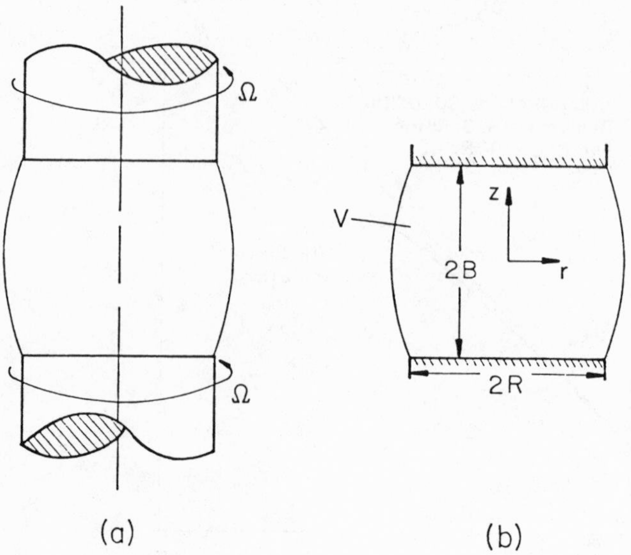

Figure 4. A rotating captive liquid drop (a) of volume V with its dimensions (b).

Because the forces acting on the drop are conservative, the drop shape is found as an extremum of the effective potential energy $\tilde{E}$ of the system:

$$E \equiv \int_0^{\pi}\int_{-B}^{+B} (\sigma \sqrt{f^2+f^2+f_{\tilde{z}}^2+f_\theta^2} \quad - \quad \rho\Omega^2\tilde{f}^4/8)dzd\theta \,. \quad (22)$$

This shape is stable if the extremum of (22) is a local minimum with respect to all infinitesimal perturbations that leave the drop fixed to the confining faces and unchanged in volume.

The shape $\tilde{f}(\theta,\tilde{z})$ is approximated in a reduced quadratic finite-element basis and the extremum problem is solved by Newton iteration for a sequence of values of rotational Bond number $\Sigma \equiv \rho\Omega^2 R^3/8\sigma$, beginning with no rotation ($\Sigma = 0$). Stability of the calculated shapes is found from the sign of the lowest eigenvalue of the constrained algebraic eigenproblem associated with the variation of the potential energy with respect to all perturbations that can be represented in the reduced quadratic basis and which satisfy the constraint of fixed volume. The lowest several eigenvalues and eigenvectors are computed with the block-Lanczos method developed by Golub and Underwood [7]. The projection method suggested by Golub [8] is used to ensure that eigenvectors satisfy the volume constraint.

As discussed in Section 3, a zero eigenvalue signals not only neutral stability to a certain shape perturbation but also bifurcation of another family of equilibrium shapes of the same or different symmetry. The eigenvector of neutral stability is used to force the Newton iteration to converge to a shape in the bifurcating family.

The finite-element algorithm for shape and stability was tested for the special case of a drop with exactly the volume of the right-circular cylinder that intersects the edges of the solid faces; for this case a cylinder is an equilibrium shape for all values of rotational Bond number Σ and traditional analysis of shape multiplicity, and stability leads to closed-form results. As Figure 5 shows, multiple equilibrium drop shapes are possible. Both other axisymmetric shapes with axial undulations and non-axisymmetric shapes are possible equilibrium shapes. The bifurcation points of the families and the stability of each family predicted by finite-element analysis match well with the closed-form results; for details see [4].

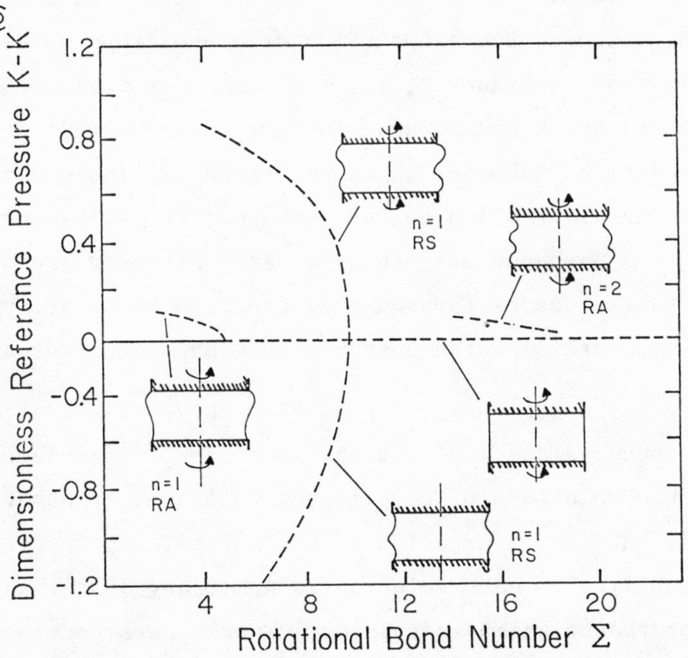

Figure 5. Axisymmetric shape families of a drop with cylindrical volume calculated by finite element analysis. The drop volume is $2\pi B$ and the length is $2B=1$.

Drops with volumes greater or less than the cylindrical drop are axisymmetric for small rotation rates (small Σ), but each reaches a maximum rotation rate beyond which axisymmetric equilibrium shapes do not exist; the family turns back to lower Σ. Since no simple representation is known for these shapes, traditional analysis of bifurcation and stability is almost impossible and only computer-aided results are feasible. Two possibilities are found for the members of these axisymmetric shape families. Either its members first become unstable to asymmetric, C-shaped perturbations or the family passes through a limit point at a maximum Σ and becomes unstable to an axisymmetric perturbation. Which fate besets stable axisymmetric forms as Σ becomes large depends on the volume and length of the drop: long skinny drops run out of existence, whereas short fat ones lose stability to C-shaped perturbations.

When the axisymmetric drops become unstable to a C-shaped perturbation there is a bifurcation to a family of C-shaped drops. Finite-element results show that the bifurcation of merely fat drops to C-shapes is subcritical, i.e. to lower Σ and the C-shapes are themselves unstable. In contrast, the bifurcation of obese drops to C-shapes is supercritical, i.e. to higher Σ, and the family of C-shapes is stable up to a limit point, which occurs as Σ is further increased. An example of this supercritical behavior is demonstrated in Figure 6 for an obese drop. Beyond the bifurcation to the C-shapes, the axisymmetric shapes are unstable to a C-shaped disturbance. At a higher value of Σ the axisymmetric family passes through a bifurcation to a family of two-lobed drops which are likewise unstable to C-shaped perturbations.

6. Summary. The two examples presented above illustrate the practicality of computer-aided analysis based on the finite-element method. Not only can the solutions of complicated nonlinear transport problems be calculated, but the bifurcation of the solution into multiple families, the stability of each solution to arbitrary disturbances, and the sensitivity of the solution to changes in parameters and input data can be found. Moreover, computer-aided analysis is applicable to a much wider range of problems than traditional analysis.

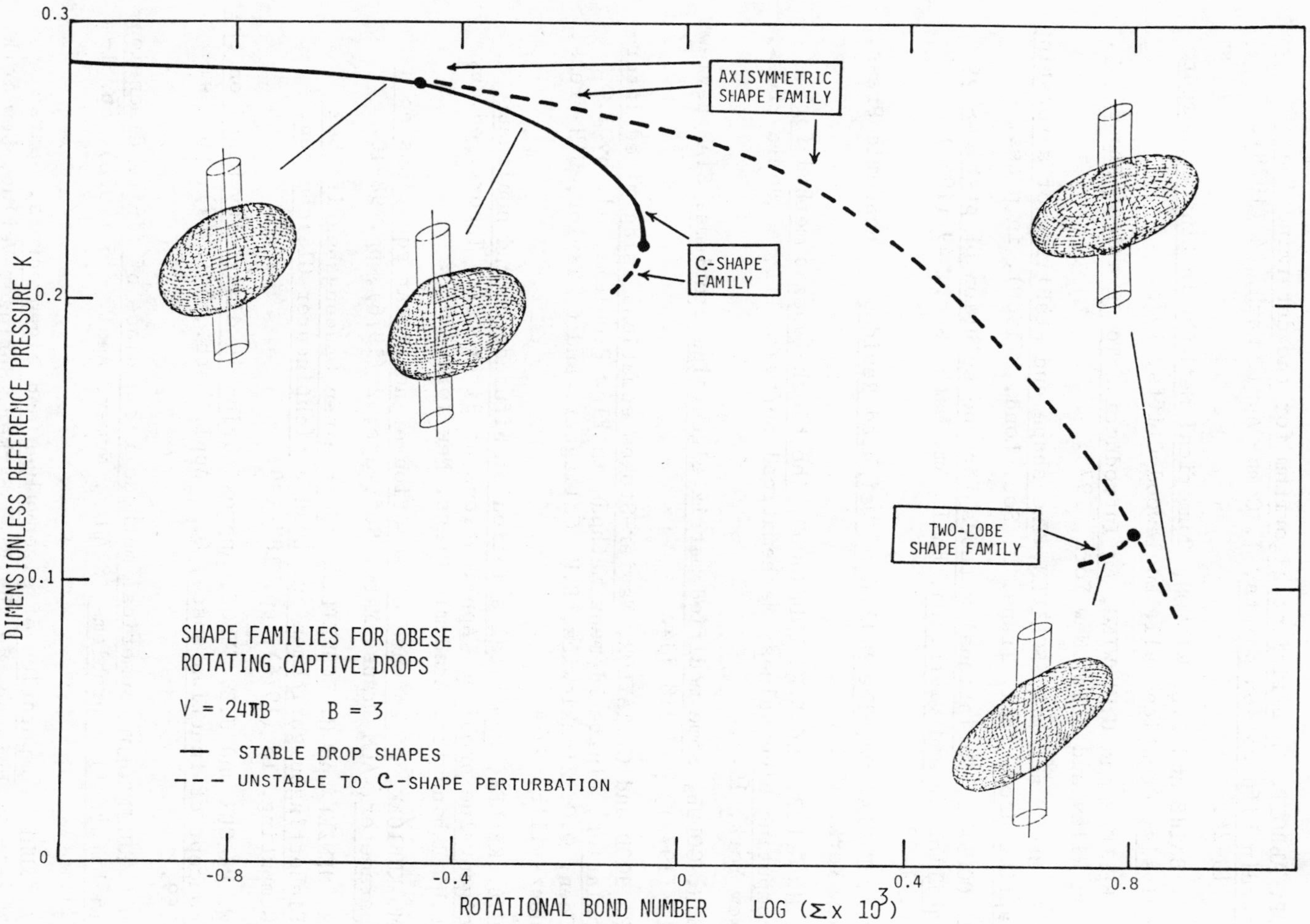

Figure 6. Axisymmetric, C-shaped, and two-lobed families of obese rotating liquid drops. $V = 24\pi B$ and $B = 3$. Stable (——) and unstable (----).

REFERENCES

[1] J.P. ABBOTT, An efficient algorithm for the determination of certain bifurcation points, J. Comp. Appl. Math., 4 (1978), pp. 19-26.

[2] K.J. BATHE and E.L. WILSON, Numerical Methods in Finite Element Analysis, Prentice-Hall, Englewood Cliffs, 1976.

[3] R. BELLMAN and G.M. WING, An Introduction To Invariant Embedding, John Wiley and Sons, New York, 1975.

[4] R.A. BROWN and L.E. SCRIVEN, The shape and stability of a rotating captive drop, Phil. Trans. R. Soc. Lond., (1980), in press.

[5] R. COURANT, Variational methods for the solution of problems of equilibrium and vibration, Bull. Am. Math. Soc., 49 (1943), pp. 1-23.

[6] B.A. FINLAYSON, The Method of Weighted Residuals, Academic Press, New York, 1972.

[7] G.H. GOLUB, and R.R. UNDERWOOD, The block-Lanczos method for computing eigenvalues, Mathematical Software III, Academic Press, New York, 1977.

[8] G.H. GOLUB, Some modified matrix eigenvalue problems, SIAM Review, 15 (1973), pp. 318-334.

[9] P. HOOD and C. TAYLOR, Navier-Stokes equations using mixed interpolation, Finite Element Methods in Flow Problems, ed. by J.T. Oden, O.C. Zienkiewicz, R.H. Gallagher, and C. Taylor, UAH Press, Huntsville, 1974.

[10] H.B. KELLER, Numerical solution of bifurcation and nonlinear eigenvalue problems, Applications of Bifurcation Theory, ed. by P.H. Rabinowitz, Academic Press, New York, 1977.

[11] M. KUBIČEK, Dependence of solution of nonlinear equations on a parameter, ACM Trans. Math. Software, 2 (1976), pp. 98-107.

[12] R. MENZEL and H. SCHWETLICK, Über einen Ordnungsbegriff bei Einbettingsulgorithmen zur Lösung nichtinearer Gleichungen, Computing, 16 (1976), pp. 187-199.

[13] W. PRAGER and J.L. SYNGE, Approximation in elasticity based on the concept of function space, Q.J. Appl. Math., 5 (1947), pp. 241-269.

[14] W.C. RHEINBOLDT, Numerical methods for a class of finite dimensional bifurcation problems, SIAM J. Numer. Anal., 15 (1978), pp. 1-11.

[15] A. RUHE, Computation of eigenvalues and eigenvectors, Sparse Matrix Techniques, ed. by V.A. Barker, Springer-Verlag, New York, 1977.

[16] S. RICHARDSON, A stick-slip problem related to motion of a free jet at low Reynolds number, Proc. Cambridge Philos. Soc., 67 (1970), pp. 477-489.

[17] E. RIKS, The application of Newton's method to the problem of elastic stability, J. Appl. Mech., 39 (1972), pp. 1060-1065.

[18] W.J. SILLIMAN, Viscous film flows with contact lines, Ph.D. thesis, University of Minnesota, 1979.

[19] W.J. SILLIMAN and L.E. SCRIVEN, Separating flow near a static contact line: slip at a wall and shape of a free surface, J. Comp. Phys., (1980), in press.

[20] W.J. SILLIMAN and L.E. SCRIVEN, Computer-aided analytics, presented at A.I.Ch.E. Meeting, Boston, 19-22 August, 1979.

[21] R.B. SIMPSON, A method for the numerical determination of bifurcation states of nonlinear systems of equations, SIAM J. Numer. Anal., 12 (1975), pp. 439-451.

[22] W.E. STEWART and J.P. SØRENSEN, Sensitivity and regression of multicomponent reactor models, 4th International/6th European Symposium on Chemical Reaction Engineering, Dechema, Frankfurt, 1976.

[23] G. STRANG and G.J. FIX, An Analysis of the Finite Element Method, Prentice-Hall, Englewood Cliffs, 1973.

[24] J.M.T. THOMPSON and G.W. HUNT, A General Theory of Elastic Stability, Wiley, London, 1973.

[25] H. WACKER, Continuation Methods, Academic Press, New York, 1978.

[26] O.C. ZIENKIEWICZ, The Finite Element Method, McGraw-Hill, New York, 1977.

SECTION FOUR
ELECTRICAL AND CIVIL ENGINEERING

Steady Motions Exhibited by Duffing's Equation: A Picture Book of Regular and Chaotic Motions

Yoshisuke Ueda*

Abstract. Various types of steady states take place in the system exhibited by Duffing's equation. Among them harmonic, higher harmonic and subharmonic motions are popularly known. Then ultrasubharmonic motions of different orders are fairly known. However chaotic motions are scarcely known. By using analog and digital computers, this report makes a survey of the whole aspect of steady motions exhibited by Duffing's equation.

1. Introduction. Duffing's equation appears in various physical and engineering problems. It is one of the simplest and the most important nonlinear differential equations. The aim of this report is to give the whole aspect of steady states exhibited by the equation. Throughout this paper the term steady state or steady motion means physical state which continues infinitely after the transient has vanished.

There are various types of steady motions exhibited by Duffing's equation. Among them deterministic or regular motions are generally known, e.g., harmonic, higher harmonic and subharmonic motions. However, owing to the perfectly deterministic nature of the equation, no reference has been made to the possibility of the existence of chaotic motions for a long time. The occurrence of chaotic motions was originally studied by the author [1, 2, 3, 4]. Holmes has also observed chaotic behavior in analog computer solutions of Duffing's equation [5, 6]. Further Moon has performed experiments for the forced vibrations of a buckled beam and showed the existence of chaotic motions [7, 8].

*Department of Electrical Engineering, Kyoto University, Kyoto, Japan.

The purpose of this report is to make a survey of the steady motions exhibited by Duffing's equation which takes the form

$$\frac{d^2x}{dt^2} + k\frac{dx}{dt} + x^3 = B\cos t \qquad (1)$$

Since the solution of Eq. (1) cannot be obtained analytically, we have relied on analog and digital computers. Thus computer solutions are examined and summarized in this report. Therefore, from the mathematical point of view, they may raise new questions, yet it is of value and interest to introduce them to many researchers in various fields.

2. Preliminaries.

2.1 Discrete Dynamical System. Equation (1) is rewritten as

$$\frac{dx}{dt} = y, \quad \frac{dy}{dt} = -ky - x^3 + B\cos t \qquad (2)$$

Let us here introduce a diffeomorphism on the xy plane into itself by using the solutions of Eqs. (2). Let $x = x(t, x_0, y_0)$, $y = y(t, x_0, y_0)$ be a solution of Eqs. (2) which starts from a point $p_0 = (x_0, y_0)$ at $t = 0$. Let $p_1 = (x_1, y_1)$ be the location of the solution at the instant of $t = 2\pi$, i.e., $x_1 = x(2\pi, x_0, y_0)$, $y_1 = y(2\pi, x_0, y_0)$; then a C^∞-diffeomorphism

$$f_\lambda : R^2 \to R^2, \quad p_0 \mapsto p_1, \quad \lambda = (k, B) \qquad (3)$$

of the xy plane into itself is defined.

A periodic solution of Eqs. (2) is represented by a fixed or n-periodic point of f_λ, i.e., $p = f_\lambda^n(p)$, $(n \in Z^+)$. A fixed point p is characterized by the eigenvalues m_1, m_2 of $Df_\lambda(p)$, the derivative of f_λ evaluated at the point. A simple fixed or periodic point is classified into: (i) completely stable fixed or periodic point or sink (S), (ii) completely unstable point or source (U), (iii) directly unstable point or saddle (D) and (iv) inversely unstable point or saddle (I).

The steady motion exhibited by Eqs. (2) is represented by an attractor of the diffeomorphism f_λ. If an attractor is composed of a single periodic group, the corresponding motion turns out to be periodic and hence deterministic or regular. But if it is composed of a closed, invariant set of f_λ containing infinitely many unstable periodic groups, chaotic motion appears resulting from the small uncertain factors in the real systems.

2.2 Chaotically Transitional Processes and Strange Attractors. A periodic motion is represented by asymptotically stable periodic solution. It corresponds to a sink of the diffeomorphism f_λ having a wide basin as compared with random noise in the real system. On the other hand a chaotic motion is represented by a bundle of solutions in the txy space which is asymptotically orbitally stable and contains infinitely many unstable periodic solutions. The representative point of the physical state wanders chaotically among the solutions of this bundle under the influence of small uncertain factors in the real systems. Considering this nature, we have called the phenomenon chaotically transitional process [1].

The set of points on the xy plane consisting of the cross section of the bundle at $t = 2n\pi$ ($n \in Z$) is called a strange attractor. We have emphasized that the strange attractor is identical with a closure of unstable manifolds of a saddle of the diffeomorphism f_λ.

The complex structure of the solution bundle for this problem is not fully understood, but it should be noted that in other systems, such as the Lorenz attractor, the presence of chaotic motion is reported even in the absence of noise [9]. Of course, in real physical systems one does have such noise.

no hardware, however.

3. Experimental Results on the Steady Motions.

3.1 kB Chart for Different Types of Steady States. In the forced oscillatory system exhibited by Eq. (1), various types of steady states are sustained depending on the system parameters $\lambda = (k, B)$ as well as on the initial conditions. Figure 1 shows the regions on the kB plane in which different steady motions are observed. These regions are obtained by using analog and digital computers. The roman numerals I, II, II', II", III and IV characterize 2π periodic motions. The fractions m/n (m = 1, 3, 4, 5, 6, 7, 11 and n = 2, 3) indicate the regions in which subharmonic or ultrasubharmonic motions of order m/n are sustained. An ultrasubharmonic motion of order m/n is a periodic motion

$f = \frac{m}{n} f_0 \quad \left(\frac{m = 1, 3, 4, 5, 6, 7, 11}{n = 2, 3}\right) =$

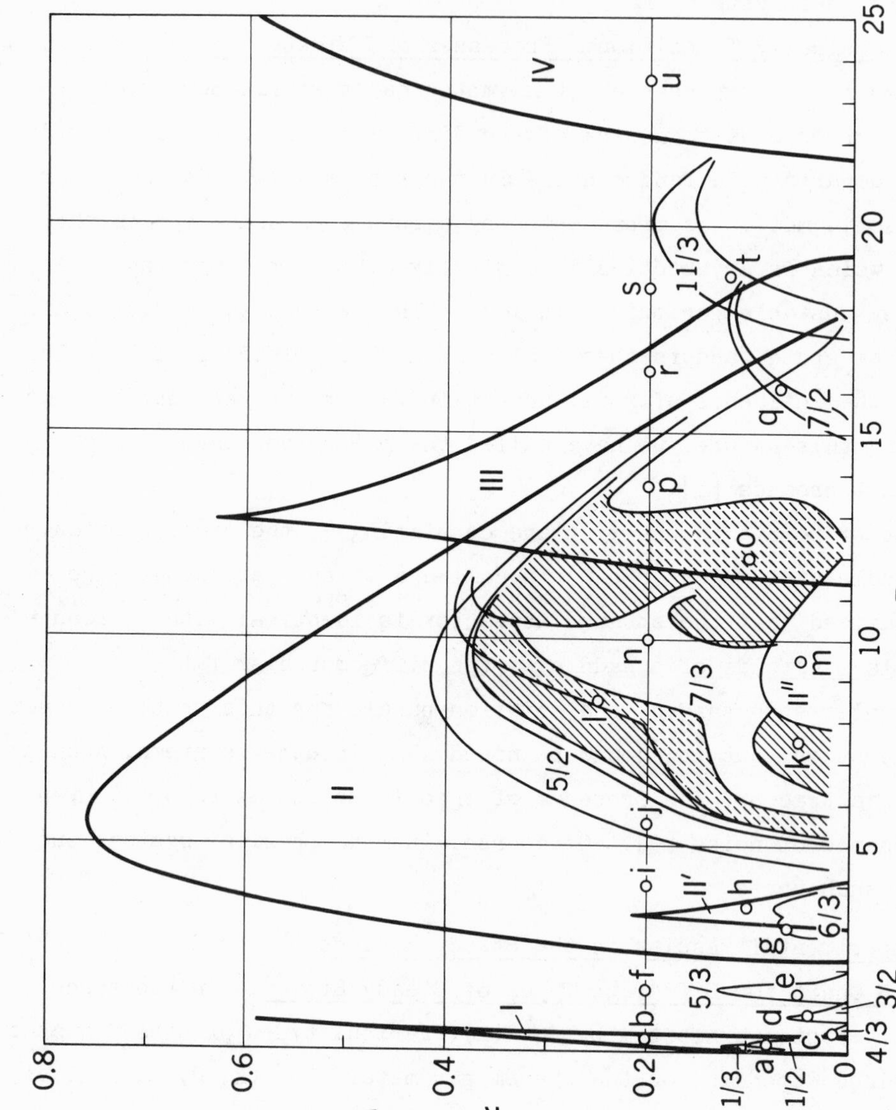

Figure 1: Regions of different steady states for the system exhibited by Eq. (1).

whose principal frequency is m/n times the frequency of external force. Chaotic motions take place in the shaded regions. In the area hatched by full lines, chaotic motion occurs uniquely, while in the area hatched by dotted lines, two different steady states take place, i.e., chaotic and regular motions. Which one occurs depends on the initial conditions.

Ultrasubharmonic motions of higher orders ($n = 4, 5, \ldots$) can occur naturally in the system, but they are omitted in Fig. 1. Further we should like to add that, though we performed the experiment carefully and repeatedly, the chart is far from perfect. In particular, the regions which lie between B = 5 and 15 are regarded as very serious problems. Details are doubtful and further investigations will be required.

3.2 A Collection of Steady Motions. In order to illustrate the regions of kB chart, we choose a set of parameters $\lambda = (k, B)$ from every region. The locations of these parameters are indicated by alphabets from a to u in Fig. 1. Figure 2 shows the trajectories of the steady motions on these points. Almost all steady motions which will occur for these parameters are supposed to be collected. In the figure, positions of the representative point at the instant $t = 2n\pi$ ($n \in Z$) are marked x. So the marks x on the periodic trajectories are the completely stable fixed or periodic points of the diffeomorphism f_λ. The three cases (k), (l_1) and (o_1) show chaotic motions, in which the trajectories are drawn after the transients have vanished and hence the marks x appear on the strange attractors.

3.3 Chaotically Transitional Processes. The outlines of the strange attractors for the cases (k), (l_1) and (o_1) are shown in Fig. 3. They are plotted after the transients have vanished. As mentioned before, the strange attractors are identical with the closures of unstable manifolds of some saddles of f_λ. The waveforms which are the realizations of these chaotically transitional processes are given in Fig. 4. The global phase plane structure of f_λ for the case (o_1) is shown in Ref. [2]. The transition of the strange attractors and average power spectra are also reported in Ref. [3].

3.4 Remarks on the Experimental Results. Here we briefly summarize

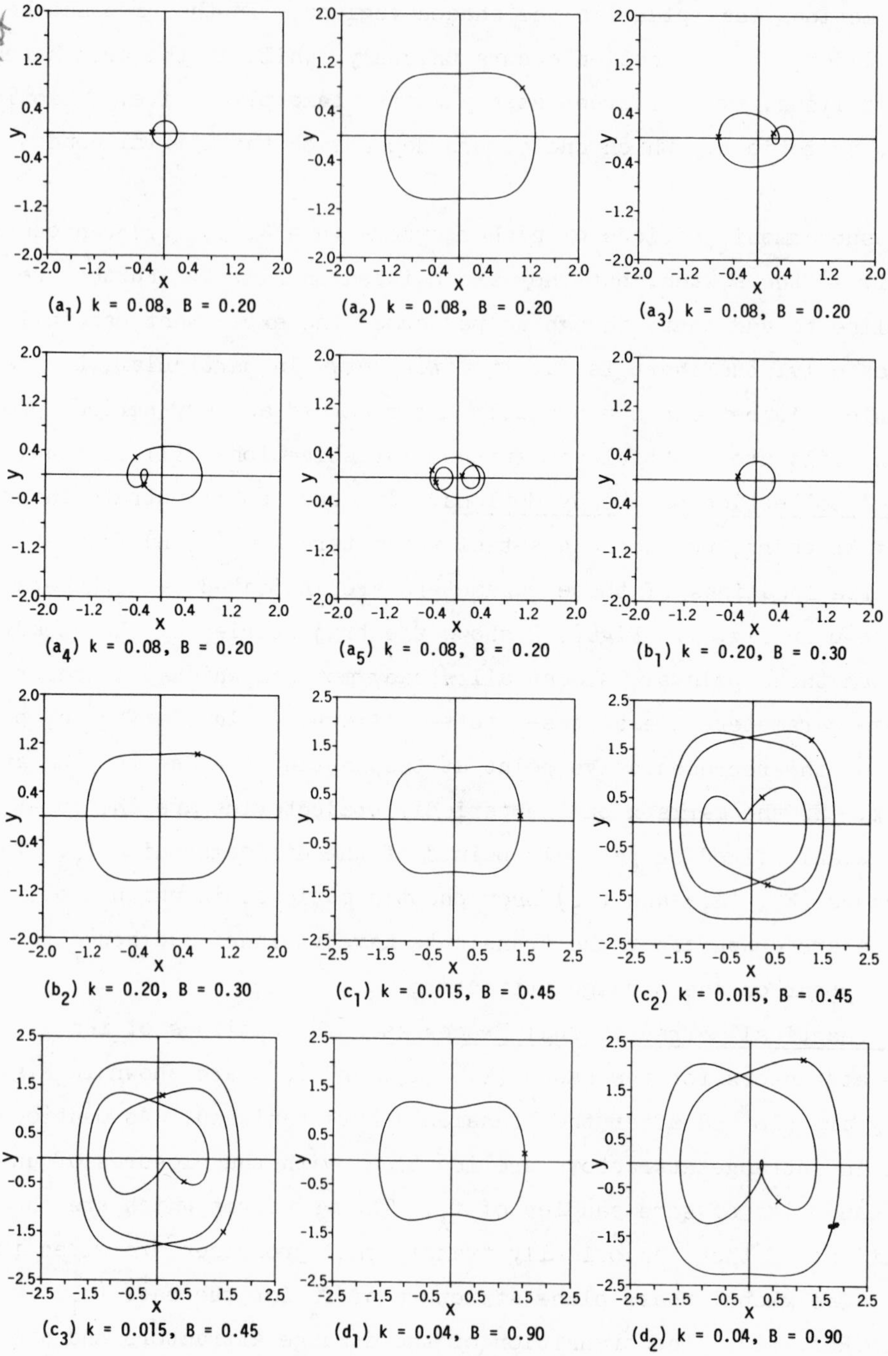

Figure 2: Trajectories of various types of steady motions.

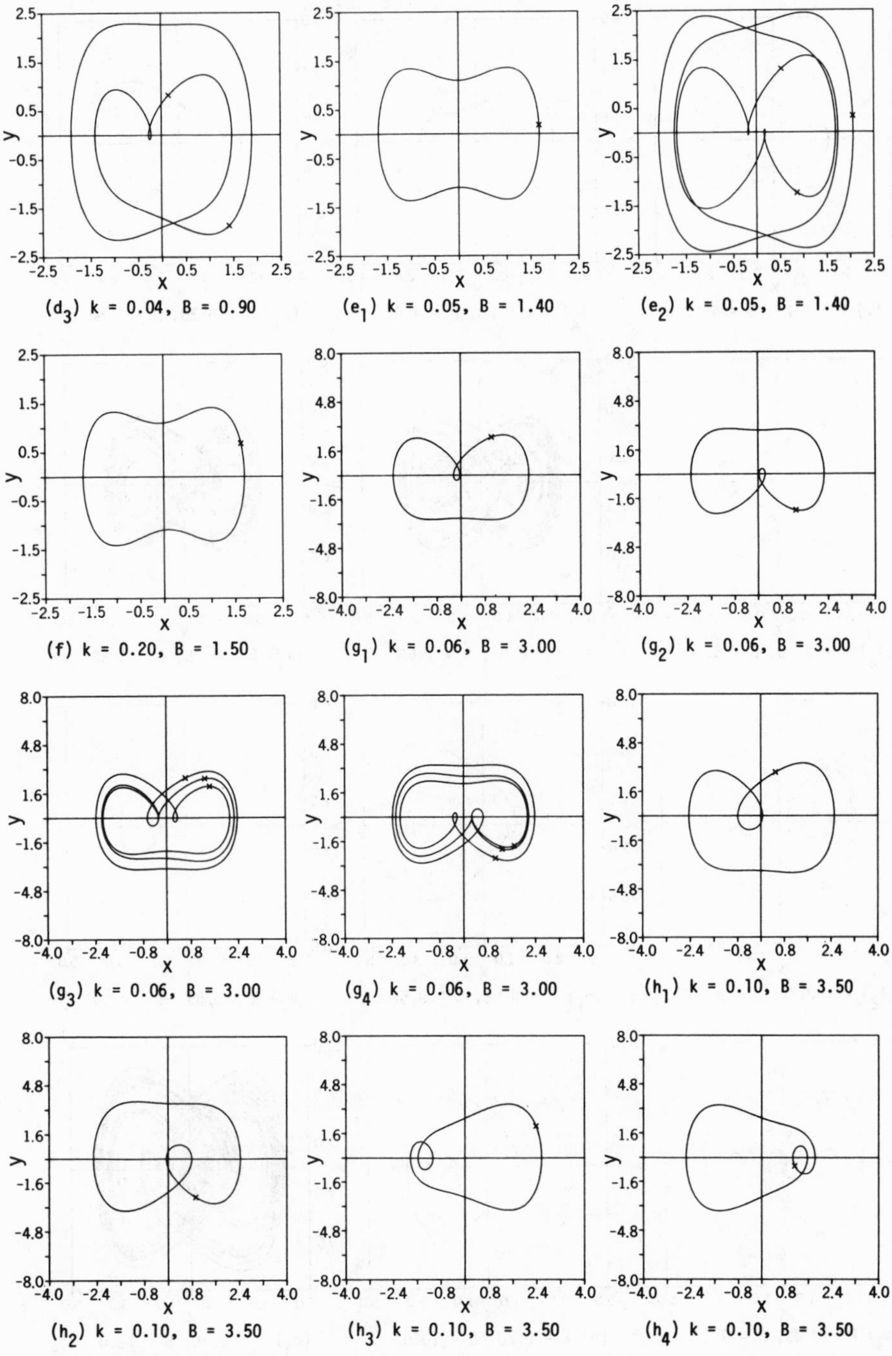

Figure 2: Continued.

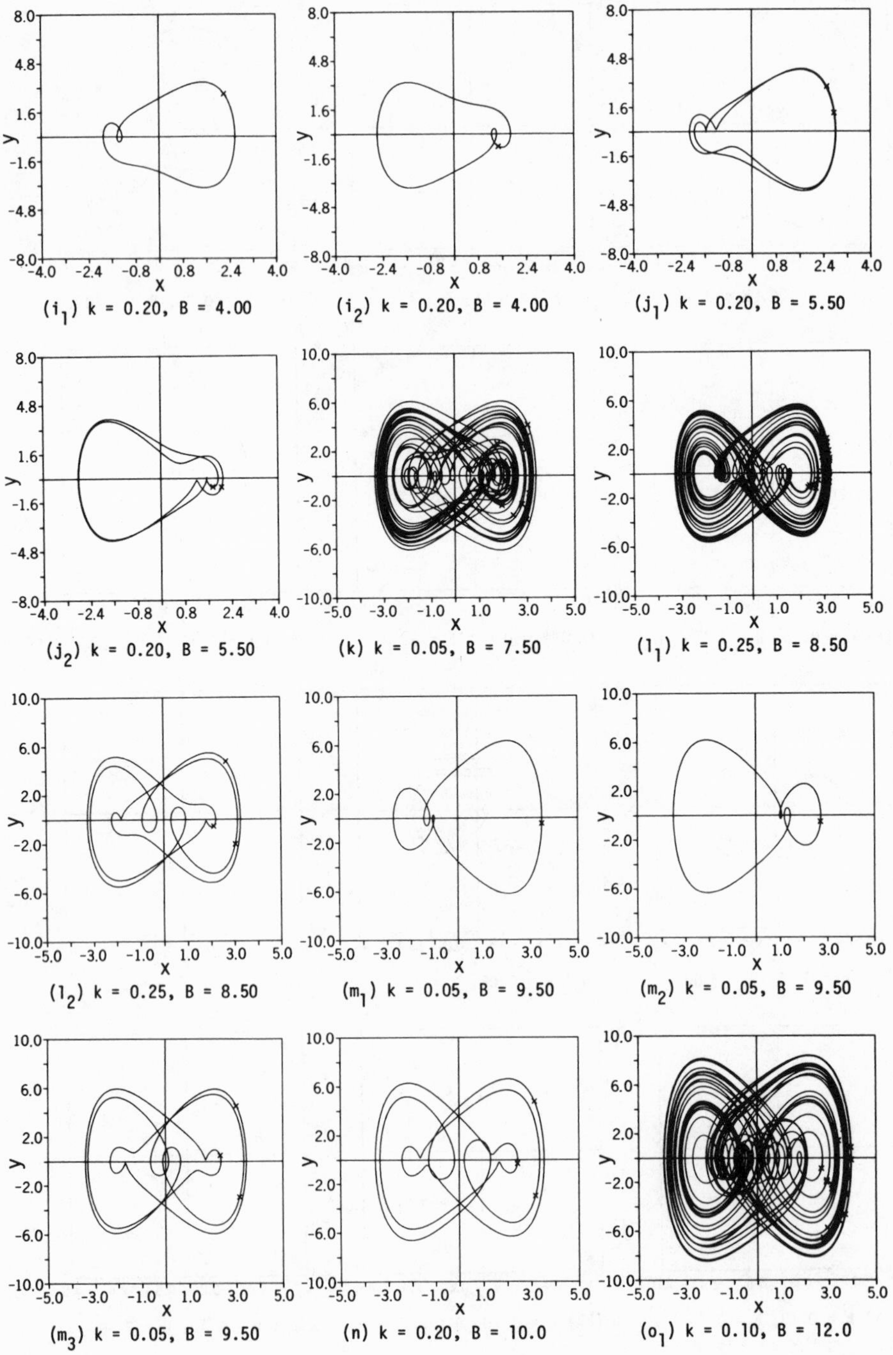

Figure 2: Continued.

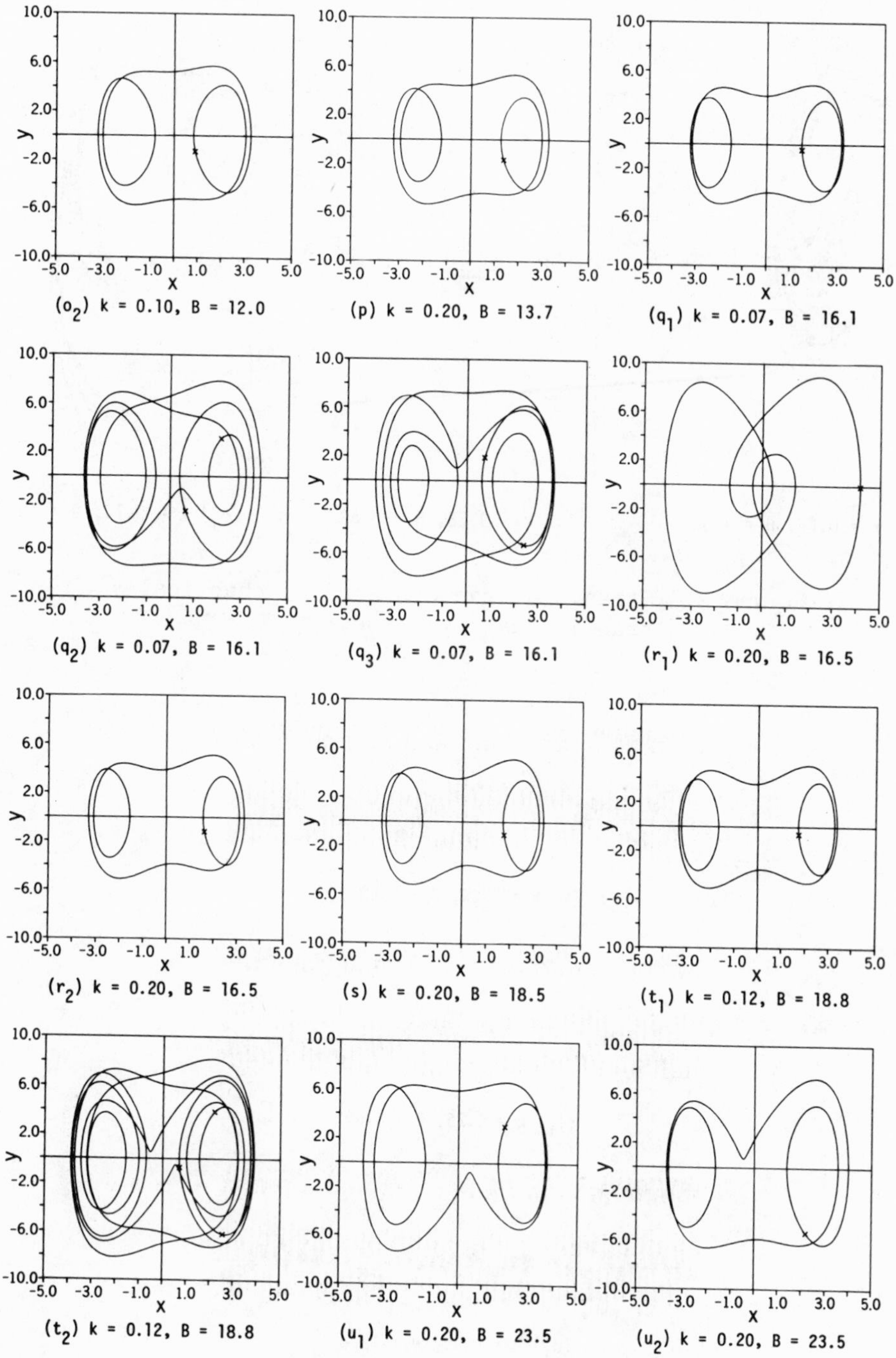

(o_2) k = 0.10, B = 12.0

(p) k = 0.20, B = 13.7

(q_1) k = 0.07, B = 16.1

(q_2) k = 0.07, B = 16.1

(q_3) k = 0.07, B = 16.1

(r_1) k = 0.20, B = 16.5

(r_2) k = 0.20, B = 16.5

(s) k = 0.20, B = 18.5

(t_1) k = 0.12, B = 18.8

(t_2) k = 0.12, B = 18.8

(u_1) k = 0.20, B = 23.5

(u_2) k = 0.20, B = 23.5

Figure 2: Continued.

(k) k = 0.05, B = 7.50   (1₁) k = 0.25, B = 8.50   (o₁) k = 0.10, B = 12.0

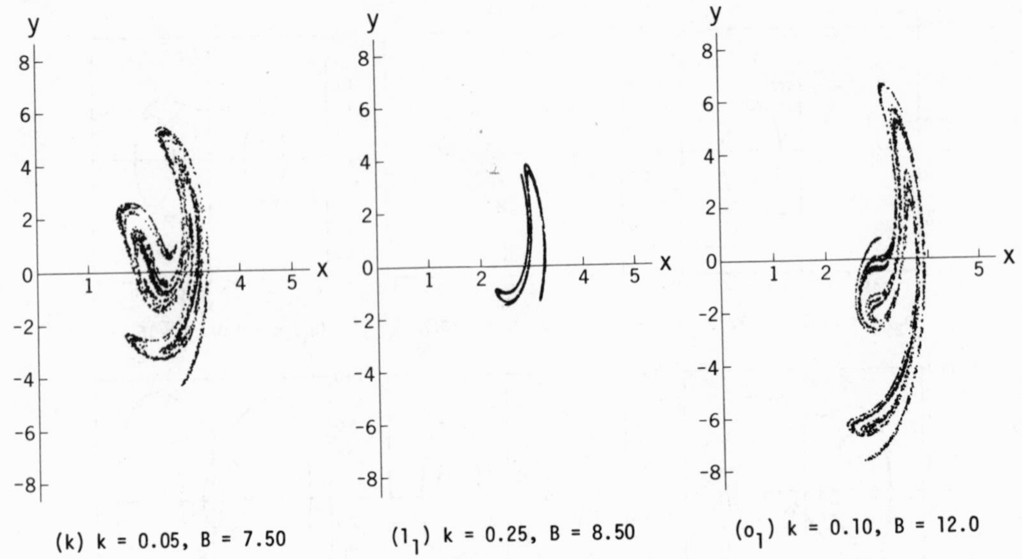

Figure 3:   Strange attractors for the chaotically transitional processes.

(k) k = 0.05, B = 7.50

(1₁) k = 0.25, B = 8.50

(o₁) k = 0.10, B = 12.0

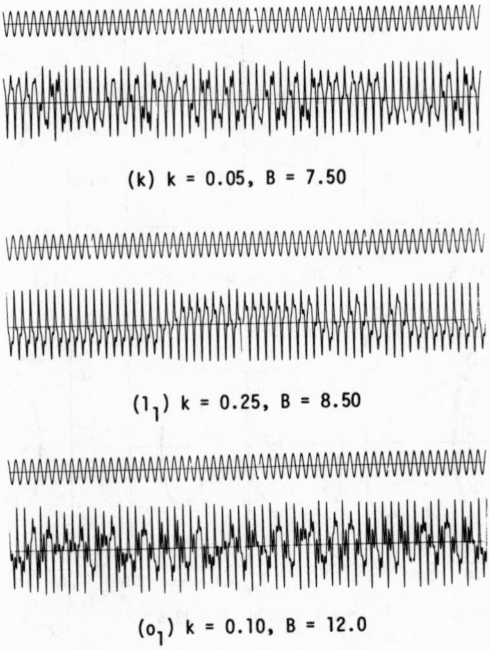

Figure 4:   Waveforms of the chaotically transitional processes.

the experimental results obtained in the preceding sections.

(1) Periodic motions with period 2π, i.e., harmonic or higher harmonic ones, occur almost everywhere except in the regions surrounded by 5/2 harmonic region. In particular, two types of 2π periodic motions are observed in the regions I, II, II", III and IV, and four types of them in the region II'. On the boundaries of I, II' and III, jump phenomenon takes place. In other words, SD coalescence (SD extinction or generation) occurs on them, while on the boundaries II and IV, SDS branching occurs.

(2) As B increases, the order of ultrasubharmonics grows larger. Ultrasubharmonic regions of all orders except 5/2 and 7/3 lie within the limits of k less than 0.2.

Ultrasubharmonic regions of order 5/2 and 7/3 participate closely in the chaotic regions [3]. This nature is related with the cubic characteristic of the nonlinear restoring term of Eq. (1).

(3) Though ultrasubharmonic motions of higher orders appear in the chaotic regions, they are omitted in Fig. 1. Similar circumstances are discussed in detail in Ref. [4]. The main results and remaining unsolved problems are also summarized there relating to chaotically transitional processes exhibited by Duffing's equation.

4. Conclusion. By using analog and digital computers, the whole aspect of steady motions exhibited by Duffing's equation has been surveyed experimentally. It is hoped that the results will be applied to various physical problems and will deserve attention as material for mathematical study.

In conclusion I should like to express my profound gratitude to Professor Chikasa Uenosono of Kyoto University, the President of the Institute of Electrical Engineers of Japan, for his constant support and encouragement during the preparation of this report.

I am particularly indebted to Professor Philip Holmes of Cornell University who gave me the opportunity to present this work at the Conference. I have also benefited greatly from conversation with Professor Francis Moon of Cornell University when he visited Kyoto. At the Conference, they gave me valuable suggestions for improvements of this report.

This work has been carried out in part under the Collaborating Research Program at the Institute of Plasma Physics, Nagoya University. The author wishes to express his sincere thanks to the staffs of the Institute.

REFERENCES

[1] Y. UEDA et al., Computer simulation of nonlinear ordinary differential equations and nonperiodic oscillations, Trans. IECE Japan, 56-A(1973), pp. 218-225; English Translation, Scripta Publ. Co., pp. 27-34.

[2] Y. UEDA, Random phenomena resulting from nonlinearity: In the system described by Duffing's equation, Trans. IEE Japan, 98-A(1978), pp. 167-173.

[3] Y. UEDA, Randomly transitional phenomena in the system governed by Duffing's equation, J. Statistical Physics, 20(1979), pp. 181-196.

[4] Y. UEDA, Explosion of strange attractors exhibited by Duffing's equation, Int. Conf. on NONLINEAR DYNAMICS, New York, Dec. 17-21, 1979, to appear in the Annals of The New York Academy of Sciences.

[5] P. HOLMES, Strange phenomena in dynamical systems and their physical implications, Appl. Math. Modelling, 1(1977), pp. 362-366.

[6] P. HOLMES, A nonlinear oscillator with a strange attractor, Proc. of Royal Soc. London, 292(1979), pp. 419-448.

[7] F. C. MOON and P. J. HOLMES, A magnetoelastic strange attractor, J. Sound and Vibration, 65(2)(1979), pp. 275-296.

[8] F. C. MOON, Experiments on chaotic motions of a forced nonlinear oscillator: strange attractor, Theoretical and Applied Mechanics Preprint, Cornell University, April 1979.

[9] J. GUCKENHEIMER, A strange, strange attractor, Chap. 12 of The Hopf bifurcation and its applications, Springer-Verlag New York, 1976.

Periodically Forced Relaxation Oscillations

Mark Levi*

Abstract. We give a qualitative analysis of the periodically forced relaxation oscillations described by the system

$$\varepsilon\ddot{x} + \phi(x)\dot{x} + \varepsilon x = bp(t) \tag{1}$$

of the Van der Pol type. An equation of this form was analyzed by Cartwright, Littlewood and Levinson. We derive a simple geometrical representation of (1). The resulting picture is applied to study of the limit behaviour of (1). Namely, combining this picture with the use of symbolic dynamics, we obtain a complete analysis of the qualitative behaviour of (1) for "most" values of b. In particular, the set of rotation numbers and the attractor are described, the latter in terms of subsifts of finite type. Another consequence of the derived geometrical picture is the proof of structural stability of (1) for "most" values of b, by application of the structural stability theorem due to Robbin. Applying the recent results of Newhouse and Palis we show that there exist uncountably many values of b for each of which the system (1) possesses infinitely many stable periodic solutions, and that as b ranges in certain (short) intervals, system (1) undergoes a sequence of bifurcations, in which it passes through infinitely many intervals of structural stability. Finally, the derived geometrical picture of (1) gives a simple explanation of the classical results of Cartwright, Littlewood and Levinson.

*Department of Mathematics, Northwestern University, Evanston, Illinois 60201.

§1. Introduction. Our aim is to apply some recent results and methods of the theory of dynamical systems to qualitative analysis of a Van der Pol-type system with forcing

$$\varepsilon\ddot{x} + \phi(x)\dot{x} + \varepsilon x = bp(t), \tag{1}$$

where ε is a small but fixed parameter, $\phi(x)$ (the damping) is negative for $|x| < 1$ and positive elsewhere, $p(t)$ is a periodic forcing of period T and b belongs to some finite interval $[b_1,b_2]$ of length of order 1 (independent of ε), to be specified later. One can choose ϕ,p close (in some sense) to $\phi_0 = \mathrm{sgn}(x^2-1)$, $p_0(t) = \mathrm{sgn}\sin\frac{2\pi}{T}t$,* see Figure 1.

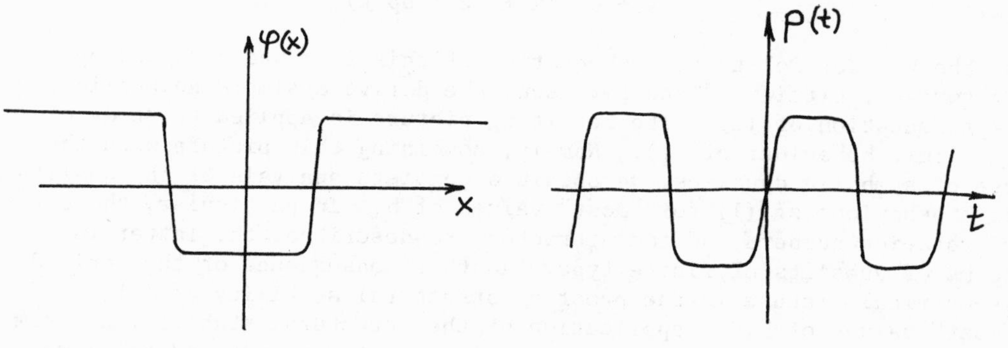

Figure 1

In the 1940's the system (1) arose as a model of the oscillations of the potential in the cavity magnetron — a device used to generate the high frequency beams in the radars. Equation (1) describes oscillations of the current in a triode circuit with a feedback and a periodic external forcing. Some biological systems also obey such an equation.

For $b = 0$ we have an autonomous system whose behaviour has been well-known for over half a century.

Making $b \neq 0$, however, complicates the behaviour drastically. In the early 1940's it was observed experimentally that the equation (1) has

---

*This specific choice of ϕ,p is inessential for the qualitative behaviour of the system; however, it allows a rigorous and complete analysis for "most" b. The choice $\phi(x) = x^2-1$ corresponds to the classical Van der Pol left-hand side.

a periodic solution of a period much larger than that of a forcing term, namely, an integer multiple (around 400 for certain ε) of T. This effect was used in electronics to obtain low-frequency oscillations. An interest in this problem was stimulated by a puzzling observation: for some values of b the system possessed two periodic solutions of <u>different</u> periods; in fact, the experiments showed that the intervals of b for which there is one or two observable periodic solutions, alternate, i.e. for b increasing, the system admits alternately only one or only two stable periodic regimes.

The significance of two periodic solutions with different periods was noticed by Cartwright and Littlewood [1], who observed that it implies existence of the so-called strange attractor* — an attractor which is neither a point nor a curve, previously not known to arise in differential equations.

Most interestingly, Cartwright and Littlewood had discovered a subfamily of solutions, which exhibits a "random" behaviour**([1,4,5]).

Their analysis, quite involved, was considerably simplified by Levinson [7], who chose $\phi(x)$ so as to make (1) piecewise linear, so that the solutions could be analyzed using the explicit formulae on each linearity interval.

These classical results described a certain subfamily of solutions of (1). It remained unclear, however, how the other solutions behave, and, most importantly, what is the geometrical reason for such a behaviour, and how is this geometry deduced from the form of equation (1). Also, what kind of bifurcations occur as b changes?

These questions will be answered in the following order. First, we state the results (§2); their informal justification is given in §3 and §4. More specifically, in §3 we reduce the study of Eq. (1) to that of an annulus map and describe its qualitative behaviour, bifurcations,

---

*The term "strange attractor" was not used by Cartwright and Littlewood, but was introduced by Ruelle and Takens in a different context about two decades later. The term attractor is used in the sense of Conley since the whole invariant set is not nonwandering.

**We will see later how this family fits within the attractor.

etc., assuming that this map has a certain simple form. This assumption is justified on intuitive level in §4; that is, we describe the behaviour of the flow of (1). We point out that the main difficulty of the analysis lies in the determination of the geometry of the flow. Here we give only the loosely stated intuitive results (§4); the exact formulation and proofs can be found in [3].

Our attack on the problem consists therefore of two main parts: 1) determination of the form of Poincaré map associated with Eq. (1); 2) deduction of the properties of the high iterates of this map using the form found in 1).

Much of the analysis in part 2) uses some recent results in the theory of dynamical systems — notably, the concept of horseshoe map of Smale [14,15]; see also Moser [8], bifurcation theory of Newhouse and Palis [9,10,11], etc. It should be noted that Eq. (1), which arose in electronics, was a major incentive in the development of the theory. We apply this theory back to the equation.

Acknowledgment. This note is a digest of the author's thesis. I would like to thank my advisor Jürgen Moser for his help and encouragement, without which this work would not have been possible.

§2. Qualitative properties of the system — the results.

2.1. Assumptions.

Prior to stating the results, we indicate the assumptions and introduce some notations.

We assume that $\phi(x)$ is even: $\phi(x) = \phi(-x)$, and that $p(t)$ satisfies a symmetry property $p(t+\frac{T}{2}) = -p(t)$, T being the period of $p(t)$. To be specific, we take for ϕ, p the functions $\operatorname{sgn}(x^2-1)$, $\operatorname{sgn}\sin\frac{2\pi}{T}t$ correspondingly, smoothed near their discontinuities so as to preserve their symmetry properties (and periodicity of p). Any functions uniformly close to these will do too.* Assume also that $p(t) > 0$ for $0 < t < T/2$. Introduce $\bar{p} = \int_0^{T/2} p(t)dt$, $\Phi(x) = \int_0^x \phi(\xi)d\xi$ (see Figure 2).

*This class can be extended (see remark in [3]).

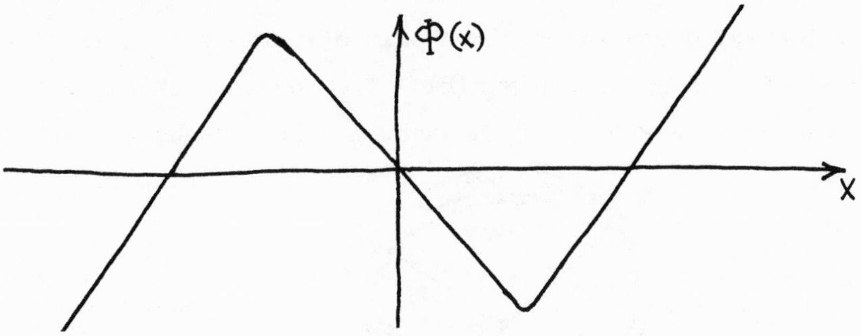

Figure 2

To specify the interval $[b_1, b_2]$, we fix $b_1 > 0$ to be a small constant (say, $b_1 = 1/100$), choose $b_2 = (2m - b_1)/\bar{p}$ and assume that period T is long enough — as it turns out in the proof (given in [3]) it suffices to have $Tb_1/2 > (2m - b_1)/\bar{p} = b_2$. Finally, instead of looking at Eq. (1) we consider an equivalent system

$$\text{(2)} \qquad \begin{aligned} \dot{x} &= \frac{1}{\varepsilon}(y - \Phi(x)) \\ \dot{y} &= -\varepsilon x + bp(t) \end{aligned}$$

and describe the nonautonomous flow (2) by sampling the positions of the solutions $(x(t), y(t))$ at discrete times nT. In other words, we look at the Poincaré map D: $(x,y)_{t=0} \to (x,y)_{t=T}$.

2.2 Qualitative properties of the system.

If the above assumptions hold, then for $\varepsilon > 0$ small enough, the following (including the classical results) holds.

The range $[b_1, b_2]$ of b-values consists of the alternating subintervals A_k, B_k separated by thin gaps g_k of small (with ε) total length, such that the qualitative behaviour of the map D throughout each interval A_k, B_k is preserved, while g_k are the bifurcation intervals.* Here is a detailed description of what happens in (A), (B), (g). For all $b \in [b_1, b_2]$, D has one totally unstable fixed point z_0; moreover,

---

*For a simple geometrical explanation of such alternating behaviour see §3.2 (Figure 6), and beginning of §3.3.

(A) for $b \in A_k$, the map D is of so-called Morse-Smale type; more specifically, D has exactly one pair of periodic points of period $(2n-1)$ with an integer $n = n(k) \sim 1/\varepsilon$ constant throughout each A_k. One of these points is a sink, another a saddle; see Fig. 3.

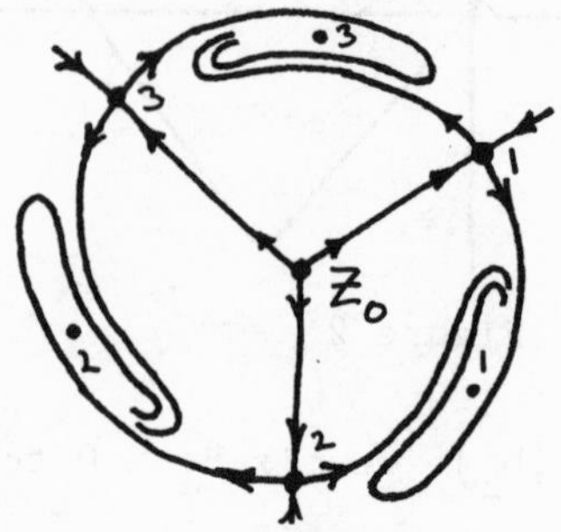

Figure 3

Any point which lies off the stable manifold of the saddle (except for z_0) tends to the sink.

A more interesting case is

(B) for $b \in B_k$, the invariant set of D consists (besides z_0) of two sink-saddle pairs of periods $2n+1$, $2n-1$ correspondingly, and of an invariant hyperbolic Cantor set C, to which the saddles belong; symbolically the situation is depicted in Figure 4. The set C can be thought of as the set of the points which are undecided to which of the two sinks to tend for future iterates, and which stay away from z_0 and ∞ for all negative iterates by D.

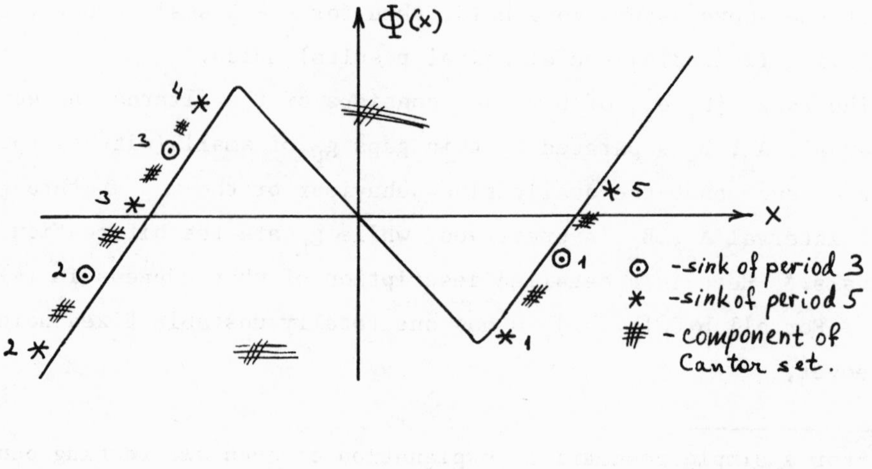

Figure 4

A more precise description of this Cantor set is given in §3 in terms of a certain return map. Here we indicate only that each point z in C can be represented by a certain sequence $\sigma = \sigma(z) = (\ldots\sigma_{-1}\sigma_0\sigma_1\ldots)$ of symbols $\sigma_i = 0,1,2$ or 3 with some pairs $\sigma_i\sigma_{i+1}$ forbidden. Each sequence $\sigma = \sigma(z)$ contains the information on the behaviour of the point $z \in C$ under the iterations by D: namely, the jth symbol determines the approximate position of jth iterate $D^j z$. In particular, our freedom to choose σ_j arbitrarily (within the restriction mentioned above) reflects in the "random" behaviour of a sequence of iterates $D^j z$, where $z = z(\sigma)$ corresponds to σ.

As a consequence of this description, we obtain infinitely many periodic points of D, since there are infinitely many periodic sequences.

We remark that the measure of C is zero — this is the answer to a conjecture of Littlewood. In particular, iterates $D^j z$ of almost all points z tend to one of the two sinks, which explains why the Cantor set is not observed experimentally.

Attractor

The attractor consists of the Cantor set C with its unstable manifolds, and two sinks. As we remarked before, the measure of C is zero.

Rotation numbers*

An interesting phenomenon related to the stochasticity is the existence of the full interval of rotation numbers; namely, the set of rotation numbers is exactly a closed interval $\left[\frac{2\pi}{2n+1}, \frac{2\pi}{2n-1}\right]$. In other words, for any number r in this interval there is a point $z = z(r) \in$ Cantor set, whose rotation number is r. Here n is the integer in the expression for the periods of the two sinks. (Here b still belongs to B_k.)

---

*Definition. A real number r is called a rotation number of a map $D: R^2 \to R^2$ with respect to a fixed point z_0 of D if for some $z \neq z_0$ $r = \lim_{n\to\infty} \left(\arg(D^n z - z_0)\right)/n$. In other words, r is an average angle (if exists) by which a point is rotated by application of D.

Structural Stability

Both cases, $b \in A_k$ and $b \in B_k$, correspond to D structurally stable. The above described behaviour is not pathological in that it cannot be destroyed by small perturbations of the system (1).

This follows by application of the structural stability theorems of Palis [12] for $b \in A_k$ and Robbin [13] for $b \in B_k$.

(g) $g \in g_k$: Bifurcations

As b crosses the gap g_k, a complicated sequence of bifurcations occurs. Perhaps the most interesting feature of these bifurcations is occurrence (for some b) of infinitely many stable periodic points. Despite their stability, they would be very hard to detect by computer due to their high period and small basin of attraction.

Classical results

We point out that the existence of alternately one and two sink-saddle pairs was shown by Cartwright, Littlewood and Levinson. The family of the solutions, found by Levinson for $b \in B_k$, corresponds to the sequences containing no 0's and 2's. The above description shows, that in addition to Levinson's periodic solutions, there are infinitely many others.

Remark 2.1. In analyzing the case $b \in B_k$ we use the concept of the horseshoe map (Smale); for its description see [8].

The bifurcations ($b \in g_k$) are analyzed by applying recent results of Newhouse and Palis [9,10,11].

§3. Reduction to the annulus map; its analysis.

3.1. Reduction to the annulus map.

It is proven in [3] that there is a rectangular region r in the (x,y)-plane (see Figure 5), such that an iterate of each point $z \neq z_0$ enters r repeatedly for the future iterations. (For an intuitive explanation see §4.) It suffices, therefore, to study map D restricted to r. In fact, we make an additional simplification: instead of studying map D, we analyze the return map $M: r \to r$, defined for each

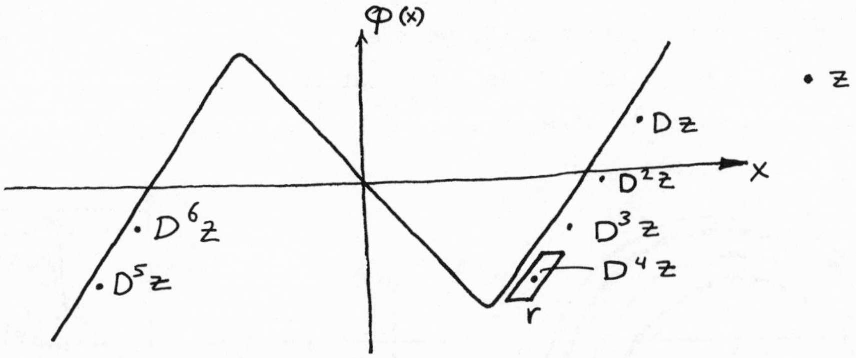

Figure 5

z ∈ r as $D^j z$, with j > 0 being the first integer for which $D^j z \in r$ again. Clearly, j depends on z, which makes the return map M discontinuous: two nearby points may require a different number of iterations to come back to r. This discontinuity is removed, however, if within r we identify any pair of points z, Dz into one. Now, r is chosen in such a way that its upper side is mapped onto the lower one (and no two points inside correspond to each other under D). Identification of the two sides of r makes M continuous, while r becomes an annulus. Summarizing, we have reduced D to an annulus map M. The only information lost by this reduction is the number of steps it takes to come back to r under iterations by D. This information is easily recoverable from some additional properties of D; the details can be found in [3].

Our aim now is to describe the form of the map M and then use it to analyze the behaviour of its high iterates. As it turns out, the symmetry properties of the damping $\phi(x)$ and of the forcing p(t) (see Sec. 2.1) reflect in the fact that M: r → r can be represented as a second iterate of another map N: r → r of a simpler form than M: $M = N \cdot N = N^2$. It suffices, therefore, to study N: r → r.

3.2. Properties of the annulus map N.

Analysis shows that r is an extremely thin ($\sim e^{-1/\varepsilon^2}$) annulus, i.e. is nearly a circle. Therefore, a two-dimensional mapping N: r → r can be represented by a circle map f (one-dimensional) to a high degree of accuracy.*

To describe properties of N, we treat r as a circle and normalize its length to be 1. Map N is such that there exists a short (for ε small) arc Δ which is stretched by N to the length between 1 and 2; say, it is 1.5; see Figure 6.

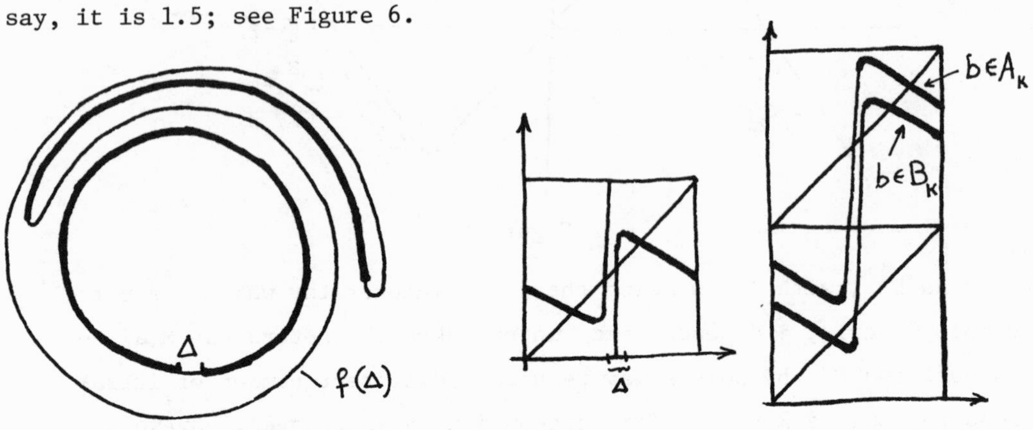

Figure 6

The rest of the circle is deformed in the simplest possible way: it is reversed in direction and somewhat contracted. In addition to having this simple form, N depends nicely on the amplitude b: in essence, increasing b causes the image N(r) to rotate clockwise.

3.3. Analysis of N.

As an immediate consequence of the above description we recover the classical result on the alternating appearance of one and two sink-saddle pairs. Namely, as b grows, N(r) rotates clockwise, i.e. the graphs on Figures 6b, c move downwards, which causes alternately one and two pairs of intersections of the graph with the bisectors (which correspond to the fixed points). The intersections where the slope of the graph is > 1 correspond to saddles, while the ones with the $|\text{slope}| < 1$ correspond to sinks of map N. These fixed points of N are the periodic points of the Poincaré map D and can be shown to have

*In the case of our map N, the properties of N can be recovered completely for most values of b from the 1-dimensional information; see [3].

periods $2n \pm 1$ ($n \sim 1/\varepsilon$) correspondingly for each pair.

Below we state without proof the results of the analysis of N. They are easily seen to imply the results of §2.

Range $[b_1, b_2]$ consists of the alternating intervals A_k, B_k separated by short gaps g_j, such that the qualitative behaviour of N persists as b ranges in A_k or B_k, while g_j are the bifurcation intervals. More precisely,

(A) if $b \in A_k$, N is a Morse-Smale type map. More exactly, every point not on the stable manifold of the saddle tends to the sinks (Figure 7a). This picture translates into Figure 3 for map D.

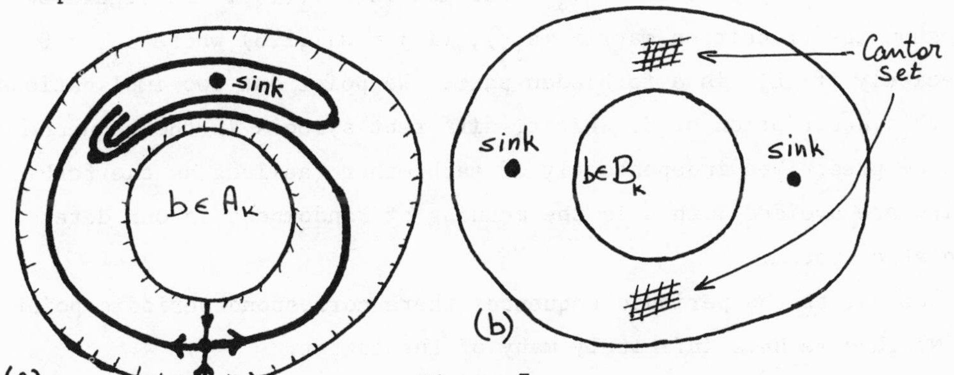

(a) Figure 7

A more interesting case is

(B) for $b \in B_k$ the map N has an invariant Cantor set S (to which the two fixed saddle points belong) and two sinks (Figure 7b). S can be thought of as a very complicated watershed — it separates the basins of attraction of the two sinks.

To completely describe N, it remains to specify its behaviour on the set S. Here is this description.

Each point $z \in S$ can be represented uniquely by a biinfinite sequence $\sigma = (\ldots\sigma_{-1}\sigma_0\sigma_1\ldots)$ of symbols σ_i which can take on one of four values 0,1,2 or 3. Also, any combination of these symbols except for 00,10,21,22,23,30 can occur. The jth symbol σ_j ($j > 0$) determines in which of the four vertical strips v_{σ_j} on Figure 8 $N^j z$ lies. We note that the fact that a pair st is forbidden means that a point in V_s cannot map into V_t — for example, no point in V_0 remains in V_0 (i.e.,

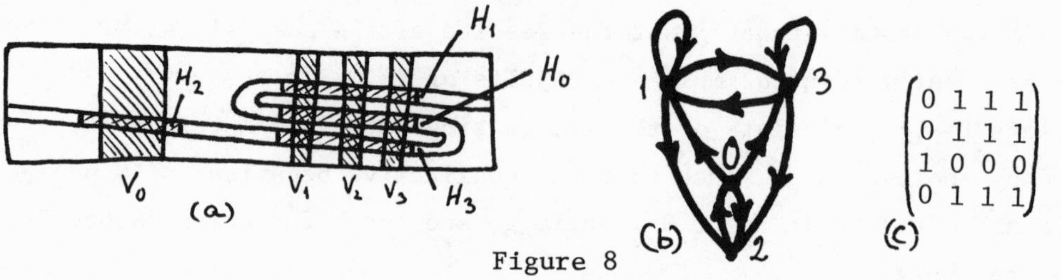

Figure 8

$V_0 \cap H_0 = V_0 \cap N(V_0) = \emptyset$), and no point in V_2 maps into V_1, V_2 or V_3. The permitted transitions between the strips are conveniently shown on graph in Figure 8b. For example, the ears of the "mouse" indicate that there are points in $V_1(V_3)$ that map into $V_1(V_3)$. On Figure 8c we show the transition matrix (a_{ij}), $(i,j = 0,1,2,3)$ where $a_{ij} = 0$ precisely if ij is a forbidden pair. We point out two implications of this description of S. First, different symbols in the sequence σ can be prescribed independently of each other (as long as the forbidden pairs are avoided); this is the meaning of randomness in our deterministic system.

Second, to the periodic sequences there correspond periodic points of N; thus we have infinitely many of the latter.

Remark 3.1. The sequences consisting of symbols 1 and 3 only correspond to the family of solutions of equation (1) described by Levinson. Note that 1 and 3 can occur in an arbitrary combination, just as in Levinson's work [7].

Remark 3.2. Sequences $\sigma^1 = (\ldots 111 \ldots)$ and $\sigma^3 = (\ldots 333 \ldots)$ correspond to the two fixed saddle points of N.

Remark 3.3. Map N is structurally stable for both $b \in B_k, A_k$.

(g) When b passes through a gap g_j from A_k to B_k, the simple situation of (A) undergoes a complicated sequence of bifurcations. Its onset can be seen from Figure 7: as b is increasing, the fold α of the unstable manifold of the saddle will move clockwise and will become tangent to the stable manifold of the saddle. This leads to the bifurcations which had been studied by Newhouse and Palis. In particular, for some values of b $\in g_j$ there are infinitely many stable periodic points of N. Another implication of their results is the existence of infinitely many intermediate open subintervals of g_j where N is structurally stable.

§4. Analysis of the flow.

Recall that our system is of the form

$$\dot{x} = \frac{1}{\varepsilon}(y - \Phi(x)), \tag{2a}$$

$$\dot{y} = -\varepsilon x + bp(t). \tag{2b}$$

We describe the flow heuristically, using pictures. Unfortunately, the rigorous description is considerably more complex. It can be found in [3].

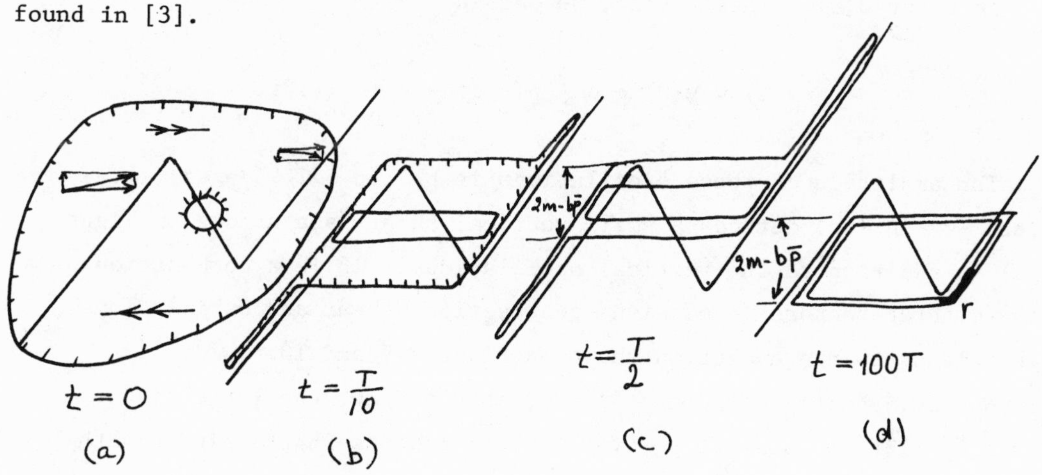

Figure 9

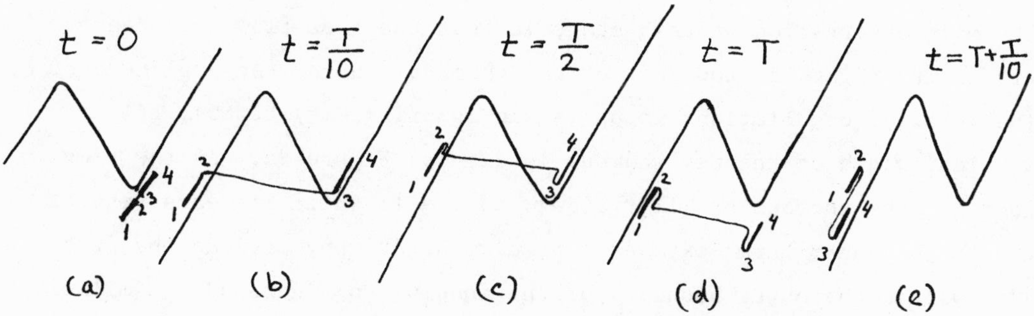

Figure 10

We start ($t = 0$) with a fat annulus on the plane — a large disc with a small disc deleted (Figure 9a). If chosen properly, the inner disc expands, whereas the outer disc contracts.

The evolution of the fat annulus shown on Figures 9, 10 (which we explain in a moment) is determined by the following properties of equation (2):

By (2a) the flow contracts strongly in the horizontal direction towards the two parts of the curve $y = \Phi(x)$ with the positive slope, and strongly expands near the part of $\Phi(x)$ with negative slope. This explains transition from (a) to (b) on Figure 9.

By (2b), the points oscillate in the vertical direction, up to an error $O(\varepsilon)$. Integrating, we obtain

$$y(t) = y(0) + b\int_0^t p(\tau)d\tau - \varepsilon\int_0^t x(\tau)d\tau.$$

The amplitude of these oscillations is $b\bar{p}$ ($\bar{p} = \int_0^{T/2} p\,dt$). The term $-\varepsilon x$ in (2b) defines a small vertical shear (see arrows on Figure 9a) in addition to the vertical oscillations. It is a combination of these three factors (expansion - contraction, shear and oscillations) that leads to the evolution shown on Figures 9 and 10.

Namely, shortly after $t = 0$, say, at $t = T/10$, the large disc quickly flattens against $y = \Phi(x)$ into a rhombus-shaped figure with two "ears," Figure 9b. This is due to the fastness of the flow away from Φ.

At $t = T/2$, the disc is in its highest position (loosely speaking). The main observation at this stage is that the fast flow had sheared off to the right the top part of the rhombus. Since the amplitude of the vertical oscillations is $b\bar{p}$ (again ignoring $O(\varepsilon)$-terms), the vertical width of the new rhombus is $2m - b\bar{p}$, Figure 9c. As the time goes on, the rhombus oscillates between $y = \pm m$ while its ears retract due to the shear term $-\varepsilon x$. As a result, after many periods the disc lies within the outer boundary of the annulus on Figure 9d. Similarly the small disc expands so that its boundary lies within that annulus, and the latter oscillates up and down while the points circulate clockwise inside it.

So far we have described a crude picture — namely, the fat annulus shrinks into a thin one. The problem is to see how does this annulus deform after one period. First, since the points circulate in the thin annulus, we can cut a representative piece r through which every point passes. The evolution of r during the time $(T + \frac{T}{10})$ is shown on Figure 10. r is chosen so that its lower part (12) lies in the fast region and thus quickly falls to the left without changing its vertical orientation. The upper part (34) has a different fate — it goes up and down along the right branch of Φ and at $t = T$ ends up sufficiently low (due to the term $-\varepsilon x$ in (2b)) to be swept to the left shortly after (say, at $t = T + \frac{T}{2}$), Figure 10e.

The intermediate part (23) became elongated.

As we sample $r(jT + \frac{T}{2})$ for $j = 1,2,\ldots$, it marches upwards along the left branch of Φ until it intersects $-r$, the symmetric image of r. In fact, every point z from r passes through $-r$ at some time $t = nT + \frac{T}{2}$ with $n = n(z)$ an integer. This defines the map $N_1: r \to -r$, or $N = -N_1: r \to r$. Now the above description of the evolution of r roughly explains the form of N given on Figure 6.

In the same way we can define $N_2: -r \to r$ to take $z = z(0) \in -r$ into $z(nT + \frac{T}{2}) \in r$ with $n = n(z) > 0$ the smallest integer for which the last inclusion holds. The symmetry properties of $\phi(x)$, $p(t)$ guarantee that $N(z) = -N_1(z) = N_2(-z)$, and thus for the map $M: r \to r$ defined in §3 we have $M = N \cdot N = N^2$.

REFERENCES

[1] M. L. CARTWRIGHT and J. E. LITTLEWOOD, On nonlinear differential equations of the second order: I. The equation $\ddot{y} - k(1 - y^2)\dot{y} + y = b\lambda k \cos(\lambda t + \alpha)$, k large, J. London Math. Soc., Vol. 20 (1945), pp. 180-189.

[2] J. E. FLAHERTY and F. C. HOPPENSTEADT, Frequency entrainment of a forced Van der Pol oscillator, Studies in Applied Mathematics 18 (No. 1) (1978), pp. 5-15.

[3] M. LEVI, Qualitative analysis of the periodically forced relaxation oscillations. To appear in the Memoirs of the A.M.S.

[4] J. E. LITTLEWOOD, On non-linear differential equation of second order: III, Acta Math., Vol. 97 (1957), pp. 267-308.

[5] J. E. LITTLEWOOD, On non-linear differential equation of second order: IV, Acta Math., Vol. 98 (1957), pp. 1-110.

[6] J. E. LITTLEWOOD, Some problems in real and complex analysis, Heath, Lexington, Mass., 1968.

[7] N. LEVINSON, A second order differential equation with singular solutions, Ann. Math., Vol. 50, No. 1 (1949), pp. 127-153.

[8] J. MOSER, Stable and random motions in dynamical systems, Princeton University Press (Study 77), 1973.

[9] S. NEWHOUSE, Diffeomorphisms with infinitely many sinks, Topology 13 (1974), pp. 9-18.

[10] S. NEWHOUSE, The abundance of wild hyperbolic sets and non-smooth stable sets for diffeomorphisms, IHES, January 1977.

[11] S. NEWHOUSE and J. PALIS, Cycles and bifurcation theory, Asterisque 31 (1976), pp. 43-141.

[12] J. PALIS, On Morse-Smale dynamical systems, Topology 8, No. 4 (1969), pp. 385-404.

[13] J. ROBBIN, A structural stability theorem, Ann. Math. 94 (1971), pp. 447-493.

[14] S. SMALE, Differentiable dynamical systems, Bull. Amer. Math. Soc., Vol. 73 (1967), pp. 747-817.

[15] S. SMALE, Diffeomorphisms with many periodic points, Differential and Comb. Top. (ed., S. Cairns), Princeton University Press, 1965, pp. 63-80.

A Discrete Dynamical System with Subtly Wild Behavior

D. G. Aronson,* M. A. Chory,** G. R. Hall,* and R. P. McGehee*

Abstract. We describe the results of computer studies on a family of discrete dynamical systems which arise in theoretical ecology. In the systems under consideration there are Hopf bifurcations for certain parameter values and our main concern is the behavior of the resulting invariant sets for parameter values far from the bifurcation point. To discuss this behavior we employ the Arnol'd tongues corresponding to various resonances and follow the evolution of the invariant set in a typical tongue.

"Things are seldom what they seem, skim milk masquerades as cream."
W.S. Gilbert, H.M.S. Pinafore

It has been well known for several years that even very simple discrete dynamical systems can exhibit extremely complicated and, on occasion, even wild behavior. One need only recall the examples found by Hénon [6] and by Curry & Yorke [4], or the spectacular space filling attractors found by Stein & Ulam [10] and by Guckenheimer, Oster &

---

*School of Mathematics, University of Minnesota, Minneapolis, Minnesota 55455. This work was supported in part by National Science Foundation grants MSC 78-02158 (D.G.A.) and MSC 79-01998 (R.P. McG.), and by a grant from the Graduate School of the University of Minnesota (G.R.H.). The computations were done in the Dynamical Systems Laboratory in the University of Minnesota School of Mathematics. The Laboratory is supported by National Science Foundation grant MSC 78-02173 and a grant from the Graduate School of the University of Minnesota.

**The Analytic Sciences Corp., 6 Jacob Way, Reading, Massachusetts 01867.

Ipaktchi [5]. In this paper we shall describe some aspects of the behavior of a family of discrete dynamical systems whose dynamics are very rich in interesting phenomena, but whose attractors are not particularly striking to the naked eye. Let us emphasize at the outset, the purpose of this work is simply to describe and, where possible, to interpret heuristically the results of rather extensive computer studies. Thus this paper should be viewed as a work of science rather than as mathematics. A fuller and more technical account of our work will appear elsewhere [2].

The starting point of our investigation is the so-called delayed logistic equation

$$(1) \qquad w_{n+1} = a\,w_n(1 - w_{n-1})$$

where $a > 0$ is a parameter. Maynard Smith [8] proposes (1) as a model for the growth of an herbivorous population. In this interpretation w_n represents the population density in the n-th generation and (1) describes how the $(n+1)$-st generation is influenced by the population in the n-th generation as well as the amount of vegetation consumed by the $(n-1)$-st generation.

In order to explore the dynamics of the second order difference equation (1) it is convenient to write it as a system of two first order difference equations. For this purpose define

$$x_n = w_{n-1} \quad \text{and} \quad y_n = w_n .$$

Then (1) is equivalent to the system

(2) $X_{n+1} = f_a(X_n)$

where

(3) $X_n = \begin{pmatrix} x_n \\ y_n \end{pmatrix}$ and $f_a(X_n) = \begin{pmatrix} y_n \\ ay_n(1-x_n) \end{pmatrix}$.

The first step in an investiagtion of the dynamics of a system such as (2) is to locate the equilibria and to study their stability. The equilibria are the fixed points of the map $X \mapsto f_a(X)$ and an equilibrium $\tilde{X}$ is stable if all the eigenvalues of the Jacobian matrix $f_a'(\tilde{X})$ lie inside the unit circle in the complex plane. It is easy to verify that for the delayed logistic map (2), (3) the equilibria are

$$X^o = \begin{pmatrix} 0 \\ 0 \end{pmatrix} \quad \text{and} \quad X^* = \begin{pmatrix} \frac{a-1}{a} \\ \frac{a-1}{a} \end{pmatrix} .$$

Since

$$f_a'(X^o) = \begin{pmatrix} 0 & 1 \\ 0 & a \end{pmatrix}$$

has 0 and a as eigenvalues, it follows that X^o is stable for $a \in (0,1)$ and unstable for $a > 1$. The eigenvalues of

$$f_a'(X^*) = \begin{pmatrix} 0 & 1 \\ 1-a & 1 \end{pmatrix}$$

are

$$\lambda^{\pm}(a) = \frac{1}{2}(1 \pm \sqrt{5-4a}) .$$

For $a \in (1,5/4]$ both eigenvalues are real, and they are complex for $a > 5/4$. If $a \in (1,2)$ then $|\lambda^{\pm}(a)| < 1$ so that the equilibrium X^* is stable for $a \in (1,2)$. As a increases through the value 2 , the complex conjugate pair of eigenvalues $\lambda^{\pm}(a)$ passes out of the unit circle via the sixth roots of unity $\frac{1}{2}(1 \pm i\sqrt{3})$. Since the map $X \mapsto f_a(X)$ is a diffeomorphism in the neighborhood of X^*, it follows from the Hopf Bifurcation Theorem for maps [7] that a smooth invariant circle develops from the fixed point X^* and persists for sufficiently small $a > 2$. Moreover, it can be shown that there is an exchange of stability, that is, the invariant circle is an attractor (in the sense of Conley [3]) for sufficiently small $a > 2$. Indeed, this is the only attractor in the first quadrant.

The Hopf Bifurcation Theorem is purely local in that it gives information only about what occurs for a in some neighborhood of the critical value 2. Thus, although it tells us that there is an attracting invariant circle for sufficiently small $a > 2$, it does not give any information about how this invariant set evolves for a far from 2. By using the computer to keep track of the successive

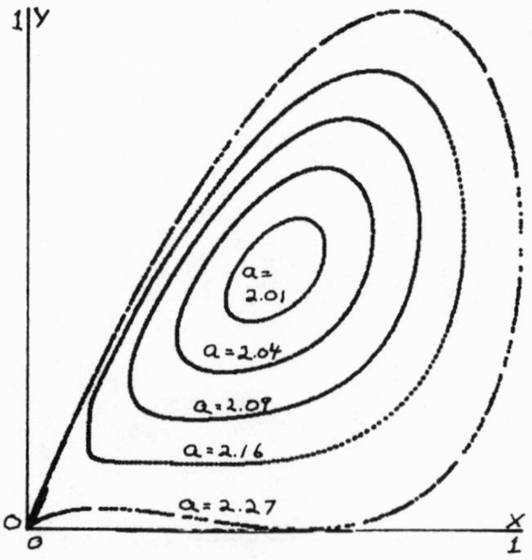

Figure 1: The visible attractor for various values of a. Each of the sets shown consists of iterates 2000 through approximately 3500 of the point $(.1, .2)^T$ under the delayed logistic map.

iterates of a single point we are able to observe the attractor or, at least, some portion of it. Figure 1 shows this visible attractor for various values of a. As a increases from the value 2 the visible attractor grows until a critical value $a^* \approx 2.271\ldots$ is reached. For $a > a^*$ there is no attractor in the first quadrant (cf. [9]). For each value of a in Figure 1 it is impossible to say of what the visible attractor consists. It might be the whole invariant set or an orbit of very high period, but it need not be either of these. For other values of a, the visible part of the attractor consists of a small number of periodic points, for example, seven points for $a \in (2.18, 2.21)$.

Figure 2a shows the visible attractor for $a \approx a^*$. Except for the loop in the lower left hand corner it appears to be a smooth curve.

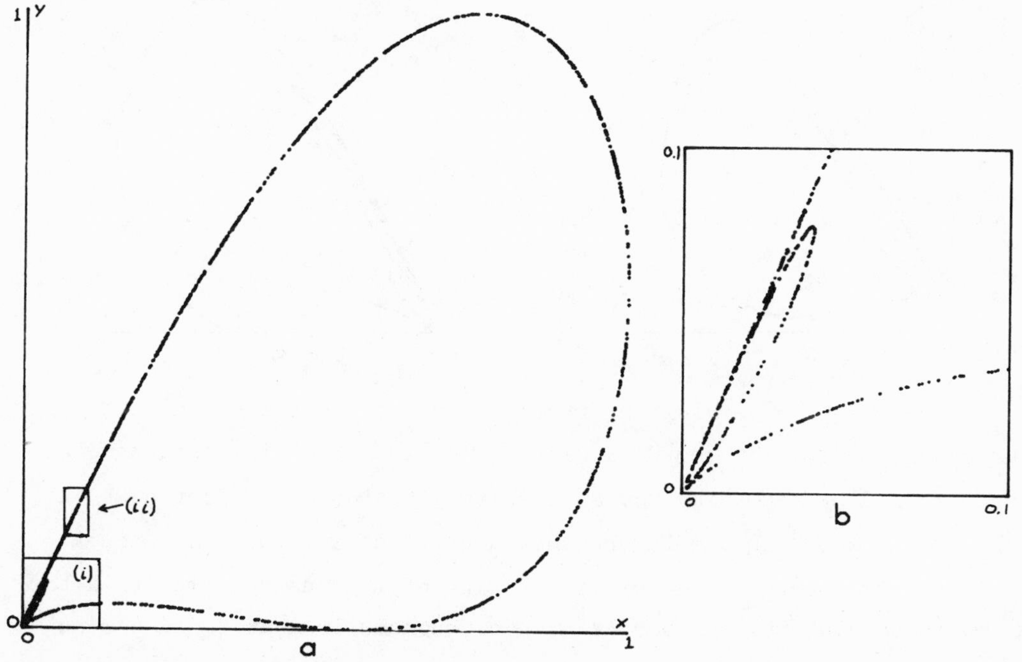

Figure 2: a. The visible attractor for $a = 2.27$. b. A detailed view of the region in box (i) in Figure 2a.

Figure 2b is a magnified view of the loop. To understand the origin and significance of this structure we must consider the saddle point

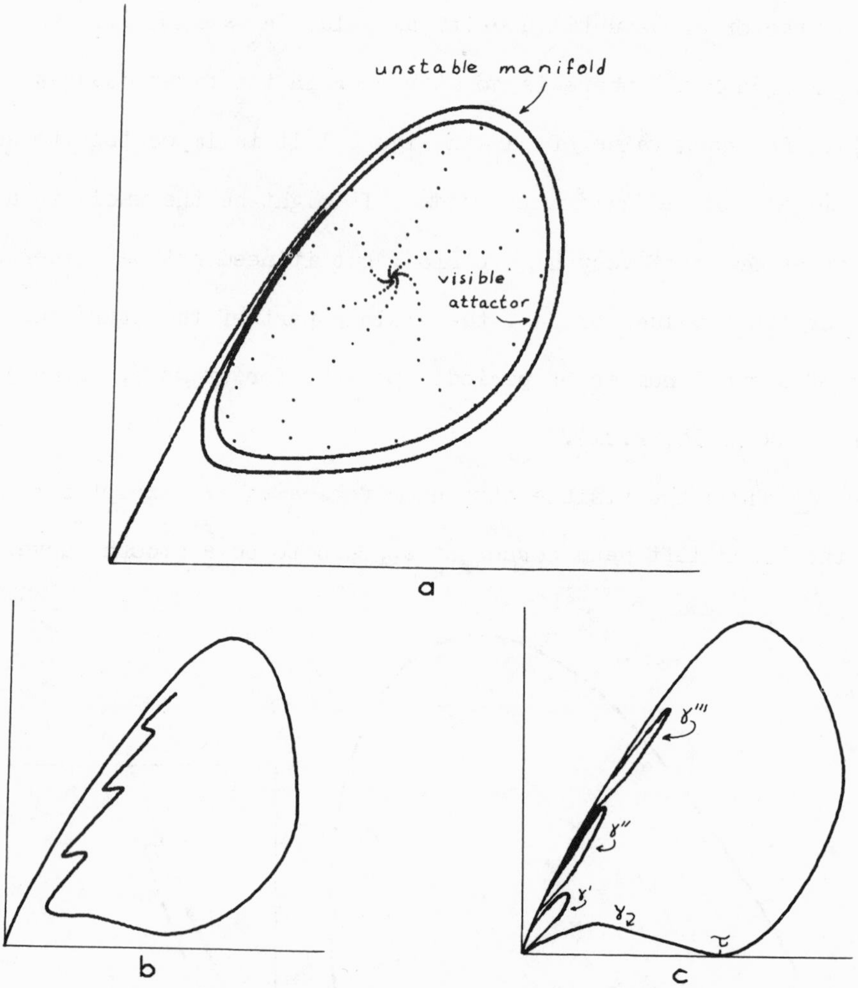

Figure 3: a. The visible attractor and the unstable manifold for the saddle point X^O computed with $a = 2.1$. The visible attractor was obtained by recording the successive iterates of a point near X^*. The "spiral arms" in the figure show how the orbit approaches the visible attractor. b. Schematic drawing of the unstable manifold for X^O with a value of a near a^*. c. Schematic drawing of the unstable manifold for X^O when $a = a^*$ showing the homoclinic tangency.

X^o . The eigenvectors of $f'_a(X^o)$ corresponding to the eigenvalues 0 and a are $(1,0)^T$ and $(1,a)^T$ respectively. It is clear from (3) that the stable manifold associated with X^o is, in fact, the x-axis. It is a very "fast" stable manifold since f_a maps the whole x-axis onto the origin. The unstable manifold associated with X^o has been studied in detail by Pounder & Rogers [9]. For $a \in (2,a^*)$ it remains in the first quadrant and spirals towards the attractor as shown in Figures 3a and b. When $a = a^*$ a homoclinic tangency occurs, that is, the unstable manifold becomes tangent to the stable manifold at a point on the positive x-axis. This homoclinic tangency is shown in Figure 3c. Since the image of the point of tangency τ is the origin, the arc γ is mapped into the loop γ' . Similarly, γ' is mapped into the loop γ'' which, in turn, is mapped into the loop γ''', ad infinitum. Thus the attractor in Figure 2a is actually composed of an infinite number of interlaced loops. Figures 4a through d show this structure graphically in a sequence of successive magnifications of a portion of the visible attractor.

As we noted above, there are values of $a \in (2,a^*)$ for which the visible attractor consists of a finite number of periodic points. With each periodic orbit we can associate a rotation number

$$\rho = \frac{\#\text{ of times around the invariant set per period}}{\text{period}} .$$

Figure 5 gives a crude summary of some of our observations on the relationship between ρ and a . On this graph we have plotted ρ against a , and the horizontal lines show the parameter interval over which we were able to detect an orbit of the corresponding rotation number. The orbits indicated by these lines are not necessarily stable. In general, the orbit of a given rotation number is stable for parameter values near the beginning and end of the interval over which it exists. In between there may be cascading bifurcations and "chaos" as indicated schematically in Figure 6. Note that some of the lines in Figure 5 overlap. Indeed, we have observed the existence

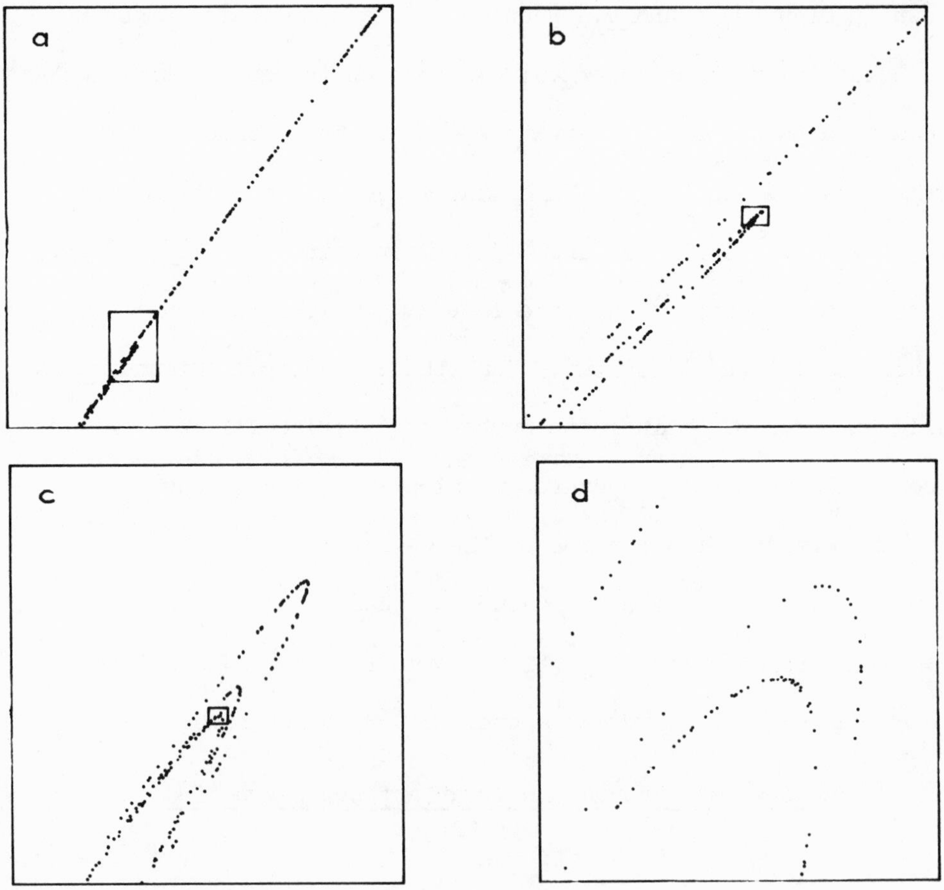

Figure 4: a. - d. Successive magnifications of the part of the visible attractor in the box (ii) in Figure 2a showing the structure described in Figure 3c. Scale. a. (.065, .105) × (.155, .224), b. (.074, .080) × (.163, .174), c. (.0773, .0776) × (.16877, .16926), d. (.0774666, .0774837) × (.16895903, .168978).

of parameter intervals for which there are two distinct stable periodic orbits. For example, stable orbits with rotation numbers 1/8 and 2/15 coexist for a near 2.2344 . As observed by Curry & Yorke, the invariant set cannot be a circle where periodic orbits of different periods coexist.

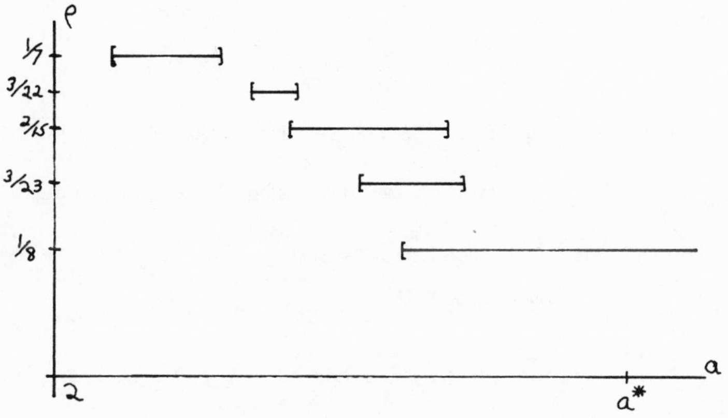

Figure 5: Rotation number plotted against a .

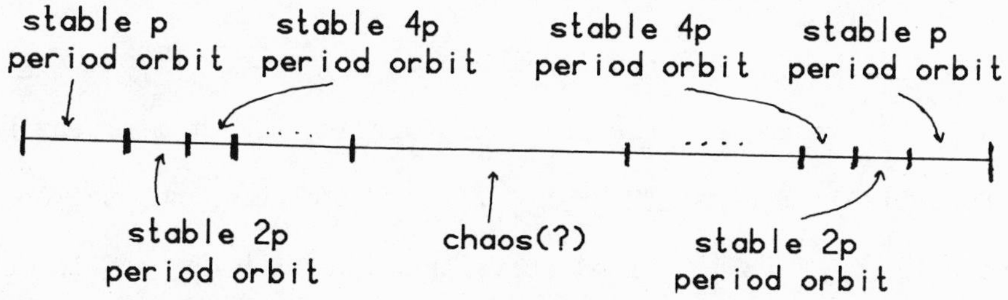

Figure 6: Schematic representation of the struction observed on a typical bar in Figure 5.

The birth and death of a periodic orbit with period p is characterized by the spontaneous appearance of p saddle-nodes (cf. Figure 7) on the invariant circle as the parameter a either increases or decreases from certain critical values. These points are fixed points of the p-th iterate of the map, $f_a^{(p)}$, and the eigenvalues of $f_a^{(p)\prime}$ at these points are both positive with one of them equal to 1. As a moves away from its critical value in the appropriate direction, the saddle-nodes split into saddle-sink pairs (cf. Figure 8). The invariant set contains the union of the fixed points and the unstable manifolds associated with the saddles. At a period doubling bifurcation the eigenvalues of $f_a^{(p)\prime}$ evaluated at the sinks are both negative with one of them equal to -1. Since the map $X \mapsto f_a(X)$ is a homeomorphism, the eigenvalues of $f_a^{(p)\prime}$ cannot be zero. Moreover, the eigenvalues of $f_a^{(p)\prime}$ evaluated at the sinks vary continuously with a. Therefore, in order to make the transition from positive to negative values there must be an interval of a-values on which the eigenvalues are complex. When the eigenvalues are complex the sinks are spiral points and the invariant set is as shown in Figure 8(b). In particular, for such values of a it is clear that the invariant set is not a globally differentiable circle.

According to the Hopf Bifurcation Theorem, given an integer $k \geq 1$ there is a $\delta_k > 0$ such that the invariant circle is k-times continuously differentiable for $a \in (2, 2+\delta_k)$. On the other hand, the observations which we have described above show that for a sufficiently far from 2 the invariant set need not be differentiable or even a circle. In order to gain some insight into the process by which the

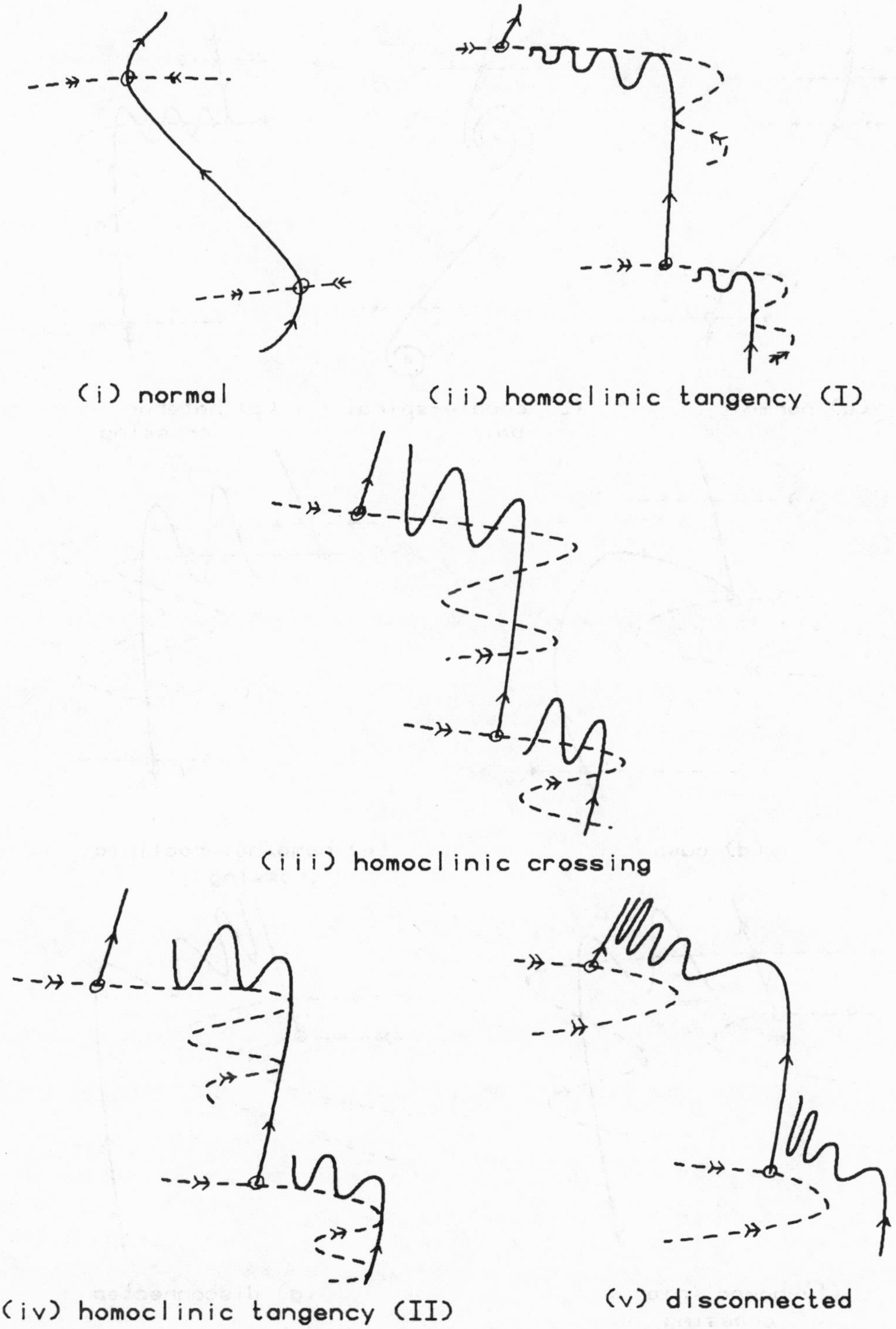

Figure 7: Saddle-nodes

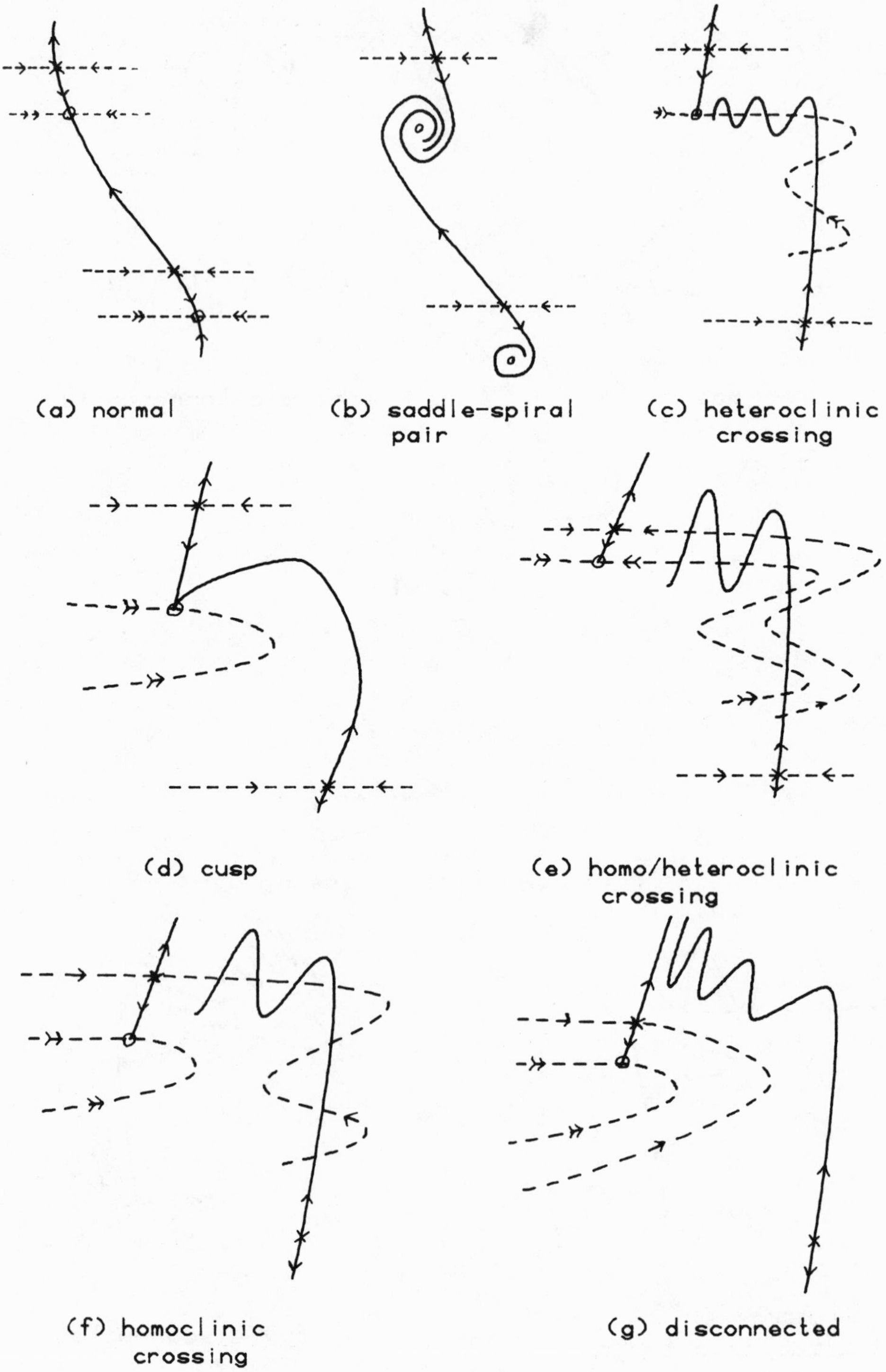

Figure 8: Saddle-sink pairs

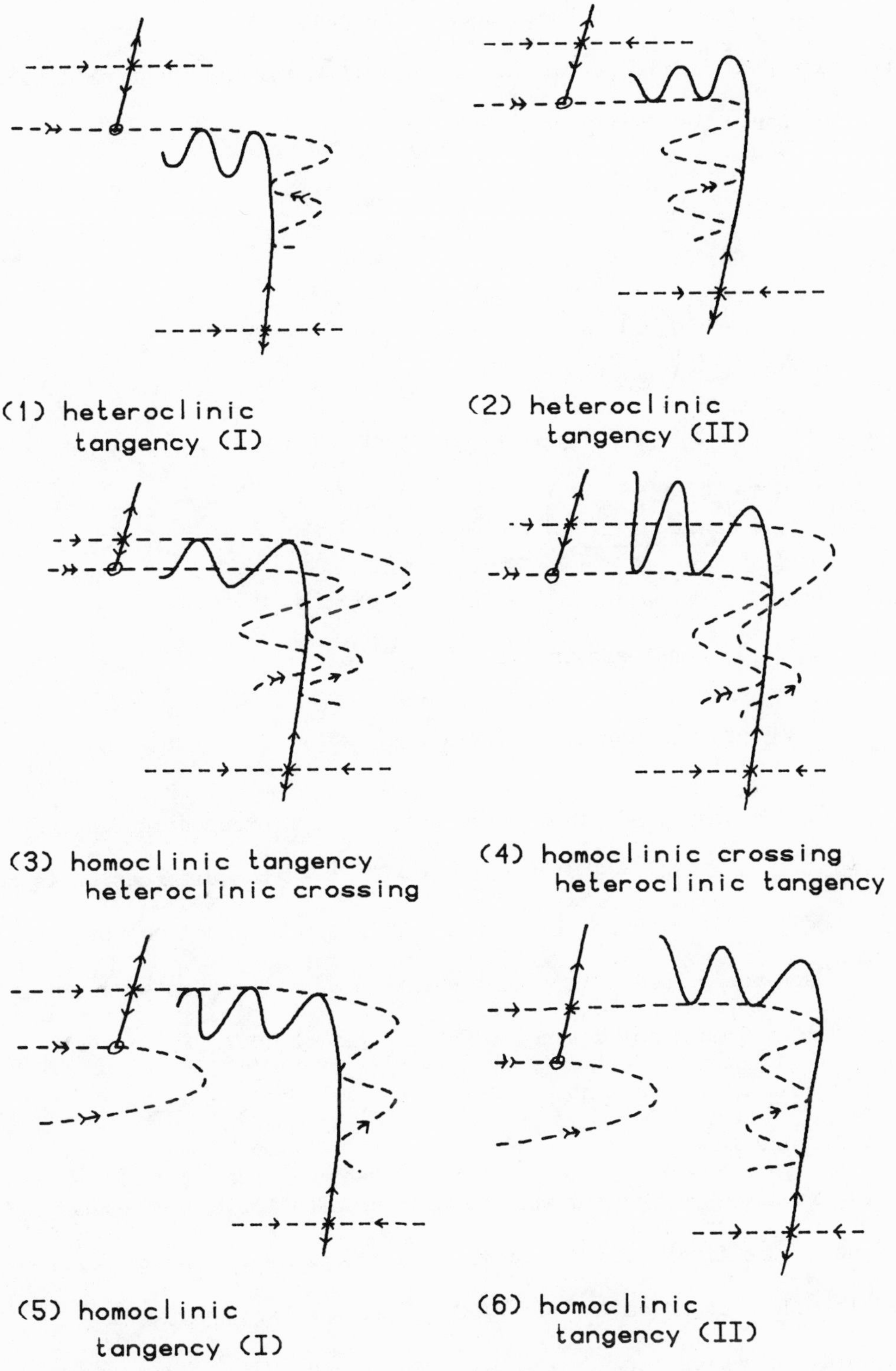

Figure 9: Homo- and heteroclinic tangencies.

invariant circle looses smoothness it is useful to embed our map in a two-parameter family of maps and to use a result due to Arnol'd [1]. The two-parameter family we choose is

$$X \mapsto f_{a\epsilon}(X)$$

where

$$f_{a\epsilon}(X) = \begin{pmatrix} a+\epsilon x \\ ay(1-x) \end{pmatrix} .$$

For all $\epsilon \in [0,1)$ the map is a diffeomorphism near the fixed point

$$X_\epsilon^* \equiv \begin{pmatrix} \frac{a-1}{a} \\ \frac{(a-1)(1-\epsilon)}{a} \end{pmatrix}$$

and the Hopf bifurcation occurs at X_ϵ^* for $a=2$.

Arnol'd [1] considers the map

$$z \mapsto \mu z + \text{higher order terms}$$

where $z \in \mathbb{C}$ and $\mu \in \mathbb{C}$ is a parameter. A Hopf bifurcation occurs on the unit circle $|\mu| = 1$. When $\mu = e^{2\pi i\, q/p}$ the bifurcation is to a periodic orbit with rotation number $\rho = q/p$. Arnol'd proves that this resonance persists in a tongue-like (cusped) region as shown in Figure 10. Some Arnol'd tongues for the map $X \mapsto f_{a\epsilon}(X)$ are shown in Figure 11. We shall now describe in some detail what occurs in each Arnol'd tongue.

Figure 12 is a highly distorted picture of a typical resonance tongue in the first quadrant of the (a,ϵ)-plane. The creation of saddle-nodes occurs on the arc ABA'. If the basic period is p, then in the region between the arcs ABA' and CC' there are p

saddle-sink pairs on the invariant set and the eigenvalues of $f_{a\varepsilon}^{(p)'}$ evaluated at the sinks are both in the interval $(0,1)$. These eigenvalues coalesce and become complex as the parameter point crosses the arc CC'.

Heteroclinic tangencies occur on the arc EF and E'F' in Figure 12. If the parameter point lies in the region bounded by the arc EBE'F'FE then the invariant circle is essentially as shown in Figure 8(a). However, if the parameter point lies on one of the arcs EF or E'F' then an unstable manifold from each saddle is tangent to a strong stable manifold of a neighboring sink. For example, on the arc EF the invariant circle is as shown in Figure 9(1). Figures 8(c) through 8(g) show the various possible configurations for the invariant set when the parameter point is inside the region bounded by the

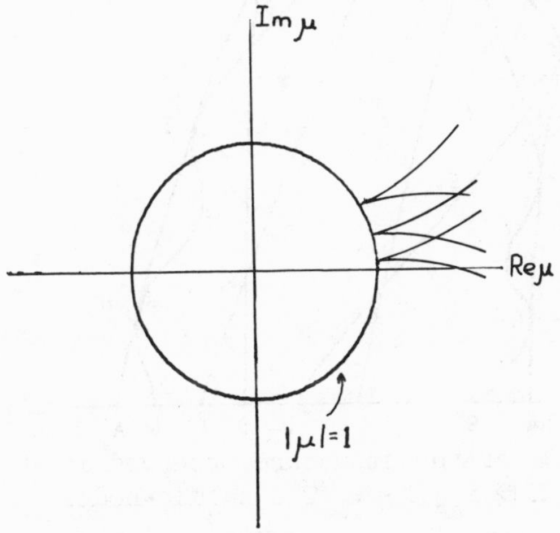

Figure 10: Schematic drawing of the Arnol'd tongues for the map $z \mapsto \mu z + \ldots$.

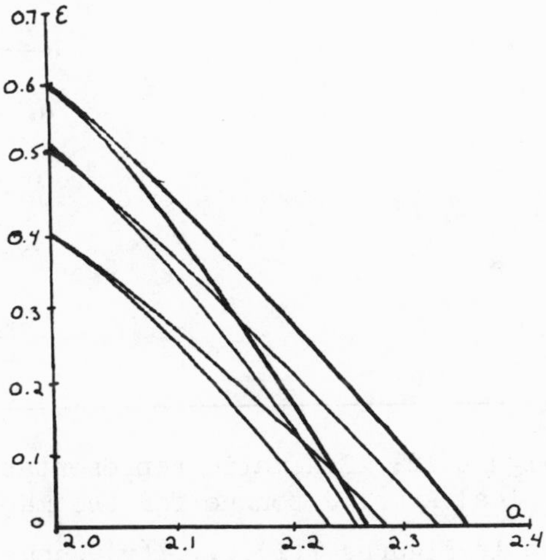

Figure 11: Arnol'd tongues for rotation numbers 1/8, 1/9 and 1/10 computed for the map $X \mapsto f_{a\varepsilon}(X)$.

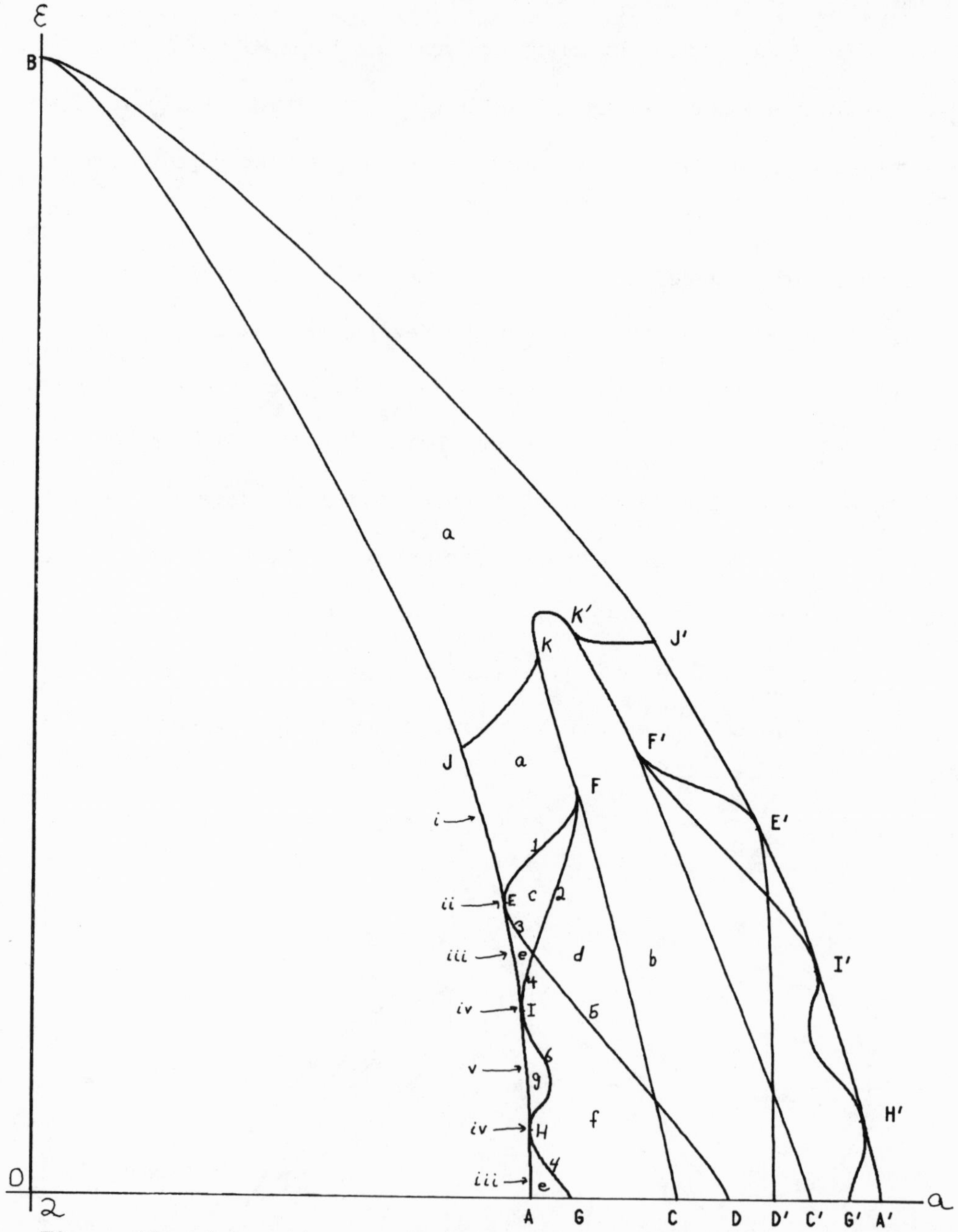

Figure 12: Schematic representation of the structure observed in a typical Arnol'd tongue for the map $X \mapsto f_{a\epsilon}(X)$. The saddle-nodes shown in Figures 7(i),...,7(v) occur on the arcs or at the points labeled i,...,v . The saddle-sink pairs shown in Figure 8(a),...,8(g) occur in the regions labeled a,...,g . (Note that the saddle-spiral pairs 8(b) occur in the region inside the arc CC' , but only in some neighborhood of the arc.) Hetero- and homoclinic tangencies shown in Figure 9(1),...,9(6) occur on the arcs labeled 1,...,6 . The arcs JK and J'K' are described in the text.

arc AEFCA in Figure 12. The transitions between these different configurations are due to the occurrence of hetero- and homoclinic tangencies on the arcs FI, IH, HG and ED as shown in Figures 9(2) through 9(6). It should be emphasized that all of the configurations drawn in Figures 7, 8 and 9 have been observed in our computer work. Figure 12 represents an attempt at a coherent synthesis of these observations. Thus, although Figure 12 is consistent with all of our computer studies to date, it must be regarded as conjectural. In particular, the structure is certainly no simpler than the one indicated in Figure 12, but it may actually be more complicated. We shall discuss some further aspects of these pictures below. The point we want to make here is that the invariant set is certainly not smooth when the parameter point is anywhere in the region bounded by the arc AEFCA. The situation is similar for the region bounded by the arc A'E'F'C'A', but we shall not discuss it here.

If the parameter point is in the region bounded by the arc EFF'E'BE and is close to the arc EF, then the invariant circle is as shown in

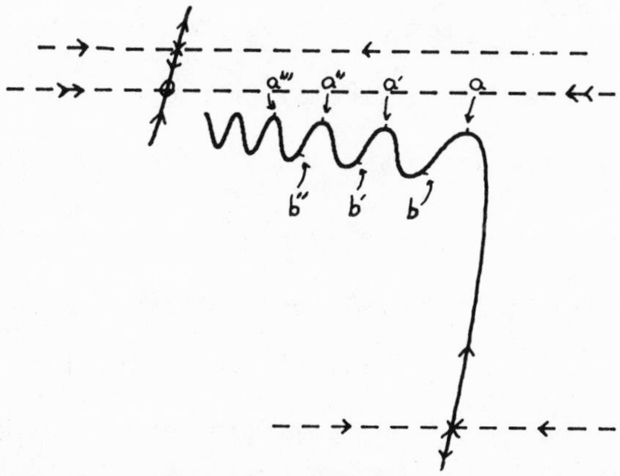

Figure 13: Schematic representation of part of the invariant circle for the map $X \mapsto f_{a\epsilon}(X)$ when the parameter point lies in the region bounded by the arc EJKFE in Figure 12.

Figure 13. On the sequence of points $a, a', a'', \dots$ the tangent to the unstable manifold from the saddle has essentially the direction of the fast stable manifold of the sink. However, on the sequence of points $b, b', b'', \dots$ the tangent to the unstable manifold from the saddle has essentially the slow direction associated with the sink. Thus, although the invariant circle has a tangent at each point, the tangent does not vary continuously when the parameter point is near the arc EF or E'F'. On the other hand, if the parameter point is far from the arcs EF and E'F' the invariant circle is as shown in Figure 8(a) and is certainly at least C^1. According to our computer observations, the transition from a C^1 invariant circle to one which is simply differentiable occurs on arcs JK and J'K' as shown in Figure 12. A more detailed description of this loss of continuous differentiability in terms of splitting in the tangent bundle will be given elsewhere [2].

In Figure 12, homoclinic tangencies occur on the arcs ED, E'D', IH, and I'H'. We have observed that these arcs are the set theoretic limits of the boundaries of the various Arnol'd tongues which intersect the given one. Specifically, the upper boundaries of tongues which impinge from below accumulate on the arc ED, while their lower boundaries accumulate on the arc IH. Periodic orbits with different rotation numbers coexist when the parameter point is in the region bounded by the arc AEDA. Indeed, at each point in the region bounded by the arc AHIEDA one can find coexisting orbits with a continuum of rotation numbers. In the region bounded by the arc DEBE'D'D the system is in phase lock, that is, all orbits have the same rotation number.

Figure 14 shows the visible attractor for $a = 2.252\ldots$ and $\epsilon = 0$. In this case the parameter point is somewhere on the segment DD' of the Arnol'd tongue for $\rho = 1/8$ (cf. Figure 12). Indeed, the parameter point in this case is between the two regions of cascading bifurcation indicated in Figure 6. Although in Figure 14, the visible attractor appears to consist of eight arcs, a closer examination shows that this is not the case. Suppose that the part of the attractor labeled i in Figure 14 were an arc. The fate of that arc under eight iterations of the map is shown schematically in Figure 15. In the first seven iterations the arc is compressed and stretched, while in the eighth iteration it is folded over on itself as shown. The results of the continuation of this process are shown in Figures 16 a through 16 d.

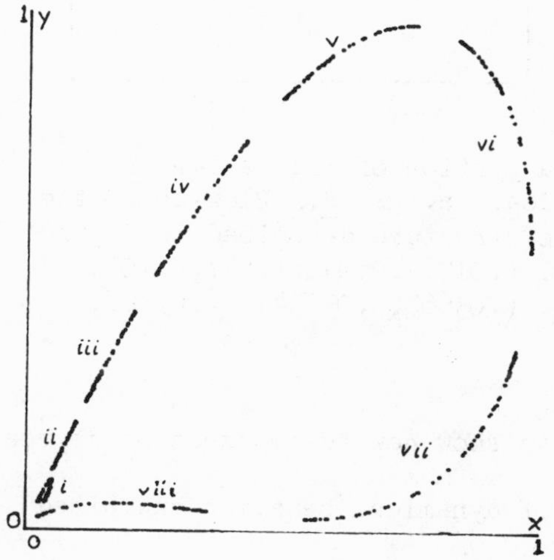

Figure 14: The visible attractor for $a = 2.2521185$, $\epsilon = 0$.

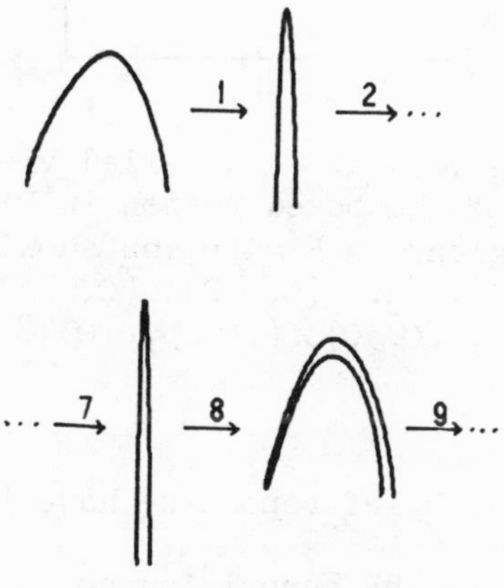

Figure 15: Schematic representation of the action of eight iteration of the map $X \mapsto f_{a\epsilon}(X)$ applied to the part of the attractor labeled i in Figure 14.

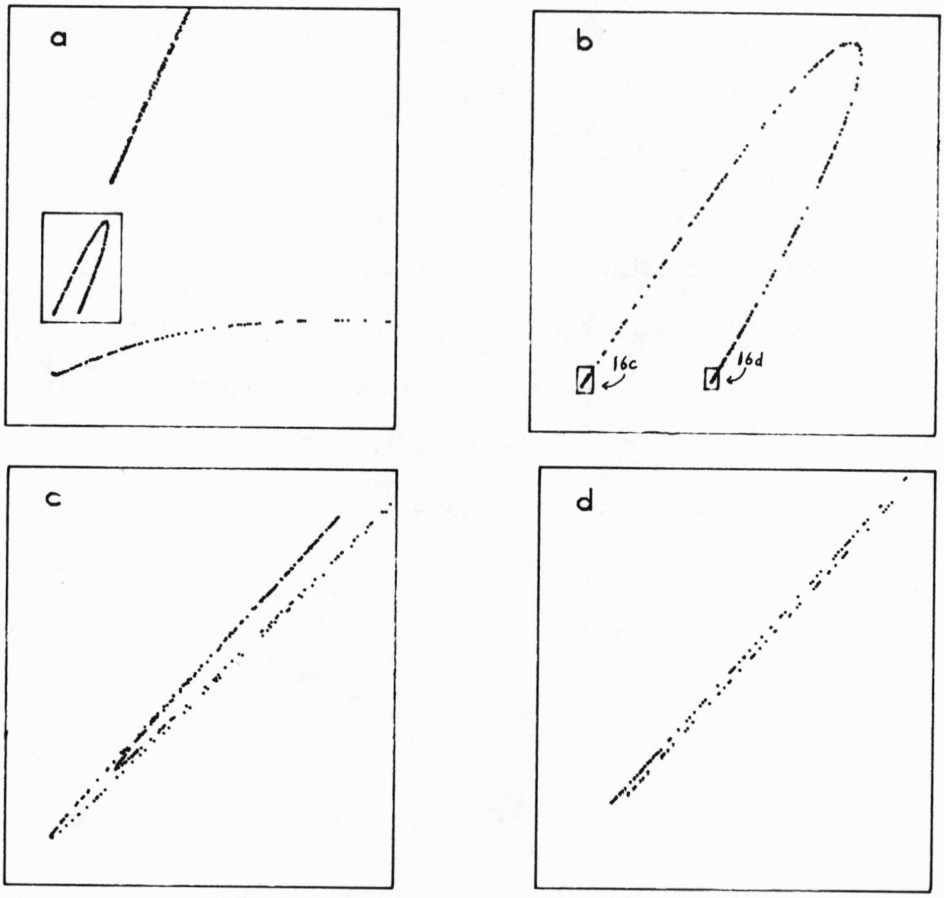

Figure 16: a. A detailed view of a portion of Figure 14. b. A blow-up of the boxed region in Figure 16a. c. & d. Blow-up of the boxed regions in Figure 16b showing the structure described in Figure 15. Scale. a. (0, .2) × (0, .2), b. (.019, .059) × (.047, .101), c. (.0239, .0250) × (.0525, .0548), d. (.03668, .03734) × (.05328, .05465).

In this brief report we have tried to show how to impose some degree of order on the surprising complexity of dynamical behavior exhibited by a very simple map of the plane. We freely admit that we do not fully understand many of the phenomena which we have observed, and it is quite clear that we have not observed all of the phenomena which need to be understood. Our work in this area is continuing.

REFERENCES

[1] V.I. ARNOL'D, Loss of stability of self-oscillations close to resonance and versal deformations of equivalent vector fields, Functional Analysis and its Applications, 11 (1977), pp. 1-10.

[2] D.G. ARONSON, M.A. CHORY, G.R. HALL and R.P. MCGEHEE, manuscript in preparation.

[3] C. CONLEY, Isolated Invariant Sets and the Morse Index, Conference Board of the Mathematical Sciences Regional Conference Series in Mathematics Number 38, American Mathematical Society, Providence, 1978.

[4] J.H. CURRY and J.A. YORKE, A transition from Hopf bifurcation to chaos: computer experiments on maps in R^2, The Structure of Attractors in Dynamical Systems, Lecture Notes in Mathematics, Vol. 668, pp. 48-68, Springer-Verlag, Berlin, 1978.

[5] J. GUCKENHEIMER, G.F. OSTER and A. IPAKTCHI, The dynamics of density dependent population models, J. Math. Biol, 4 (1976), pp. 101-147.

[6] M. HÉNON, A two-dimensional mapping with a strange attractor, Comm. Math. Phys., 50 (1976), pp. 67-77.

[7] J.E. MARSDEN and M. MCCRAKEN, The Hopf Bifurcation and its Applications, Springer Verlag, Berlin, 1976.

[8] J. MAYNARD SMITH, Mathematical Ideas in Biology, Cambridge U. Press, Cambridge, 1971.

[9] J.R. POUNDER and T.D. ROGERS, The geometry of chaos: dynamics of a nonlinear second-order difference equation, Bull. Math. Biol, to appear.

[10] P.R. STEIN and S.M. ULAM, Nonlinear transformation studies on electronic computers, Rozprawy Matem., 39 (1964), pp. 3-65.

Buckling of Shallow Elastic Structures

Raymond H. Plaut*

Abstract. This paper reviews some recent work on the snap-through instability of shallow elastic arches and trusses. Both static and dynamic loads are considered. The influence of load position on the instability of an arch is treated first. Then the interactive behavior of multiple, independent loads is analyzed. Stability boundaries are constructed in the load-space, and their properties are investigated.

1. Introduction. When shallow elastic structures, such as arches or spherical caps, are subjected to downward loads, they may buckle suddenly into an inverted (or partially inverted) configuration. This type of instability is called snap-through. In general, the governing equations are nonlinear, and numerical techniques are required to obtain solutions. For quasi-static loading, critical loads are determined with the use of the equilibrium equations or the principle of minimum potential energy. For dynamic loading, a stability criterion must be defined, and in most cases the equations of motion must be integrated numerically.

The work reviewed in this paper primarily involves the effect of the spatial distribution of loading on critical values of the loads. In §2, a single load is applied to an arch at an arbitrary location, and plots of critical load versus load position are presented. Arches and trusses under three independent, static loads are considered in §3 and §4, respectively. Critical loading combinations form stability boundaries (interaction surfaces) in the three-dimensional load-space.

---

*Department of Civil Engineering, Virginia Polytechnic Institute and State University, Blacksburg, Virginia 24061. This work was supported by Grant No. ENG77-17847 awarded by the National Science Foundation.

Dynamic loads are then discussed in §5 and §6. For a rigid-bar model of an arch, stability boundaries are determined for step loads and impulse loads in §6.

2. Arch with single static load. Consider first a shallow elastic arch subjected to a single concentrated load. The case of clamped-pinned boundary conditions is shown in Fig. 1 (in nondimensional terms). The unloaded configuration $y_o(x)$ (dash-dot line) is assumed to be circular with height c. Under the load p at location x_p, the equilibrium shape y(x) induces an axial thrust $\eta(y)$ given by

$$\eta(y) = 2 \int_0^1 [(y_o')^2 - (y')^2] \, dx, \tag{1}$$

and the equilibrium equation is assumed to be [1]

$$y'''' - y_o'''' + \eta(y) \, y'' = - p \, \delta(x - x_p) \tag{2}$$

where δ is the Dirac delta function. For the case in Fig. 1, the boundary conditions are $y(0) = 0$, $y'(0) = y_o'(0)$, $y(1) = 0$, and $y''(1) = y_o''(1)$.

For given values of c, p, and x_p, (1) and (2) can be solved numerically for the equilibrium shape (or shapes) y(x). Typical load-deflection equilibrium paths are depicted in Fig. 2, where dashed lines denote the unstable portions. If c is sufficiently small, the path is monotonically increasing, as in Fig. 2(i). Otherwise, snap-through instability occurs as the load is increased, either at a limit (maximum) point, as in Fig. 2(ii) and Fig. 2(iii), or at a bifurcation point, as in Fig. 2(iv).

Plots of critical load versus load position are presented in Figs. 3-5 for pinned-pinned, clamped-clamped, and clamped-pinned boundary conditions, respectively. All critical loads correspond to limit points except at the cusps (at $x_p = 0.5$ for $c \geq 3$ in Fig. 3, and at $x_p = 0.53$ for c=5 and $x_p = 0.54$ for c=6 in Fig. 5). For the clamped-clamped, shallow, circular arch under a single load, bifurcation can only occur after the limit point has been reached, as shown in Fig. 2 (iii), no matter how large the value of c [2].

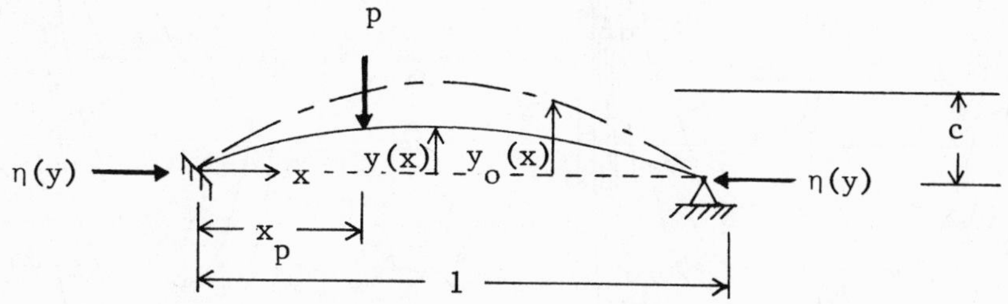

Fig. 1. Clamped-pinned arch

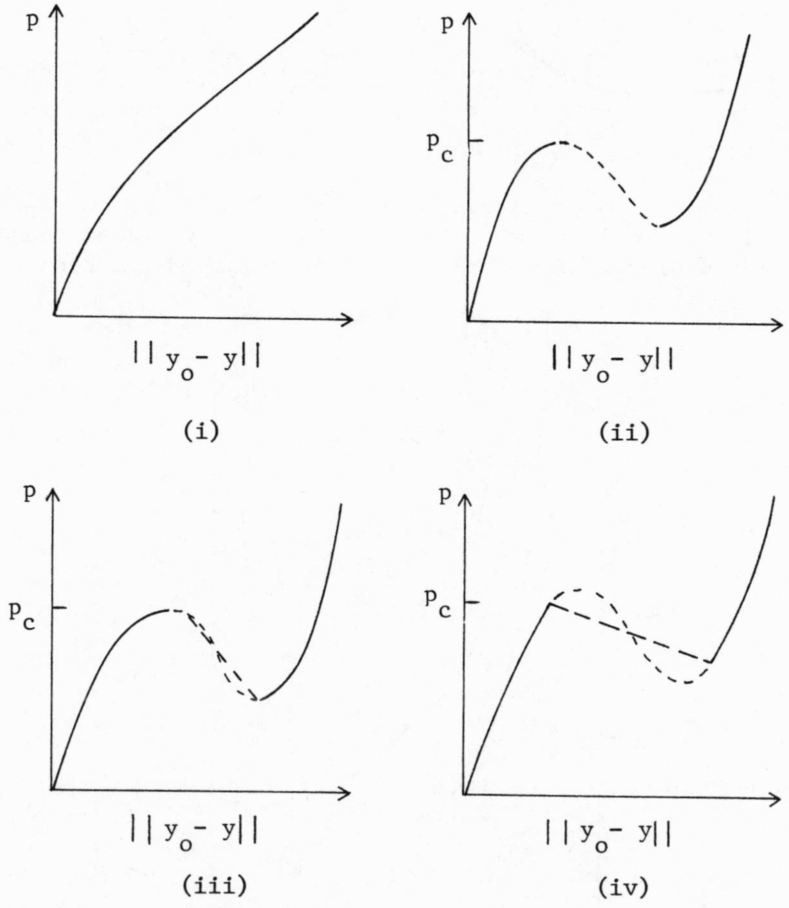

Fig. 2. Typical load-deflection equilibrium paths

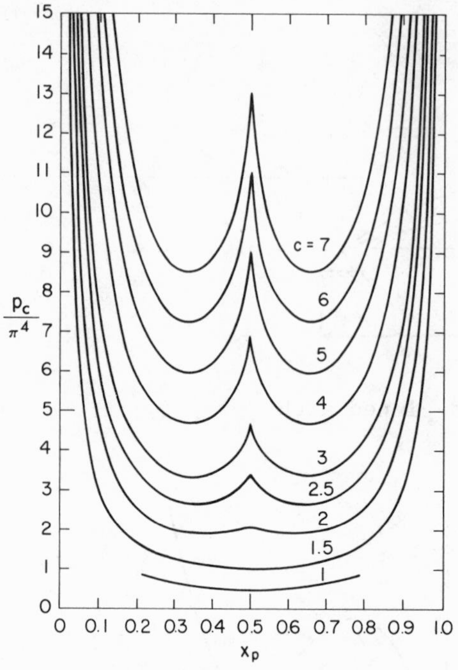

Fig. 3. Critical loads for pinned-pinned arch

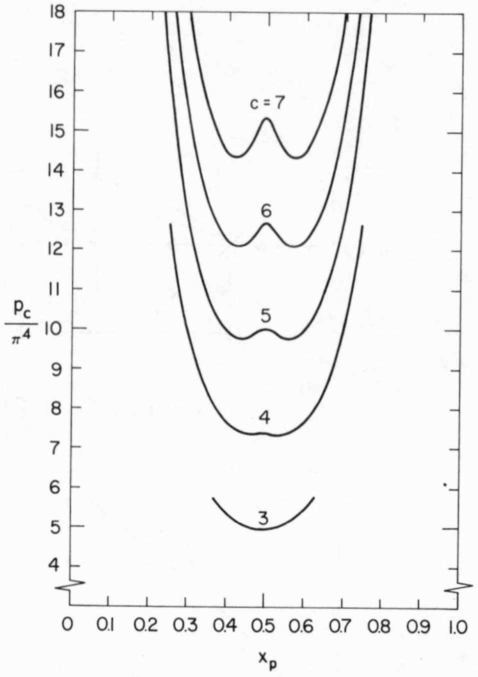

Fig. 4. Critical loads for clamped-clamped arch

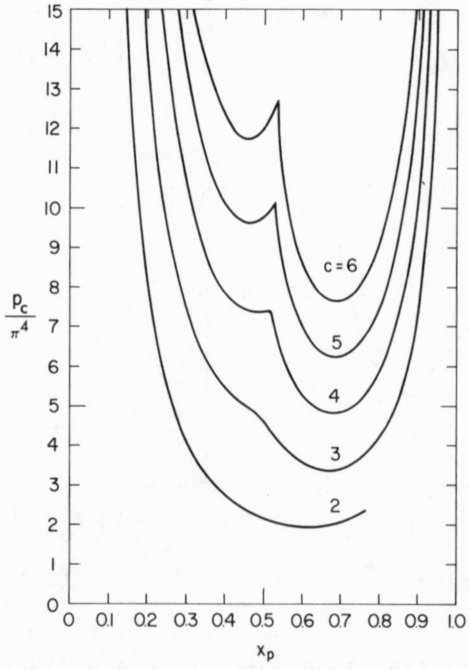

Fig. 5. Critical loads for clamped-pinned arch

Similar problems, with the load at or near the center, have been analyzed by Troger [3] and Zeeman [4] in terms of catastrophe theory and by Golubitsky and Schaeffer [5] in terms of the theory of singularities.

3. Arch with multiple static loads. Now consider the pinned-pinned, circular arch shown in Fig. 6, subjected to independent loads p_1, p_2, and p_3. Critical loading combinations are computed numerically by fixing the load ratios and increasing the load magnitudes [6]. For a given arch height c, the locus of these critical points forms a surface, called the stability boundary, in the (p_1, p_2, p_3) load-space. If c is sufficiently small, there are no critical loads. As c increases, the stability boundary takes on the shapes sketched in Fig. 7.

The stability boundaries are symmetric with respect to the plane $p_1 = p_3$. For small values of c, loading rays near the p_1 and p_3 axes do not yield snap-through instability, and the stability boundary has cut-off values as in Fig. 7(i). All critical points in Fig. 7(i) and Fig. 7(ii) are limit points, and these stability boundaries are concave toward the origin.** As c increases, bifurcation instability (denoted B) occurs for some symmetric loading combinations (up to the x in Fig. 7(iii)) and then for all critical points in the plane of symmetry (Fig. 7(iv)). The stability boundary has a cusp at bifurcation points. Imperfections in the arch tend to shift the stability boundary in the load-space.

If the ends of the arch are clamped, the critical load on the p_2 axis always corresponds to a limit point (see §2), and Fig. 7(iv) is not obtained. If one end is clamped and the other is pinned, the stability boundaries are not symmetric and the plane containing bifurcation points is not the plane of symmetry.

The postbuckling analysis of continuous elastic systems under multiple, independent, static loads is presented in [8], with the arch in

**This concavity property is also satisfied by multiply-loaded, conservative, nongyroscopic systems which do not exhibit prebuckling deformations, such as columns loaded axially and plates under in-plane loading [7]; in those cases, however, buckling corresponds to bifurcation from a trivial equilibrium state.

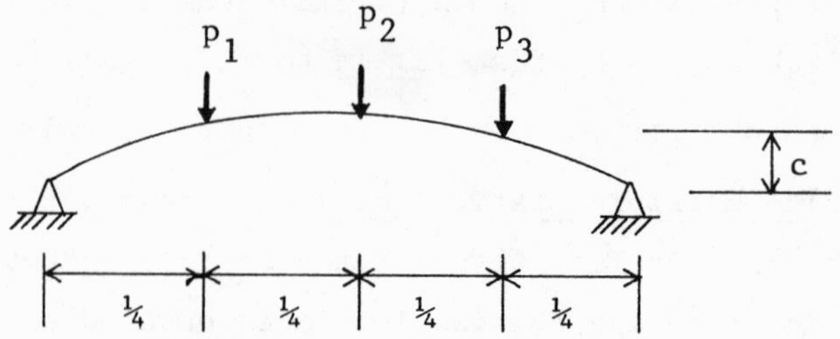

Fig. 6. Arch with multiple loads

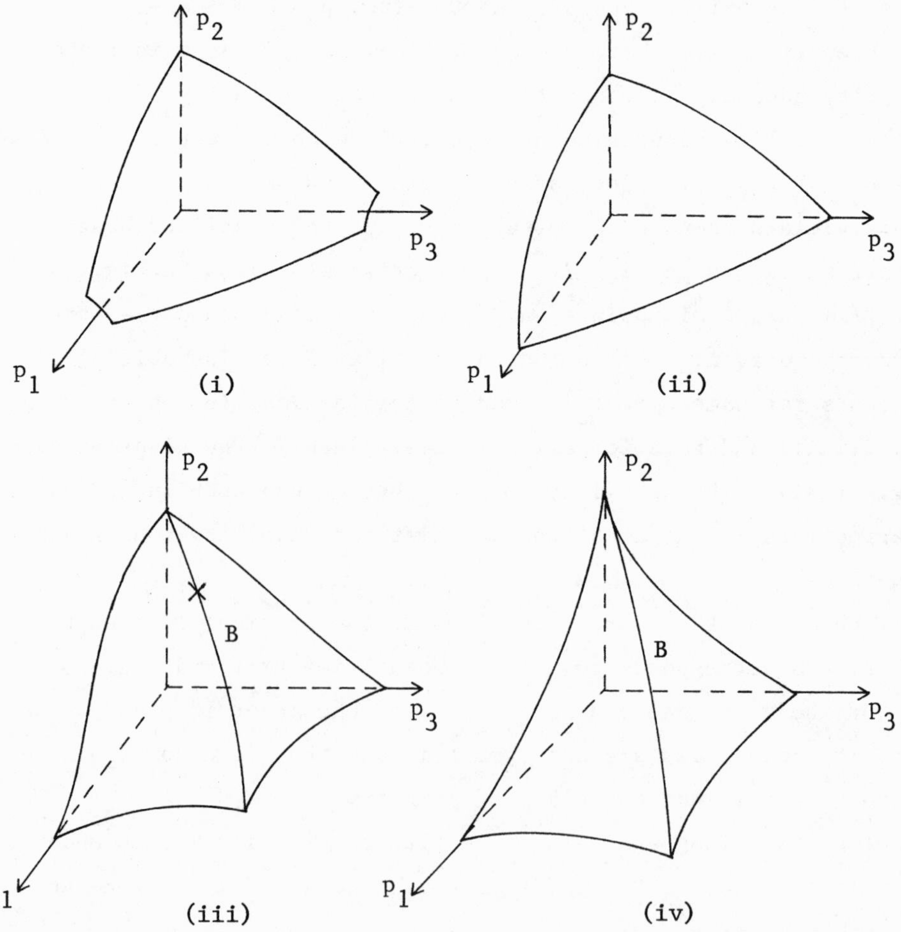

Fig. 7. Typical stability boundaries for arch

Fig. 6 treated as an example. A multiple-parameter perturbation technique is utilized in the neighborhood of critical points.

4. Truss with multiple loads. A truss is comprised of straight, elastic bars which are connected by pin joints and only deform axially. Consider the shallow, planar truss shown in Fig. 8. As the loads are increased, the initial, stable part of the equilibrium path can be obtained by minimizing the potential energy, which is a function of six generalized coordinates, the vertical deflections q_1, q_2, q_3 (downward) and the horizontal deflections q_4, q_5, q_6 of the interior joints. The potential energy has the form [9]

$$V = \frac{1}{2} \sum_{i=1}^{7} k_i \Delta_i^2 - \sum_{j=1}^{3} p_j q_j \tag{3}$$

where k_i represents the stiffness of bar i and Δ_i is its change of length (in terms of the generalized coordinates). The critical loading combinations can then be determined, and the resulting stability boundary has the form sketched in Fig. 9. There are three curves corresponding to bifurcation instability, one of which lies in the plane of symmetry, and the surface is divided into three smooth sections by these curves. A similar type of stability boundary has also been obtained for a shallow space truss (a lattice, or reticulated, dome) in [10].

5. Arch with single dynamic load. Consider a shallow arch subjected to a dynamic load p(x,t) which has a specified spatial distribution q(x). Typical examples are a step load $p(x,t) = p_o q(x)H(t)$, where H is the Heaviside step function and p_o is a measure of the load magnitude, and an impulse load $p(x,t) = p_o q(x)\delta(t)$, which imparts an initial velocity to the arch. Defining a critical value of the load is not as straightforward in the dynamic case as it was for static loads (based on Fig. 2).

The most commonly used criterion is that of Budiansky and Roth [11]. For a given value of p_o, the motion of the arch is determined numerically and the maximum value d of a norm of the deflection is computed.

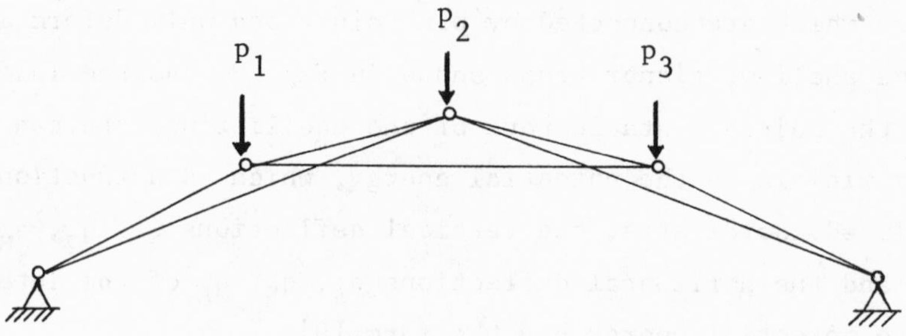

Fig. 8. Truss with multiple loads

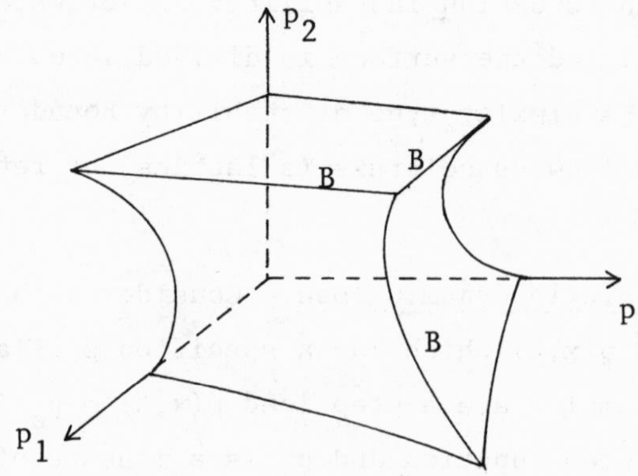

Fig. 9. Typical stability boundary for truss

A plot of d versus p_o is then drawn. Two typical plots are sketched in Fig. 10, where c denotes the height of the unloaded arch. If a sharp increase in slope occurs at a certain value of p_o, as in Fig. 10 (i), that value is said to be "critical". In some cases, however, the maximum response shows a gradual increase with p_o. It also may behave erratically, as in Fig. 10(ii), and some sharp peaks may be skipped if the load increments are not chosen sufficiently small [12]. Another interesting characteristic is that the motion may oscillate for a while close to the original configuration before snapping to an inverted shape [13,14]. The numerical integration must be carried out for a sufficiently long time period in order to be confident of obtaining the maximum response.

Another stability criterion was proposed by Hsu [15]. For impulse loads, the static equilibrium configurations of the unloaded arch are determined. In a phase space, surfaces of constant total energy (potential plus kinetic) are constructed about the unloaded state, expanding until another equilibrium state is reached. The arch is said to be stable under a given impulse load if the initial state (corresponding to the initial velocities) lies within the final surface. For step loads, similar surfaces are constructed for the arch under the corresponding static loading, and stability exists if the unloaded configuration lies within the final surface [16]. These definitions are sufficient to prevent snap-through instability in the sense of Budiansky and Roth.

6. Arch model with multiple dynamic loads. The arch model shown in Fig. 11 consists of four rigid bars of length L and mass m, three rotational springs of stiffness C, and a translational spring of stiffness K. The unloaded model has angles $\theta_i = \alpha_i$, and from geometry one can write

$$(4) \qquad \theta_4 = \sin^{-1} (\sin \theta_1 + \sin \theta_2 - \sin \theta_3).$$

The kinetic energy has the form [17]

$$(5) \qquad T = \sum_{i=1}^{4} [\tfrac{1}{2} m (\dot{a}_i^2 + \dot{b}_i^2) + \tfrac{1}{24} m L^2 \dot{\theta}_i^2],$$

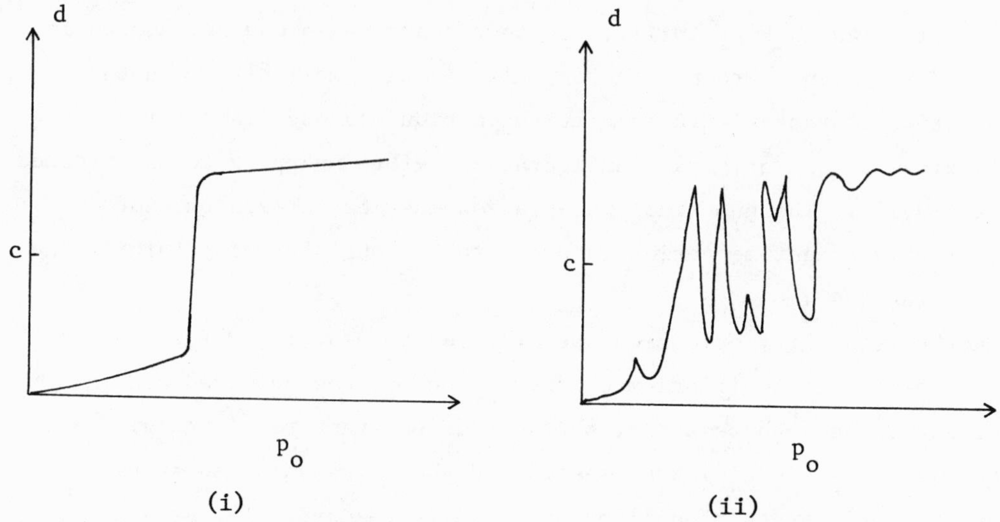

Fig. 10. Maximum response of arch

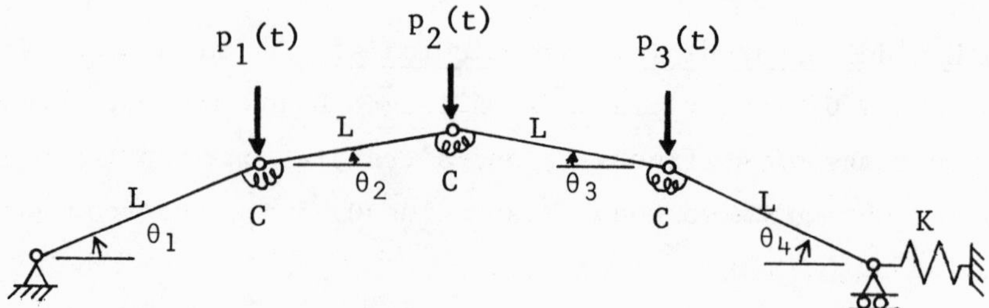

Fig. 11. Arch model

where a_i and b_i are the horizontal and vertical positions, respectively, of the centroid of bar i relative to the left hand exterior joint, and the potential energy (in the case of step loads) is given by

$$V = \frac{1}{2} C \left[(\alpha_1 - \theta_1 - \alpha_2 + \theta_2)^2 + (\alpha_3 - \theta_3 + \alpha_2 - \theta_2)^2 + (\alpha_4 - \theta_4 - \alpha_3 + \theta_3)^2\right] + \frac{1}{2} K \gamma^2 - \sum_{j=1}^{3} p_j q_j, \tag{6}$$

where γ is the change in length of the translational spring and q_j are the vertical deflections of the joints. (For impulse loads, the terms $p_j q_j$ in (6) are deleted and initial velocities are imparted to the bars.) After putting T and V in terms of the generalized coordinates θ_1, θ_2, θ_3, and their first derivatives, Lagrange's equations of motion are determined and the motion is computed numerically. The parameters are chosen as $\alpha_1 = \alpha_4 = 0.30$, $\alpha_2 = \alpha_3 = 0.15$, and $C = 0.001KL^2$.

For the sake of clarity, the results are presented in Figs. 12 and 13 as interaction curves in the following three load-planes: (i) p_2 vs. p_3 when $p_1 = 0$; (ii) p_1 vs. p_3 when $p_2 = 0$; and (iii) p_2 vs. p_3 when $p_1 = p_3$. For static loading, the critical load combinations are given by the dashed lines in Fig. 12. The static stability boundary in the three-dimensional load-space can be constructed with the use of these curves, and has the shape depicted in Fig. 7(iv).

For step and impulse loads, the maximum response of the model increases gradually as the load magnitude increases, for many load combinations. Therefore, the Budiansky-Roth criterion is not applicable. Two other criteria are defined. In the first, which yields the solid curves in Fig. 12 (step loads) and Fig. 13 (impulse loads), the critical load is chosen as the lowest magnitude which causes the outer bars to simultaneously deflect below the horizontal (i.e., $\theta_1<0$ and $\theta_4<0$) at some time during the motion. The second, which yields the dash-dot curves, gives the load combinations at which one of the angles θ_i, i=1,2,3,4, becomes negative during the motion. Dynamic stability boundaries in the (p_1, p_2, p_3) load-space can be constructed with the use of these curves.

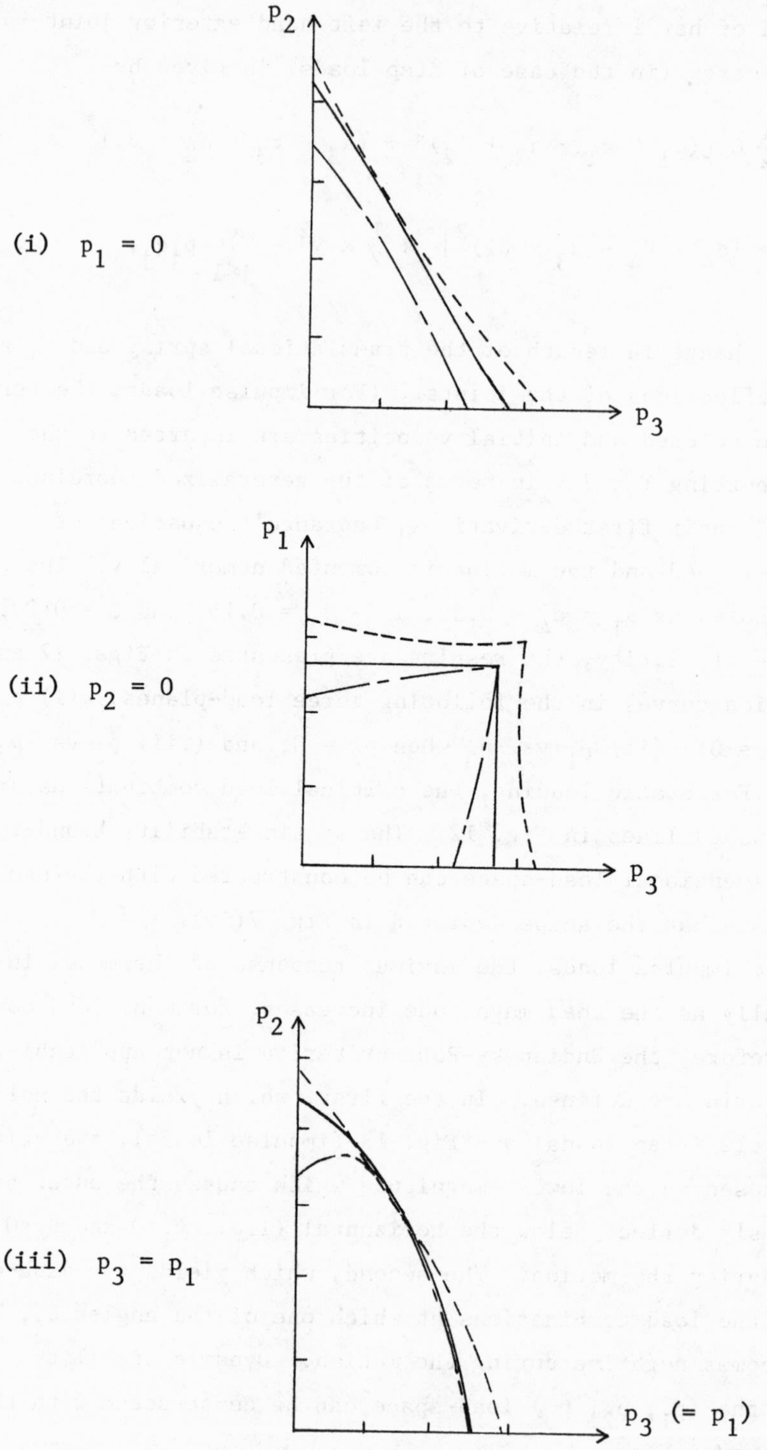

Fig. 12. Interaction curves for static and step loads

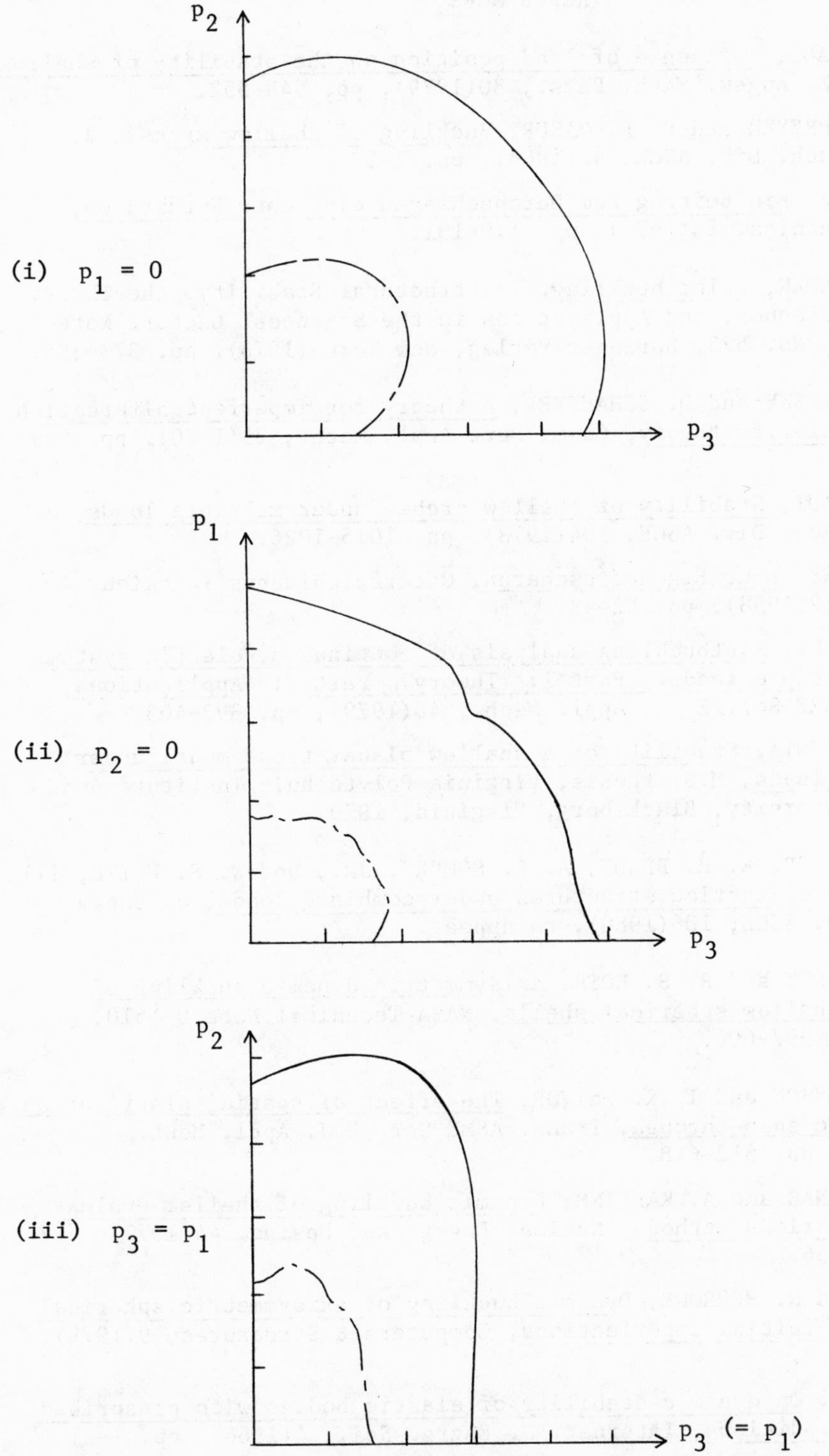

Fig. 13. Interaction curves for impulse loads

REFERENCES

[1] R. H. PLAUT, Influence of load position on the stability of shallow arches, Z. Angew. Math. Phys., 30(1979), pp. 548-552.

[2] H. L. SCHREYER and E. F. MASUR, Buckling of shallow arches, J. Engrg. Mech. Div. ASCE, 92(1966), pp. 1-17.

[3] H. TROGER, Ein Beitrag zum Durchschlagen einfacher Strukturen, Acta Mechanica, 23(1975), pp. 179-191.

[4] E. C. ZEEMAN, Euler buckling, in Structural Stability, the Theory of Catastrophes, and Applications in the Sciences, Lecture Notes in Math., No. 525, Springer-Verlag, New York (1976), pp. 373-395.

[5] M. GOLUBITSKY and D. SCHAEFFER, A theory for imperfect bifurcation via singularity theory, Comm. Pure Appl. Math., 23(1970), pp. 529-568.

[6] R. H. PLAUT, Stability of shallow arches under multiple loads, J. Engrg. Mech. Div. ASCE, 104(1978), pp. 1015-1026.

[7] F. BUCKENS, Über Eigenwertscharen, Österreichisches Ingenieur-Archiv, 12(1958), pp. 82-93.

[8] R. H. PLAUT, Postbuckling analysis of continuous, elastic systems under multiple loads. Part 1: Theory. Part 2: Applications, Trans. ASME Ser. E J. Appl. Mech., 46(1979), pp. 393-403.

[9] S. J. VALENTA, Stability of a shallow planar truss model under multiple loads, M.S. thesis, Virginia Polytechnic Institute and State University, Blacksburg, Virginia, 1979.

[10] S. M. HOLZER, R. H. PLAUT, A. E. SOMERS, JR., and W. S. WHITE, III, Stability of lattice structures under combined loads, J. Engrg. Mech. Div. ASCE, 106(1980), to appear.

[11] B. BUDIANSKY and R. S. ROTH, Axisymmetric dynamic buckling of clamped shallow spherical shells, NASA Technical Note D-1510, 1962, pp. 597-606.

[12] E. R. JOHNSON and I. K. McIVOR, The effect of spatial distribution on dynamic snap-through, Trans. ASME Ser. E J. Appl. Mech., 45(1978), pp. 612-618.

[13] V. SVALBONAS and A. KALNINS, Dynamic buckling of shells: evaluation of various methods, Nuclear Engrg. and Design, 44(1977), pp. 331-356.

[14] R. KAO and N. PERRONE, Dynamic buckling of axisymmetric spherical caps with initial imperfections, Computers & Structures, 9(1978), pp. 463-473.

[15] C. S. HSU, On dynamic stability of elastic bodies with prescribed initial conditions, Internat. J. Engrg. Sci., 4(1966), pp. 1-21.

[16] C. S. HSU, Stability of shallow arches against snap-through under timewise step loads, Trans. ASME Ser. E J. Appl. Mech., 35(1968), pp. 31-39.

[17] J. M. FITZGERALD and R. H. PLAUT, Snap-through of a shallow arch model under the interaction of multiple dynamic loads, in Developments in Theoretical and Applied Mechanics, Vol. 10, 1980, to appear.

SECTION FIVE
REVIEW: MATHEMATICAL METHODS AND MECHANICS

Remarks on Bifurcation Theory in Differential Equations

Jack K. Hale*

Abstract. This paper discusses primarily the idea of genericity in bifurcation theory for one and two parameter problems. Recent results on homoclinic points are mentioned as well as a connection between the bifurcation function and dynamic behavior.

My objective in this lecture is to point out some trends in the theory of bifurcation in differential equations and the manner in which very abstract theory has an influence on the basic understanding of specific examples. The examples will be in low dimension - never more than three. This choice is not made because examples never arise in higher dimensions. In fact, some of the most interesting applications of bifurcation theory today concern infinite dimensional systems - partial differential equations, functional differential equations or various types of integral equations. Even though the original problem is of higher dimension, the essential ingredient to bifurcation often is determined by a vector field in a low dimensional space. The appropriate space is obtained from the theory of integral manifolds

---

*Lefschetz Center for Dynamical Systems, Division of Applied Mathematics, Brown University, Providence, R.I. 02912. This work was supported in part by the National Science Foundation under MCS-79-05774, in part by the United States Army under AROD DAAG 27-76-G-0294, and in part by the United States Air Force under AF-AFOSR 76-3092C.

which is discussed by Marsden [16] in this volume.

Even though I will not attempt to give an historical perspective on bifurcation theory, a careful study of the literature shows that Poincaré [26], [27] and Lyapunov [14], [15] are responsible for the basic philosophy as well as several of the fundamental ideas of the methods that we presently employ. One can find a direct link with the importance of exchanges of stability, the reduction principle to lower dimensional problems, the philosophy of genericity and the transformation theory so relevant to obtaining approximations of the center manifold and the flow on the center manifold. In many respects, we are still exploiting the ideas and methods of these two giants.

A fundamental step toward modern bifurcation theory in differential equations was made by Andronov and Pontryagin [3] when they gave a definition of structural stability in the plane. To avoid the difficulties that arise from the noncompactness of the plane, they restricted the discussion to the interior of a closed curve without contact to any of the vector fields to be considered. Two vector fields X, Y in C^r, $r \geq 1$, are equivalent if the trajectories of one are homeomorphic to the other. This is an equivalence relation among vector fields. X is structurally stable if every Y in a neighborhood of X is equivalent to X. The set of structurally stable systems is open and dense and characterized by the following properties: every critical point and periodic orbit is hyperbolic and no orbit connects saddles.

One can now say X is a bifurcation point (a vector field for which a perturbation could lead to a bifurcation) if X is structurally unstable; that is, not structurally stable. It is impossible to study the behavior of all vector fields in the neighborhood of an arbitrary bifurcation point; for example, $X = 0$ is a bifurcation point and any flow in the plane can be obtained by choosing an appropriate Y near zero. This is where the idea of genericity enters bifurcation theory. One must find those bifurcation points X which have the property that the simplest possible bifurcations occur near X. Andronov and Leontovich [1] made this precise by defining structural instability of degree k (or bifurcation point of degree k).

The vector field X is a bifurcation point of degree 0, if it is structurally stable. X is a bifurcation point of degree 1, if it is not of degree zero and there is a neighborhood of X which has only bifurcation points of degree 0, or, ones which are equivalent to X. It is a bifurcation point of degree 2, if it is not of degree zero or one and there is a neighborhood containing only bifurcation points of degree 0 or 1, or, ones which are equivalent to X. Similarly, one defines degree k.

It can be shown (see Andronov et al [2], Sotomayor [29]) that $r \geq 3$ implies X being a bifurcation point of degree 1 is equivalent to the fact that there is a neighborhood U of X such that $U = U_1 \cup \Gamma \cup U_2$ where Γ is a smooth submanifold of codimension one, U_1, U_2 are open sets belonging to distinct equivalence classes of structurally stable systems. Furthermore, X is a bifurcation point of degree 1, if and only if exactly one of the following alternatives hold as one passes through X from U_1 to U_2:

(i) Two limit cycles coalesce and disappear

(ii) A focus loses its hyperbolicity and a periodic orbit appears (elementary Hopf bifurcation)

(iii) A saddle and node coalesce and disappear

(iv) There is a smooth invariant curve containing a saddle and node which coalesce, disappear and the invariant curve becomes a periodic orbit (this is actually a special case of (iii)).

(v) $\mathrm{tr}\,\partial X(0)/\partial x \neq 0$, a periodic orbit and saddle merge to form a homoclinic orbit (the stable and unstable manifolds of the saddle intersect) at X and then the homoclinic orbit disappears leaving only the saddle.

The fact that only two possibilities arise in a neighborhood of a bifurcation point of degree 1 suggests that this is the typical or generic situation that arises in the discussion of one parameter families of vector fields. Sotomayor [29] has proved this is the case. In fact, in the family of smooth one parameter families of vector fields in C^r, $r \geq 3$, the set which contains only bifurcation points of degree 0 or 1 is residual.

Extensive applications of these results to the theory of nonlinear oscillations were made in the late 1930's (see Andronov, Vitt and Khaikin [4]). Most of the results in the literature on one parameter problems are a consequence of these results.

The characterization of bifurcation points of degree two has recently been completed, but the detailed results will not be given here (see Andronov et al [2], Sotomayor [30], [31], Takens [32], Carr [5]). This theory of bifurcation points of degree two can be considered the typical or generic situation for two parameter families of vector fields. Genuine two parameter problems frequently arise in applications and often, the tendency is to scale the parameters in terms of a single small parameter in order to use some classical perturbation procedure. Sometimes it is possible (but with considerable effort) to show that the information obtained is exactly the same as one would obtain by allowing the original parameters to vary independently. In other cases, information is lost by scaling. The simplest illustration of this is the Hopf bifurcation when the degenerate term is fifth degree rather than cubic. For any one parameter family of vector fields for which the eigenvalues cross the imaginary axis with non-zero velocity, it can be shown that only one periodic orbit will be obtained (see Chafee [6]). The same result is true when the degeneracy is $(2k+1)$ (see Negrini and Salvadori [35]). If the eigenvalues are allowed to cross with zero velocity, two orbits sometimes appear (see Flockerzi [10]). If this problem is discussed in the general setting, no confusion or loss of information occurs. This result for Hopf bifurcation has recently been applied by Chow, et al [8].

Interesting applications of the two parameter methods for the equation

$$\ddot{x} + \lambda_1 x + \lambda_2 \dot{x} + f(x,\dot{x}) = 0$$

have been given by Takens [32], Holmes and Marsden [11], Carr [5] for the case when $f(x,0) = \beta x^3$, $\beta \neq 0$ and by Howard and Koppell [12] for $f(x,0) = \beta x^2$, $\beta \neq 0$.

For a discussion of bifurcation points of degree greater than two, see Andronov et al [2], Sotomayor [30], [31].

When the dimension of the system exceeds 2, the class of structurally stable systems is small in the sense that it is not a residual set. In addition, many new phenomena occur which make a classification of such systems extremely difficult. Much of the modern theory of dynamical systems is an attempt to discover a residual set of "simple" dynamical systems - ones which can be classified and which will preserve their essential features when subjected to perturbations (see Smale [28], Nitecki [23], Peixoto [25], Newhouse [20], Palis and Melo [24]).

Bifurcation theory is not as well understood. In dim $m > 2$, much attention has been devoted to the appearance of strange attractors through bifurcation from a homoclinic orbit. For a hyperbolic fixed point of a diffeomorphism, a point $q \neq p$ is <u>homoclinic</u> to p if q is in the intersection of the stable and unstable manifolds of p. It is <u>transverse homoclinic</u> to p if the intersection is transversal. A special case of the results of Newhouse and Palis [21], [22] is concerned with one parameter families of diffeomorphisms $f(x,\lambda)$ which have the stable and unstable manifolds of p at $\lambda = 0$ tangent at q. Under certain conditions, they show there are families with the property that there is an infinite nonwandering set for each $\lambda < 0$ and the map is structurally stable for each $\lambda < 0$. Furthermore, there are infinitely many changes in the topological structure as $\lambda \to 0$. Newhouse [19] shows each of these systems also has infinitely many sinks.

The abstract theory can be realized for a planar differential equation with periodic coefficients

$$\ddot{y} + g(y) = -\lambda_1\dot{y} + \lambda_2 f(t)$$

where $f(t+1) = f(t)$ and the equation for $\lambda_1 = \lambda_2 = 0$ has a homoclinic orbit. The appearance of homoclinic orbits for (λ_1,λ_2) in a neighborhood of zero has been discussed by several persons including Mel'nikov [17], Morozov [18], Holmes (see [11]), Chow, Hale, Mallet-Paret [7] and Hayashi and Ueda (see [34]). The most complete discussion is in [7] and relies only on classical concepts in differential equations. It is shown there are two curves in (λ_1,λ_2)-space which correspond to bifurcation to homoclinic points. In a neighborhood of any point on these curves, there are infinitely many subharmonic bifurcations, half being saddles and the others sinks. This example thus gives a concrete illustration of the abstract results in diffeomorphisms.

One also can observe this type of bifurcation in a more generic way by considering a planar equation

$$\dot{x} = X(x,\lambda) + \mu f(t,x)$$

where $f(t+1,x) = f(t,x)$ and the function $X(x,\lambda)$ at $\lambda = 0$ is a bifurcation point of degree one satisfying property (v) where a homoclinic orbit joins a saddle. For a generic class of f, one can adapt the methods in [7] to get the curve of bifurcation to homoclinic points. The theory of integral manifolds will imply the existence of a hyperbolic invariant torus $M_{\lambda,\mu}$ near the periodic orbit which bifurcated from the homoclinic orbit of $\dot{x} = X(x,\lambda)$. One can show there are subharmonic bifurcations on $M_{\lambda,\mu}$ as λ,μ vary.

As a final topic, we mention a recent result on the relationship between the dynamic behavior of a system and the bifurcation function obtained by the method of Liapunov-Schmidt. Consider the n-dimensional equation

$$\dot{x} = Ax + f(t,x,\lambda) \tag{1}$$

where $f(t+2\pi,x,\lambda) = f(t,x,\lambda)$, $f(t,0,0) = 0, \partial f(t,0,0)/\partial x = 0$, λ is a vector parameter, the matrix A has zero as a simple eigenvalue and the remaining ones with negative real parts. If we apply the method of Liapunov-Schmidt for the existence of 2π-periodic solutions for (x,λ) near zero, we obtain a scalar bifurcation function $G(a,\lambda)$ depending on a scalar a and λ whose zeros correspond to 2π-periodic solutions. If $x = (y,z)$, $f = (g,h)$, $y,g \in \mathbb{R}$, then the function $G(a,\lambda)$ is defined as follows. Let $x(a,\lambda,t) = (y(a,\lambda,t),z(a,\lambda,t))$ for $a \in \mathbb{R}$, λ small, be the 2π-periodic function satisfying

$$\dot{y} = g(t,y,z,\lambda) - \frac{1}{2\pi}\int_0^{2\pi} g(t,y(t),z(t),\lambda)dt$$

$$\dot{z} = Bz + h(t,y,z,\lambda)$$

$$\frac{1}{2\pi}\int_0^{2\pi} y(t)dt = a$$

and define

$$G(a,\lambda) = \frac{1}{2\pi}\int_0^{2\pi} g(t,y(a,\lambda,t),z(a,\lambda,t),\lambda)dt .$$

Consider the scalar equation

$$\dot{a} = G(a,\lambda). \tag{2}$$

It is shown by deOliveira and Hale [9] that the stability properties of the equilibrium points of (2) determine the stability properties of the 2π-periodic solutions of (1). The bifurcation function is relatively easy to approximate since it only requires equating Fourier series. It is certainly much easier than averaging techniques. More importantly, it permits a complete discussion of several parameter problems which averaging alone cannot do since it requires scaling of the parameters and an application of the implicit function theorem. The results apply directly to Hopf bifurcation since one can arrive at (1) by a change of variables with t representing an angle. In fact, if one is interested in periodic orbits of

$$\dot{u} = Cu + U(u,v,\alpha)$$
$$\dot{v} = Bu + V(u,v,\alpha)$$

where $u \in \mathbb{R}^2$, C has eigenvalues $\pm i, \mathrm{Re}\lambda B < 0$, U,V are $O((|u|+|v|)(|u|+|v|+|\alpha|))$ at zero, let $u_1 = y\cos\theta$, $u_2 = -y\sin\theta$, $v = yz$, and express the derivatives of y,z in terms of θ to obtain a special case of the above equation. Certain types of infinite dimensional evolutionary equations also can be discussed (see [9]).

The extent to which the above remarks are valid where A has a double eigenvalue zero certainly deserves extensive study. The problem is difficult because of the possibility of the appearance of invariant tori and homoclinic points. When one considers a three dimensional autonomous equation near zero for which the linear part has a pair of purely imaginary eigenvalues and one zero eigenvalue, a change of variables leads to equation (1) in the plane with $A = 0$ and t an angle variable. More specifically, consider the equation

$$\dot{u}_1 = U_1(u_1,u_2,u_3,\lambda)$$

(3)
$$\dot{u}_2 = u_3 + U_2(u_1,u_2,u_3,\lambda)$$

$$\dot{u}_3 = -u_2 + U_3(u_1,u_2,u_3,\lambda)$$

where $U_j(u,\lambda) = O(|u|(|u|+|\lambda|))$ as $u,\lambda \to 0$. If we assume that U_2, U_3 vanish for $u_2 = u_3 = 0$, then it is legitimate to introduce the new coordinates $u_1 = x_1$, $u_2 \cos\theta$, $u_3 = -x_2 \sin\theta$ and eliminate t in the equations to obtain

$$\frac{dx}{d\theta} = X(\theta,x,\lambda) \tag{4}$$

for the two vector $x = (x_1,x_2)$. The function X is 2π-periodic in θ and satisfies $X(\theta,x,\lambda) = O(|x|(|x|+\lambda))$ as $x,\lambda \to 0$.

Let $x(a,\lambda,\theta)$ for $a \in \mathbb{R}^2$, λ small, be the 2π-periodic function satisfying the equation

$$\frac{dx}{d\theta} = X(\theta,x,\lambda) - \frac{1}{2\pi}\int_0^{2\pi} X(\theta,x(\theta),\lambda)d\theta$$

$$\frac{1}{2\pi}\int_0^{2\pi} x(\theta)d\theta = a$$

and define the bifurcation function

$$G(a,\lambda) = \frac{1}{2\pi}\int_0^{2\pi} X(\theta,x(a,\lambda,\theta),\lambda)d\theta .$$

The problem is to relate the solutions of the second order equation (2) to the solutions of (3).

Under some hypotheses, Langford [13] has obtained specific results on this problem. More specifically, he related the stability properties of the equilibrium points of (2) to the stability properties of the periodic orbits of (3). At this conference, Guckenheimer has given a more complete solution of the problem considered by Langford and has shown how a periodic orbit of (2) yields a torus of (3) and how a homoclinic orbit of (2) leads to transverse homoclinic points for (3) and thus strange attractors. These latter results were independently discovered by Chow and Hale in some work not yet published. Also, they were able to eliminate the symmetry hypothesis that U_2, U_3 vanish for $u_2 = u_3 = 0$.

Also, Chow and Hale have obtained some other new types of bifurcations in their consideration of more general problems. We only state one result. Suppose the equation (2) depends on a scalar parameter λ and has the property that there is a hyperbolic stable periodic orbit for $\lambda < 0$, there is a stable periodic orbit for $\lambda = 0$ governed by the cubic terms of the Poincaré map and there are three hyperbolic periodic orbits (two stable, one unstable) for $\lambda > 0$. Then for the three dimensional equations (3), there is a hyperbolic stable invariant torus for $\lambda < 0$, a stable invariant torus for $\lambda = 0$ and three hyperbolic invariant torii (two stable, one unstable) for $\lambda > 0$. The proofs are complicated and will appear elsewhere.

REFERENCES

[1] A. A. ANDRONOV and F. A. LEONTOVICH, Sur la théorie de la variation de la structure qualitative de la division du plan en trajectories. Dokl. Akad. Nauk 21(1938), pp. 427-430.

[2] A. A. ANDRONOV, F. A. LEONTOVICH, I. I. GORDON and A. G. MAIER, Theory of Bifurcations of Dynamical Systems on a Plane, Wiley, 1973.

[3] A. A. ANDRONOV and L. S. PONTRJAGIN, Grubye sistemy. Dokl. Akad. Nauk SSR 14(1937), no. 5.

[4] A. A. ANDRONOV, A. A. VITT, S. E. KHAIKIN, Theory of Oscillations, Pergamon 1966.

[5] J. CARR, Applications of Centre Manifold Theory, Lecture Notes, Lefschetz Center for Dynamical Systems, Division of Applied Mathematics, Brown University, June, 1979.

[6] N. CHAFEE, Generalized Hopf bifurcation and perturbation in a full neighborhood of a given vector field, Indiana Univ. Math. J. 27(1978).

[7] S-N. CHOW, J. K. HALE and J. MALLET-PARET, An example of bifurcation to homoclinic orbits, J. Diff. Eqn. Submitted.

[8] S-N. CHOW and R. WHITE, On the transition from supercritical to subcritical Hopf bifurcation. To appear.

[9] J. C. deOLIVEIRA and J. K. HALE, Dynamic behavior from bifurcation equations. Tohoku Math. J. To appear.

[10] D. FLOCKERZI, Bifurcation of Periodic Solutions from an Equilibrium Point. Dissertation, Wurzburg, 1979.

[11] P. HOLMES and J. E. MARSDEN, Qualitative techniques for bifurcation analysis of complex systems. Bifurcation Theory and Applications to Scientific Disciplines, N. Y. Acad. Sci. 1979.

[12] L. HOWARD and N. KOPPELL, Bifurcations and trajectories joining critical points. Adv. Math. 18(1976), pp. 306-358.

[13] W. F. LANGFORD, Periodic and steady state mode interactions lead to torii. SIAM J. Appl. Math. 37(1979), pp. 22-48.

[14] A. M. LYAPUNOV, Problème Général de la Stabilité du Mouvement, Princeton, 1949.

[15] A. M. LYAPUNOV, Sur les figures d'equilibre peu différentes des ellipsoides d'une masse liquide homogène donnée d'un mouvement de rotation. Zap. Akad. Nauk, St. Petersburg (1906).

[16] J. E. MARSDEN, Dynamical systems and invariant manifolds. Proc. New Approaches to Nonlinear Problems in Dynamics, Monterey, Calif. Dec. 9-14, 1979, SIAM Publications.

[17] V. K. MEL'NIKOV, On the stability of the center for time periodic solutions. Trans. Moscow Math. Soc. (Trudy) 12(1963), pp. 3-56.

[18] A. D. MOROZOV, On the complete qualitative investigation of the equation of Duffing. Diff. Uravn. 12(1976), pp.241-255.

[19] S. NEWHOUSE, Diffeomorphisms with infinitely many sinks. Topology 12(1974), pp.9-18.

[20] S. NEWHOUSE, Lectures on dynamical systems, C.I.M.E. Summer Session on Dynamical Systems, June 19-27, 1978.

[21] S. NEWHOUSE and J. PALIS, Bifurcations of Morse-Smale dynamical systems. Dynamical Systems, pp. 303-365, Academic Press, 1973.

[22] S. NEWHOUSE and J. PALIS, Cycles and bifurcation theory. Asterisque 31(1976), pp.43-141.

[23] Z. NITECKI, Differentiable Dynamics, M.I.T. Press, 1971.

[24] J. PALIS and W. MELO, Introducão aos Sistemas Dinamicos, IMPA, Rio de Janeiro, 1978.

[25] M. PEIXOTO, Dynamical Systems, Academic Press, 1973.

[26] H. POINCARÉ, Les Méthodes Norwelles de la Méchanique Céleste, Vol. 3, Gauthier-Villars, 1892.

[27] H. POINCARÉ, Sur l'équilibre d'une masse fluide animés d'un mouvement de rotation. Acta Math. 7(1885), pp.259-380.

[28] S. SMALE, Differentiable dynamical systems, Bull. Am. Math. Soc. 73(1967), pp.747-817.

[29] J. SOTOMAYOR, Generic one parameter families of vector fields on two dimensional manifolds. Pub. I.H.E.S. 43(1973).

[30] J. SOTOMAYOR, Structural stability and bifurcation theory, Dynamical Systems, pp. 549-560, Academic Press, 1973.

[31] J. SOTOMAYOR, Generic bifurcations and dynamical systems. Dynamical Systems, pp. 561-582, Academic Press, 1973.

[32] F. TAKENS, Forced oscillations and bifurcations. Communication 3 Inst. Rijksuniversiteit, Utrecht (1974).

[33] F. TAKENS, Singularities of vector fields. Publ. I.H.E.S. 43 (1974), pp. 47-100.

[34] Y. UEDA, Mapped forced oscillations by analog computer, Proc. New Approaches to Nonlinear Problems in Dynamics, Monterey, Calif. Dec. 9-14, 1979, SIAM Publications.

[35] P. NEGRINI and L. SALVADORI, Attractivity and Hopf bifurcation, Nonlinear Anal. 3(1979), pp.87-99.

Bifurcation Theory and Averaging in Mechanical Systems

P. R. Sethna*

Abstract. Ideas relating to center manifold theory, the method of integral manifolds and theory of normal forms are shown to have application in the study of nonlinear mechanical systems. Specific applications to the study of the plane motion of a tube carrying a fluid, the motion of ships exhibiting internal resonance and bifurcations more general than Hopf bifurcations in three dimensional motions of tubes are discussed.

I. Introduction. Several new mathematical ideas have recently had significant influence on the study of nonlinear dynamical systems. The ideas related to the center manifold theorem [1], method of averaging and its generalization to infinite dimensional systems [2], [3] and the theory of normal forms [4] have all made it possible to study bifurcation phenomena in mechanical systems that have been hitherto either intractable or studied by less rigorous means.

We will discuss the relevance of these ideas in the study of mechanical systems by means of three examples. In section two is given the study of the plane motions of a tube carrying a fluid [5]. The tube exhibits Hopf bifurcations in a system modeled by a coupled system of partial and ordinary differential equations. The method of integral averaging and the ideas based on the center manifold theorem are central to this study. In section three we study forced motions

* Department of Aerospace Engineering and Mechanics, 107 Aerospace Engineering Bldg., University of Minnesota, Minneapolis, Minnesota 55455. The work was supported by funds from National Science Foundation under Grant NSF-ENG-76-00030. Portions of this paper have appeared in the ASME Journal of Applied Mechanics, 45(1978), pp. 895-902 and the Proceedings of the VIIIth International Conference on Nonlinear Oscillations.

of dynamical systems with quadratic nonlinear terms in the case when internal resonance occurs [6]. Such systems are known to model the motion of ships. In this problem we use the results of the method of averaging and give some new mathematical results for the case when the averaged equations themselves have a Hopf bifurcation [7] and show that the forced periodic solutions bifurcate into tori. In section four we discuss some work in progress on general classes of autonomous dynamical systems of finite dimension, which exhibit bifurcations due to the passage of more than one pair of complex eigenvalues into the right half plane. Concepts related to the center manifold theory, the theory of normal forms and the theory of bifurcations under a symmetry group are all important concepts in this study.

II. Hopf bifurcations in tubes carrying a fluid. Consider an elastic tube with a circular cross section which is long in comparison to the outer diameter. Its center line is assumed inextensible and initially straight. It carries an incompressible fluid. The flow velocity is assumed uniform at any cross section and parallel to the center line of the tube. The fluid enters at the fixed end at a constant pressure and exits at the free end. The motions are assumed to occur in a plane. The mathematical statement of the problem depends on three parameters, ρ, the flow velocity, β, the mass ratio of the tube and the fluid and α a parameter that is essentially related to the length to area ratio of the tube.

If x is the dimensionless length coordinate and τ the dimensionless time, if $u_2(x,\tau)$ is the lateral deflection of the tube and if $\tilde{v}(\tau)$ is the fluctuation of the fluid flow velocity about a mean flow velocity, then the following formidable boundary value problem describes the system:

$$\frac{\partial \underline{u}}{\partial \tau} = \underline{\underline{L}}\underline{u} + \varepsilon \underline{N}_1(\underline{u}) + \varepsilon \underline{N}_2(\underline{u},\tilde{v}) + 0(\varepsilon^2)$$

$$(2.1)\qquad \frac{\partial \tilde{v}}{\partial \tau} = -2\rho_0 \tilde{\alpha}\tilde{v} + g(\underline{u}) + 0(\varepsilon)$$

where ε is a small positive scaling parameter,

$$\underline{u} \equiv \begin{pmatrix} u_1 \\ u_2 \end{pmatrix}, \quad \underline{L} \equiv \begin{pmatrix} 0 & -\rho_0^2(\ )'' - (\ )^{iv} \\ 1 & -2\rho_0\beta(\ )' \end{pmatrix},$$

$$\underline{N}_1(\underline{u}) \equiv \begin{pmatrix} -\frac{3}{2}\{(u_2'')^3 + 2u_2'u_2''u_2'''\} \\ +[u_2'\{\int_x^1\{\int_0^{x_1}\{(u_1')^2 - u_2''(2\rho_0\beta u_1' - \rho_0^2 u_2^{iv})\}dx_2\}dx_1]' \\ 0 \end{pmatrix}$$

$$\underline{N}_2(\underline{u},\tilde{v}) \equiv \begin{pmatrix} -(1-x)\beta u_2''\dot{\tilde{v}} + 2\rho_0\tilde{v}(u_2^{iv} + \rho_0^2 u_2'' + 2\rho_0\beta u_1' + 4\rho_0^2\beta^2 u_2'') \\ -2\tilde{v}\beta(u_1' - 2\rho_0\beta u_2'') \\ 0 \end{pmatrix}$$

and

$$g(\underline{u}) \equiv \beta \int_0^1 \left[\int_0^x \left\{ \left(\frac{\partial^2 u_2}{\partial x \partial \tau}\right)^2 + \frac{\partial u_2}{\partial x} \cdot \frac{\partial^3 u_2}{\partial x \partial \tau^2} \right\} dx_1 - \frac{\partial^2 u_2}{\partial \tau^2} \cdot \frac{\partial u_2}{\partial x} \right] dx,$$

where primes denote derivatives with respect to x.

The boundary conditions are

(2.2) $\quad u_2(0,t) = u_2'(0,t) = u_2''(1,t) = u_2'''(1,t) = 0$.

The object of the analysis is to study the motion of the system for small nonlinear oscillations, i.e. when ε is small, as they depend on the parameters ρ, β, α and in particular as the flow parameter ρ goes through a critical value $\rho = \rho_0$ when the stable solution $\underline{u}(x,t) = 0$ bifurcates and the system has for the same value of ρ the solution $\underline{u}(x,t) = 0$ and another periodic solution.

Linearized system

For $\varepsilon = 0$ the boundary value problem reduces to

$$\frac{\partial \underline{u}}{\partial \tau} = \underline{L}\ \underline{u}$$

with the same boundary conditions (2.2). If $\underline{u} = \underline{w}(x)e^{\lambda\tau}$ then the eigenvalues satisfy a very complicated transcendental equation. The eigenvalues obviously depend on ρ and β. For $\rho = 0$, the problem is self-adjoint and all the eigenvalues are pure imaginary. For $\rho > 0$ and small, it can be shown that all the eigenvalues are in the left half plane and thus the flow damps out the elastic motions. For larger values of ρ, we postulate that there is a value of ρ, $\rho = \rho_0$, a critical value of ρ, when one pair of complex eigenvalues cross the pure imaginary axis with $\lambda(\rho_0) = i\omega_0$, $\bar{\lambda}(\rho_0) = -i\omega_0$. Furthermore, we assume that for $0 < \rho < \rho_0$ all other eigenvalues remain in the left half plane and a Hopf bifurcation occurs. There is some numerical evidence for these assumptions. In Fig. 1 is shown, for $\beta^2 = 0.60$, the behavior of the first three eigenvalues corresponding to the self adjoint problem. The behavior is seen to be very complicated and it is seen that it is the second mode that becomes unstable.

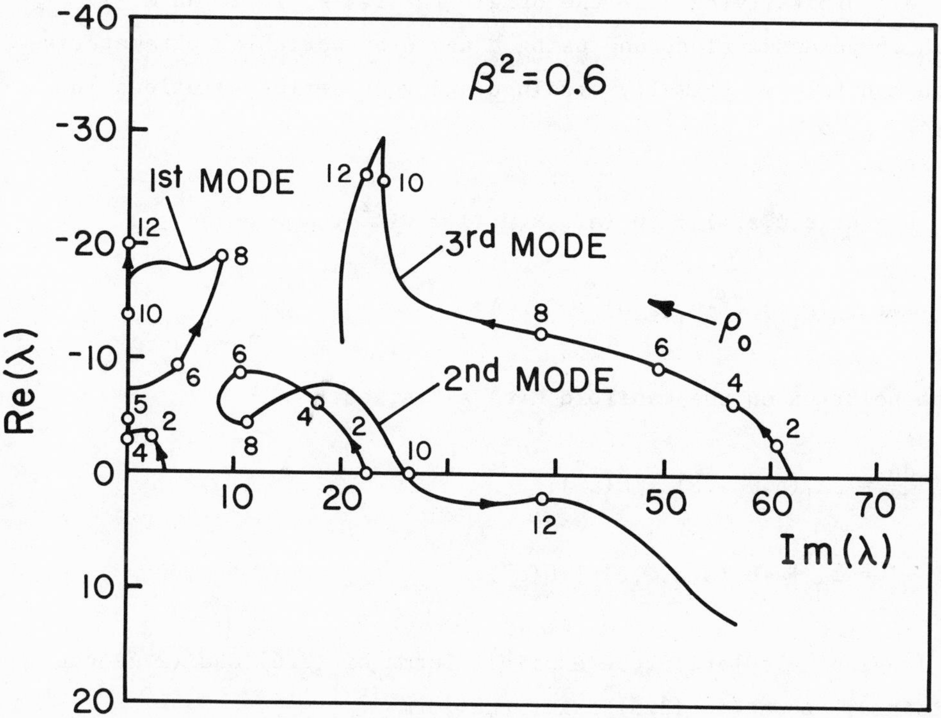

Figure 1: Behavior of eigenvalues corresponding to the lowest modes of the system as a function of the flow velocity ρ_0

Bifurcation into a periodic solution

We follow here the general ideas in [3]. Let $\rho = \rho_0 + \varepsilon\mu$. For $\varepsilon = 0$ we have $\underline{u} = \underline{u}_0 = a[\underline{w}(x)e^{i\psi} + \bar{\underline{w}}(x)e^{-i\psi}]$ where $\underline{w}(x)$ is the vector valued eigenfunction corresponding to the critical eigenvalue, "a" is the amplitude and $\phi = \psi - \omega_0\tau$ is the phase of the motion.

For $\varepsilon \neq 0$, let $\underline{u}$ be decomposed into $\underline{u} = \underline{u}_0 + \underline{y}$ where $\underline{u}_0$ lies in the subspace spanned by $\underline{w}$ and $\bar{\underline{w}}$ for all "a" and ϕ while $\underline{y}$ lies in a space orthogonal to the space of eigenvectors adjoint to $\underline{w}$ and $\bar{\underline{w}}$ for all "a" and ϕ. Following [3], for $\varepsilon \neq 0$ we assume that

(i) there exists a smooth invariant center manifold $M(\varepsilon)$: $\underline{y} = \underline{y}^*(a,\phi,\varepsilon\mu)$, $\underline{y}^*(0,\phi,\varepsilon\mu) = \frac{\partial y^*}{\partial a}(0,\phi,0) = 0$ passing through the origin $\underline{u} = 0$ and tangent to the $(a,\phi,\varepsilon\mu)$ space.

(ii) the differential equation induces a smooth flow on $M(\varepsilon)$

(iii) all orbits lying near the origin for $t \in (-\infty,\infty)$ lie on $M(\varepsilon)$.

With these assumptions and using a and ϕ as variables parameterising the manifold we formally obtain $\underline{u}$ and $\tilde{v}$ as series solutions in powers of ε.

(2.4) $$\underline{u} = \underline{u}_0(a,\psi,x,\tau) + \varepsilon\underline{u}_1(a,\psi,x,\mu,\tau) + O(\varepsilon^2),$$

(2.5) $$\tilde{v} = \tilde{v}_0(a,\psi) + \varepsilon\tilde{v}_1(a,\psi) + O(\varepsilon^2)$$

and the solution on the manifold $M(\varepsilon)$ satisfies:

(2.6) $$\frac{da}{d\tau} = \varepsilon A_1(a,\mu,\alpha,\beta) + O(\varepsilon^2),$$

(2.7) $$\frac{d\psi}{d\tau} = \omega_0 + \varepsilon B_1(a,\mu,\alpha,\beta) + O(\varepsilon^2).$$

After lengthy calculation the explicit forms of (2.6) and (2.7) can be obtained. Equation (2.6) takes the form

$$\frac{da}{d\tau} = \varepsilon[C_1(\beta)\mu + C_2(\alpha,\beta)a^2]a + O(\varepsilon^2)$$

and the periodic solution has amplitude

$$(2.8)\quad a_0^2 = -\frac{\mu C_1(\beta)}{C_2(\alpha,\beta)}$$

In (2.8), $C_1(\beta) > 0$ for all β and if $C_2(\alpha,\beta) < 0$ we have the "supercritical" case and we have a stable periodic solution for $\mu > 0$. If $C_2(\alpha,\beta) > 0$, we have the subcritical case and, as a first approximation, we have an unstable periodic solution. For a given β, the mass ratio, the length parameter α determines the nature of the bifurcation. For long tubes α is large and bifurcations for all β are supercritical. For α small the bifurcation is subcritical for all β. These results are shown in Fig. 2. The anamolous behavior at $\beta^2 \simeq 0.3$ and $\beta^2 \simeq 0.7$

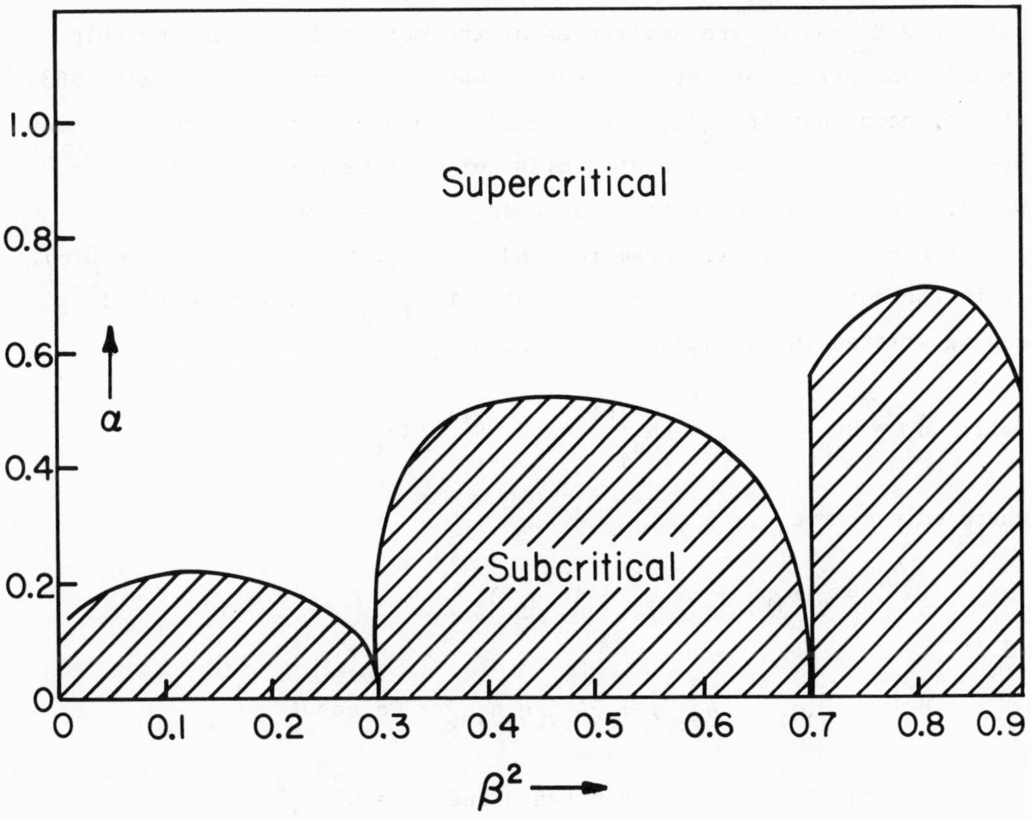

Figure 2: Critical "slenderness parameter" α vs. β^2

is due to the fact that the critical eigenvalue as a function of ρ does not cross the pure imaginary axis but tangents that axis at these values of ρ and thus Hopf bifurcations are not possible.

III. Bifurcation of the Averaged Equations in Study of Motion of Ships. The motion of a ship can be modeled by a system of equations of the form [8]:

$$\ddot{\phi} + \omega_1^2\phi + k_1\phi\theta + c_\phi\dot{\phi} = M_\phi\cos(\nu t+\tau_1)$$

(3.1)

$$\ddot{\theta} + \omega_2^2\theta + k_2\phi^2 + c_\theta\dot{\theta} = M_\theta\cos(\nu t + \tau_2)$$

where ϕ is the angular motion in roll, θ is the angular motion in pitch and M_ϕ and M_θ are amplitudes of the moment acting on the ship in roll and pitch respectively due to waves. Very early (Froude 1863) it was known that if $\omega_2/\omega_1 \simeq 2$ special phenomena occur in such systems. System (3.1) is an example of a class of systems discussed in [6]. We will be interested in the case when $\nu = \omega_1$ and $\omega_2/\omega_1 \simeq 2$.

Let $\varepsilon > 0$ be a small parameter and let, M_ϕ, M_θ, c_ϕ, c_θ all be $0(\varepsilon)$. If furthermore $\phi = \varepsilon q_1$ and $\theta = \varepsilon q_2$ and if $k_1/k_2 = 2$, system (3.1) can be written in the quasi-Hamiltonian form

(3.2) $$\dot{q}_i = H,_{p_i}^{(0)}, \quad \dot{p}_i = -H_{q_i}^{(0)} - \varepsilon H,_{q_i}^{(1)} - \varepsilon dp_i$$

where $\varepsilon d = c_\phi = c_\theta$,

$$H^{(0)} = \frac{1}{2}(p_ip_i + \nu^2q_1^2 + 4\nu^2q_2^2),$$

(3.3)

$$H^{(1)} = \frac{\beta}{2}(q_1^2 + 4q_2^2) + \frac{1}{3}A_{ijk}q_iq_jq_k + Gq_i\cos\nu t$$

where A_{ijk} and G are system constants and $\varepsilon\beta = \nu^2 - \omega_1^2$.

Then using appropriate variation of constant transformations [6] the system equations can be reduced to the vector form:

$$(3.4)\qquad \frac{d\underline{x}}{dt} = \varepsilon \underline{f}(t,\underline{x},\varepsilon,\alpha)$$

where $\underline{x}$ is a four-vector, $\underline{f}$ is T-periodic in t with $T = \frac{2\pi}{\nu}$ and α is some parameter to be discussed below.

We first give below, very briefly, a result for the general case of system of the form (3.4) [7]. Let $\underline{x} \in R^n$ and $\underline{f}$ an n-vector function, with the appropriate number of derivatives, and almost periodic in t uniformly with respect to $\underline{x}$ and α in some bounded sets. Let

$$\underline{f}_0(\underline{\xi},\alpha) = \lim_{T\to\infty} \frac{1}{T} \int_0^T \underline{f}(t,\underline{\xi},0,\alpha)dt$$

and consider the averaged system

$$(3.5)\qquad \frac{d\underline{\xi}}{dt} = \varepsilon \underline{f}_0(\underline{\xi},\alpha)$$

Then if $\underline{\xi}^0(\alpha)$ is a constant solution of (3.5) and if $C(\underline{\xi}^0(\alpha),\alpha) \equiv \partial \underline{f}_0(\underline{\xi}^0(\alpha),\alpha)/\partial\underline{\xi}$ has eigenvalues with non-zero real parts, then as is well known, there exist almost periodic solutions of (3.4) with the same stability properties as those of C. We will be interested in the solution of (3.4) as they depend on the parameter α. If all the eigenvalues of C, for $\alpha < 0$ have negative real parts and for $\alpha = 0$ the eigenvalues go through a change of stability, then, in the generic case, either (i) one eigenvalue goes through zero or (ii) a pair of complex eigenvalues with non-zero imaginary parts $\omega(\alpha)$, go through the pure imaginary axis as α goes through zero and becomes positive and the autonomous system (3.5) has a Hopf bifurcation. System (3.5) then has a unique non-zero periodic solution $\underline{\xi}^*_\alpha(\varepsilon\omega(\alpha)t)$, 2π-periodic in $\varepsilon\omega(\alpha)$ t and the variational equation of (3.5) for the periodic solution has n-1 characteristic exponents with non-zero real parts.

We now appeal to a theorem in Hale [2], which we will not state in full here for lack of space. Then based on the above we have the following theorem:

Theorem. There exist constant ε_α and σ_α such that for all ε, $0 < \varepsilon \leq \varepsilon_\alpha$, system (3.4) has a unique integral manifold M lying for all real t in σ_α neighborhood of the cylinder $\underline{\xi}^*_\alpha(\theta)$, $0 \leq \theta \leq 2\pi$, $-\infty < t < \infty$ and M has a parametric representation of the form

$$\underline{x} = \underline{g}(t,\theta,\varepsilon,\alpha), \quad 0 \leq \theta \leq 2\pi, \quad -\infty < t < \infty,$$

where $\underline{g}$ is periodic in θ of period 2π, almost periodic in t uniformly with respect to $\theta \in R^1$ for each fixed ε, $0 < \varepsilon \leq \varepsilon_\alpha$, and has uniformly continuous derivatives with respect to θ up through order two. Furthermore,

$$\lim_{\varepsilon \to 0} ||\underline{g}(t,\theta,\varepsilon,\alpha) - \underline{\xi}^*_\alpha(\theta)|| = 0$$

for $-\infty < t < \infty$, $0 \leq \theta \leq 2\pi$ and fixed α.

The manifold M is stable if and only if $\underline{\xi}^*_\alpha$ is stable.

Corollary. In the special case when $\underline{f}$ is periodic in t with period T we have that

M: $\underline{x} = \underline{g}(t,\theta,\varepsilon,\alpha)$ is a two dimensional surface in the $(\underline{x},t)$ space in the form of a torus.

We now apply the theorem to system (3.2). If the amplitude of the role motion of the ship is b_1, then b_1 satisfies

$$(3.6) \quad b_1^6 - 2(2\gamma^2-\delta^2)b_1^4 + (\delta^2+\gamma^2)(\delta^2+4\gamma^2)b_1^2 - (\delta^2+4\gamma^2) = 0$$

$$\text{where } \gamma = \beta\left(\frac{2}{GU}\right)^{1/2}, \quad U = -A_{112}, \quad \delta = \left(\frac{2}{GU}\right)^{1/2} d\omega_1$$

In Fig. 3 we show the response curve, i.e., amplitude b_1^2 vs. γ, the detuning parameter. The motions become unstable, as mentioned before, in two ways. If the eigenvalue goes through zero we call the transitions from stable to unstable motions transitions of the first kind; these are indicated by Loci AB, A'B'; CD and C'D'. The case of interest is the one when Hopf bifurcations occur in the averaged equations. They occur for γ in the neighborhood of zero and they depend on δ as shown. If $\alpha = \gamma - \gamma_{cr}$ where γ_{cr} is the value of γ at which Hopf bifurcations occur, and if α is small, the theorem assures

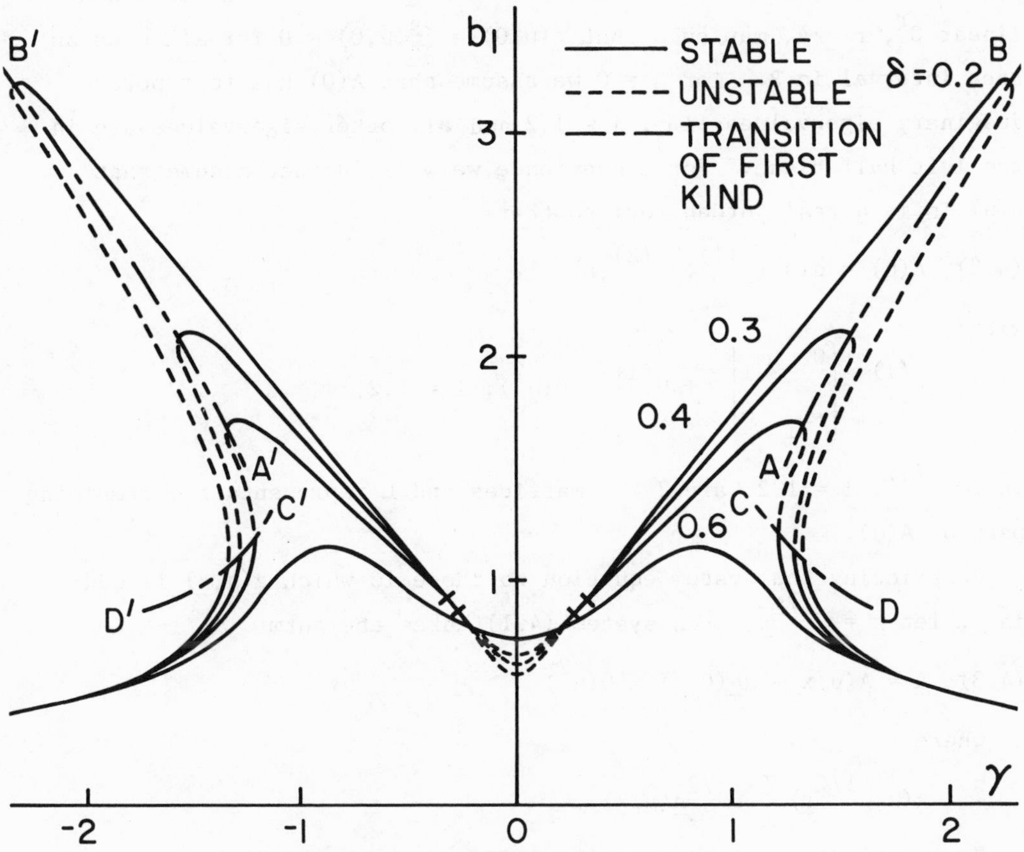

Figure 3: Response curves

the existence of the integral manifold, and gives the estimate of its shape, size and location explicitly. Furthermore, it can be shown that the manifold is stable. The physical motion occurs in the form of amplitude modulated motions, basically at the natural frequency, with a very slow modulation that appears to be periodic. In the case of the ship the modulation of the roll and pitch motions transfers energy from the roll to the pitch motion and back again to the roll motion and so forth at a very slow rate.

IV. <u>Bifurcations with multiple eigenvalues</u>. Consider a dynamical system described by

$$(4.1)\quad \dot{\underline{z}} = A(\mu)\underline{z} + \underline{g}(\mu,\underline{z})$$

where $A(\mu)$ is a $n \times n$ constant matrix and $\underline{g}$: $R \times R^n \to R^n$ is a non-linear C^r, $r \geq 4$, map such that $\underline{g}(\mu,0) = \frac{\partial \underline{g}}{\partial \underline{z}}(\mu,0) = 0$ for all μ on an open interval in R. For $\mu = 0$ we assume that A(0) has four pure imaginary eigenvalues $\pm i\omega_j$, $j = 1,2$ and all other eigenvalues are in the left half plane. For convenience we will further assume that $A(\mu)$ is in a real Jordan form so that

$$(4.2) \quad A(\mu) = \mathrm{diag}(L^{(1)}, L^{(2)}, L)$$

with

$$L^{(i)} = \begin{pmatrix} 0 & \omega_i \\ -\omega_i & 0 \end{pmatrix} + \mu U^{(i)} + 0(\mu^2), \; i = 1,2$$

where $U^{(i)}$, $i = 1,2$, are 2×2 matrices and L represents the remaining part of $A(\mu)$.

Restricting the system equation to those in which $\underline{g}(\mu,\underline{z})$ is odd in $\underline{z}$, let $\underline{z} = \mu^{1/2}\underline{x}$, then system (4.1) takes the form

$$(4.3) \quad \dot{\underline{x}} = A(\mu)\underline{x} + \mu\underline{n}(0,\underline{x}) + 0(\mu^2)$$

where

$$\underline{g}(\mu,\mu^{1/2}\underline{x}) \equiv \mu^{3/2}\underline{n}(\mu,\underline{x}).$$

System (4.3) has fundamentally different behavior depending on the ratio of ω_2/ω_1. If ω_2/ω_1 is near 1, 3 or 1/3 then it has internal resonance, and behaves very differently from the case when this ratio is not near these numbers. In the latter case, using the center manifold theorem, and the theory of normal forms, the nth ($n \geq 4$) order system (4.3) can be studied for its essential qualitative behavior near $\mu = 0$ by studying the following much simpler 4th order system in the new variables ρ_1, ρ_2, θ_1 and θ_2:

$$(4.4) \quad \dot{\rho}_1 = \mu\rho_1(C_1 + \beta_1\rho_1^2 + \beta_2\rho_2^2) + 0(\mu^2)$$

$$(4.5) \quad \dot{\rho}_1 = \mu\rho_2(C_2 + \beta_3\rho_1^2 + \beta_4\rho_2^2) + 0(\mu^2)$$

$$(4.6) \quad \dot{\theta}_1 = \omega_1 + \mu(C_3 + \gamma_1\rho_1^2 + \gamma_2\rho_2^2) + 0(\mu^2)$$

$$(4.7) \quad \dot{\theta}_2 = \omega_2 + \mu(C_4 + \gamma_3\rho_1^2 + \gamma_4\rho_2^2) + 0(\mu^2)$$

where C_j, β_j, γ_j, $j = 1,2,3,4$ are real constants. We note that (4.4) and (4.5) are completely independent of (4.6) and (4.7) up to $O(\mu)$ and can be studied independently by methods of analysis in the plane (ρ_1, ρ_2). The motion lies for μ small on an integral manifold of dimension three in $R \times R^n$ space which is the product of the state space and time.

If on the other hand ω_2/ω_1 is equal to 1, 1/3 or 3, the reduction is again to a four dimensional system but with three coupled equations and integral manifold is of dimension two.

Some physical problems of this type have additional symmetry properties and this forces modifications in the above mentioned results. Consider for example a finite dimensional specialization of the problem discussed is section 2. The system consists of rigid circular tubes with elastic joints. Suppose the system of tubes hangs vertically and has polar symmetry about a vertical axis and suppose the tubes are carrying a fluid. In this case we have possibility of identical motion of the linearlized system in two orthogonal vertical planes. Thus $\omega_2/\omega_1 = 1$. Furthermore, due to the polar symmetry the location of these planes is arbitrary and thus the system has certain symmetry properties. A study of the system equations shows that if $B(\phi)$, $\phi \in [0, 2\pi]$, is a rotation matrix parameterized by ϕ, then if

$$\underline{x} = B(\phi)\underline{\xi}$$

then $A(\mu)B = BA(\mu)$ and $\underline{g}(\mu,B\underline{\xi}) = B\underline{g}(\mu,\underline{\xi})$ and thus the system equations in the variable $\underline{\xi}$ remain in the form

$$\dot{\underline{\xi}} = A(\mu)\underline{\xi} + \underline{g}(\mu,\underline{\xi}).$$

In problems of this type the standard methods of analysis based either on the implicit function theorem or the method of averaging fail and modified approaches are necessary.

REFERENCES

[1] J.E. MARSDEN, M. McCRAKEN: The Hopf bifurcation and its applications, Springer-Verlag, New York, 1976.

[2] J.K. HALE: Oscillations in nonlinear systems, McGraw-Hill, New York, 1963.

[3] S.N. CHOW, J. MALLET-PARET: Integral averaging and bifurcations, Journal of Differential Equations, Vol. 26, No. 1, (1976), pp. 112-159.

[4] F. TAKENS: Singularities of vector fields, I.H.E.S., Publications Mathematiques, No. 43, (1974), pp. 47-100.

[5] A.K. BAJAJ, P.R. SETHNA, T.A. LUNDGREN: Hopf bifurcation phenomena in tubes carrying a fluid, SIAM Journal on Applied Mathematics (to appear).

[6] P.R. SETHNA, A.K. BAJAJ: Bifurcations in dynamical systems with internal resonance, ASME Journal of Applied Mechanics, Vol. 45, No. 4, (1978), pp. 895-902.

[7] P.R. SETHNA, A.K. BAJAJ: Bifurcation in nonlinear oscillatory systems, presented at VIIIth International Conference on Nonlinear Oscillations, Prague, 1978.

[8] A.H. NAYFEH, D.T. MOOK, L.R. MARSHALL: Nonlinear coupling of pitch and roll modes in ship motions, Journal of Hydronautics, Vol. 7, No. 4, (1973), pp. 145-152.

On Some Global Results of Point Mapping Dynamical Systems

C. S. Hsu*

Abstract. Two investigations are reported here. One is concerned with a class of nonlinear dynamical systems under impulsive, periodic, and parametric excitation which permits the integration of the differential equations of motion over one period to yield a point mapping. This mapping can aid considerably the study of the global behavior of the systems. Moreover, since the point mapping is exact, one has complete confidence that the behavior of the original system is entirely governed by the point mapping.

The second topic reported here is a discussion of the generalization of Poincaré's theory of index for two-dimensional systems to systems of higher orders. The concept of the degree of a map well-known in topology is used in the generalization. The intention here is to present the extension of Poincaré's theory to higher order systems in such a manner that hopefully the results of this theory might become more readily accessible to researchers outside mathematics for application to physical problems. In the paper the theory of index for vector fields of any dimensions is presented first. Its application to point mapping dynamical systems is then discussed in detail.

1. Introduction. In this paper we shall report on two directions of investigation we have pursued in our study of nonlinear dynamical systems at the University of California, Berkeley. Although our interest is naturally concentrated on dynamical systems of mechanical origin, the basic theoretical developments are of course equally applicable to dynamical systems encountered in other fields of science and engineering.

---

*Department of Mechanical Engineering, University of California, Berkeley, California 94720. This work was supported by a grant from from the National Science Foundation.

First we consider the broad class of nonlinear dynamical systems under parametric excitation. These systems are governed by non-autonomous differential equations. When the excitation is periodic, then a useful approach is to recast the problem in the form of a point mapping by integrating the equation of motion over one period. If one can obtain the corresponding point mapping in an exact manner, then one can first study the dynamic behavior of the point mapping. After having ascertained the behavior of the point mapping system, one can go back to the original differential equation to obtain the continuous time history of motion of the original system. The basic difficulty of this approach is that in general we are unable to carry out the integration. In this paper we discuss one special class of problems for which it is possible to obtain the point mappings exactly. For these problems this approach allows us to study the nonlinear problems in a new and yet clear-cut and complete manner without approximation.

The second topic we shall report to this Conference is the development of a generalized theory of index for dynamical systems. Here we recall that one of the most interesting global results in the theory of nonlinear oscillations is the celebrated Poincaré's theory of index for two-dimensional systems. This theory is now classic and is discussed in practically all the textbooks on nonlinear oscillations [1-4]. However, one rarely finds any discussion of the extension of the theory to higher order systems. One of the places where one can find a brief discussion of the extension is in [5].

On the other hand, in topology there is the concept of the <u>degree of a map</u> which is valid for manifolds of any dimension. This concept has been used to discuss the index of a vector field. Unfortunately, the theory does not seem to have been sufficiently developed into a form which will make it available to engineers for application to physical problems. Thus, the potential of the concept is largely unexploited. In this paper we shall use the concept of the degree of a map to present a theory of index for higher order systems in a form which is a parallel to Poincaré's for two-dimensional systems and in mathematical terms with which the researchers outside mathematics are perhaps more familiar. Inasmuch as the basic idea involved here is one concerning mapping, the theory of index presented here could have very wide applications. In this paper we shall concentrate on its application to point mapping systems.

2. Systems Under Impulsive Periodic Parametric Excitations. Consider a nonlinear mechanical system of N' degrees of freedom

$$\underset{\sim}{M}\,\ddot{\underset{\sim}{z}} = \underset{\sim}{F}(\underset{\sim}{z}, \dot{\underset{\sim}{z}}, t) \tag{2.1}$$

where $\underset{\sim}{z}$ is an N'-vector, $\underset{\sim}{M}$ an inertia matrix and $\underset{\sim}{F}$ a vector-valued nonlinear function of $\underset{\sim}{z}$, $\dot{\underset{\sim}{z}}$ and t. Let us assume that the system is under a periodic excitation. In principle, for such a periodic system one can integrate the equation of motion over one period to obtain a period-to-period point mapping. But in reality this cannot be done analytically for most cases. For this reason let us confine our attention to a special class of systems which are under impulsive, periodic, parametric excitations of the following type.

$$\underset{\sim}{M}\,\ddot{\underset{\sim}{z}} + \underset{\sim}{C}\,\dot{\underset{\sim}{z}} + \underset{\sim}{K}\,\underset{\sim}{z} + \sum_{m=1}^{M} \underset{\sim}{p}^{(m)}(\underset{\sim}{z}) \sum_{j=-\infty}^{\infty} \delta(t-t_m-j) = \underset{\sim}{0} \tag{2.2}$$

where $\underset{\sim}{C}$ and $\underset{\sim}{K}$ are N'xN' constant matrices and $\underset{\sim}{p}^{(m)}(\underset{\sim}{z})$ is a vector-valued nonlinear function of $\underset{\sim}{z}$. The parametric excitation is periodic of period 1 and it consists of M impulses within one period. The time instant t_m at which the m-th impulsive parametric excitation takes place satisfies the following ordering.

$$0 \leqq t_1 < t_2 < \ldots < t_M < 1 \tag{2.3}$$

The excitation strength of the m-th impulsive action is given by $\underset{\sim}{p}^{(m)}$.

Each of the impulsive excitation can be expected to produce a discontinuity in the velocity $\dot{\underset{\sim}{z}}$ with the displacement $\underset{\sim}{z}$ remaining continuous. Thus, at $t=t_m$ one finds by integration

$$\dot{\underset{\sim}{z}}(t_m+) - \dot{\underset{\sim}{z}}(t_m-) = -\underset{\sim}{M}^{-1}\,\underset{\sim}{p}^{(m)}(\underset{\sim}{z}(t_m)) \tag{2.4}$$

where t_m+ and t_m- denote respectively the instant just after and the instant just before t_m. For the time periods between impulses the system is linear. Let

$$z_i = x_i, \qquad \dot{z}_i = x_{N'+i}. \tag{2.5}$$

(2.2) may now be written as

$$\dot{\underset{\sim}{x}} = \underset{\sim}{A}\,\underset{\sim}{x} + \sum_{n=1}^{M} \underset{\sim}{b}^{(m)}(\underset{\sim}{x}) \sum_{j=-\infty}^{\infty} \delta(t-t_m-j) \tag{2.6}$$

where

$$\underset{\sim}{A} = \begin{pmatrix} \underset{\sim}{0} & \underset{\sim}{I} \\ -\underset{\sim}{M}^{-1}\underset{\sim}{K} & -\underset{\sim}{M}^{-1}\underset{\sim}{C} \end{pmatrix} \qquad \underset{\sim}{b}^{(m)}(\underset{\sim}{x}) = \begin{pmatrix} \underset{\sim}{0} \\ -\underset{\sim}{M}^{-1}\underset{\sim}{p}^{(m)}(\underset{\sim}{z}) \end{pmatrix} \tag{2.7}$$

Let $\underset{\sim}{X}(t)$ be the fundamental matrix of the linear equation $\dot{\underset{\sim}{x}}=\underset{\sim}{A}\underset{\sim}{x}$ with $\underset{\sim}{X}(0)=\underset{\sim}{I}$. If $\underset{\sim}{x}(0)$ is the initial state, the solution of (2.6) may be written as

$$\begin{aligned} \underset{\sim}{x}(t) &= \underset{\sim}{X}(t)\,\underset{\sim}{x}(0), \qquad 0 \leq t < t_1 \\ \underset{\sim}{x}(t_1-) &= \underset{\sim}{X}(t_1)\,\underset{\sim}{x}(0), \\ \underset{\sim}{x}(t_1+) &= \underset{\sim}{x}(t_1-) + \underset{\sim}{b}^{(1)}(\underset{\sim}{x}(t_1-)), \\ \underset{\sim}{x}(t) &= \underset{\sim}{X}(t-t_1)\,\underset{\sim}{x}(t_1+), \qquad t_1 < t < t_2 \\ &\cdots\cdots\cdots \\ \underset{\sim}{x}(t) &= \underset{\sim}{X}(t-t_M)\,\underset{\sim}{x}(t_M+), \qquad t_M < t \leq 1. \end{aligned} \tag{2.8}$$

This process allows us to express the state at the end of a period of excitation in terms of the state at the end of the preceding period, leading therefore to a point mapping of the form

$$\underset{\sim}{x}(n+1) = \underset{\sim}{G}(\underset{\sim}{x}(n)) \tag{2.9}$$

In this case because the point mapping is obtained in an exact manner the nonlinear mapping $\underset{\sim}{G}$ completely governs the behavior of (2.2).

Consider now two simple examples. First consider the nonlinear plane motion of a rigid bar hinged at one end and subjected to a periodic impact load of period 1 at the other free end. Let A be the hinged point and I_A be the moment of inertia about A. Let the bar be restrained at the hinged end by a linear rotational spring of modulus k and a linear rotational damper with a damping constant c. Let x_1 and x_2 be the angular displacement and velocity of the bar with x=0 coinciding with the natural equilibrium position of the bar. The periodic impact load P_0 has a fixed direction which is in that of x=0. Let ℓ be the length of the bar. This problem falls within the class of problems we have just discussed. Therefore, the exact point mapping $\underset{\sim}{G}$ can be found. Let

$$(2.10) \qquad \mu = \frac{c}{2I_A}, \qquad \omega^2 = \frac{k}{I_A} - \mu^2, \qquad \alpha = \frac{P_0\ell}{I_A}$$

The mapping $\underset{\sim}{G}:(g_1,g_2)$ is given by

$$(2.11) \qquad \begin{aligned} x_1(n+1) &= g_1(\underset{\sim}{x}) = E[(C + \frac{\mu}{\omega}S)x_1(n) - \frac{\alpha S}{\omega}\sin x_1(n) + \frac{S}{\omega}x_2(n)] \\ x_2(n+1) &= g_2(\underset{\sim}{x}) = E[-(\omega + \frac{\mu^2}{\omega})Sx_1(n) - (C - \frac{\mu}{\omega}S)\alpha \sin x_1(n) \\ &\quad + (C - \frac{\mu}{\omega}S)x_2(n)] \end{aligned}$$

where

$$(2.12) \qquad E = e^{-\mu}, \qquad C = \cos\omega, \qquad S = \sin\omega.$$

Next consider a two degrees of freedom system which consists of two identical rigid bodies free to rotate about a common axis. These bodies are coupled to each other as well as connected to stationary support through linear rotational dampers. Let the bodies be subjected to synchronized periodic impact loads of fixed direction and of period 1. For this system we can again determine the exact period-to-period mapping. Let q_1 and q_2 be the angular displacements of the bodies. Let c_0 and c_{12} be the damping coefficients of the anchoring and coupling dampers. Let P_0 again denote the magnitude of the impact load. Let

$$x_1=\frac{1}{2}(q_1+q_2),\quad x_2=\frac{1}{2}(q_1-q_2),\quad x_3=\frac{1}{2}(\dot{q}_1+\dot{q}_2),\quad x_4=\frac{1}{2}(\dot{q}_1-\dot{q}_2)$$

(2.13)

$$\mu_1=\frac{c_0}{2I_A},\quad \mu_2=\frac{(c_0+2c_{12})}{2I_A},\quad \alpha=\frac{P_0\ell}{I_A}.$$

The nonlinear mapping $\underset{\sim}{G}:(g_1,g_2,g_3,g_4)$ is given by

(2.14)

$$x_1(n+1)=g_1(\underset{\sim}{x})=x_1(n)+C_1[x_3(n)-\alpha \sin x_1(n) \cos x_2(n)]$$

$$x_2(n+1)=g_2(\underset{\sim}{x})=x_2(n)+C_2[x_4(n)-\alpha \cos x_1(n) \sin x_2(n)]$$

$$x_3(n+1)=g_3(\underset{\sim}{x})=D_1[x_3(n)-\alpha \sin x_1(n) \cos x_2(n)]$$

$$x_4(n+1)=g_4(\underset{\sim}{x})=D_2[x_4(n)-\alpha \cos x_1(n) \sin x_2(n)]$$

where

(2.15)

$$C_1=\frac{1-e^{-2\mu_1}}{2\mu_1},\quad D_1=e^{-2\mu_1},\quad C_2=\frac{1-e^{-2\mu_2}}{2\mu_2},\quad D_2=e^{-2\mu_2}.$$

These point mappings have been studied in detail. Many interesting global results concerning the existence of various periodic solutions and the bifurcation phenomena have been obtained. For the details of these studies the reader is referred to [6,7,8].

3. A Theory of Index for Vector Fields. When one comes to the study of the global behavior of dynamical systems, one is naturally reminded of the remarkable index theory of Poincaré's for two-dimensional systems. The ideal tool needed to generalize Poincaré's theory to higher order systems is the concept of the degree of a map. It is a well-known concept in topology. Unfortunately, in literature this concept is invariably presented and discussed in topological terms. It is probably because of this that its meaning and its significance are not widely known outside mathematics and its potential use in other fields has hardly been explored at all. In this section we shall use this concept of the degree of a map to present a theory of index for vector fields in a form which might lend itself more readily for applications to problems outside mathematics.

3.1. The Degree of a Map. The concept of the degree of a map [9-13] may be stated in the following form. Let $\underset{\sim}{M}^m$ and $\underset{\sim}{N}^m$ be two compact and oriented manifolds of dimension m. Let f be a mapping from $\underset{\sim}{M}^m$ to $\underset{\sim}{N}^m$.

$$f: \underset{\sim}{M}^m \rightarrow \underset{\sim}{N}^m \tag{3.1.1}$$

Then there is an integer called the degree of f given by

$$\int_{\underset{\sim}{M}^m} f^*\omega = \deg(f) \int_{\underset{\sim}{N}^m} \omega \tag{3.1.2}$$

where ω is any m-form on $\underset{\sim}{N}^m$ and f* is the pull-back (or reciprocal image) induced by f. This degree of a map has a very simple geometrical meaning. It is the number of times $\underset{\sim}{N}^m$ is covered by the image $f(\underset{\sim}{M}^m)$, each covering being counted as positive or negative depending upon whether the mapping is locally orientation-preserving or orientation-reversing.

3.2. Mapping of a Vector Field. Consider now an n-dimensional Euclidean space $\underset{\sim}{X}$. Let $\underset{\sim}{F}$ be a continuously differentiable real-valued vector field defined on a bounded open set $\underset{\sim}{B}_0$ of $\underset{\sim}{X}$. A point in $\underset{\sim}{B}_0$ at which $\underset{\sim}{F}=\underset{\sim}{0}$ is called a singular point of $\underset{\sim}{F}$. We assume that all the singular points of $\underset{\sim}{F}$ are isolated. Let $\underset{\sim}{S}$ be a compact, connected and oriented hypersurface of dimension (n-1) in $\underset{\sim}{B}_0$ which divided $\underset{\sim}{X}$ into two parts, its interior and its exterior, and which passes through no singular points of $\underset{\sim}{F}$. At each point $\underset{\sim}{x}$ on $\underset{\sim}{S}$, $\underset{\sim}{F}(\underset{\sim}{x})$ is defined. A unit vector $\underset{\sim}{y}(\underset{\sim}{x})$ can then be constructed from this $\underset{\sim}{F}(\underset{\sim}{x})$ according to

$$\underset{\sim}{y}(\underset{\sim}{x}) = \frac{1}{|\underset{\sim}{F}(\underset{\sim}{x})|} \underset{\sim}{F}(\underset{\sim}{x}) \tag{3.2.1}$$

Evidently, in an n-dimensional Euclidean space $\underset{\sim}{Y}$, this vector will terminate on the hypersurface of a unit sphere. This hypersurface will be denoted here by $\underset{\sim}{\Sigma}$. Thus this construction defines a mapping f which maps $\underset{\sim}{x}$ on $\underset{\sim}{S}$ to $\underset{\sim}{y}$ on $\underset{\sim}{\Sigma}$.

$$f: \underset{\sim}{x}(\underset{\sim}{S}) \rightarrow \underset{\sim}{y}(\underset{\sim}{\Sigma}) \quad \text{or} \quad \underset{\sim}{S} \rightarrow \underset{\sim}{\Sigma}. \tag{3.2.2}$$

We now apply the degree of a map to the present case by identifying $\underset{\sim}{S}$ with $\underset{\sim}{M}^m$, $\underset{\sim}{\Sigma}$ with $\underset{\sim}{N}^m$ and the dimension (n-1) with m.

To go further, let u^i (i=1,2,...,n-1) be the local coordinates on $\underset{\sim}{S}$ and let v^i (i=1,2,...,n-1) be those on $\underset{\sim}{\Sigma}$. Let the m-form ω of the degree formula (3.1.2) be chosen to be the volume form of $\underset{\sim}{\Sigma}$.

(3.2.3) $$\omega = \tau(\underset{\sim}{v})\, dv^1 \wedge dv^2 \wedge \ldots \wedge dv^{n-1}$$

(3.1.2) may now be written as

(3.2.4) $$\deg(f) = \frac{A(\underset{\sim}{S})}{B(\underset{\sim}{\Sigma})}$$

where

(3.2.5) $$A(\underset{\sim}{S}) = \int_{\underset{\sim}{S}} \tau(f(\underset{\sim}{u}))\, \frac{\partial(v^1,v^2,\ldots,v^{n-1})}{\partial(u^1,u^2,\ldots,u^{n-1})}\, du^1\wedge du^2\wedge\ldots\wedge du^{n-1}$$

(3.2.6) $$B(\underset{\sim}{\Sigma}) = \frac{n\pi^{n/2}}{\Gamma(\frac{n}{2}+1)}.$$

<u>3.3.</u> <u>A Theory of Index</u>. We are now ready to present a theory of index for vector fields. We shall only state the results. The detailed development may be found in [14]. In what follows a surface $\underset{\sim}{S}$ will always mean a hypersurface of dimension (n-1) possessing those properties as stated in Section 3.2.

<u>Definition 1. The index of a compact, connected and oriented surface $\underset{\sim}{S}$ with respect to $\underset{\sim}{F}$, to be denoted by</u> $I(\underset{\sim}{S},\underset{\sim}{F})$, <u>is defined as</u>

$$I(\underset{\sim}{S},\underset{\sim}{F}) = \deg(f)$$

<u>where</u> deg(f) <u>is the degree of the map</u> f: $\underset{\sim}{S}\to\underset{\sim}{\Sigma}$.

<u>Theorem 1. Let $\underset{\sim}{S}_1$ and $\underset{\sim}{S}_2$ be two compact, connected and oriented surfaces which have disjointed interiors but have a common part $\underset{\sim}{S}_c$. Let $\underset{\sim}{S}$ be $\underset{\sim}{S}_1$ plus $\underset{\sim}{S}_2$ but with $\underset{\sim}{S}_c$ deleted. Then</u>

$$I(\underset{\sim}{S},\underset{\sim}{F}) = I(\underset{\sim}{S}_1,\underset{\sim}{F}) + I(\underset{\sim}{S}_2,\underset{\sim}{F}).$$

<u>Theorem 2. If $\underset{\sim}{S}$ is a compact, connected and oriented surface con-</u>

taining no singular points of F on it and in its interior, $I(S,F)=0$.

Theorem 3. If S_1 is a compact, connected and oriented surface which is contained in the interior of another compact, connected and oriented surface S_2, and if no singular points lie between S_1 and S_2, then $I(S_1,F)=I(S_2,F)$.

Definition 2. The index of an isolated singular point P with respect to a vector field F is defined as the index of any compact, connected and oriented surface S the interior of which contains only P and no other singular points of F; it is denoted by $I(P,F)$, i.e. $I(P,F)=I(S,F)$.

Let $DF(P)$ denote the Jacobian matrix of F at P. A singular point P of F is said to be non-degenerate with respect to F if $\det[DF(P)]\neq 0$.

Theorem 4. The index of an isolated non-degenerate singular point P of F is given by

$$I(P,F) = \begin{matrix} +1 \\ -1 \end{matrix} \quad \text{if} \quad \det[DF(P)] \begin{matrix} > \\ < \end{matrix} 0.$$

Theorem 5. If S is a compact, connected and oriented surface containing a finite number $P_1, P_2, \ldots, P_N$ of singular points of F in its interior, then

$$I(S,F) = \sum_{i=1}^{N} I(P_i,F)$$

Once the index theory for vector fields has been formulated, its application to autonomous dynamical systems governed by differential equations such as

$$\dot{x} = F(x) \tag{3.3.1}$$

is immediate. We shall not dwell on this.

4. Application of the Theory of Index to Point Mapping Systems. Consider now a dynamical system governed by a point mapping such as (2.9). Consider the periodic solutions of the mapping. Let us denote by G^k the mapping G applied k times, with G^0 understood to be an identity mapping. A periodic solution of period K of a mapping G is a set

of K distinct points $\underset{\sim}{x}^*(j)$, $j=1,2,\ldots,K$, such that

(4.1) $$\underset{\sim}{x}^*(m+1) = \underset{\sim}{G}^m(\underset{\sim}{x}^*(1)), \quad m = 1, 2, \cdots, K-1$$
$$\underset{\sim}{x}^*(1) = \underset{\sim}{G}^K(\underset{\sim}{x}^*(1)).$$

Since we will refer to periodic solutions of this kind time and again, it is convenient to adopt an abbreviated name. We call a periodic solution of period K a P-K solution and any of its elements x*(j), j=1, 2,···,K, a periodic point of period K, or, in abbreviation, a P-K point. The P-1 points can, of course, be interpreted as the equilibrium states of the point mapping system. In general, however, if the mapping is in fact derived from a system of periodic differential equations by integrating over one period, then a P-1 point could be either an equilibrium state or a periodic solution of period 1 of the original differential dynamical system and the P-K points will correspond to subharmonic solutions of period K.

4.1. The Index of a P-1 Point. First consider a P-1 point $\underset{\sim}{x}^*$ of $\underset{\sim}{G}$. If we identify the vector field $\underset{\sim}{F}$ of Section 3 with $\underset{\sim}{G}-\underset{\sim}{I}$, where $\underset{\sim}{I}$ is the identity mapping,

(4.1.1) $$\underset{\sim}{F}(\underset{\sim}{x}) = (\underset{\sim}{G}-\underset{\sim}{I})(\underset{\sim}{x}) = \underset{\sim}{G}(\underset{\sim}{x}) - \underset{\sim}{x},$$

then it is obvious that all the P-1 points of $\underset{\sim}{G}$ are singular points of $\underset{\sim}{F}$ and all the singular points of $\underset{\sim}{F}$ are the P-1 points of $\underset{\sim}{G}$. By Theorem 4 we have the following.

Theorem 6. If $\underset{\sim}{x}^*$ is a P-1 point of $\underset{\sim}{G}$ non-degenerate with respect to $(\underset{\sim}{G}-\underset{\sim}{I})$, then its index with respect to $(\underset{\sim}{G}-\underset{\sim}{I})$, to be denoted by $I(\underset{\sim}{x}^*,\underset{\sim}{G}-\underset{\sim}{I})$, is equal to +1 or -1, according as $\det[\underset{\sim\sim}{DG}(\underset{\sim}{x}^*)-\underset{\sim}{I}]>0$ or <0.

Theorem 7. Given a point mapping $\underset{\sim}{G}$, if a hypersurface S of dimension n-1 contains a finite number $\underset{\sim}{P}_1,\underset{\sim}{P}_2,\ldots,\underset{\sim}{P}_N$ of P-1 points of $\underset{\sim}{G}$ in its interior, and if p_1 is the number of points having positive $\det[\underset{\sim\sim}{DG}(\underset{\sim}{P}_i)-\underset{\sim}{I}]$ and q_1 is the number of points having negative $\det[\underset{\sim\sim}{DG}(\underset{\sim}{P}_i)-\underset{\sim}{I}]$, with $p_1+q_1=N$, then the index of S with respect to $(\underset{\sim}{G}-\underset{\sim}{I})$, to be denoted by $I(\underset{\sim}{S},\underset{\sim}{G}-\underset{\sim}{I})$, is p_1-q_1.

Next consider a vector field defined as

$$\mathbf{F}(\mathbf{x}) = (\mathbf{G}^2 - \mathbf{I})(\mathbf{x}) = \mathbf{G}^2(\mathbf{x}) - \mathbf{x}. \tag{4.1.2}$$

A P-1 point of $\mathbf{G}$ is again a singular point of $\mathbf{F}$. With respect to this vector field, the index of the P-1 point is determined by the following theorem.

Theorem 8. If $\mathbf{x}^*$ *is a P-1 point of* $\mathbf{G}$ *non-degenerate with respect to* $(\mathbf{G}^2 - \mathbf{I})$, *then its index with respect to* $(\mathbf{G}^2 - \mathbf{I})$, *to be denoted by* $I(\mathbf{x}^*, \mathbf{G}^2 - \mathbf{I})$, *is +1 or -1, according to* $\det[\mathbf{DG}^2(\mathbf{x}^*) - \mathbf{I}] > 0$ *or* < 0.

If the vector field $\mathbf{F}$ is chosen to be

$$\mathbf{F}(\mathbf{x}) = (\mathbf{G}^k - \mathbf{I})(\mathbf{x}) = \mathbf{G}^k(\mathbf{x}) - \mathbf{x}, \tag{4.1.3}$$

where k is a positive integer, a P-1 point of $\mathbf{G}$ is again a singular point of $\mathbf{F}$. The index of the point with respect to this $\mathbf{F}$ turns out to depend only upon whether k is even or odd.

Theorem 9. If k is odd and if $\mathbf{x}^*$ *is a P-1 point of* $\mathbf{G}$ *non-degenerate with respect to* $(\mathbf{G}^k - \mathbf{I})$, *then the index of* $\mathbf{x}^*$ *with respect to* $(\mathbf{G}^k - \mathbf{I})$ *is equal to its index with respect to* $(\mathbf{G} - \mathbf{I})$.

Theorem 10. If k is even and if $\mathbf{x}^*$ *is a P-1 point of* $\mathbf{G}$ *non-degenerate with respect to* $(\mathbf{G}^k - \mathbf{I})$, *then the index of* $\mathbf{x}^*$ *with respect to* $(\mathbf{G}^k - \mathbf{I})$ *is equal to its index with respect to* $(\mathbf{G}^2 - \mathbf{I})$.

4.2. The index of a P-K Point. Consider now a P-K point of $\mathbf{G}$. Let $\mathbf{x}^*(j)$ be such a point. Let us take $\mathbf{F}$ to be $(\mathbf{G}^K - \mathbf{I})$. Then a P-K point of $\mathbf{G}$ is evidently a singular point of $\mathbf{F}$.

Theorem 11. If $\mathbf{x}^*(j)$ *is a P-K point of* $\mathbf{G}$ *non-degenerate with respect to* $(\mathbf{G}^K - \mathbf{I})$, *then the index of* $\mathbf{x}^*(j)$ *with respect to* $(\mathbf{G}^K - \mathbf{I})$, *to be denoted by* $I(\mathbf{x}^*(j), \mathbf{G}^K - \mathbf{I})$, *is* +1 *or* -1, *according as* $\det[\mathbf{DG}^K(\mathbf{x}^*(j)) - \mathbf{I}] > 0$ *or* < 0.

Similar to Theorems 9 and 10 we have the following.

Theorem 12. If k is odd and if $\mathbf{x}^*(j)$ *is a P-K point of* $\mathbf{G}$ *non-degenerate with respect to* $(\mathbf{G}^{kK} - \mathbf{I})$, *then the index of* $\mathbf{x}^*(j)$ *with respect to* $(\mathbf{G}^{kK} - \mathbf{I})$ *is +1 or -1, according to whether* $\det[\mathbf{DG}^K(\mathbf{x}^*(j)) - \mathbf{I}]$

>0 *or* <0.

Theorem 13. *If* k *is even and if* $x^*(j)$ *is a* P-K *point of* G *non-degenerate with respect to* $(G^{kK}-I)$, *then the index of* $x^*(j)$ *with respect to* $(G^{kK}-I)$ *is* +1 *or* -1, *according to whether* $\det[(DG^K(x^*(j)))^2-I]$ >0 *or* <0.

4.3. The Index of a Hypersurface S *with Respect to* G^L-I. We are now ready to present a global result of the index theory for point mapping dynamical systems. Consider a surface S and a vector field F defined by

$$F(x) = (G^L-I)(x) = G^L(x) = x \tag{4.3.1}$$

where L is a positive integer. Let us denote the complete set of positive integer factors of L by $f_1=1$, f_2, $f_3,\ldots$, $f_Q=L$. Obviously, all P-1, P-f_2, P-$f_3,\ldots$, P-L points of G are singular points of F and vice versa. Let N of these points be located in the interior of S and let them be labelled as $P_i(K_i)$, i-1,2,...,N. Here K_i is equal to one of the factors f_q, q=1,2,...,Q, and is also the periodicity of P_i. To each $P_i(K_i)$ is associated a number $k_i=L/K_i$. We assume that all $P_i(K_i)$ points are non-degenerate with respect to F. The index of $P_i(K_i)$ with respect to F can be determined by Theorem 12 if k_i is odd or by Theorem 13 if k is even. The global result is the following.

Theorem 14. *The index of* S *with respect to* (G^L-I) *is*

$$I(S,G^L-I) = \sum_{i=1}^{N} I(P_i(K_i),G^{k_iK_i}-I).$$

For some indication of possible applications of this theory of index to specific systems the reader is referred to [14,15].

REFERENCES

[1] J. J. STOKER, *Nonlinear Vibrations*, Interscience Publishers, New York, 1950.

[2] N. MINORSKY, *Nonlinear Oscillations*, D. Van Nostrand Co., Princeton, New Jersey, 1962.

[3] E. A. CODDINGTON and N. LEVINSON, *Theory of Ordinary Differential Equations*, McGraw-Hill, New York, 1955.

[4] S. LEFSCHETZ, Differential Equations: Geometric Theory, Interscience Publishers, New York, 1957.

[5] V. I. ARNOLD, Ordinary Differential Equations, translated and edited by R. A. Silverman, The MIT Press, Cambridge, Massachusetts, 1973.

[6] C. S. HSU, H. C. YEE, and W. H. CHENG, Steady-state response of a nonlinear system under impulsive periodic parametric excitation, J. of Sound and Vibration, 50(1977), pp. 95-116.

[7] C. S. HSU, On nonlinear parametric excitation problems, Advances in Applied Mechanics, 17(1977), pp. 245-301.

[8] C. S. HSU, Nonlinear behavior of multibody systems under impulsive parametric excitation, Dynamics of Multibody Systems, Editor: K. Magnus, Springer-Verlag, 1978. Proceedings of the IUTAM Symposium on Dynamics of Multibody Systems held in Munich, Germany, August 29-September 3, 1977, pp. 63-74.

[9] S. LEFSCHETZ, Introduction to Topology, Princeton University Press, Princeton, New Jersey, 1964.

[10] S. STERNBERG, Lectures on Differential Geometry, Prentice-Hall Inc., Englewood Cliffs, New Jersey, 1964.

[11] V. GUILLEMIN and A. POLLACK, Differential Topology, Prentice-Hall Inc., Englewood Cliffs, New Jersey, 1974.

[12] H. FLANDERS, Differential Forms, Academic Press, New York, 1963.

[13] Y. CHOQUET-BRUHAT, C. DEWITT-MORETTE, and M. DILLARD-BLEICK, Analysis, Manifolds and Physics, North-Holland Publishing Company, Amsterdam, 1977.

[14] C. S. HSU, Theory of index for dynamical systems of order higher than two, to appear in the Journal of Applied Mechanics.

[15] C. S. HSU, A theory of index for point mapping dynamical systems, To appear in the Journal of Applied Mechanics.

Jets and Geneticity in Qualitative Dynamics

Lawrence Markus*

Abstract. The k - jet of a real function is the k - th degree polynomial obtained upon truncating the Taylor power series; and a traditional problem of mathematical analysis concerns the extent to which this k - jet genetically determines the qualitative features of the given function, or the qualitative behavior of various mathematical systems involving this function. In order to illustrate such concepts of determinacy, we shall investigate the qualitative behavior of a nonlinear oscillator with one-degree of freedom along a line, and interpret this behavior in terms of the appropriate k - jet of the force function.

For conservative oscillators it is proved that the force k - jet always determines the qualitative behavior of the dynamical system, that is, each such k - jet is dynamically-determinate. For nonconservative oscillators various results on jet-determinacy are illustrated by examples.

1. Jet determination. The k - jet of a real function is the k - th degree polynomial obtained upon truncating the Taylor power series; and a traditional problem of mathematical analysis concerns the extent to which this k - jet genetically determines the qualitative features of the given function, or the qualitative behavior of various mathematical systems involving this function. In order to illustrate such concepts of determinacy, we shall investigate the qualitative behavior of a nonlinear oscillator with one-degree of freedom along a line, and

---

*Mathematics Department
University of Minnesota, Minneapolis, Minnesota 55455
This research was partially supported by NSF grant MCS 79-01998.

interpret this behavior in terms of the appropriate k-jet of the force function.

In more detail, we shall consider the nonlinear oscillator

$$\frac{d^2x}{dt^2} + g(x, \frac{dx}{dt}) = 0 ,$$

or rather the corresponding differential system in the real (x,y)-plane $\mathbb{R}^2$,

$$\dot{x} = y$$

$$\dot{y} = -g(x,y) .$$

Here $g(x,y)$, which represents a force depending on both position x and velocity y, is assumed to be a real differentiable function (by this we mean of class C^∞) in a neighborhood of the origin, which is to be a critical point so $g(0,0) = 0$. We shall be concerned with the qualitative nature of the function $g(x,y)$, and the resulting differential system, only near the origin; and we seek to describe such local behavior in terms of the k-jet

$$g_1(0,0)x + g_2(0,0)y + \frac{1}{2!}[g_{11}(0,0)x^2 + 2g_{12}(0,0)xy + g_{22}(0,0)y^2] + \\ + \dots + \frac{1}{k!}P_k(x,y) ,$$

for some specified integer $k \geq 1$. Here $P_k(x,y)$ is the usual k-th degree homogeneous polynomial arising in the Taylor power series for the function $g(x,y)$, expanded about the origin where $g(0,0) = 0$. Naturally we refer only to non-trivial jets that are not identically vanishing.

In order to consider the structure of the qualitative behavior of $g(x,y)$, or the differential system $\ddot{x} + g(x,\dot{x}) = 0$, near the origin in $\mathbb{R}^2$ we must introduce the appropriate concepts of qualitative equivalence.

Two real functions $h(x,y)$, and $\hat{h}(x,y)$, differentiable near the origin of $\mathbb{R}^2$ where both vanish, are qualitatively equivalent about $x = y = 0$ in case: there exists an invertible differentiable coordinate transformation Φ , in some neighborhood of the origin, which carries the function h to $\hat{h}$. This means $h \circ \Phi = \hat{h}$ so that the function germs of h and $\hat{h}$ are right-equivalent, according to the standard definition [2 , p. 89].

For the corresponding differential systems

$$\mathcal{S})\ \ \ddot{x} + h(x,\dot{x}) = 0 \quad \text{and} \quad \hat{\mathcal{S}})\ \ \ddot{x} + \hat{h}(x,\dot{x}) = 0$$

the requirement of qualitative equivalence near the origin demands only that a topological map Φ carry the solution curve family (sensed but not parametrized solution curves) of $\mathcal{S}$ in some neighborhood N , onto the solution curve family of $\hat{\mathcal{S}}$ in the neighborhood $\hat{N} = \Phi(N)$ of the origin.

Definition. The k-jet of the differentiable function $g(x,y)$ is determinate in case: each differentiable function $h(x,y)$ with the same k-jet must be qualitatively equivalent to $g(x,y)$ near the origin.

The k-jet of $g(x,y)$ is dynamically-determinate in case: each differentiable function $h(x,y)$ with the same k-jet must define a differential system $\ddot{x} + h(x,\dot{x}) = 0$ that is locally qualitatively equivalent to the given differential system $\ddot{x} + g(x,\dot{x}) = 0$ near the origin of $\mathbb{R}^2$.

Of course, we consider only k-jets which are not identically vanishing (have some non-zero coefficients) when dealing with the problem of jet determinacy.

Remark. In an analogous manner we can define the concepts of qualitative equivalence and jet-determinacy for real differentiable functions of one real variable x near the origin of $\mathbb{R}$. That is, we study functions such as $g(x)$, with $g(0) = 0$, and the corresponding differential systems $\ddot{x} + g(x) = 0$, or $\dot{x} = y$, $\dot{y} = -g(x)$ in $\mathbb{R}^2$. In this case the comparison functions are all of the corresponding form $h(x)$ defined near the origin of $\mathbb{R}$. Also the differential systems are all conservative, for instance, $\ddot{x} + g(x) = 0$ with a potential function $G(x) = \int_0^x g(s)\,ds$ and total energy

$$E(x,y) = \frac{y^2}{2} + G(x) \quad \text{in} \quad \mathbb{R}^2 .$$

2. Jet determinacy for conservative oscillators. For simplicity we first study the problem of jet-determinacy for conservative oscillators with one-degree of freedom, say $\ddot{x} + g(x) = 0$, or the system in the (x,y)-plane $\mathbb{R}^2$

$$\dot{x} = y , \quad \dot{y} = -g(x) .$$

As usual $g(x)$ is differentiable near the origin in $\mathbb{R}$, so the differential system has differentiable coefficients near the origin in $\mathbb{R}^2$, which is a critical point. Moreover, the solution curves lie along the energy levels $E(x,y) = y^2/2 + G(x) = E_0$ (constant), where the potential energy is given by $G(x) = \int_0^x g(s)\,ds$.

Let us suppose that the lowest order non-vanishing jet of $g(x)$ is $c\,x^k$, for some integer $k \geq 1$ and constant $c \neq 0$. Then, as is well-known [2 , p. 111], $g(x)$ is determined by its k - jet $c\,x^k$. Namely write the new coordinate $\bar{x}$, near the origin on $\mathbb{R}$,

$$\bar{x} = x(1 + \varphi(x))^{1/k} ,$$

for differentiable $\varphi(x)$ with $\varphi(0) = 0$, where $\varphi(x)$ is defined by

$$g(x) = cx^k(1 + \varphi(x)) .$$

Then, in terms of coordinate $\bar{x}$, we write

$$g(x(\bar{x})) = c\,\bar{x}^k .$$

Thus $g(x)$, or any differentiable function with the k - jet $c\,x^k$, (or any jet whose lowest degree term is $c\,x^k$) is qualitatively equivalent to the monomial $c\,x^k$. Hence every non-vanishing k - jet in $\mathbb{R}$ is determinate. The corresponding result for dynamical-determinacy is not so obvious, but still remains valid as we show below using a case-by-case analysis.

Consider the nonlinear conservative oscillator

$$\ddot{x} + g(x) = 0$$

where $g(x) = -c\,x^k + \ldots$, that is, the first nonvanishing jet of $g(x)$ is $-c\,x^k$, for $k \geq 1$ and $c \neq 0$. Then by a linear scaling of t and y , we can take $c = \pm 1$, that is, the given differential system is qualitatively equivalent to

$$\dot{x} = y \ , \ \dot{y} = \pm x^k + \ldots = \frac{1}{|c|} g(x) \ .$$

We shall show that any such differential system is qualitatively equivalent to the k-jet system (and hence, to a linear or quadratic model):

$$\begin{cases} \dot{x} = y \ , \\ \dot{y} = x^k \end{cases} \text{or} \quad \begin{cases} \dot{x} = y \\ \dot{y} = -x^k \end{cases}$$

Case 1: $k = 2\ell - 1$ odd, and $c > 0$, so consider

$$\dot{x} = y \ , \ \dot{y} = -x^{2\ell - 1} + \ldots \quad \text{for some integer} \quad \ell \geq 1 \ .$$

The solution curves are the energy levels

$$E(x,y) = \frac{y^2}{2} + G(x) = \frac{y^2}{2} + \frac{x^{2\ell}}{2\ell} + \ldots ,$$

where the potential is defined by $G(x) = \int_0^x g(s)\,ds$. Then in some small neighborhood of the origin, the origin is the unique critical point and every other solution is periodic along some energy locus $E(x,y) = E_o > 0$. Thus the differential system is qualitatively equivalent to the center $\dot{x} = y \ , \ \dot{y} = -x^{2\ell - 1}$; in fact, qualitatively equivalent to the linear center $\dot{x} = y \ , \ \dot{y} = -x$ near the origin in R^2 .

Case 2: $k = 2\ell - 1$ odd, and $c < 0$ so consider

$$\dot{x} = y \ , \ \dot{y} = x^{2\ell - 1} + \ldots \quad \text{for some integer} \quad \ell \geq 1 \ .$$

The solution curves are along the evergy levels for

$$E(x,y) = \frac{y^2}{2} + G(x) = \frac{y^2}{2} - \frac{x^{2\ell}}{2\ell} + \ldots \ .$$

An easy graphical analysis (using $\dot{x} = y$, $\dot{y} = 0$ on $x = 0$; and $\dot{x} = 0$, $\dot{y} = x^{2\ell - 1} + \ldots$ on $y = 0$) shows that the solution curve family is qualitatively equivalent to the linear saddle $\dot{x} = y$, $\dot{y} = x$.

Case 3: $k = 2\ell$ even, and $c \neq 0$ so consider

$$\dot{x} = y \; , \; \dot{y} = x^{2\ell} + \ldots$$

(if $c < 0$, reverse sense of both x and y , to obtain an equivalent system with $c > 0$). On the axis $x = 0$ we have $\dot{x} = y$, $\dot{y} = 0$, and on the horizontal axis we find $\dot{x} = 0$, $\dot{y} > 0$. The energy locus $E(x,y) = 0$ consists of the origin, one solution curve in $y > 0$, and one solution in $y < 0$ (the uniqueness of the solution leaving the origin follows from the inequality, $\frac{\partial E}{\partial y}(0,y) = 2y > 0$ if $y > 0$). Thus the differential system in this case is qualitatively equivalent to the system $\dot{x} = 0$, $\dot{y} = (x^2 + y^2)$ which has a unique critical point at the origin and otherwise has solutions that are vertical lines.

Theorem 1. Every k - jet in one variable $x \in \mathbb{R}$ is determinate and also dynamically-determinate. That is, $c\,x^k$, for $k \geq 1$ and $c \neq 0$, determines the local qualitative behavior of each function $g(x)$, and also each differential system $\ddot{x} + g(x) = 0$, whenever $g(x)$ has the k -jet $c\,x^k$. Thus the k - jet of a conservative force genetically determines the qualitative dynamics.

3. Jet determinacy for nonconservative oscillators.

We now consider the more difficult problem of the qualitative behavior of the damped oscillator

$$\ddot{x} + g(x,\dot{x}) = 0$$

or the differential system in R^2

$$\dot{x} = y \ , \ \dot{y} = -g(x,y) \ .$$

Here, as earlier, $g(x,y)$ is a real differentiable function near the origin of the (x,y) - plane, and $g(0,0) = 0$ so the origin is a critical point for the corresponding differenital system. In this situation we find that a k - jet may or may not be function-determinate, or quite independently, dynamically-determinate.

Examples.

1) 1 - jet $\xi = (x+y)$.

This 1 - jet is both determinate and dynamically-determinate.

This 1 - jet certainly determines the qualitative nature of any function

$$g(x,y) = x + y + \dots \ ,$$

by direct observations based on the implicit function theorem. Also we can apply the sufficiency test for determinacy of a k - jet ξ [2 , p. 113].

Sufficiency Test. If every homogeneous polynomial q , of degree $(k+1)$ in (x,y) , can be expressed as below, then ξ is determinate:

$$q = q_1(2) \frac{\partial\xi}{\partial x} + q_2(2) \frac{\partial\xi}{\partial y} + r(k+2) \ ,$$

where the notation indicates that $q_1(2)$ and $q_2(2)$ are polynomials with no terms of degree less than 2 , and $r(k+2)$ has no terms of degree less than $(k+2)$, or else they vanish identically.

Remark. If ξ is homogeneous we can further require that q_1 and q_2 are homogeneous quadratic polynomials and that $r \equiv 0$. In this case the test is necessary and sufficient for determinacy (and the analogous assertion holds for homogeneous k-jets in n real variables).

The corresponding differential system

$$\dot{x} = y \ , \ \dot{y} = -x - y + \ldots$$

displays a stable focus about the origin, regardless of the higher order perturbation terms. Thus $\xi = (x+y)$ is dynamically-determinate and leads to the qualitative behavior of the linear damped oscillator

$$\dot{x} = y \ , \ \dot{y} = -x - y \quad \text{or} \quad \ddot{x} + \dot{x} + x = 0 \quad .$$

2) 3-jet $\xi = x^3$.

This 3-jet is not determinate, and not dynamically-determinate.

Since $\xi = x^3$ can be perturbed to either $x^3 - y^6 = (x - y^2)(x^2 + xy^2 + y^4)$ or $x^3 - xy^3 = x(x^2 - y^3)$, and since the loci of zeros for these two functions are topologically different, we note that ξ cannot be determinate. Also we can refer to the necessity test for determinacy of a k-jet ξ [2, p. 114].

Necessity Test. If ξ is determinate, then every homogeneous polynomial q , of degree (k+1) in (x,y) , can be expressed:

$$q = q_1(1) \frac{\partial \xi}{\partial x} + q_2(1) \frac{\partial \xi}{\partial y} + r(k+2) \quad .$$

In this case $q = y^4$ cannot be expressed as a multiple q_1 of

$\frac{\partial \xi}{\partial x} = 3x^2$, as required in the test.

The corresponding differential system is a center

$$\dot{x} = y \ , \ \dot{y} = -x^3 \ ,$$

with solution curves along the energy levels of $E(x,y) = \frac{y^2}{y} + \frac{x^4}{4}$. But arbitrarily little damping can modify the qualitative behavior to a stable focus

$$\dot{x} = y \ , \ \dot{y} = -x^3 - y^5$$

or

$$\ddot{x} + (\dot{x})^5 + x^3 = 0 \ .$$

3) 3 - jet $\xi = -x^3$.

This 3 - jet is not determinate, but it is dynamically-determinate.

The necessity test for determinancy would require that

$$y^4 = q_1 x^2 + r(5) \ ,$$

which is impossible.

Next, consider the differential system

$$\dot{x} = y \ , \ \dot{y} = x^3 + O_4(x,y) \ ,$$

where $O_4(x,y)$ is differentiable near the origin with a vanishing 3 - jet. We seek to prove that such a differential system must be determined as qualitatively equivalent to a linear saddle

$$\dot{x} = y \ , \ \dot{y} = x \ .$$

First note that on $y = 0$ we have $O_4(x,0)$ as a quartic in x, so $\dot{y} > 0$ for $x > 0$ (and $\dot{y} < 0$ for $x < 0$) within some small neighborhood of the origin, which must therefore be an isolated critical point. Also in the upper half-plane $\dot{x} = y > 0$, with the reverse horizontal velocity in the lower half-plane. By elementary geometrical analyses we find that there exist separatrix solution curves approaching the origin from the first quadrant (as $t \to -\infty$), and similarly from each of the quadrants.

We next show that, in the first quadrant, there is only one such separatrix solution curve, which we write $x(t), y(t)$ or else $y = \psi(x)$ since $\dot{x} = y \neq 0$. That is, there cannot be a "fan" of such solution curves approaching the origin in the first quadrant. From this it is easy to demonstrate that the differential system is topologically equivalent to a linear saddle system in some neighborhood of the origin.

On the parabola curve $y = (1/2)x^2$ in $x > 0$ we find $y' = x$, but the slope of the differential system is

$$\frac{dy}{dx} = \frac{x^3 + O_4}{y} = \frac{x^3 + O_4(x, \frac{1}{2}x^2)}{(x^2/2)} > \frac{3}{2}x \quad \text{for small } x > 0 .$$

Similarly on $y = 2x^2$ where $y' = 4x$, we find

$$\frac{dy}{dx} = \frac{x^3 + O_4(x, 2x^2)}{2x^2} < \frac{3}{4}x \quad \text{for small } x > 0 .$$

Thus the solution curve $y = \psi(x)$ lies within the sector bounded by the two parabolas for $0 < x < x_o$, for some small $x_o > 0$.

We define $\psi(x)$ to be continuous on $[0,x_o]$ with $\psi(0) = 0$, and $\psi(x) \in C^1$ in $(0,x_o]$. But inside the parabolic sector, the slope of any solution lies between $\frac{3}{4}x$ and $\frac{3}{2}x$. Thus $\psi(x) \in C^1$ on the closed interval, with $\psi'(0) = 0$.

Now suppose there were a different such separatrix curve $y = \psi_1(x)$, say $\psi_1(x) < \psi(x)$ on $(0,x_o]$. Let $\Delta(x) = \psi(x) - \psi_1(x)$ in C^1 on $[0,x_o]$, so $\Delta(x) > 0$ on $(0,x_o]$ with $\Delta(0) = 0$. Then compute

$$\frac{d\Delta(x)}{dx} = f(x,\psi(x)) - f(x,\psi_1(x)) = f_y(x,\tilde{y}(x))\ \Delta(x) ,$$

where the slope function $f(x,y) = (x^3 + O_4)/y$, and $\tilde{y}(x)$ lies between $\psi_1(x)$ and $\psi(x)$, as usual. Note that

$$f_y = \frac{y\frac{\partial O_4}{\partial y} - (x^3 + O_4)}{y^2} < 0$$

inside the parabolic sector $\frac{1}{2}x^2 < y < 2x^2$ for small $x > 0$. This means that $\frac{d\Delta(x)}{dx} < 0$ and so $\Delta(x)$ is nonincreasing with x . But if $\Delta(x_o) > 0$, this contradicts the supposition that $\Delta(0) = 0$. Thus we conclude that there is just one separatrix curve in the first quadrant.

By standard techniques it now follows that every such differential system

$$\dot{x} = y\ ,\ \dot{y} = x^3 + O_4(x,y)$$

is qualitatively equivalent to a linear saddle system in some neighborhood of the origin, see [3].

4) 1 - jet $\xi = x$.

This 1 - jet is determinate, but not dynamically determinate.

Since $\frac{\partial\xi}{\partial x} = 1$ is nonzero at the origin, the 1 - jet is determinate.

However the differential system

$$\dot{x} = y \ , \ \dot{y} = -x + \ldots$$

is not qualitatively determined, since any frictional perturbation changes the center to a stable focus, for instance,

$$\dot{x} = y \ , \ \dot{y} = -x - y^3 \ .$$

REFERENCES

[1] JANE AUSTEN, Collected Novels, London 1811, 1813.

[2] D. CHILLINGWORTH, Differential Topology with a View to Applications, Pitman, London, 1976.

[3] L. MARKUS, Global structure of ordinary differential equations in the plane, Trans. Am. Math. Soc. Vol. 76, (1954) pp. 127-148.

SECTION SIX
BIFURCATION WITH SYMMETRY

An Application of Singularity Theory to Convection in a Spherical Shell

David Schaeffer* and Martin Golubitsky**

Abstract. The problem of convection in a spherical shell is explored using singularity theory methods. The rotational symmetry of the problem plays a crucial role. Theoretical and numerical results are compared.

Singularity theory provides a useful language in which to discuss bifurcation problems. In particular it can be very helpful in organizing experimental data by relating them to known theoretical results. As an illustration of this power, we consider the Bénard problem which arises in plate tectonics. This problem has been considered by, among others, Chossat [1] from a theoretical point of view and Young [2] in a numerical simulation. Chossat considers the selfadjoint case - this means the gravity field is proportional to the temperature gradient. He shows that in this case the only bifurcating solutions are axisymmetric and they appear supercritically. Young, on the other hand, considers a nonselfadjoint case, and in his data nonaxisymmetric solutions play an important role. In the process of trying to reconcile this disparate information, one is

*Department of Mathematics, Duke University, Durham, NC 27706. This work was sponsored in part by NSF Grant MCS-79-02010.
**Department of Mathematics, Arizona State University, Tempe, AZ 85281. This work was sponsored in part by NSF Grant MCS 79-5799.

led to a rather convincing plausibility argument that the basic bifurcation in the case Young considers must be subcritical, a point on which he makes no comment. Of course this involves no contradiction, as he considers a nonself-adjoint case. Nevertheless, it is a little surprising that an apparently small change in certain auxiliary parameters can reverse the orientation of the primary bifurcation in this problem. It should be remarked that the complexity of this problem renders a direct analysis exceedingly difficult; for example, the calculations of Young required integration of a time-dependent system of nonlinear equations with non-trivial dependence on all three space coordinates.

The singularity theory approach to bifurcation considers only steady state solutions (so far, at least), and it is restricted to problems for which the Lyapunov-Schmidt reduction to a finite dimensional problem is possible. The central concept in the theory is the notion of contact equivalence which formalizes the idea of two systems having the same qualitative behavior. (See the article of M. Golubitsky in these proceedings for more details concerning the general theory.) In many cases it is possible to find a finite-parameter family of distinguished perturbations of a bifurcation problem which enumerates all perturbations of the given bifurcation problem up to contact equivalence. In such cases one calls the family of distinguished perturbations the <u>universal unfolding</u> of the original bifurcation problem, and for technical reasons the number of perturbation parameters in this family is called the <u>codimension</u> of the singularity. Furthermore, there is often a particularly simple, preferred form of the bifurcation equations which is contact equivalent to the original problem, and one refers to this as a <u>canonical form</u> of the equations. Judicious use of canonical forms can simplify certain bifurcation calculations enormously.

The notion of codimension provides a measure of the complexity of a bifurcation problem. Although this measure does not lead to a linear ordering, it does suggest a rough order in which to consider canonical forms to match the data of a given experiment. Indeed one expects the canonical form which actually governs a given experimental situation to have as low a codimension as possible, for otherwise there are accidental degeneracies to explain. In applying this principle, however, it is of paramount importance to consider the structure of the equations governing the phenomena at hand. Most obviously, for convection in a spherical shell all equations are equivariant under the action of the orthogonal group $O(3)$. For a physically useful classification one must define codimension in a context which only considers perturbations preserving the $O(3)$ symmetry. Even given this qualification, care must still be exercised in the application of this principle.

For example, in the convection problem Chossat [1] shows that the canonical form becomes more singular in the selfadjoint case.

The nature of our plausibility argument that Young's case [2] is one with subcritical bifurcation may be summarized as follows. We have found the universal unfolding (in the O(3) context) of the bifurcation problem describing the selfadjoint case. This unfolding depends on only one parameter, and as this parameter varies all small, O(3)-symmetric perturbations of the original problem are enumerated. We have examined all the problems so enumerated, and <u>none agrees with the data of Young</u> - specifically for no such problem is there a range of Rayleigh numbers where <u>both</u> the axi-symmetric and nonaxi-symmetric solutions are stable. We have also examined the next few cases on the classification list looking for this feature, but none possesses it. However, a simple case exhibiting subcritical bifurcation does possess it; rather than postulate many accidental degeneracies we conjecture that the latter canonical form governs Young's case.

REFERENCES

[1] P. Chossat, "Bifurcation and Stability of Convective Flows in a Rotating or Nonrotating Shell," preprint, Univ. of Minn. (1978).

[2] R. Young, "Finite Amplitude Thermal Convection in a Spherical Shell," JFM <u>63</u> (1974), pp. 695-721.

The Analysis of Nonlinear Equations with Symmetries

Hans G. Othmer*

Abstract. When a physical theory embodies invariance under a group of transformations, such invariance must be reflected in the structure of the transient and steady-state equations that describe the theory and in the solutions to these equations. Analysis of the structure of the symmetry group shows what modes are excited by nonlinear coupling and leads to self-consistent subsets of the amplitude equations in nonlinear problems. This is illustrated with the simplest non-trivial example, a group of order two.

Group invariance may also lead to degeneracy in the spectrum of an invariant linear operator and thus to bifurcation problems for which there is no general theory that guarantees bifurcation. We derive the bifurcation equations for a class of problems with a degenerate imaginary eigenvalue at criticality and illustrate a new phenomenon that arises. We also show how the structure of the solution set at a degenerate eigenvalue changes under small perturbations.

Introduction. In many problems in physics and engineering, the evolution of the state u is governed by an abstract autonomous differential equation of the form

$$\frac{du}{dt} = F(u,p), \tag{1}$$

where X and Y are Banach spaces $(X \subseteq Y)$, $p \in I \subseteq \mathbb{R}$, and $F: X \times I \to Y$. For certain classes of equations the stability of a known steady state u_0 can be determined from the spectrum of the Frechet derivative $F_u(u_0,p)$, and bifurcation can only occur at points (u_0,p_0) at which the spectrum of F_u intersects the imaginary axis.

---

*Department of Mathematics, Rutgers University, New Brunswick, New Jersey 08903 and Department of Mathematics, University of Utah, Salt Lake City, Utah 84112. This research was partially supported by grant GM-21558 from the National Institutes of Health.

When F is linear and has a complete set of eigenfunctions, the Fourier modes all evolve independently and any subset spans an invariant linear manifold for the transient equations. Usually nonlinear problems do not have invariant linear manifolds, but they may exist when F is invariant under a group of transformations. When two functions ϕ_i and ϕ_j of a basis set are coupled nonlinearly, the modes that appear in the representation $\phi_i\phi_j(x) = \Sigma\Gamma_{ijk}\phi_k(x)$ are dictated in part by the symmetry properties of ϕ_i and ϕ_j. Which Γ_{ijk}'s vanish for symmetry reasons can be determined from knowledge of all chains of subgroups of the symmetry group, and one can thereby identify all invariant linear manifolds and all self-consistent subsets of the amplitude equations [8]. An example is given in the following section.

Now suppose that a solution u_1 bifurcates from u_0 at p_0 and that F, u_0 and u_1 are invariant under the groups G, G_0 and G_1, respectively, where $G_1 \subset G_0 \subseteq G$. If the transition $u_0 \to u_1$ occurs we say that there is a loss of symmetry at p_0, whereas the reverse transition $u_1 \to u_0$ results in a gain of symmetry. Examples of such spontaneous changes in symmetry abound. In a liquid to crystal transition in $\mathbb{R}^3$, the symmetry group drops from the full Euclidean group to the subgroup that characterizes the lattice structure, and a similar reduction of symmetry occurs in the transition to convection cells in the Benard problem. Both of these transitions are reversible and the symmetry can be regained, but this is often not the case, as for instance in the buckling of symmetric plates and shells.

The use of group-theoretic techniques in the analysis of phase transitions originated with Landau's theory of second-order phase transitions [7]. More recently, similar techniques have been used in the analysis of general invariant bifurcation problems in which the dimension of the null space of F_u is greater than one [8, 10, 11 and references therein]. In the third section we outline the analysis for bifurcating periodic solutions in the presence of symmetry and illustrate the results with a simple example for a reaction-diffusion system. The fourth section gives one example of the changes that occur in the solution set when a symmetric equation is perturbed by a term of lower symmetry.

Symmetry-adapted Amplitude Equations. Suppose that $F(0,p) = 0$ and that $F(u,p) = L_0 u + Q[u,u]$ where Q is a bilinear mapping. Let L_0 and its adjoint L_0^*, the latter defined by the duality pairing $< , >$ between X^* and X, have a complete set of eigenfunctions $\{\phi_i\}$ and eigenfunctionals $\{\psi_j\}$, respectively, normalized so that $<\psi_j, \phi_k> = \delta_{jk}$. If one writes $u = \Sigma a_k \phi_k$, the amplitudes a_k satisfy

$$\frac{da_j}{dt} = \lambda_j a_j + \sum_{k,\ell} Q_{jk\ell} a_k a_\ell \tag{2}$$

where
$$Q_{jk\ell} \equiv \overline{<\psi_j, Q[\phi_k, \phi_\ell]>} .$$

When F is invariant under a compact group G, some of the eigenvalues of L_0 may be degenerate, but a basis can be chosen for every eigenspace of L of dimension greater than one in such a way that each basis function transforms according to an irreducible representation (an irrep) of G (see [3] for the terminology). In this symmetry-adapted basis many of the components Q_{ijk} vanish solely for symmetry reasons, and those that do can be determined once the product rules for G and all its subgroups are known. Furthermore, once the Q_{ijk}'s are known one can identify all amplitudes with the property that $a_k(t) = 0 \;\; \forall\, t > 0$ if $a_k(0) = 0$, and can thereby identify self-consistent subsets for the full system of equations [2,9]. An example will illustrate this.

Suppose that λ is doubly degenerate with eigenfunctions ϕ_1 and ϕ_2 and consider the two-mode approximation $u = a_1\phi_1 + a_2\phi_2$. Further, suppose that (1) is defined on $x \in (-\infty,\infty)$ and that F is even in x, i.e. it is invariant under the group $G = \{e,\sigma\}$, where $\sigma x = -x$ and $T_\sigma u(x) \equiv u(\sigma^{-1}x) = u(-x)$. Let ϕ_1 and ψ_1 be even and let ϕ_2 and ψ_2 be odd. Then $\{\phi_1,\phi_2\}$ and $\{\psi_1,\psi_2\}$ are symmetry-adapted, and because the Q_{ijk}'s are scalars,

$$<\psi_j, Q[\phi_k,\phi_\ell]> = <T_\sigma\psi_j, Q[T_\sigma\phi_k, T_\sigma\phi_\ell]>.$$

It follows that $Q_{ijk} = 0$ for $(ijk) = (121), (112), (211)$ and (222), and so half of the Q_{ijk}'s vanish for symmetry reasons. The amplitude equations are

$$\frac{da_1}{dt} = \lambda a_1 + Q_{111}a_1^2 + Q_{122}a_2^2 \tag{3}$$

$$\frac{da_2}{dt} = \lambda a_2 + 2Q_{212}a_1a_2$$

and it is clear that if $a_2(0) = 0$ then $a_2(t) = 0 \ \forall t > 0$. Thus ϕ_2 cannot be excited by nonlinear coupling of ϕ_1 with itself, i.e., the linear manifold $a_2 = 0$ is invariant under (3). Consequently one can study (3) on this manifold and because it is invariant, we say that the first equation, with $a_2 = 0$, forms a self-consistent set.

A similar procedure with similar results applies to other groups and general polynomial nonlinearities that are invariant. For larger groups one must know the structure of all subgroups of G and the multiplication properties of functions that transform according to their representations, and from these one can determine which coefficients vanish for symmetry reasons and find all self-consistent subsets of equations. An example in which the group is C_{4v} is given in [8].

Bifurcation of Periodic Solutions. If $F \in C^3$ we can write

$$F(u,p) = (L_0 + pL_1(p))u + Q[u,u,p] + C[u,u,u,p] + R(u,p), \tag{4}$$

where L_0 is as before and Q, C and R are bilinear, trilinear and higher-order functions of u, respectively. Suppose that the spectrum of L_0 consists of a part $\sigma_1(L_0)$ for which $\mathrm{Re}\lambda < -\gamma$, and an n-fold degenerate pair of eigenvalues $\pm i\omega_0$. By the hypotheses on L_0, there are n linearly independent solutions for each of the problems

$$L_0\Phi_k = i\omega_0\Phi_k \qquad L_0^*\Psi_k = -i\omega_0\Psi_k$$

and $\langle\psi_j,\Phi_k\rangle = \delta_{jk}$. We set $\tau = \omega t$ and look for solutions of (1) in the form

$$u = \varepsilon u_0 + \varepsilon^2 u_1 + \varepsilon^3 u_2 + O(\varepsilon^4)$$

$$p = \varepsilon p_1 + \varepsilon^2 p_2 + \varepsilon^3 p_3 + O(\varepsilon^4)$$

$$\omega = \omega_0 + \varepsilon\omega_1 + \varepsilon^2\omega_2 + O(\varepsilon^3).$$

This leads to the sequence of problems

$$\omega_0 \frac{du_0}{d\tau} - L_0 u_0 = 0$$

$$\omega_0 \frac{du_1}{d\tau} - L_0 u_1 = -\omega_1 \frac{du_0}{d\tau} + p_1 L_1(0) u_0 + Q[u_0,u_0,0]$$

$$\omega_0 \frac{du_2}{d\tau} - L_0 u_2 = -\omega_2 \frac{du_0}{d\tau} - \omega_1 \frac{du_1}{d\tau} + p_1 L_1(0) u_1 + p_2 L_1(0) u_0$$

$$+ p_1^2 \frac{\partial L_1(0)}{\partial p} u_0 + 2Q[u_0,u_1,0] + p_1 \frac{\partial Q}{\partial p}[u_0,u_0,0] + C[u_0,u_0,u_0,0].$$

It follows that for initial data in the null space of $L_0 \pm i\omega_0$

$$u_0 = \sum_{k=1}^{n} a_k \left\{ e^{i(\tau+\theta_k)} \Phi_k + e^{-i(\tau+\theta_k)} \overline{\Phi}_k \right\},$$

where ε is fixed by the normalization $\Sigma a_k^2 = 1$. One of the phase differences can be set equal to zero by translating the origin of τ. One can prove a Fredholm alternative for equations of the above form and when it is applied to the second equation, the result is that $\omega_1 = p_1 = 0$, just as in the case $n = 1$ [5]. At $O(\varepsilon^3)$ the solvability condition yields the bifurcation equations

$$i\omega_2 a_j e^{-i\theta_j} + p_2 \sum_{k=1}^{n} \langle\psi_j, L_1(0)\Phi_k\rangle a_k e^{-\theta_k} + \int_0^{2\pi} \langle\psi_j, 2Q[u_0,u_1,0]\rangle e^{i\tau} d\tau$$

$$+ \sum_{p,q,r=1}^{n} \left\{ \Gamma^j_{pqr} e^{i(\theta_p-\theta_q-\theta_r)} + \Gamma^j_{pqr} e^{i(-\theta_p+\theta_q-\theta_r)} + \Gamma^j_{rqp} e^{i(-\theta_p-\theta_q+\theta_r)} \right\} a_p a_q a_r = 0$$

$$\sum_{k=1}^{n} a_k^2 = 1 \qquad j = 1, \ldots, n,$$

where

$$\Gamma^j_{pqr} \equiv \langle \psi_j, C[\Phi_p, \overline{\Phi}_q, \overline{\Phi}_r, 0] \rangle .$$

To illustrate the types of solutions that are possible, consider the following example. Let (1) represent a system of reaction-diffusion equations, in which case $L_0 = \mathcal{D}\Delta + K(0)$, where $\mathcal{D}$ is the diffusion matrix, $K(0)$ the Jacobian of the kinetic terms, and Δ is the Laplacian on a domain Ω with associated boundary conditions [8]. In this case $\langle\ ,\ \rangle = \langle\ ,\ \rangle_1 \langle\ ,\ \rangle_2$ where '1' and '2' denote the Euclidean and L_2 inner products, respectively. For simplicity, assume that both Q and R vanish in (4). When the degeneracy is due to the symmetry of $\Omega, \Phi_j = y\phi_j$, where y is a fixed eigenvector of $K(0) + \alpha_j \mathcal{D} - i\omega_0$ and $\Delta\phi_j = \alpha_j \phi_j$. Suppose, for instance, that Ω is a square of length π centered at the origin, with zero Dirichlet boundary conditions. Then (1) is invariant under $G = C_{4v} \times \{1,-1\}$, where the action is defined by $T^{\pm}_g f = \pm T_g f$ for any $g \in G$, and when m and n are odd and unequal, the symmetry-adapted eigenfunctions are $\phi_{1,2} = \frac{\sqrt{2}}{\pi}$ (cos mx cos ny ± cos nx cos my). ϕ_1 is invariant under C_{4v} while ϕ_2 is invariant under a group G_1 of order 8 consisting of C_{2v} and the four operations $f(x,y) \to -f(\pm y,x), -f(\pm y,-x)$.

The bifurcation equations are

$$(i\omega_2 + p_2\Omega_1)a_1 + \Omega_2[21a_1^3 + 5(1 + 2e^{2i(\theta_1-\theta_2)})a_1a_2^2] = 0$$

$$(i\omega_2 + p_2\Omega_1)a_2 + \Omega_2[5(2 + e^{-2i(\theta_1-\theta_2)})a_1^2a_2 + 21a_2^3] = 0$$

where

$$\Omega_1 = \langle y^*, L_1(0)y\rangle_1 \quad \text{and} \quad \Omega_2 \equiv \frac{3\langle y^*, C[y,\overline{y},\overline{y},0]\rangle_1}{8\pi^2} .$$

These equations have the following solutions.

	I	II	III	IV	V
a_1	± 1	0	$\pm 1/\sqrt{2}$	$\pm 1/\sqrt{2}$	$\pm\sqrt{13/21}$
a_2	0	± 1	$\pm 1/\sqrt{2}$	$\pm 1/\sqrt{2}$	$\pm\sqrt{8/21}$
$\theta_1 - \theta_2$	0	0	0	$\pm\pi$	$\pm\pi/2$

Solutions in the first four classes represent standing waves, while in the fifth class the ith component of u_0 is proportional to one of the following pair, either of which represents a rotating wave

$$\cos(\tau + \delta_i)\phi_1 \pm \sqrt{\frac{8}{13}}\ \sin(\tau + \delta_i)\phi_2 .$$

The orbits under G are (I_+, I_-), (II_+, II_-) and $(III_+, III_-, IV_+, IV_-)$, where the subscripts denote the sign on the amplitudes. The fifth solutions have C_{2v} symmetry and if they were stationary they would lie on an orbit of length four, but since these solutions rotate periodically, all points on the orbit appear periodically. One can easily show that the above solutions all exist either for $p > 0$ or for $p < 0$, i.e. either all bifurcate supercritically or all bifurcate subcritically, and the direction is determined by Ω_1 and Ω_2. These quantities depend on the specific reaction mechanism used and on the geometry of the domain, and some results on the existence and stability of the wave solutions for the Field-Noyes model of the Zhabotinskii-Belousov reaction will be presented elsewhere. Others have predicted the coexistence of standing and rotating waves previously [5], but in the case studied there the waves are necessarily unstable, since it is known that under Neumann boundary conditions at least three species are needed to ensure that all bifurcating periodic solutions have a non-zero wave number. Other results on bifurcating waves in axisymmetric domains are given in [1].

Perturbations That Reduce the Symmetry. The nontrivial wave solutions in the preceding section all have lower symmetry than does the trivial solution and in case any of the former are stable, spontaneous reversible symmetry changes of the type discussed in the Introduction will occur at $p = 0$. Permanent symmetry changes (usually reductions) may be externally imposed or may be due to slight imperfections, and when they are small one can determine how the structure of the solution set near bifurcation points of the nonlinear problem is affected. Let μ be a measure of the perturbation and consider the two-parameter family or mappings $F(u,p,\mu)$, where it is assumed that $F(0,p,\mu) = 0$ for (p,μ) near $(0,0)$. The first step is to determine the effect of μ on the spectrum of F_u, and if $d^k F_u(0,0,0)/d\mu^k \neq 0$ for some k, the perturbation may remove some or all of the degeneracy in the critical eigenvalues of $F_u(0,0,0)$. Whether or not splitting occurs is determined in part by the symmetry properties of $F_u(0,0,0)$, $F_u(0,0,\mu)$ and the degenerate eigenfunctions Φ_j [3]. For instance, if $F_u(0,0,0)$ and $F_u(0,0,\mu)$ are both invariant under G and the degenerate Φ_j's all transform according to the same irrep of G, then Schur's lemma [3] implies that no splitting occurs for $\mu \neq 0$. In any case, the effect of perturbations on the solutions of the nonlinear problem depends heavily on the structure of the higher-order terms and therefore we turn to an example.

As in the preceding section, consider a system of reaction-diffusion equations, here having non-vanishing quadratic terms. For the same region Ω and boundary conditions as before, the problem is now invariant under C_{4v} but the symmetry-adapted ϕ's remain unchanged. ϕ_1 now transforms according to the identity representation of C_{4v} while ϕ_2 transforms according to a different one-dimensional irrep, that labelled Γ_3 in [9]. Let $F_{u\mu} = B\Delta_1$ where B is a constant matrix and Δ_1 is a differential operator. We look for steady state solutions of (1) in which a_1 and a_2 are $\mathcal{O}(\varepsilon)$, $0 < \varepsilon \ll 1$, and therefore scale p and μ by setting $p = \varepsilon p_1$, $\mu = \varepsilon\mu_1$, where p_1 and μ_1 can vary in some $\mathcal{O}(1)$ neighborhood of $(0,0)$. The bifurcation equations at $\mathcal{O}(\varepsilon^2)$ are

$$(p_1\Omega_1 + \mu_1\Omega_2 M_{11})a_1 + \mu_1\Omega_2 M_{12}a_2 + \Omega_3(\Gamma_{111}a_1^2 + \Gamma_{122}a_2^2) = 0$$

$$\mu_1\Omega_2 M_{21}a_1 + (p_1\Omega_1 + \mu_1\Omega_2 M_{22})a_2 + 2\Omega_3\Gamma_{122}a_1 a_2 = 0$$

where $\Omega_1 = \langle y^*, L_1(0)y\rangle_1$, $\Omega_2 = \langle y^*, By\rangle_1$, $\Omega_3 = \langle y^*, Q[y,y,0]\rangle_1$, $\Gamma_{ijk} =$

$\int\phi_i\phi_j\phi_k d\Omega$ and $M_{ij} = \langle\phi_i,\Delta_1\phi_j\rangle_2$. After rescaling these can be written

$$(5)\qquad \begin{aligned} M_1 a_1 + c_3\mu_1 a_2 \pm a_1^2 + a_2^2 &= 0 \\ c_4\mu_1 a_1 + M_2 a_2 + 2a_1 a_2 &= 0 \end{aligned}$$

where $M_1 \equiv c_1 p_1 + c_2\mu_1$ and $M_2 = c_1 p_1 + c_5\mu_1$.

First suppose that $c_3 = c_4 = 0$. The Jacobian of (5) is

$$J^{\pm} = \begin{bmatrix} M_1 \pm 2a_1 & 2a_2 \\ 2a_2 & M_2 + 2a_1 \end{bmatrix}$$

and therefore

$$\operatorname{trace} J^{\pm} = M_1 + M_2 + 2a_1 \pm 2a_1$$

and

$$\det J^{\pm} = M_1 M_2 + 2a_1(M_1 \pm M_2) \pm 4a_1^2 - 4a_2^2 .$$

The solutions of (5) and the corresponding values of the invariants are as follows.

(I) $\quad (a_1,a_2) = (0,0) \qquad \operatorname{trace} J = M_1 + M_2$, $\quad \det J = M_1 M_2$

(II) $\quad (a_1,a_2) = (\mp M_1, 0) \qquad \operatorname{trace} J^+ = M_2 - 3M_1$, $\quad \operatorname{trace} J^- = M_1 + M_2$, $\quad \det J^{\pm} = -M_1(\mp 2M_1 + M_2)$

(III) $\quad (a_1,a_2) = \left(-\frac{M_2}{2}, \pm\sqrt{M_2(2M_1 \mp M_2)/2}\right) \qquad \operatorname{trace} J^{\pm} = M_1 \mp M_2$, $\quad \det J^{\pm} = M_2(\pm M_2 - 2M_1)$.

Thus the Jacobian is singular along the lines $M_1 = 0$ and $M_2 = 0$ for (I), along $M_1 = 0$ and $M_1 = \pm M_2/2$ for (II), and along $M_2 = 0$ and $M_1 = \pm M_2/2$ for (III). These are the bifurcation curves and, since

$J^{\pm}$ is symmetric, they are the only curves along which the stability of solutions can change. Solutions of (5) are isolated except on the bifurcation curves and therefore, the implicit function theorem implies that these are also solutions of the complete bifurcation equations. Since the quadratic terms are nonvanishing this accounts for all solutions of the complete bifurcation equations when $c_3 = c_4 = 0$.

The foregoing information can be used to obtain a complete picture of existence and stability of all solutions in a neighborhood of $(p_1, \mu_1) = (0,0)$ for this case. The various bifurcation diagrams that result are given in [6]. When c_3 and/or c_4 are nonzero but small, there are no qualitative changes in the bifurcation diagrams. The major change that occurs for $c_4 \neq 0$ is the disappearance of the nontrivial single-mode solution.

Conclusion. It is apparent from the examples given that group invariance of a nonlinear operator restricts the structure of the differential equations that govern the evolution of the amplitudes in a modal decomposition and correspondingly restricts the structure of the bifurcation equations. If all existing symmetries are built into the equations, all consistent subsets of the equations can be determined and often one can establish *a priori* that certain types of nontrivial solutions exist. This is especially useful when the degeneracy is large, because it is usually difficult to determine when all solutions of a large nonlinear system have been found.

REFERENCES

[1] J. F. G. AUCHMUTY, Bifurcating waves, Ann. N.Y. Acad. Sci., 316 (1979), pp. 263-278.

[2] J.I. GMITRO, Concentration patterns generated by reaction and diffusion, Thesis, Univ. of Minnesota, (1969).

[3] M. HAMERMESH, Group Theory and its Application to Physical Problems, Addison-Wesley, Reading, Mass., 1962.

[4] M. HERSCHKOWITZ-KAUFMAN and T. ERNEUX, The bifurcation diagram of model chemical reactions, Ann. N.Y. Acad. Sci., 316(1979), pp. 296-313.

[5] D.D. JOSEPH and D. SATTINGER, Bifucating time-periodic solutions and their stability, Arch. Rat. Mech. Anal., 45 (1972), pp. 79-109.

[6] J.P. KEENER, Secondary bifurcation in nonlinear diffusion reaction equations, Stud. Appl. Math., 55(1976), pp. 187-211.

[7] L.D. LANDAU and E.M. LIFSCHITZ, Statistical Physics, Pergamon Press, London, 1958.

[8] H.G. OTHMER, Applications of bifurcation theory in the analysis of spatial and temporal pattern formation, Ann. N.Y. Acad. Sci., 316(1979), pp. 64-77.

[9] H.G. OTHMER and L.E. SCRIVEN, Nonlinear aspects of dynamic pattern in cellular networks, J. Theor. Biol., 43(1974), pp. 87-112.

[10] D.H. SATTINGER, Spontaneous symmetry breaking in nonlinear problems, Ann. N.Y. Acad. Sci., 316(1979), pp. 49-63.

[11] A.L. VANDERBAUWHEDE, Alternative problems and symmetry, J. Math. Anal. Applics., 62(1978), pp. 485-494.

SECTION SEVEN
STOCHASTIC PROBLEMS

Statistical Performance Analysis of Nonlinear Stochastic Systems

James H. Taylor*

Abstract. This paper presents and contrasts two methods that can be used to generate approximate time histories of the second-order statistics of the state variables of nonlinear stochastic systems: the monte carlo method [1] and covariance analysis based on statistical linearization [2]. In both cases, it will be shown that these statistics can be determined quite accurately if the variables are reasonably gaussian, where the measure of the gaussianness of a random variable is its kurtosis, or the ratio of the fourth central moment to the square of the variance. The first technique has the advantage of better error diagnostic capability, while the second generally is much less expensive to use, in terms of computer time consumption. If the variables of a stochastic system are quite nongaussian, then second-order statistics are most generally not meaningful; the monte carlo method can then be used to generate approximate cumulative distributions. The utilization and efficacy of the two methods are illustrated in a number of applications.

1. Introduction and Problem Statement. The dynamics of many nonlinear continuous-time stochastic systems can be written in the form of a state vector differential equation,

$$\dot{\underline{z}} = \underline{f}(\underline{z},\underline{y},t) \tag{1}$$

where $\underline{z}$ is the vector of system states, $\underline{y}$ is a vector of random inputs, and $\underline{f}(\underline{z},\underline{y},t)$ represents the nonlinear time-varying dynamic relationships in the system. Often the elements of $\underline{y}$ are correlated random processes with deterministic components that may be nonzero, so that a system model of the form

---

*School of Mechanical and Aerospace Engineering, Oklahoma State University, Stillwater, Oklahoma 74078. This work was supported by contract N00014-73-C-0213 from the Office of Naval Research, Vehicle Technology Program.

(2) $$\dot{\underline{x}} = \underline{f}(\underline{x},t) + G(t)\ \underline{w}(t)$$

may be used, where $\underline{x}$ is an augmented state vector, $\underline{x}^T = [\underline{z} \vdots \underline{y}]^T$, and $\underline{w}$ is the sum of a vector of white noise processes and a deterministic vector which serves as the input to generate the random vector $\underline{y}$. Henceforth, we treat (2) as the basic system model.

The initial condition of the state vector is specified by assuming that the state variables are jointly normal. Thus, the initial condition is completely specified by an initial mean vector and covariance matrix,

(3) $$E\ [\underline{x}(0)1] = \underline{m}_0,\ E\ [(x(0) - m_0)(x(0) - m_0)^T] = S_0$$

As stated above, the input vector $\underline{w}$ is assumed to be composed of elements that are gaussian white noise processes, plus an additive deterministic component or mean; thus

(4) $$E\ [\underline{w}(t)] = \underline{b}(t),\ E\ [\underline{w}(t)-\underline{b}(t))\ (\underline{w}(\tau)-\underline{b}(\tau))^T] = Q(t)\ \delta(t-\tau)$$

where $Q(t)$ is the input spectral density matrix. Given the initial condition and input statistics defined above, and the system model formulated in (2), our usual goal is to determine the state variable second-order statistics $\underline{m}(t)$, $S(t)$ over a time period of interest.

2. The Monte Carlo Method. This technique provides a straightforward approach to the statistical analysis of the performance of a nonlinear system with random inputs, based on direct simulation. It entails determining the system response to a finite number of "typical" initial conditions and random input functions which are generated according to their specified statistics. Given the information outlined in Section 1, monte carlo analysis requires a large number, say q, of representative simulations of the system response. Performing q independent simulations yields an ensemble of state trajectories, denoted $\underline{x}^{(i)}$, i=1, 2, ..., q. The mean $\underline{m}(t)$ and covariance $S(t)$ of the state vector are estimated by averaging over the ensemble of trajectories using the relations

(5)
$$\underline{\hat{m}}(t) \triangleq \frac{1}{q} \sum_{i=1}^{q} \underline{x}^{(i)}(t) \cong \underline{m}(t),$$
$$\hat{S}(t) \triangleq \frac{1}{q-1} \sum_{i=1}^{q} (\underline{x}^{(i)} - \underline{\hat{m}})(\underline{x}^{(i)} - \underline{\hat{m}})^T \cong S(t)$$

where $\underline{\hat{m}}(t)$ and $\hat{S}(t)$ denote the resulting sample statistics.

In order to assess the accuracy of the approximate statistics given in (5), it is necessary to consider the statistical properties of the estimates $\underline{\hat{m}}(t)$ and $\hat{S}(t)$. To simplify the notation, consider a scalar random variable u (e.g., the value of some system state variable at some time of interest), and let m and σ represent the true values of the mean and standard deviation of u,

(6)
$$m = E[u], \quad \sigma^2 = E[(u - m)^2]$$

By performing one set of q monte carlo trials, we obtain a single estimate of m and σ, denoted $\hat{m}$ and $\hat{\sigma}$. These estimates are also random variables; that is, if another set of q monte carlo trials were performed independently of the first set, then different estimates for the mean and variance would be obtained. Thus, the accuracy of the sample statistics must be characterized statistically.

The usual approach is to determine <u>confidence</u> <u>limits</u>, $\underline{m}$, $\overline{m}$, $\underline{\sigma}$ and $\overline{\sigma}$ so that

(7)
$$\text{Prob}\,[\underline{m} \leq m \leq \overline{m}] = \psi, \quad \text{Prob}\,[\underline{\sigma} \leq \sigma \leq \overline{\sigma}] = \psi$$

where ψ is generally nearly unity (0.95 or 0.99, for example). The confidence limits for m are obtained directly from the sample statistics [1]:

(8)
$$\underline{m} = \hat{m} - n_\sigma \frac{\hat{\sigma}}{\sqrt{q}}, \quad \overline{m} = \hat{m} + n_\sigma \frac{\hat{\sigma}}{\sqrt{q}}$$

where n_σ is chosen to correspond to the desired degree of confidence; since $\hat{m}$ is nearly gaussian for q reasonably large*, n_σ satisfies:

---

*The best rule-of-thumb for determining what q is reasonably large is that $q > 25(\lambda - 1)$; refer to next page.

(9) $$\frac{1}{\sqrt{2\pi}}\int_{-n_\sigma}^{n_\sigma} \exp\left(-\frac{1}{2}\zeta^2\right)d\zeta = \psi \;\rightarrow\; \boxed{\begin{array}{l}\psi = 0.90, n_\sigma = 1.645\\ \psi = 0.95, n_\sigma = 1.960\\ \psi = 0.99, n_\sigma = 2.576\end{array}}$$

The confidence limits for σ are not completely determined by the sample statistics unless the kurtosis, λ, is known:

$$\lambda \triangleq \mu_4/\sigma^4, \quad \mu_4 = E\,[(u-m)^4]$$

Then, for q reasonably large* [1]:

(10) $$\underline{\sigma} = \frac{\hat{\sigma}}{\sqrt{1 + n_\sigma\sqrt{\frac{\lambda - 1}{q}}}} \triangleq \underline{\rho}\,\hat{\sigma}, \quad \overline{\sigma} = \frac{\hat{\sigma}}{\sqrt{1 - n_\sigma\sqrt{\frac{\lambda - 1}{q}}}} \triangleq \overline{\rho}\,\hat{\sigma}$$

Typical values of λ for various density functions of u are: gaussian, $\lambda = 3$; exponential, $\lambda = 6$; triangular, $\lambda = 2.4$; uniform, $\lambda = 1.8$; and bipolar, $\lambda = 1.0$. Observe that the confidence limits for σ are expressed as dimensionless multipliers, $\underline{\rho}$ and $\overline{\rho}$, times the sample standard deviation. Example: For 95% confidence limits of about $\pm$ 10% ($\underline{\rho} = 0.92$, $\overline{\rho} = 1.10$) 256 trials are required for gaussian random variables ($\lambda = 3$). A more complete discussion of the variation of the limit multipliers with varying q, ψ, and λ with accompanying tables, may be found in [1].

3. Covariance Analysis Using Statistical Linearization. The differential equations that govern the propagation of the mean vector and covariance matrix for the system described by (2) can be derived directly [3]:

(11) $$\begin{aligned}\dot{m} &= E\,[\underline{f}(\underline{x},t)] + G(t)\underline{b} \triangleq \hat{\underline{f}} + G(t)\underline{b}\\ \dot{S} &= E\,[\underline{f}\,\underline{r}^T] + E\,[\underline{r}\,\underline{f}^T] + G(t)QG^T(t)\end{aligned}$$

The equation for S can be put into a form analogous to the covariance equations corresponding to $\underline{f}$ being linear, by defining the auxiliary matrix N through the relationship

(12) $$NS \triangleq E\,[\underline{f}(\underline{x},t)\,\underline{r}^T]$$

*See footnote on the preceding page.

Then (11) may be written as

$$\dot{\underline{m}} = \hat{\underline{f}} + G(t)\underline{b}, \quad \dot{S} = NS + SN^T + G(t)QG^T(t) \tag{13}$$

The quantities $\hat{\underline{f}}$ and N defined in (11) and (12) must be determined before (13) can be solved. Evaluating the indicated expected values exactly requires knowledge of the joint probability density function (joint pdf) of the state variables. The fact that the pdf is generally not available thus precludes the exact solution of (13).

The procedure for obtaining an approximate solution to (13) is to assume the form of the joint probability density function of the state variables in order to evaluate $\hat{\underline{f}}$ and N according to (11) and (12). Although it is possible to use any joint pdf, it is both reasonable and convenient to assume that the state variables are jointly normal.

As a consequence of the gaussian assumption, which is discussed more fully in [2] and [4], the elements of $\hat{\underline{f}}$ and N are random-input describing functions [5,6], which for a given nonlinearity are dependent only upon the mean and the covariance of the system state vector. Furthermore, it can be proved [7] that

$$N(\underline{m},S,t) = \frac{d}{d\underline{m}} \hat{\underline{f}}(\underline{m},S,t) \tag{14}$$

Since calculating $\hat{\underline{f}}$ is required for the propagation of the mean, it is generally much easier to employ (14) than to evaluate N directly using (12). The random-input describing functions (RIDF's) for a wide variety of nonlinearities are immediately available from [5, 6, 8], so generally little effort is required in obtaining $\hat{\underline{f}}$ and N.

The combination of covariance analysis with describing function theory to provide approximate solutions to (13) was originated in [9, 10]. For succinctness, this approach is called CADET™ -- the Covariance Analysis DEscribing Function Technique [10].

A comparison of statistical linearization with the classical Taylor series or small-signal linearization technique, wherein $\underline{f}(\underline{m})$ and

*The term CADET is a trademark of The Analytic Sciences Corporation.

the Jacobian $J = d\underline{f}/d\underline{m}$ are used instead of $\hat{\underline{f}}$ and N, provides a great deal of insight into the success of RIDF's in capturing nonlinear effects. If a saturation or limiter is present in a system and its input x is zero-mean, for example, the Taylor series approach leads to replacing f(x) with a unity gain regardless of the input amplitude, while the corresponding RIDF [5,6] decreases as the rms value of x, σ_x, increases, thus accurately reflecting the nonlinear effect that is neglected in the small-signal linear model. The fact that RIDF's retain an essential characteristic of system nonlinearities -- input-amplitude dependence -- provides the basis for the proven accuracy of statistical linearization.

4. Illustrative Examples. The following results demonstrate the two techniques of statistical performance analysis, and illustrate many of the accompanying comments about their use and accuracy.

4.1 A missile/target intercept problem. A variable of particular interest in the planar missile-target intercept problem (cf. [11]) during the terminal homing phase is the cross-range (lateral) separation between the missile and target, denoted y. In a typical analysis, y (and all other system variables) is assumed to be gaussian at the initiation of the terminal homing phase, and y remains quite gaussian until the last few seconds of the engagement. Figure 1 shows the variation of σ_y with time during a six-second engagement ensemble, where a quite highly nonlinear system model [11] has been used for simulation purposes. The solid curve is obtained by CADET, and the results of a 500-trial monte carlo study are indicated with circled data points to indicate $\hat{\sigma}_y$ and vertical I-bars to indicate the 95% confidence limits. The estimated value of kurtosis is also indicated near each data point; as observed above, $\hat{\lambda}$ is nearly 3 until the last second, while at the final time, t=6 sec, $\hat{\lambda} \cong 15$, which is indicative of the quite highly gaussian character of the final lateral separation ("miss distance").

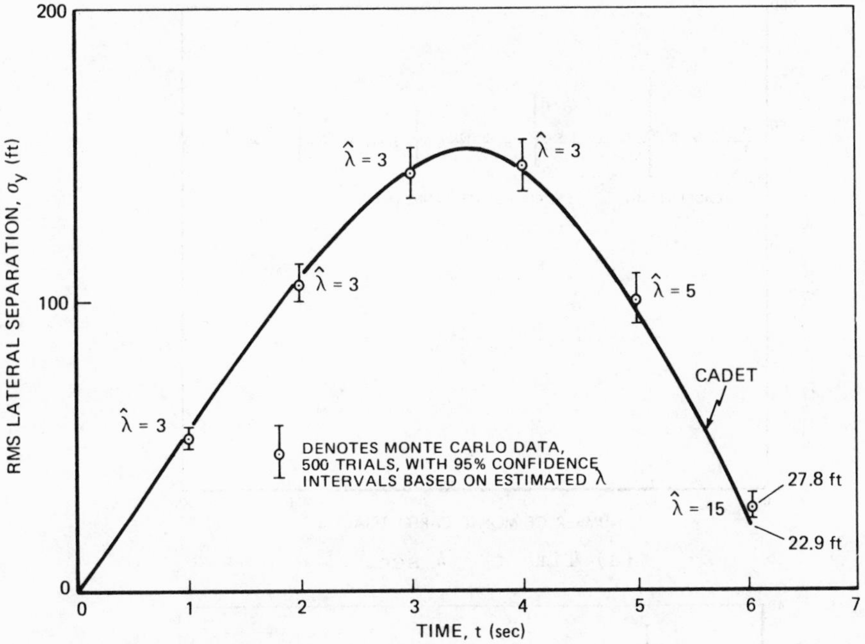

Figure 1. Time History of rms Missile-Target Lateral Separation.

Figure 2 gives a more detailed view of the CADET and monte carlo analyses depicted in Figure 1; for two values of time the estimated σ_y is shown as a function of the number of trials performed, q. We note in Figure 2 that the estimated value of σ_y at t=4 appears to "settle" to about 145 ft after a few hundred trials, while $\hat{\sigma}_y(6)$ converges much more slowly, as (10) predicts. The CADET result appears to be excellent at t=4 sec, and is certainly as accurate as several hundred monte carlo trials at t=6 where y is quite highly nongaussian.

4.2 An antenna pointing and tracking problem. The purpose of the antenna pointing and tracking system modeled in Fig. 3 is to follow a target line-of-sight (LOS) angle, θ_t. Assume that θ_t is a deterministic ramp, with an angular rate of ω degrees per second. The pointing error, $e = \theta_t - \theta_a$, where θ_a is the antenna centerline angle, is the input to a nonlinearity $f(\cdot)$ which represents the limited beamwidth of the antenna; for the present discussion,

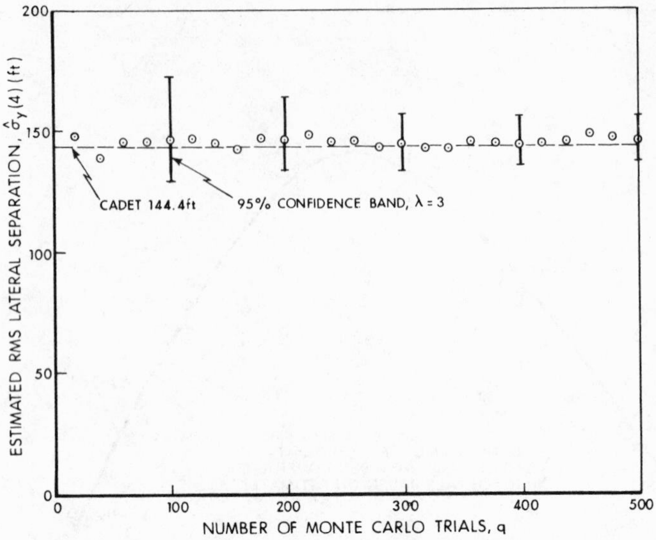

(a) Time t = 4 sec.

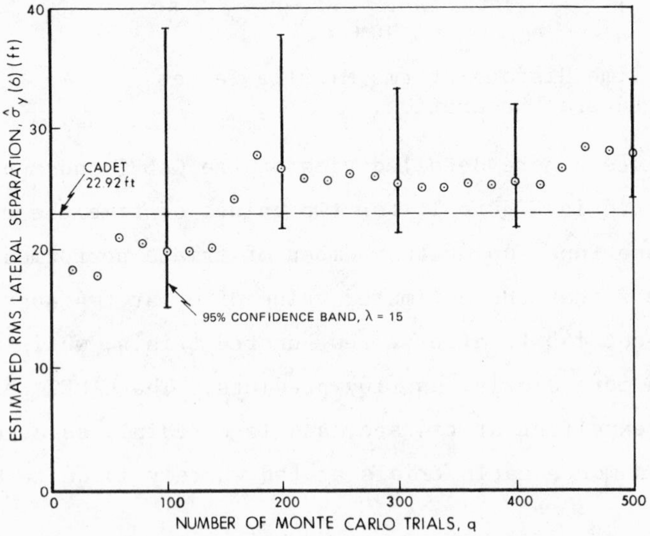

(b) Time t = 6 sec.

Figure 2. Comparison of CADET and Monte Carlo rms Lateral Separation

$$f(e) = e(1-k_a e^2)$$

where k_a is suitably chosen to represent the antenna characteristic.

The noise w(t) injected at the receiver is a white noise process having zero mean and spectral density q. This problem formulation is taken directly from [12]; a more thorough discussion of the approach and results in [12] vis-à-vis the current treatment is given in [13]. In the form given in (2), the system is governed by $\dot{\underline{x}} = \underline{f}(\underline{x}) + \underline{w}$, where

$$\underline{f}(\underline{x}) = \begin{bmatrix} -kx_2 \\ \\ a[f(x_1) - x_2] \end{bmatrix} , \qquad \underline{w} = \begin{bmatrix} \dot{\theta}_t \\ \\ aw(t) \end{bmatrix}$$

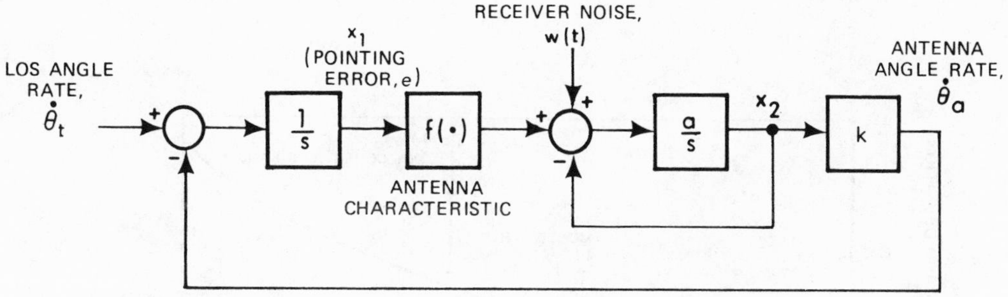

Figure 3. Antenna Pointing and Tracking Model

The solutions depicted in Fig. 4 are based on the assumption that e(0) and w(t) are gaussian. The system parameters are: $a = 50\ \text{sec}^{-1}$, $k = 10\ \text{sec}^{-1}$, $k_a = 0.4\ \text{deg.}^{-2}$, $m_{e0} = 0.4$ deg, $\sigma_{e0} = 0.1$ deg, $q = 0.004\ \text{deg}^2$. The goal is to determine tracking capability for various values of ω; for brevity, only $\omega = 6$ deg/sec is shown here. Four solutions are presented: CADET results; ensemble statistics from a 200-trial monte carlo simulation, with 95% confidence intervals based on the gaussian assumption; results using the linear approximation (corresponding to $k_a = 0$); and the second-order Volterra series solution based on a technique given in [12]. The CADET solution provides a significantly better agreement with the monte carlo data than either of the other analytical approximations, which failed to capture the instability that occurs when tracking error is large, thus clearly establishing the superiority of statistical linearization.

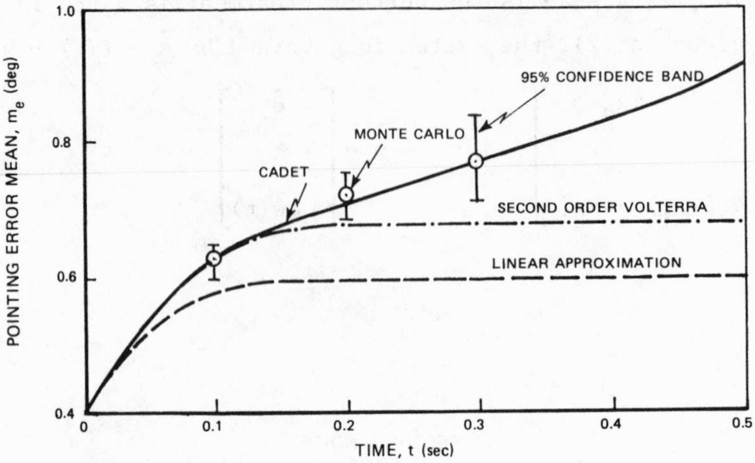

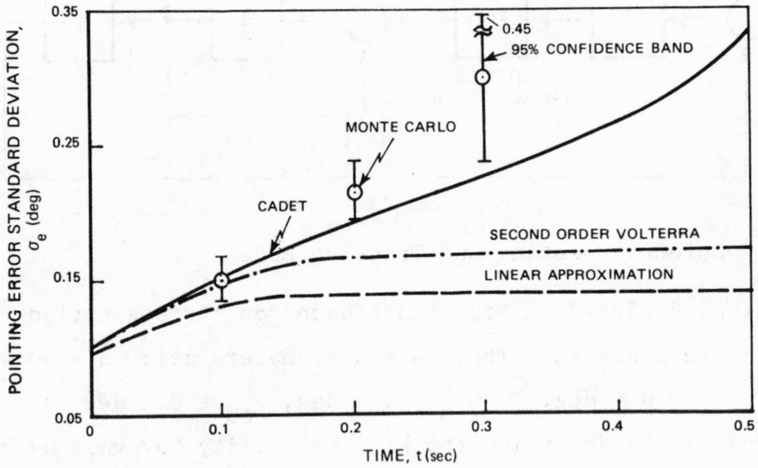

Figure 4. Pointing error statistics for ω = 6 deg/s.

5. Conclusion. The two best methods now available for the statistical performance analysis of nonlinear systems have been outlined. Both are approximate, although the monte carlo method can be made arbitrarily accurate by performing enough trials. Since it is necessary to estimate kurtosis quite well in order to assess this accuracy, the trials required may be excessive [1]. If kurtosis is reasonable (experience suggests that λ should be less than 10 to 15), then CADET produces results that are comparable to 250- to 500-trial monte carlo studies with a small fraction of the computer time expenditure. (Savings of 90 to 98% have been achieved.)

For larger values of λ, the significance of second-order statistics is largely lost, unless one knows the pdf of the variable; one's only recourse is to use the monte carlo method to generate histograms to approximate the cumulative distribution of the random variable. The meaninglessness of $\hat{\sigma}$ as a performance measure is illustrated in Figure 5, which shows an empirical miss distance distribution plot with standard deviation σ = 1018 ft and kurtosis of about 25; clearly the missile under consideration performs much better than this statistic suggests, as the "gaussian fit" curve for this value of rms miss distance shows.

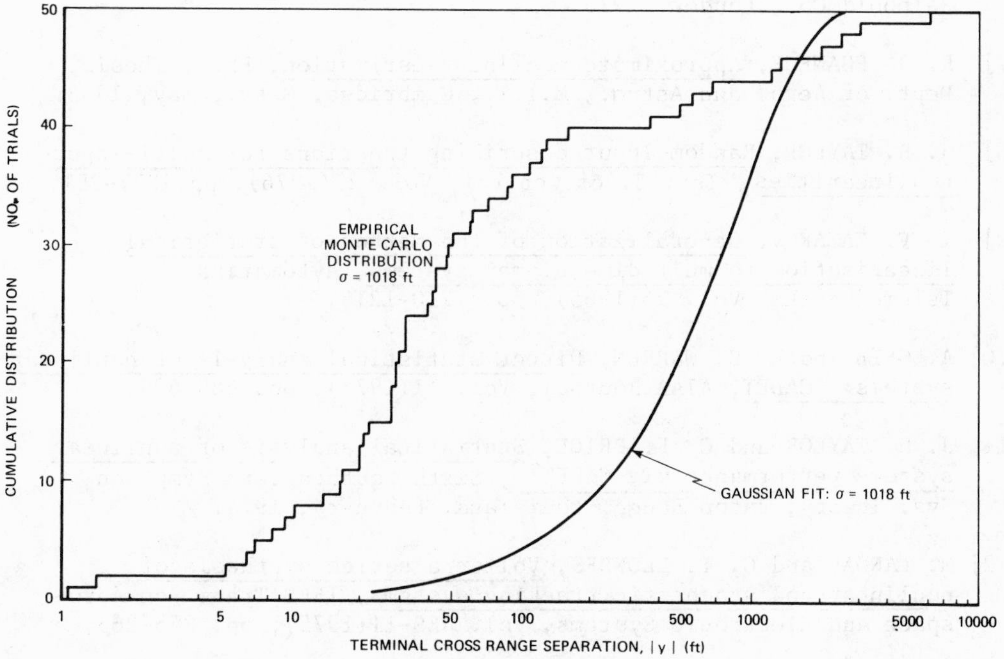

Figure 5. Empirical Probability Distribution Plot for Tactical Missile Miss Distance.

REFERENCES

[1] J. H. TAYLOR, On the credibility of the monte carlo method for nonlinear systems analysis, Tenth Pittsburgh Modeling and Simul. Conf., Vol. 10(1979), pp. 569-575.

[2] J. H. TAYLOR, C. F. PRICE, J. SIEGEL, and A. GELB, Covariance analysis of nonlinear stochastic systems via statistical linearization, ASME Winter Ann. Meeting 1976 (To appear in J. K. HEDRICK, and H. M. PAYNTER (Eds.), Nonlinear system analysis and synthesis: Vol. 2, American Society of Mechanical Engineers, New York, 1980).

[3] A. H. JAZWINSKI, Stochastic Processes and Filtering Theory, Academic Press, New York, 1970.

[4] J. BEAMAN, Accuracy of statistical linearization methods, These proceedings.

[5] A. GELB and W. E. VANDER VELDE, Multiple-input Describing Functions and Nonlinear System Design, McGraw-Hill Book Co., New York, 1968.

[6] D. P. ATHERTON, Nonlinear Control Engineering, Van Nostrand Reinhold Co., London, 1975.

[7] R. J. PHANEUF, Approximate nonlinear estimation, Ph.D. Thesis, Dept. of Aero. and Astro., M.I.T., Cambridge, Mass., May, 1968.

[8] J. H. TAYLOR, Random-input describing functions for multi-input nonlinearities, Int. J. of Control, Vol. 23(1976), pp. 277-281.

[9] I. E. KAZAKOV, Generalization of the method of statistical linearization to multidimensional systems, Avtomatika i Telemekhanika, Vol. 26(1965), pp. 1210-1215.

[10] A. GELB and R. S. WARREN, Direct statistical analysis of nonlinear systems: CADET, AIAA Journal, Vol. 11(1973), pp. 689-694.

[11] J. H. TAYLOR and C. F. PRICE, Statistical analysis of nonlinear systems performance via CADET™, Sixth Southeastern Symp. on Sys. Theory, Baton Rouge, Louisiana, February, 1974.

[12] M. LANDAU and C. T. LEONDES, Volterra series synthesis of nonlinear and stochastic tracking systems, IEEE Trans. on Aerospace and Electronic Systems, Vol. AES-11(1975), pp. 245-265.

[13] J. H. TAYLOR, Comment on 'Volterra series synthesis of nonlinear stochastic tracking systems,' IEEE Trans. on Aerospace and Electronic Systems, Vol. AES-14 (1978), pp. 390-393.

Experimentally Derived Reformulation of the Wheelset Nonlinear Hunting Problem

Larry M. Sweet*

Abstract. Recent measurements of the limit cycle stability of a single railroad vehicle wheelset reveal behavior that is different from that expected from presently available mathematical analysis. In the presence of track irregularity inputs only, the wheelset exhibits stable limit cycle oscillations at velocities above the linear critical speed. In contrast to behavior predicted neglecting nonlinearities in the creep forces, limit cycles are observed at subcritical velocities when preceeded by large initial conditions. Non-Gaussian amplitude probability distributions are presented which are useful in quasi-linearization analysis of wheelset hunting.

Introduction. The analysis of nonlinear oscillations of a railroad vehicle wheelset known as hunting has become the focus of numerous recent investigations [1-8]. From an engineering viewpoint, this interest results from the fact that understanding of the dynamic characteristics of this primitive component of rail vehicle systems yields insights into more complex complete vehicle phenomena. The problem is potentially interesting in the system theory and applied mathematics contexts in that it provides a classic practical case study for which prediction of global or limit cycle stability is of significant importance. This short paper briefly reviews results of recent experimental research on wheelset dynamics conducted at Princeton University that may have a significant input on future analytical approaches to the wheelset nonlinear hunting problem.

---

*Associate Professor and Associate Director, Transportation Program, Department of Mechanical and Aerospace Engineering, Princeton University, Princeton, N.J. 08544. Supported under DOT Contract DOT-OS-60147.

The system under study, shown in Figure 1, consists of a single wheelset (two wheels rigidly mounted on a solid axle) with lateral, vertical, roll, and yaw degrees of freedom, in addition to axle rotation. Linear springs restrain the wheelset in the lateral and yaw directions, with negligible bearing friction. The apparatus used is designed to reproduce, in scale, all significant nonlinear forces due to wheel/rail interactions, including contact geometry, gravitational stiffness, and creep forces [1]. In related research, the wheelset scale model has been used to measure the nonlinear wheel/rail contact forces in steady state and criteria for predicting derailment [2].

Limit Cycle Phenomena Observed in Experiments.

The dynamic behavior of the wheelset is represented by the following equations [3]:

$$M\ddot{y} + 2 \cdot F_L \left[\frac{\dot{y}}{V}, \psi\right] + K_y\, y + F_g\, [y - \delta] = 0 \tag{1}$$

$$I\ddot{\psi} + 2\,\ell \cdot F_T \left[\alpha_e\, (y-\delta), \ell\frac{\dot{\psi}}{V}\right] + K_\psi\, \psi = 0 \tag{2}$$

The lateral force F_g due to wheel/rail contact angle differences and the effective conicity α_e are nonlinear functions of the relative position y of the wheelset with respect to the rail δ, and the forces F_L and F_T are nonlinear cross-coupling terms due to creepage, or slip, between wheels and rails. The dependence of the $\dot{y}$ and $\dot{\psi}$ terms on forward velocity V leads to an instability in the form of a Hopf bifurcation at critical speed V_c [8].

While the critical speed may be predicted from a purely linear analysis, understanding of the behavior of the nonlinear system is important, since the existence of stable limit cycles above the critical speed or instabilities below it are of major practical importance. Results of experiments with a single wheelset model on very straight track at Princeton University demonstrate the possibilities. In Figure 2 time histories of the wheelset lateral displacement relative to the rail are shown for experiments over the same track section. At velocities below the linear critical speed of 9 m/sec, the rms

Figure 1 - Single Wheelset Experimental Apparatus. One Fifth Dynamically Scaled Model in 200 m Track.

amplitude of the response of the wheelset to the broadband track input remains fairly constant (Case A). Above 9m/sec a stable limit cycle of substantially larger amplitude results (Case B). The dominance of the limit cycle behavior over the response due to track input alone is shown in the phase plane plot for the stable limit cycle in Figure 3.

The time history shown in Figure 4 is startlingly different. In this experiment, conducted at 6.1 m/sec where the wheelset was found to be stable when subjected to track inputs alone, the limit cycle found previously at higher velocity is found when the wheelset is released from a state with large values for initial lateral displacement (Case C). This type of response might be encountered at the exit from a curve, where the wheels would be released from flange contact. At 3 m/sec, the wheelset is globally stable, regardless of initial condition (Case D).

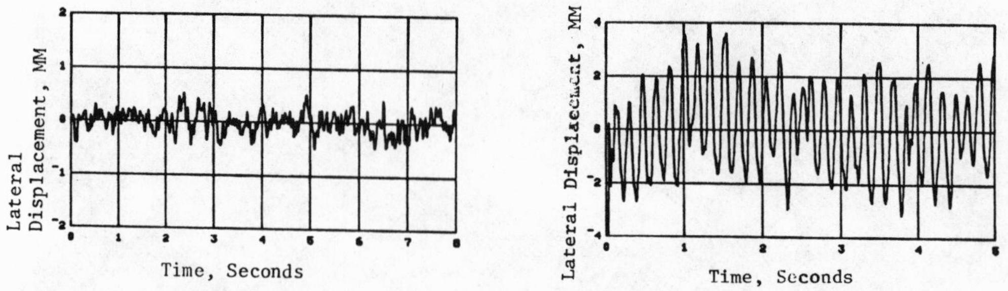

Figure 2 - Wheelset Lateral Motion for Case A (left) and Case B (right).

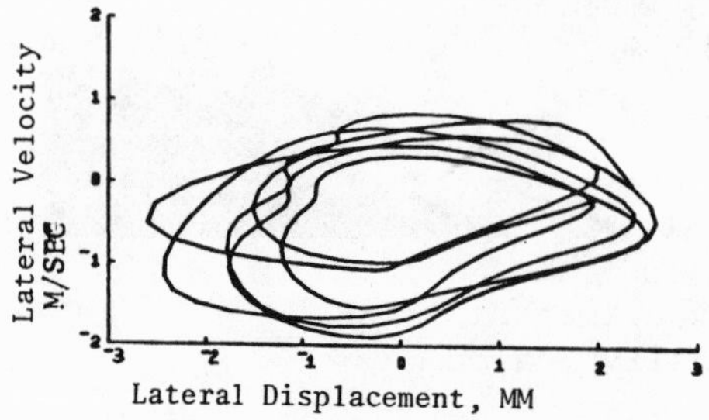

Figure 3 - Phase Plane Trajectory During Limit Cycle Oscillation. Frequency of Oscillation about 6 Hz.

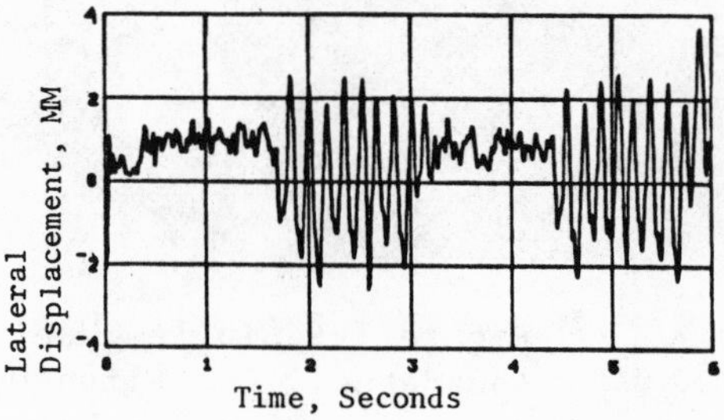

Figure 4 - Limit Cycles at Sub-Critical Velocity Due to Large Initial Condition (Case C).

The behavior represented by these cases is summarized in Figure 5. The solid lines indicate stable equilibria or periodic orbits; the dashed lines are solutions to Eqs. (1) and (2) which are unstable to perturbations in system parameters or track disturbance inputs. The predicted limit cycle amplitudes result from application of describing function analysis to the F_g and α_e terms in Eqs. (1) and (2), and assuming that the creep forces functions are linear [6,7],

$$F_L = f_L \left(\frac{\dot{y}}{V} - \psi\right) \qquad (3)$$

$$F_T = f_T \left(\frac{\ell\dot{\psi}}{V} + \alpha_e\right) \qquad (4)$$

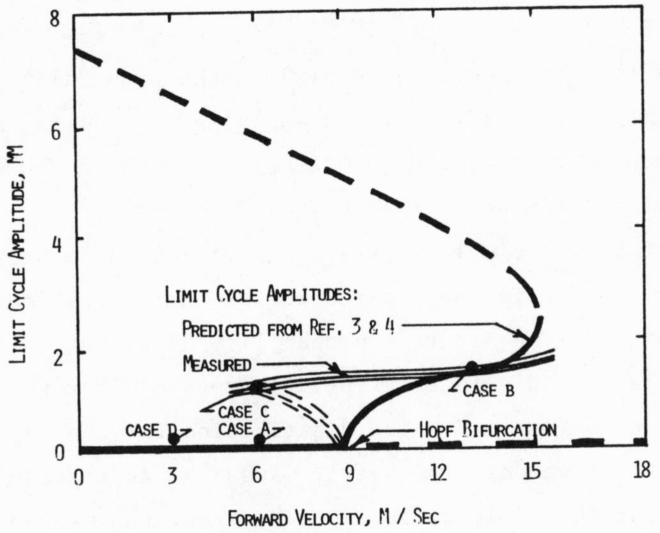

Figure 5 - Behavior of Stable (Solid Lines) and Unstable (Dotted Lines) Limit Cycles.

These widely used approximations to the complete nonlinear problem yield limit cycle behavior which is qualitatively different from that measured in the subcritical velocity range, as shown in Figure 5. Recent results for wheelsets with different wheel profiles (and hence different F_g and α_e functions), but including the F_T and F_L nonlinearities, do exhibit limit cycle behavior below V_c [3,4]. The experimental evidence indicates, therefore, that the nonlinear creep force terms should not be neglected in prediction of limit cycle behavior.

Random Variable Probability Distributions.

As seen in the responses shown in Figures 3 through 5, the presence of the random input results in displacements that are of the same order of magnitude as those of the limit cycle. Statistical linearization methods have been developed [5] to provide improved accuracy in contrast to sinusoidal describing function or Galerkin methods, by minimizing the error due to linearization over the broad range of frequencies exhibited by the signal. In this approach, the linear gain k is found from,

$$k = \frac{E[xf(x)]}{E[x^2]} = \frac{\int_{-\infty}^{\infty} xf(x)p(x)dx}{\int_{-\infty}^{\infty} x^2(p(x))dx} \tag{5}$$

where f(x) is the nonlinearity and p(x) is the probability density function of x(t). The difficulty in applying Eq. (5) is that p(x) may not be known *a priori*, and is often assumed to be Gaussian in the absence of experimental or simulated data.

The results of the single wheelset experiments help provide a better foundation for future application of the statistical linearization techniques. In Figure 6 probability distribution functions for relative lateral displacement p(y-δ) are shown for the time histories shown in Figure 3. The distribution for the experiment below the critical velocity is nearly Gaussian, as expected from the literature. For the limit cycle case, however, the probability distribution is more uniform over the range of displacements. Once p(x) is known, the non-Gaussian nature presents no conceptual difficulty in applying the technique, since Eq. (5) may be integrated numerically.

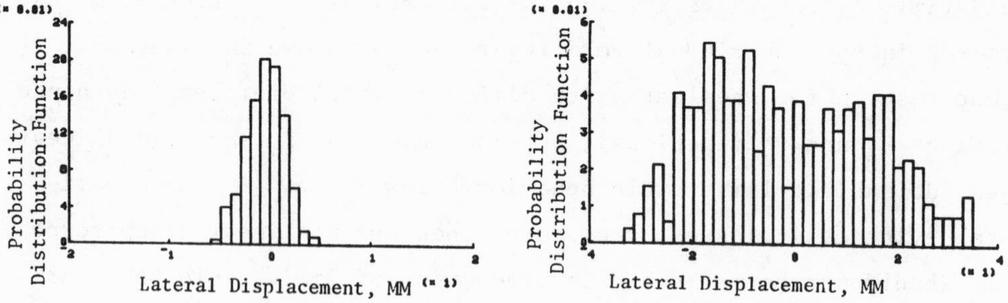

Figure 6 - Amplitude Probability Distribution Functions for Subcritical (left) and Limit Cycle (right) cases.

Summary

Limit cycle behavior of a single wheelset subject to track disturbance inputs above and below the critical velocity (Hopf bifurcation point) has been studied. The results suggest that analysis of the nonlinear wheelset hunting problem should include nonlinear creep force terms and non-Gaussian amplitude probability distributions.

REFERENCES

[1] Sweet, L.M., Sivak, J.A., and Putman, W.F., "Nonlinear Wheelset Forces in Flange Contact - Part 2: Experimental Measurements Using Dynamically Scaled Models". ASME TRans., J. DSMC, Vol. 101 (1979), pp. 247-255.

[2] Sweet, L.M. and Karmel, A. "Wheelset Derailment Criteria Under Steady Rolling and Dynamic Loading Conditions." Dynamics of Vehicles on Roads and Tracks, Swets and Zeitlinger, B.V., Amsterdam, Neth., 1980.

[3] Moelle, D., Steinborn, H., and Gasch, R., "Computation of Limit Cycles of a Wheelset Using a Galerkin Method." Ibid, 1980.

[4] Hauschild, W., "The Application of Quasilinearization to the Limit Cycle Behaviour of the Nonlinear Wheel-Rail System." Ibid., 1980.

[5] Hedrick, J.K. and Arslan, A.V., Nonlinear Analysis of Rail Vehicle Forced Lateral Response and Stability". ASME Trans., J. DSMC, Vol. 101 (1979), pp. 230-237.

[6] Hannelbrink, D.N., Lee, H., Weinstock, H., and Hedrick, J.K., "Influence of Axle Load, Track Gage, and Wheel Profile on Rail-Vehicle Hunting". ASME Trans., J. Engr. Ind. Vol. 99 (1977), pp. 186-196.

[7] Garg, D.P., "Wheelset Lateral Dynamic Analysis Using the Describing Function Technique" ASME Paper No. 77-DET-149, September, 1979.

[8] Huigol, R.R., "Hopf-Friedricks Bifurcation and the Hunting of a Railway Axle." Q. of Appl. Math. (1979), pp. 85-95.

Bifurcation in Nonlinear Stochastic Systems

S. T. Ariaratnam*

Abstract. The response of a nonlinear parametrically excited stochastic system is investigated to illustrate the phenomenon of stochastic bifurcation and the possibility of errors in the use of the Gaussian closure technique for the approximate analysis of such systems.

1. Introduction. The paper deals with the response of a single degree-of-freedom nonlinear oscillator subjected to stochastic parametric excitation. The purpose of the investigation is two-fold: first, to illustrate a method of examining bifurcational instability in nonlinear stochastic systems, and second, to point out the possibility of error in the use of the Gaussian closure technique for the analysis of problems of parametric excitation.

The system considered is governed by the differential equation

$$\ddot{x} + (\alpha + \beta x^2)\dot{x} + [1 + f(t)]x = 0 \tag{1}$$

where α, β are small positive constants of the same order of smallness as that of the spectral density of the stochastic excitation process $f(t)$. An equation of this form often arises in problems in the dynamic stability of elastic structures. In the present analysis, it is assumed that $f(t)$ is a sufficiently wide-banded stationary process so that it may be approximated by a white noise process. Eq. (1) may then be replaced by the following pair of stochastic differential equations:

---

*Solid Mechanics Division, University of Waterloo, Waterloo, Ontario, Canada, N2L 3G1. The research for this paper was supported (in part) by the National Science and Engineering Research Council of Canada through Grant No. A-1815.

(2)
$$dx_1 = x_2 dt$$
$$dx_2 = -x_1 dt - (\alpha + \beta x_1^2) x_2 dt - S^{1/2} x_1 dw$$

where $x_1 = x$, $x_2 = \dot{x}$, $w(t)$ is the Wiener process, and S is the spectral density of $f(t)$. It is immaterial whether Eq. 2 is regarded in the sense of Stratonovich [1] or in the sense of Itô [3] since, in this particular case, the so-called Wong and Zakai [5] correction terms to the drift vector are identically zero. The moment stability and the steady-state mean square response of the system governed by Eq. (2) will be investigated.

2. Analysis. Treating Eq. (2) in the sense of Stratonovich and transforming to new variables a, ϕ by the relations

(3)
$$x_1 = a \cos\Phi, \quad x_2 = -a \sin\Phi, \quad \Phi = t + \phi(t)$$

leads to the pair of Stratonovich equations

(4)
$$da = -(\alpha + \beta\, a^2 \cos^2\Phi)\, a \sin^2\Phi\, dt + a \sin\Phi \cos\Phi\, dw$$
$$d\phi = -(\alpha + \beta\, a^2 \cos^2\Phi) \sin\Phi \cos\Phi\, dt + \cos^2\Phi\, dw$$

To a first approximation, the first of these equations may be replaced by the following Itô equation using the stochastic averaging method of Stratonovich [2] and Khas'minskii [4]:

(5)
$$da = \left[\frac{3S}{16} - \frac{1}{2}\left(\alpha + \frac{1}{4}\beta\, a^2\right)\right] a\, dt + \left(\frac{S}{8}\right)^{1/2} a\, dw$$

It is evident that, since the above equation for the amplitude $a(t)$ is decoupled from that for the phase $\phi(t)$, the amplitude process $a(t)$ constitutes a Markov diffusion process by itself. The transition probability density $p(a,t/a_o,t_o)$, $t>t_o$ of $a(t)$ is then governed by the Fokker-Planck equation

(6)
$$\frac{\partial p}{\partial t} = -\frac{\partial}{\partial a}\left[\left\{\frac{3S}{16} - \frac{1}{2}\left(\alpha + \frac{1}{4}\beta\, a^2\right)\right\} a\, p\right] + \frac{S}{16}\frac{\partial^2}{\partial a^2}(a^2 p).$$

The analytical solution of Eq. (6) in closed form is not available. However, the stationary solution $p(a)$ corresponding to $\partial\phi/\partial\, t = 0$ can be obtained and is found to be

(7) $$p(a) = C\, a^{2k-1} \exp(-\beta a^2/S)$$

where

(8) $$C = (\beta/S)^k\, \Gamma^{-1}(k), \quad k = 1 - (4\alpha/S),$$

provided $k>0$, i.e. $S>4\alpha$. The stationary mean square amplitude is given by

(9) $$< a^2 > = (S - 4\alpha)/2\beta$$

provided $S > 4\alpha$. For $S < 4\alpha$, the stationary solution for $< a^2 >$ is the trivial solution which can be shown to be stable in mean square. $S = 4\alpha$ corresponds to a point of bifurcation at which the trivial solution is no longer stable. This result is shown by the solid lines in the figure. It will now be compared with that obtained by the method of Gaussian closure.

3. Gaussian Closure. The following notations for the moments are used:

$$m_i = \langle x_i \rangle,\ m_{ij} = \langle x_i x_j \rangle,\ m_{ijj} = \langle x_i x_j^2 \rangle, \text{ etc., } i,j = 1,2 .$$

From Eq. (2), by the use of the Itô differential rule, the following differential equations governing the first-and the second-order moments are obtained:

(10) $$\begin{aligned}
\dot{m}_1 &= m_2 \\
\dot{m}_2 &= m_1 - \alpha m_2 + \beta m_{112} \\
\dot{m}_{11} &= 2m_{12} \\
\dot{m}_{12} &= m_{22} - m_{11} - \alpha m_{12} - \beta m_{1112} \\
\dot{m}_{22} &= - 2m_{12} - 2\alpha m_{22} - 4\beta m_{1122} + S m_{11}
\end{aligned}$$

It may be noted that the equations for the lower order moments involve the higher order ones. Hence, to obtain a closed system, a scheme for truncating the equations is often adopted. One such scheme is the Gaussian closure method in which moments of order higher than the second are expressed in terms of lower order moments using the relationships connecting the moments of a set of Gaussian random variables. This approximation yields useful results for weakly nonlinear systems subject to external Gaussian excitation, since, in the absence of the nonlinearity the response is Gaussian and hence the true response when the nonlinearity is present may be regarded as nearly Gaussian. However, its use by some authors [6], [7] in parametrically excited systems is somewhat questionable since in the linear case the parametric response is not Gaussian. To illustrate this, if the Gaussian relationships

(11) $$m_{112} = 0,\ m_{1112} = 3m_{11}m_{12},\ m_{1122} = m_{11}m_{22} + 2m_{12}^2$$

are used to truncate Eq. (9), one obtains the set of equations

(12) $$\begin{aligned} \dot{m}_1 &= m_2 \\ \dot{m}_2 &= m_1 - \alpha m_2 \\ \dot{m}_{11} &= 2m_{12} \\ \dot{m}_{12} &= m_{22} - m_{11} - \alpha m_{12} - 3\beta m_{11}m_{12} \\ \dot{m}_{22} &= -\ 2m_{12} - 2\alpha m_{22} - 4\beta m_{11}m_{22} - 8\beta m_{12}^2 + Sm_{11} \end{aligned}$$

whose non-trivial stationary solution is

(13) $$\begin{aligned} m_1 &= m_2 = m_{12} = 0 \\ m_{11} &= m_{22} = (S - 2\alpha)/4\beta\ . \end{aligned}$$

The stationary mean square amplitude is, from Eqs. (3) and (13), given by

(14) $$< a^2 > = m_{11} = m_{22} = (S - 2\alpha)/2\beta$$

This result, shown by the dotted curve in the figure, shows the solution bifurcating from a different point.

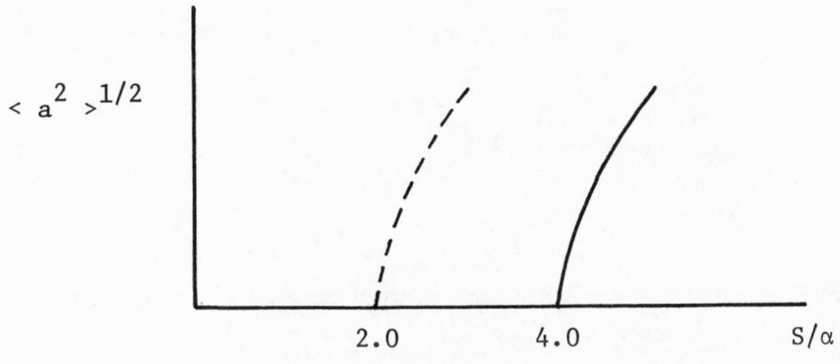

REFERENCES

[1] R. L. Stratonovich, A New representation for stochastic integrals and equations, SIAM J. Control, 4 (1966), pp. 363-371.

[2] R. L. Stratonovich, Topics in the theory of random noise, Vol I. Gordon and Breach, New York, (1967).

[3] K. Itô, On stochastic differential equations, Memoirs of the American Mathematical Society, 4, (1961).

[4] R. Z. Khas'minskii, A limit theorem for solutions of differential equations with a random right-hand side, Theory Prob. Applications, 11 (1966), pp. 390-406.

[5] E. Wong and M. Zakai, On the relation between ordinary and stochastic differential equations, Int. J. Engrg. Sci, 3 (1965), pp. 213-229.

[6] V. V. Bolotin, Stochastic stability and reliability of structures, Chapter 11 in Study No. 6, Stability, Ed: H. Leipholz, Solid Mechanics Division, University of Waterloo, Ontario, (1972), pp. 385-422.

[7] F. Weidenhammer, Biegeschwingungen des stabes unter axial pulsierender zufallslast, VDI-Berichte Nr. 135 (1969), pp. 101-107.

SECTION EIGHT
STRANGE ATTRACTORS

Chaos and Bijections Across Dimensions

Otto E. Rossler*

Abstract. Hopf's baker's transformation generates a bijection between a square and an (infinitely often cut) line, in the limit of infinite iteration number. So does the 'contracting baker's transformation.' The attractor formed under this condition is a 'non-sink attractor.' The Lorenz attractor and the non-Lorenz attractor form examples. The 'three-dimensional baker's transformation' generates a bijection between 3 and 2 dimensions and, under time reversal, a bijection between 3 dimensions and one. The latter map can be realized as an area-contracting diffeomorphism (generating a 4-solenoid). Chaos-generating bijections across dimensions can be contracting or non-contracting, maximal or non-maximal, pure or diluted, full or rarefied, regular or singular, finite or infinitesimal. The set of maximal (n-1 dimensional) non-sink attractors de fines a natural hierarchy. It may be used to classify chaotic phenomena.

1. Introduction. Bijections across dimensions were first demonstrated by Cantor (1,2). Brouwer (3) proved that such bijections are necessarily discontinuous. They may, nonetheless, arise in a continuous map in the limit of infinite iteration number.

A second result will be that a map generating a bijection across dimenions through iteration need not be area-preserving. As a consequence, a new class of attractors ('non-sink attractors') emerges.

In the following, some simple maps generating bijections across dimensions will be considered.

2. Generalized Baker's Transformation. The baker's transformation is illustrated in Figure 1. A slab of dough is being rolled out, then chopped into two pieces, then rearranged (Figure 1a, b). Alternatively,

*Institute for Physical and Theoretical Chemistry, University of Tubingen, Auf der Morgenstelle 8, 7400 Tubingen, F.R.G., and Institute for Theoretical Physics, University of Stuttgart, 7000 Stuttgart-80, F.R.G.

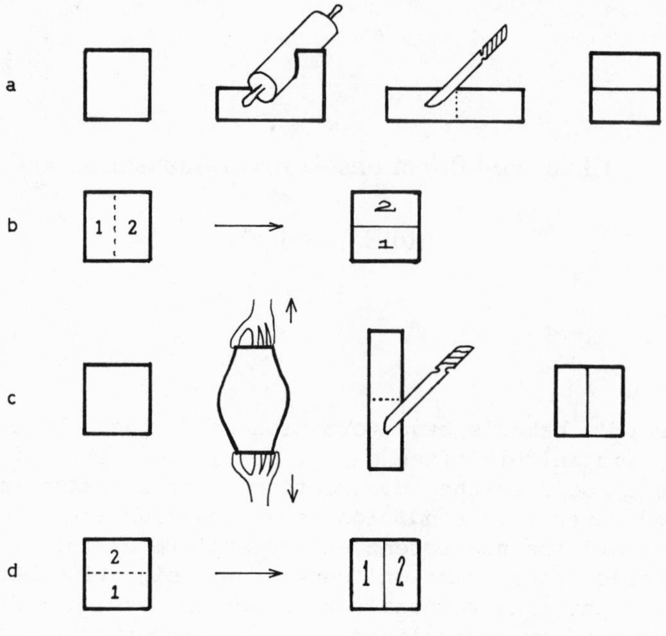

Figure 1: The baker's transformation, a) in detail, b) symbolic. c) Inverse baker's transformation, d) symbolic.

a slab of dough is being stretched, then cut into two pieces, then rearranged (Figure 1c, d). The two processes are mutually identical under time reversal. Each process may be repeated indefinitely.

The baker's transformation has been invented by Eberhard Hopf (ref. (4), p. 42) as an example of a 'mixing-transformation' in ergodic theory. The baker's transformation is strongly mixing (5). It moreover gives rise to an entropy-like observable (6).

In the present context, one particular property of the baker's transformation is of interest: it in the limit of infinite iteration number generates an uncountable stack of line segments covering the original square. The map thus in the limit generates a bijection between a plane and an (infinitely often cut) line.

This bijection between the plane and the (infinitely often cut) line is different from the trivial bijection of the same shape that is being produced by first cutting the square through the middle, then cutting the remaining halves through the middle, and so forth. The latter bijection breaks down completely as soon as the slightest amount of contraction is admitted at each step; moreover, it cannot be generated through iterating one and the same process.

The baker's transformation is, with its main properties, insensitive to the introduction of a loss of volume during each step. As shown in Figure 2, a 'contracting baker's transformation' generates two strips

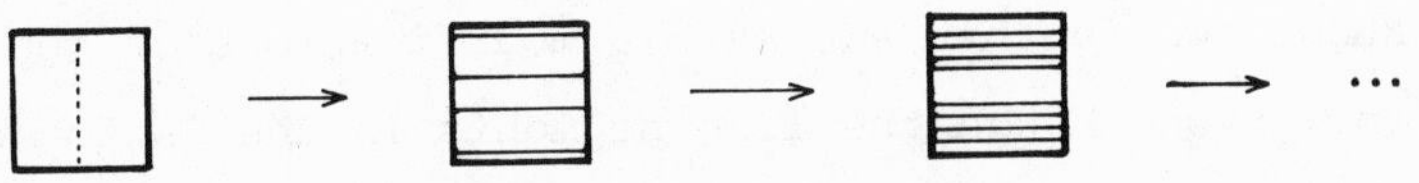

Figure 2: Contracting baker's transformation.

at the first iterate, four at the second, and so on. In the limit, still an uncountable number of line segments is being generated, namely a Cantor set. The fact that the ordinate in the limit now is a Cantor set rather than the full interval does not destroy the bijective relationship (2) between the two limiting images. Thus, the two baker's maps admit a functor (7) which is an isomorphism such that, through all iterates, the diagram of composed maps commutes. If one of the two baker's transformations is a bijection in the limit, so is the other.

Any area-contracting map which remains a bijection in the limit of infinite iteration number generates a non-sink attractor. For on the one hand, an attracting set of measure zero is being formed in the limit; on the other, the map in the limit still determines a bijection between this attracting set and the original domain (except for subsets of zero measure). The attractor formed therefore differs from 'ordinary' attractors. Ordinary attractors (point attractors, limit cycles, toroidal attractors) are sinks, meaning that 'transient' motions leading toward the attractor die out upon reaching the attractor. This happens because the latter is an invariant set for both positive and negative t (8), that is, consists wholly of 'non-transient' trajectories. In contrast, the present attractor is not a (both positively and negatively) invariant set under the map. It therefore is not a sink, eating transients, but rather leaves an attracted higher dimensional neighborhood fully invertible even after transfinite <u>t</u> . The attractor in this respect behaves just like the 'limiting configuration' formed in a strongly mixing Hamiltonian system, which too cannot be pinpointed.

The two properties of 'lack of attractors' and 'indefinite invertibility,' characteristic of conservative systems, thus need not be lost simultaneously in the transition toward nonconservativity.

Figure 3 shows two further admissible modifications of the baker's transformation. Even though the line segments in the Cantor set that in

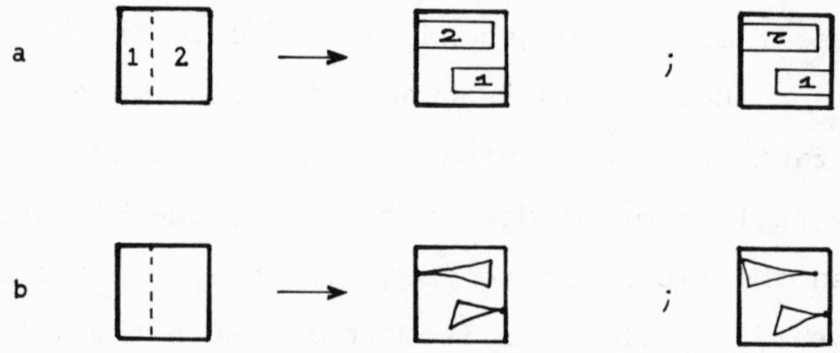

Figure 3: Contracting baker's transformation with displaced pieces, a) normal; top-inverted, b) with singularity; top-inverted.

the limit is formed in the map of Figure 3a no longer have equal lengths, the bijective relationship under the map between domain and image, discussed above, is preserved. This can be shown by choosing an appropriate line (for example, a stable manifold) along which to enumerate the pieces. The argument carries over to the two 'sandwich maps' (looking as if prepared with a dull knife (9)) of Figure 3b. The first of these (9,10) generates a Lorenz attractor (10,11), after (12); the second, top-inverted, case (13) generates a related attractor which may be termed the non-Lorenz attractor. Both attractors can be found in cross-sections through flows generated by ordinary differential equations of a realistic kind; see (14) for two abstract reaction kinetic examples.

<u>3. Higher-order Baker's Transformation</u>. There are two kinds of 3-dimensional baker's transformations. The 'trivial' kind consists of a

composition of the two-dimensional baker's transformation with an area-preserving one-dimensional map (identity map; rotation map). The non-trivial case is depicted in Figure 4. Under positive $\underline{t}$, a 'stack of

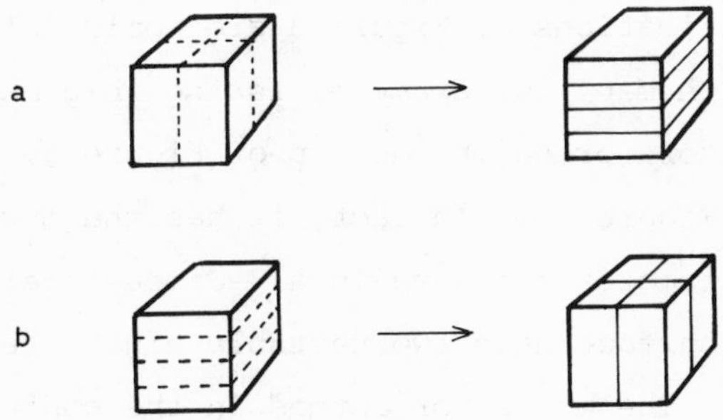

Figure 4: Three-dimensional baker's transformation, a) ordinary case, b) inverse case.

trays' is generated; under negative $\underline{t}$, a 'bunch of pencils.' The former map (Figure 4a) may also be called the pancake map; the latter, the noodle map. The pancake map indeed is in everyday use in bakeries, because fluffy pastry (cf. (4)) is being prepared according to the same principle: rolling out the dough in two directions, laying the pieces over, rolling out again, and so forth. The noodle map, in contrast, has apparently not been tried yet.

There are two nontrivial contracting three-dimensional baker's transformations (Figure 5). These two maps are, just as their two-dimension-

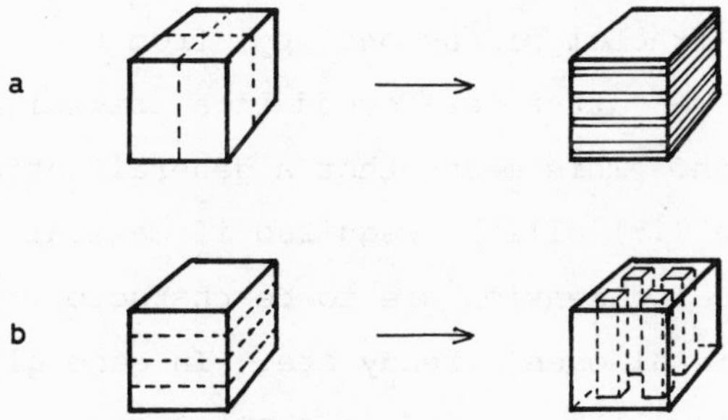

Figure 5: Contracting three-dimensional baker's transformations (compare Figure 4). a) Pancake version, b) noodle version.

al analogues (see Figure 2 for one of them), no longer mutually identical under time reversal. The former (Figure 5a) generates a bijection from three to two dimensions, the latter (Figure 5b) in the limit is a bijection between three dimensions and one.

Again, all the modifications of Figure 3 are admissible. The pancake analogue to the sandwich map, for example, is the 'big map' (11).

The non-sink attractor formed in the map of Figure 5a is 'thicker' than that produced in Figure 5b. In fact, it has the maximum dimensionality (two) that an attractor can have in a 3-dimensional map. Points inside the attracting surface have two unstable eigendirections (characteristic exponents). The attractor formed in the contracting noodle map (Figure 5b), by contrast, is not very different from the - also one-dimensional - attractor formed in the two-dimensional contracting baker's transformation of Figure 2. Its internal instability is confined to one dimension. The attractor formed in the contracting pancake map is a maximal non-sink attractor; the noodle map attractor is nonmaximal.

In terms of fractal theory (15), the attractor formed in the noodle map of Figure 5 has a fractal dimensionality between unity and two, while the attractor of the corresponding pancake map (Figure 5a) has a fractal dimensionality between two and three. The attractor formed in the noodle map thus again does not differ from that formed in a lower-dimensional map: the attractor formed in the map of Figure 2 also has a fractal dimensionality between 1 and 2 (see (15)). On the other hand, the maximal attractor formed in the pancake map of Figure 5a is indistinguishable (in its fractal dimensionality) from the - classically also two-dimensional - attractor formed in a trivial 3-dimensional baker's transformation. This means that a generalization of Mandelbrot's classification (15) will be required if maximal attractors of more than one classical dimension are to be characterized as effectively as the one-dimensional ones already are. In general, more than one real number (fractal) will be needed to characterize a maximal attractor; the attractor of Figure 5a, for example, needs two.

Proceeding to four dimensions, it is easy to see that now two pairs of nontrivial baker's transformations exist: a map generating a bijection from four to three dimensions; its inverse, generating a bijection from four dimensions to one; and two bijections from four to two dimensions which are each other's inverses. The modifications of Figure 3 may be applied again. Similarly for the fifth dimension, and so forth.

Thus, a whole hierarchy of non-sink and maximal attractors emerges.

4. Smooth Baker's Transformation. The modification of Figure 5b shown in Figure 6a (gentle twist by 90 degrees) does not affect the for-

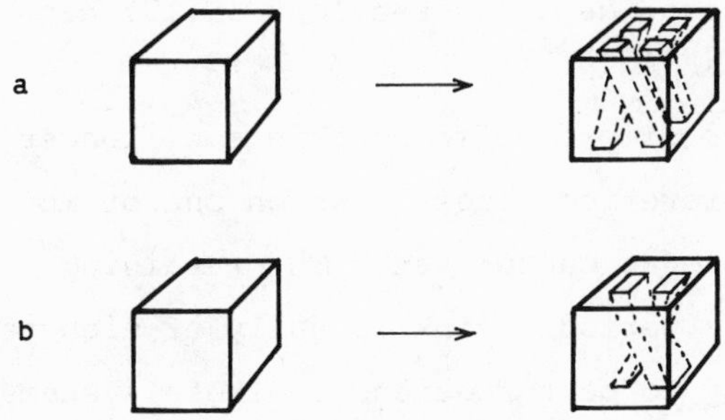

Figure 6: Smooth contracting 3-dimensional baker's transformation: a) 4-solenoid. b) Analogous classical solenoid. See text for explanation.

mation of a non-sink attractor. The new map nonetheless possesses a novel property: it is no longer necessarily discontinuous. If the top and the bottom of the cube in Figure 6a are identified (so that the cube becomes a solid 3-dimensional annulus), the map of Figure 6a is a diffeomorphism.

This diffeomorphism is closely related to the diffeomorphism described by Smale (16) as generating a 'solenoid.' Smale's map is depicted in Figure 6b. It is obtained by taking a solid annulus of dough, elongating it, thinning it, wrapping it up once, and putting it back. The diffeomorphism of Figure 6a is obtained in the same way, except for one more wrapping-up. The 4-solenoid is a non-sink attractor.

Smooth versions exist also for the pancake map of Figure 5a and its higher-dimensional analogues. They are all higher-order versions of Plykin's map (17) which is a smooth analogue to the map of Figure 2. It is shaped like a mitten and consists of 5 smoothly connected pieces which have to be considered separately in order to understand how the map is working (18). The 'hypermitten' (being analogous to Figure 5a), accordingly needs 25 pieces for its specification, and so on. Fortunately, both the original mitten and its higher version possess simpler relatives: the 'walking-stick map' (like the right-hand map of Figure 3a, but with the 2 pieces connected by a smooth bridge) and the 'folded towel map' (which is the one actually used in bakeries). For computer-generated pictures of these two maps produced by simple difference equations, see (19). A sound movie displaying simple differential equations possessing such maps as cross-sections (9,19) has been shown at this conference.

Interestingly, these simple diffeomorphisms no longer generate 'full' bijections, but only 'rarefied' ones in which one or more axis is replaced by a one-dimensional Cantor set. The remaining 'inset' of finite measure then contains a periodic sink (usually of high periodicity and with strong sensitivity to perturbations). These systems therefore act as stochastic monoflops (9,20).

5. Implications for Chaos Theory. Chaos - that is, unstable erratic trajectorial behavior - occurs in all of the maps considered above. To see this, one only has to choose an arbitrary initial point in the left-hand domain of any of the Figures above, and follow up its course.

This permits the hypothesis that deterministic chaos in general is a manifestation of a bijection across dimensions. On the one hand, Brouwer's (3) discontinuity principle explains that every point in such a bijection is bound to have a very different 'fate' as the bijection is being 'woven' step by step in time. On the other hand, it seems that all types of chaos known so far can be explained by this weaving process.

There are, of course, a number of special distinctions possible: The bijection is not in all cases 'maximal'; or 'pure' (without trivially behaving directions added to the map); or 'full' (without interwoven inset); or 'regular.' The latter distinction permits to include the well-known chaos-generating 'endomorphisms with overlap' (like the logistic difference equation (21)) as singular limiting cases, cf. (9). Another singular limiting case is the 'infinitesimally small' chaos-generating map, cf. (22).

The suggested 'reduction' allows one to define 'maximal chaos' as a function of dimensionality. Maximal attractors and their less ideal analogues define the most complicated type of chaos possible in a system of given dimensionality. One set of examples: the maximal non-sink attractor formed in the 3-dimensional map of Figure 5a; its rarefied analogue formed in the folded towel map; and the latter's 'ironed flat' version (as generated, for example, by two coupled logistic difference equations).

The set of maximal non-sink attractors thus defines a natural hierarchy of 'levels of chaos.'

REFERENCES

(1) G. CANTOR, A contribution to the theory of manifolds (in German), Journ. f. Math., 84(1878), pp. 242-258.

(2) G. CANTOR, On infinite linear point-manifolds V (in German), Math. Ann., 21(1883), pp. 545-591.

(3) L. E. J. BROUWER, Proof of the invariance of dimension number (in German), Math. Ann., 70(1911), pp. 161-165.

(4) E. HOPF, Ergodic Theory (in German), Berlin, Springer-Verlag, 1937.

(5) P. BILLINGSLEY, Ergodic Theory and Information, New York, Wiley, 1965.

(6) B. MISRA, I. PRIGOGINE, and M. COURBAGE, From deterministic dynamics to probabilistic descriptions, Proc. Natl. Acad. Sci. U.S.A., 76(1979), pp. 3607-3611.

(7) S. MAC LANE, Categories. For the Working Mathematician, Berlin, Springer-Verlag, 1972.

(8) J. P. LASALLE, The Stability of Dynamical Systems, Philadelphia, SIAM, 1976.

(9) O. E. ROSSLER, Different types of chaos in two simple differential equations, Z. Naturforsch., 31a(1976), pp. 1664-1670.

(10) J. GUCKENHEIMER, A strange strange attractor, The Hopf Bifurcation, eds. J.E. Marsden and M. McCracken, 368-381, Berlin, Springer, 1976.

(11) R. F. WILLIAMS, The structure of Lorenz attractors, Turbulence Seminar, Lecture Notes in Mathematics, Vol. 615, 94-112, Berlin, Springer-Verlag, 1977.

(12) E. N. LORENZ, Deterministic nonperiodic flow, J. Atmos. Sci., 20(1963), pp. 130-141.

(13) O. E. ROSSLER, Continuous chaos - four prototype equations, Bifurcation Theory and Applications in Scientific Disciplines, ed. by O. Gurel and O. E. Rossler, Ann. N. Y. Acad. Sci., 316(1979), pp. 376-392.

(14) O. E. ROSSLER and P. ORTOLEVA, Strange attractors in 3-variable reaction systems, Theoretical Approaches to Complex Systems, E. Pfaffelhuber Memorial Meeting, Lecture Notes in Biomathematics, Vol. 21, 67-73, Berlin, Springer-Verlag, 1978.

(15) B. B. MANDELBROT, Fractals: Form, Chance, and Dimension, San Francisco, Freeman, 1977.

(16) S. SMALE, Differentiable dynamical systems, Bull. Amer. Math. Soc., 73(1967), pp. 747-817.

(17) R. V. PLYKIN, Sources and sinks of A-diffeomorphisms of surfaces, Math. USSR Sbornik, 23(1974), pp. 233-253.

(18) O. E. ROSSLER, Chaos and strange attractors in chemical kinetics, Synergetics - Far from Equilibrium, ed. by A. Pacault and C. Vidal, 107-113, Berlin, Springer-Verlag, 1979.

(19) O. E. ROSSLER, Chaos, Structural Stability in Physics, ed. by W. Güttinger and H. Eikemeier, 290-309, Berlin, Springer-Verlag, 1979.

(20) J. L. KAPLAN and J. A. YORKE, Preturbulence: a regime observed in a fluid flow model of Lorenz, Commun. Math. Phys., 67(1979), pp. 93-108.

(21) F. HOPPENSTEADT and J. M. HYMAN, Periodic solutions of a logistic difference equation, SIAM J. Appl. Math., 32(1977), pp. 78-81.

(22) P. HOLMES and J. E. MARSDEN, Qualitative techniques for bifurcation analysis of complex systems, Bifurcation Theory and Applications in Scientific Disciplines, ed. by O. Gurel and O. E. Rossler, Ann. N.Y. Acad. Sci., 316(1979), pp. 608-622.

Experimental Models
for Strange Attractor Vibrations in Elastic Systems

Francis C. Moon*

Abstract

The harmonically forced motion of a buckled rod and beam-plate exhibit apparantly random or chaotic motions. These mechanical models simulate the solution of one and two dimension Duffing's type nonlinear differential equations with multiple equilibrium points. Poincaré maps of the motion, synchronized with the driving force, show a complex but stable structure. A chaotic motion threshold criterion for the driving amplitude as a function of frequency is discussed.

Introduction

The classical study of random vibrations in mechanical systems assumes the existence of random inputs or forces. However, in recent years it has been shown that bounded, nonperiodic, apparently chaotic vibrations can occur in deterministic mathematical models with no random inputs. These motions have been named "strange attractors" in contrast with attracting equilibrium points or limit cycle behavior. Fourier analysis of the motions has shown a broad spectrum of frequencies in spite of the fact that the inputs were single frequency or the system was autonomous. Electrical analog computer studies have been made of these motions by Ueda [1], [2] and Holmes [3].

---

*Department of Theoretical and Applied Mechanics, Cornell University, Thurston Hall, Ithaca, N.Y.14850. Supported in part by the National Science Foundation, Grant No. Eng 76-23627.

These papers have studied the nonperiodic solutions of Duffings equation

$$\ddot{A} + \gamma\dot{A} - \frac{1}{2}(1-A^2)A = f \cos\omega t \quad . \tag{1}$$

Recently, experimental observations of strange attractor motions were reported for forced vibrations of a buckled beam (Moon and Holmes [4], Moon [5]) which exhibited all the characteristics of the mathematical model analyzed by Holmes [3].

Buckled Beam - One degree of freedom

The apparatus consisted of a thin cantilevered ferromagnetic steel plate-like beam. With the beam suspended vertically, permanent magnets placed beneath the tip of the beam created forces which buckled the beam into two or three stable equilibrium positions [4]. When the clamped base of the cantilever was vibrated by a single sinusoidal force, chaotic motions appeared for certain forcing amplitudes and frequencies (Figure 1).

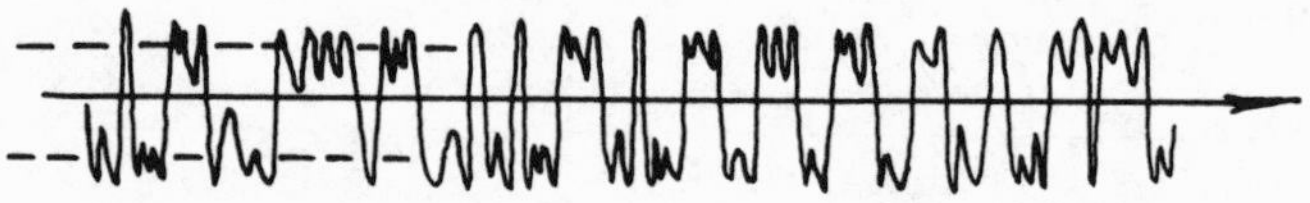

Figure 1: Time history of chaotic vibrations of a beam.

The smallest values of forcing amplitude for which the motion was chaotic were determined as a function of frequency. Experimental threshold values of the forcing amplitude, for a given driving frequency, are shown in Figure 2 for the one degree of freedom model. Holmes [3] has proposed a theoretical criterion based on a method due to Melnikov.

$$f_c = \omega^2 A_o = \frac{\gamma\sqrt{2}}{3\pi\omega} \cosh\left(\frac{\pi\omega}{\sqrt{2}}\right) \quad . \tag{2}$$

This threshold is shown in Figure 2 and gives a sufficient criterion for chaotic motion. A heuristic criterion has been proposed by the author [5] which is based on the nonlinear vibration of the beam about one of the stable equilibrium positions

$$f = \omega^2 A_o = \frac{\alpha}{2\omega}\{[(1-\omega^2) - \frac{3}{8}\frac{\alpha^2}{\omega^2}]^2 + \gamma^2\omega^2\}^{1/2} \tag{3}$$

where α is an empirical factor and $\alpha \sim O(1)$. This is also shown in Figure 2.

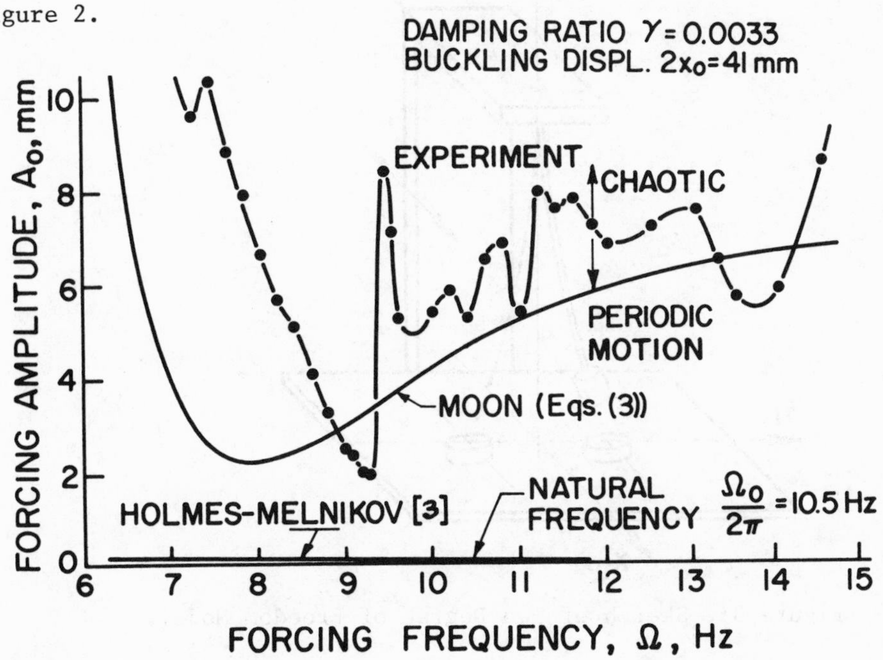

Figure 2: Threshold for Forced Chaotic Vibrations - One Degree of Freedom.

This criterion postulates a critical velocity of the mass at which the stable forced motions become chaotic. A more rigorous criterion may be related to the stability of subharmonic oscillations about one of the stable equilibrium positions which seem to occur before the chaotic motions.

Buckled Beam - Two degrees of freedom

A 5.3 gm ferromagnetic cylinder was attached to a 29.0 cm long cantilevered rod of circular cross section allowing the tip mass to move in two dimensions (Figure 3). Below the ferromagnetic tip mass, one, two or more high field permanent magnets were placed on a steel base. With two magnets, the beam could be buckled as in the one degree of freedom model for which there were two stable and one unstable equilibrium positions. The system consisting of a beam, base and magnets was vibrated in a direction normal to the line joining the three equilibrium positions in contrast to the one degree of freedom case.

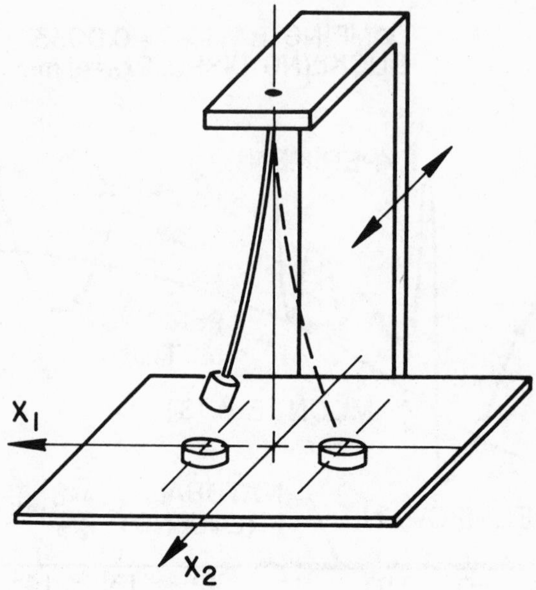

Figure 3: Sketch of Two Degree of Freedom Model.

The conservative elastic forces in this model can be represented, to a good approximation, by a quadratic potential,

$$V = \frac{1}{2} k(x^2 + y^2) \quad . \tag{4}$$

The essential features of the magnetic forces can be represented by a potential of the form

$$\mathcal{W} = \frac{1}{2} k_1 x^2 - \frac{1}{2} k_2 y^2 - \frac{1}{4} (k_3 x^4 + k_4 y^4 + 2k_5 x^2 y^2) + k_6 x + k_7 y \quad . \tag{5}$$

The combined potential is given by $V_1 = V - \mathcal{W}$. By changing the positions of the magnets, one can create a relative maximum in V_1 at the origin, leading to a static instability in the x direction. The resulting equations of motion for this multiple equilibria state, adding linear damping and an external forcing function are written below in nondimensional form.

$$(6)\qquad \begin{aligned} \ddot{x} + \gamma\dot{x} - \frac{1}{2}(1-x^2)x + \beta y^2 x &= f_2 \\ \ddot{y} + \gamma\dot{y} + \alpha(1+\varepsilon y^2)y + \beta x^2 y &= f_o + f_1\cos\omega t \end{aligned}$$

Experimentally f_o, f_2 represent components of the gravitational force on the end mass of the cantilever if the rod clamp is tilted at an angle to the vertical. The harmonic force represents the effect of vibrating the clamp and base.

When $f_o = f_2 = 0$, and f_1 is small, we may consider the y motion as sinusoidal, i.e., $y \simeq a\cos(\omega t+\phi)$ and the model represents a nonlinear Mathieu equation of the form

$$(7)\qquad \ddot{x} + \gamma\dot{x} + \frac{1}{2}[\beta a^2 - 1 + \beta a^2\cos(\omega t+\phi)]x + \frac{1}{2}x^3 = 0 .$$

Returning to the two degree of freedom equations, when $f_o = f_1 = f_2 = 0$, the equilibrium solution becomes $y = 0$, $x = 0, \pm 1$ when $\alpha, \varepsilon, \beta \geq 0$. Thus this set of equations is similar to the buckled Duffing equation (1) in that three equilibria exist. What makes (6) a nontrivial extension of (1) into two dimensions is that the forcing function does not act in the x direction, but in the y direction. Thus the x motion is parametrically excited by the y motion as shown in (7).

Unlike Duffing's equation (1), for which analytical, numerical and analog simulation have shown strange attractor behavior, there has been no comparable study of the two dimensional system (6). However the 'experimental' <u>two degree</u> of <u>freedom model</u> has <u>exhibited chaotic behavior</u> over a wide range of parameter values of f_1, ω.

The experimental natural frequencies about the stable equilibrium points were 8.3 hz in the x direction and near 6 hz in the y direction. Chaotic behavior was found over a frequency range $5.5 \leq \omega \leq 9.0$ hz. As in the one degree of freedom model, it was found that Poincaré maps of the motion showed more structure when the damping was increased. This was accomplished by vibrating the end mass in an olive oil bath. The average measured damping coefficient was

$\gamma = 0.14$.

As in the previous case, chaotic motions were observed under harmonic excitation of the base. Strain gages on the rod measured the motion normal and parallel to the excitation. A phase plane portrait of the motion normal to the base excitation, $(x,\dot{x})$, is shown in Figure 4 along with a Poincaré map synchronized with the excitation.

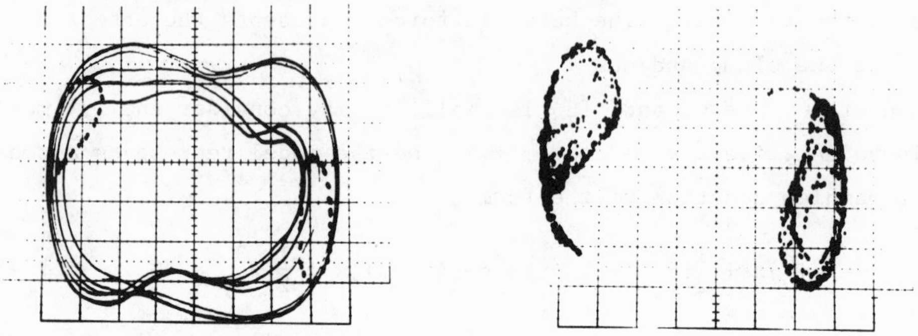

Figure 4: a) Continuous Time Traces b) Poincaré Map in the Phase Plane

While the continuous time portrait of the motion did not show any structure, Poincaré maps in the phase plane $(x,\dot{x})$ did show remarkable structure which remained stable after many thousands of excitation cycles. Of course this is only a two dimensional projection of the four dimensional phase space $(x,\dot{x},y,\dot{y})$. Poincaré maps of other projections, i.e. $(y,\dot{y})$, $(\dot{x},\dot{y})$, did not show as clear a structure as the $(x,\dot{x})$ maps.

In the two degree of freedom experiments an upper and lower chaos threshold for the driving amplitude $f_1 \equiv \omega^2 A_o$, was found for certain frequencies as shown in Figure 5. Typical Poincaré maps at the upper and lower threshold respectively are shown in Figure 6.

The Poincaré maps vary with the phase of the driving signal at which the map is synchronized. Poincaré maps for chaotic motion in two degrees of freedom at a given frequency and driving amplitude are shown in Figure 7 for 0°, 90° and 180° phase points with respect to the harmonic driving signal. By continuously changing the synchronizing phase of the Poincaré map, the structure of the strange attractor can be unravelled as sketched in Figure 8, (see e.g. [5]).

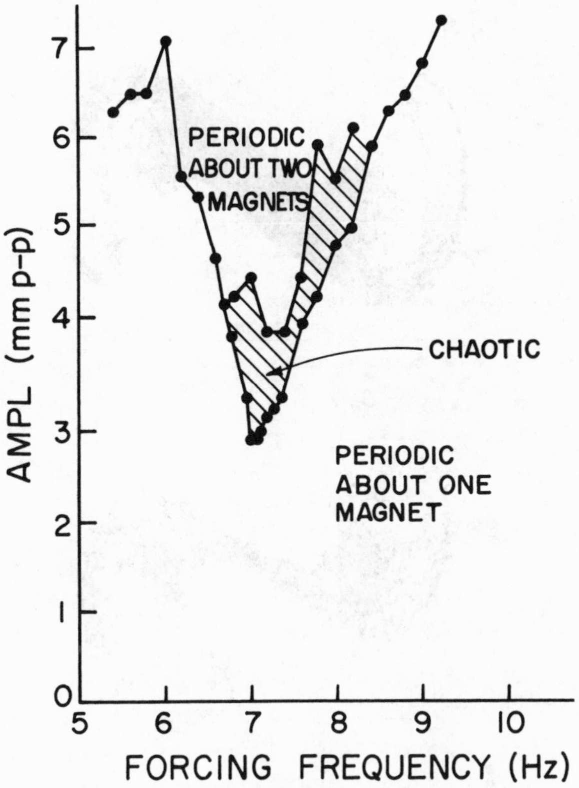

Figure 5

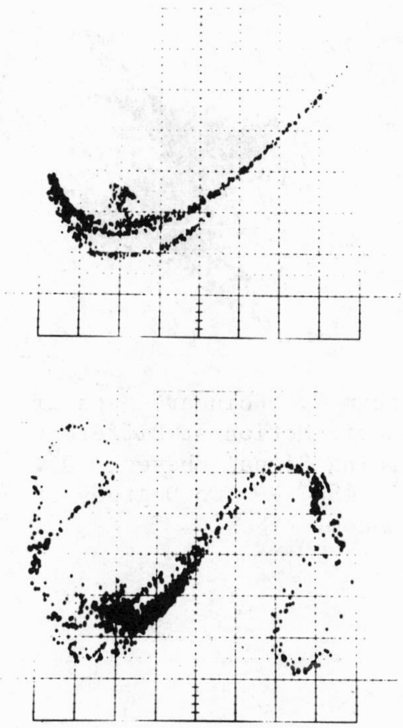

Figure 6: Poincaré Maps.

Summary

Experimental models for the study of bounded, non-periodic or chaotic ocsillations in mechanical systems otherwise known as "strange attractors" have been built. Observations of chaotic motions have been recorded for one and two degree of freedom systems under deterministic sinusoidal forces. Time varying random influences on these motions were extremely small and are not considered to be the "cause" of the resulting chaotic motions. Chaotic motions were observed under sinusoidal oscillations of the clamped end of the beam. The driving amplitude threshold for chaos was determined for different driving frequencies.

The two degree of freedom model consisted of a cantilevered beam with a soft iron cylinder at the free end. Magnets placed beneath

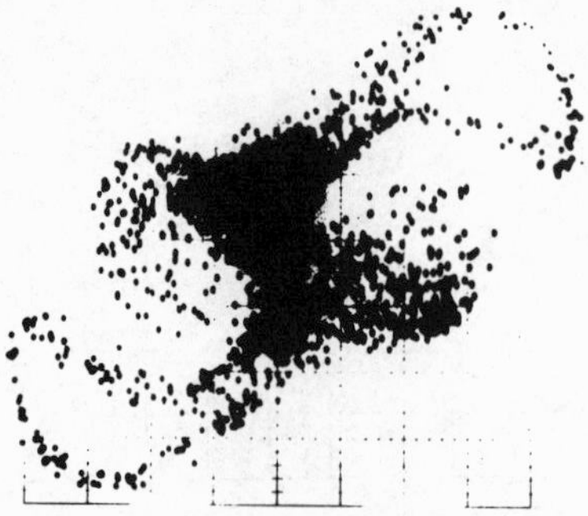

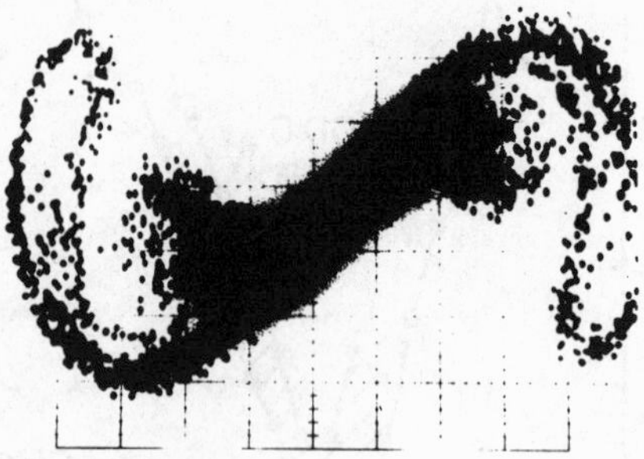

Figure 7: Poincaré Maps of Chaotic Motion at Different Forcing Signal Phases; 0°, 90°, 180° -- Two Degrees of Freedom.

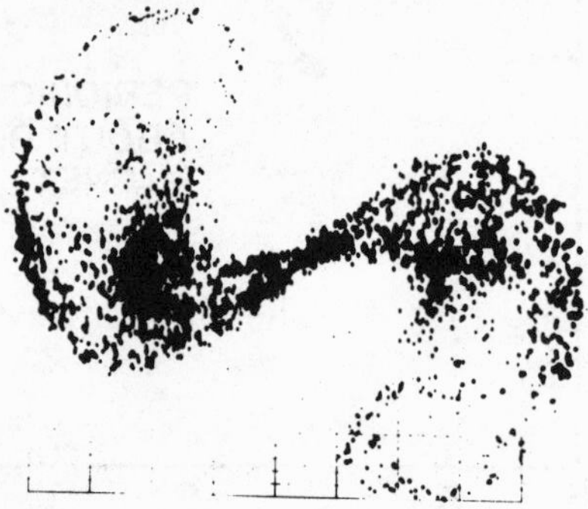

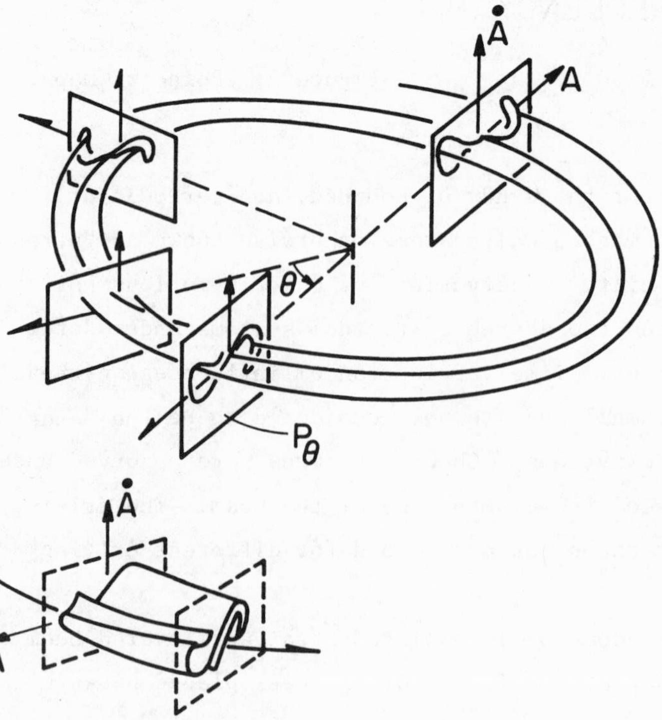

Figure 8: Proposed Structure of a Strange Attractor

the beam were used to buckle the beam creating multiple equilibrium positions. The static behavior has been shown to be similar to a butterfly catastrophe.

In both one and two degree of freedom models, experimental Poincaré maps, synchronized with the driving frequency, were used. The structure of these Poincaré maps was found to be extremely complex, but stable, even after thousands of cycles of driving frequency.

REFERENCES

[1] Ueda, Y., Hayashi, C., Akamatsu, N. Computer simulation of nonlinear ordinary differential equations and nonperiodic oscillations, Electronics and Communications in Japan, 56A, (1973) No. 4, pp. 27-34.

[2] Ueda, Y. Randomly transitional phenomena in the system governed by Duffing's equation, J. of Statistical Physics, 20, (1979) pp. 181-196.

[3] Holmes, P.J., A nonlinear oscillator with a strange attractor, Proc. of Royal Soc. London, 292, (1979) No. 1394, pp. 419-448.

[4] Moon, F.C., Holmes, P.J. "A magnetoelastic strange attractor", J. Sound and Vibration, 65(2), (1979) pp. 275-296.

[5] Moon, F.C. Experiments on chaotic motions of a forced nonlinear oscillator: strange attractors. J. Applied Mechanics, to appear.

ACKNOWLEDGMENT

The author wishes to thank Professor P. Holmes and Professor R. Rand of Cornell University and Professor Ueda of Kyoto University for helpful discussions. Thanks are also due to Dr. Stephen King, research engineer, Cornell University for the design of some of the electronic equipment.

On Coupled Cells

Igor Schreiber,* Milan Kubicek,* and Milos Marek**

Abstract. A characteristic example of apparently aperiodic time dependent concentration trajectories often observed in the system of two coupled reaction cells with mutual mass exchange is presented. The evolution of aperiodic concentration regime with "period three" intervals (in the system of two identical coupled cells with mass exchange) through the series of periodic regimes with the periods approximately equal to T, 2T, 4T, 8T is then discussed for the Brussellator model in two coupled cells. Hysteresis in periodic solutions including aperiodic regimes was found, when evolution of regimes due to slow change of the intensity of mass exchange between cells was followed. Either aperiodic or periodic regime can be reached for the same values of parameters, if the intensity of mass exchange between the cells is increased or decreased, respectively.

1. Introduction. Coupled cells with reaction and mutual mass exchange is a model system often used for description of processes in the living cell aggregates, tissues, compartmental models, distributed systems with chemical reactions and transport (including various forms of chemical reactors), to name just a few systems of chemical and biological interest.[1-8] We have studied systems of coupled cells with oscillating chemical reactions experimentally, using Zhabotinski reaction, and also explored several reaction models mostly using numerical techniques. Here we shall present report on aperiodic regimes; first we shall present a characteristic example of experimental observations and then briefly discuss some simulation results with Brusellator model.

[x]Department of Chemical Engineering, Prague Institute of Chemical Technology, Prague, Czechoslovakia.

[xx]on leave at Chemical Engr. Dept., University of Wisconsin, Madison, Wisconsin. The author is indebted to the Dept. of Chemical Engineering, University of Wisconsin, Madison, for financial support which enabled writing up of this paper.

2. Experimental observations. One of the limiting situations in the exploration of the dynamic behavior of the systems of coupled cells is the study of one cell with periodic variation of the inlet concentration of one of the reaction components. We have employed continuous flow stirred reaction cell and used Zhabotinski reaction with concentrations of reactants and temperature such, that for constant concentration of the reactants in the inlet stream undamped oscillations in the concentrations of the reaction components within the cell and at the outlet (manifested in the periodic course of the recorded redox potential) were observed.

When concentration of ceric ions in the inlet stream is periodically varied in a rectangular pattern, a number of regular (periodic) resonance regimes of increasing complexity is observed, including frequency modulation in the form of beats.[9,10] However, also aperiodic regimes are often observed. Similar phenomena are followed in a model system of coupled cells.

In Fig. 1a, experimental system of two coupled cells with mutual mass exchange through a common wall is shown.

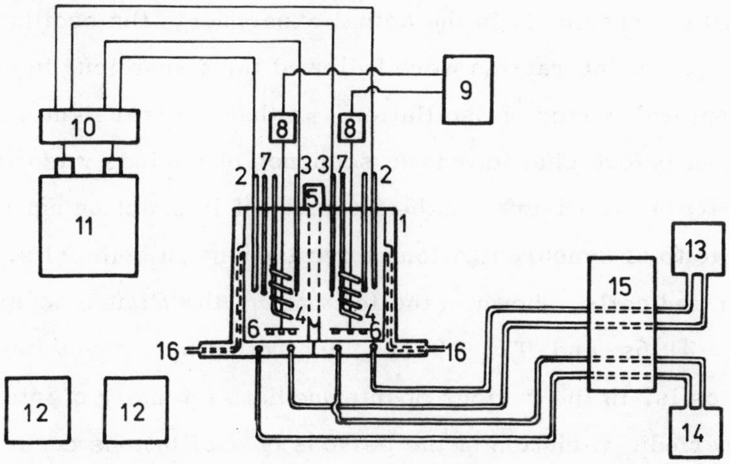

Figure 1a) Experimental system of two coupled flow-through cells with mutual mass exchange. 1-two flow-through cells; 2-calomel electrode; 3-Pt wire electrode; 4-thermostating coil; 5-movable plate; 6-stirrer; 7-temperature measurement; 8-stirrer drive; 9-electronic control of the stirrer speed; 10-signal processing; 11-4-line chart recorder; 12-thermostats; 13,14-feed solutions; 15-pump; 16-level control in the cells.

In Fig.1b typical course of relaxation oscillations in a single (uncoupled cell) is recorded.

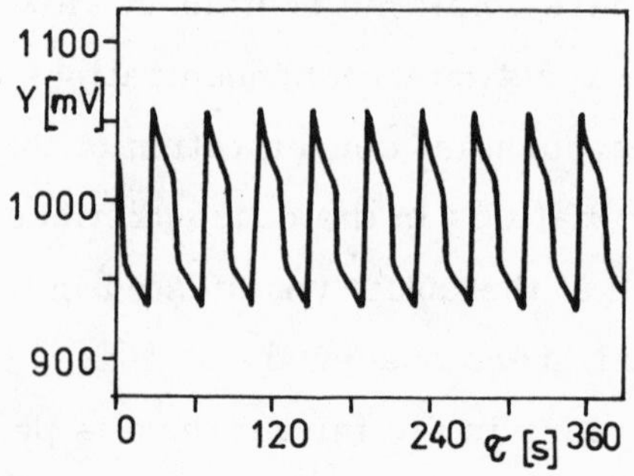

Figure 1b) Continous undamped oscillations of the redox potential (Y corresponds to $\ln(Ce^{4+}/Ce^{3+})$) in the flow-through well stirred cell. Zhabotinski reaction, concentrations: BrO_3^--0.05M, malonic acid 0.05M, Ce^{4+} 0.001M, sulfuric acid 1.5M. Temperature 32^oC, residence time 10 min. Periods of oscillations - 40 s.

In Fig.1c then the characteristics of the time courses of the periods of oscillations in both cells for various degrees of the intensity of interaction (denoted 1-5D) is presented. In the actual experiments the oscillations at a particular degree of interaction were followed for a time long in comparison with the natural period of oscillation, so that the transient effects supposedly died out before characteristic sequence of periods was evaluated. In the right part of the Figure for the intensity of interaction equal to 5D we can observe total synchronization of oscillations in both cells. Oscillations in separated cells, shown in the left part of the Figure occurred with periods $T_{p1} = 41.6s$ and $T_{p2} = 31.1s$, controlled by varying intensity of mixing in the cells. In the regions of intermediate intensity of interaction apparently aperiodic variation of the periods of oscillations occurred. In the Table in the lower part of the Fig.1c we can follow the ratios of perriods in the both cells and phase delays between the first and the second cell. We can infer from the Table, that at low degree of interaction (2D) the ratio of the periods is constant, but phase delays vary. At the degrees of interaction equal to 3D and 4D we can observe no constant trend even if the periods are in the ratios of small integers. The changes in the periods of oscillation in the first cell (higher T_p) follow the changes in the second reactor (lower period). The changes in the periods of oscillations (apparently aperiodic) are largest close to the synchronization. Synchronization occurs at the common period $T_p = 38.5s$. The above observations of aperiodic regimes are characteristic for a number of such plots obtained for various experimental conditions.[11]

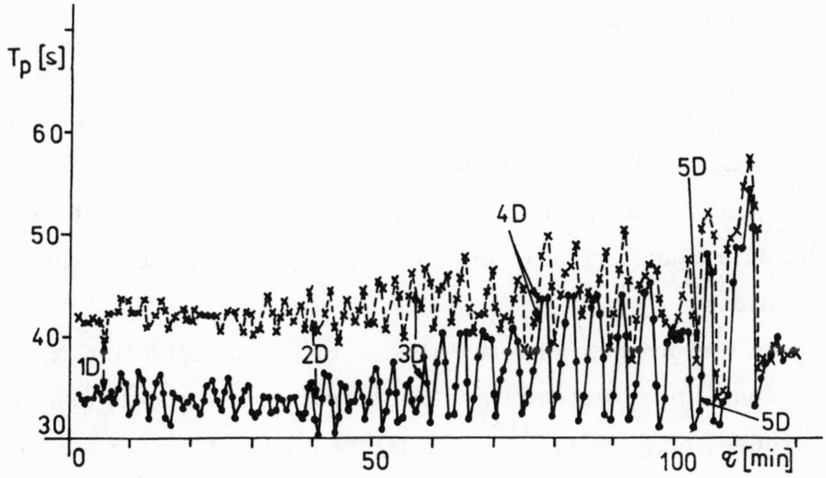

Figure 1c) Dependence of the periods of oscillations in cells on time; 1D– 5D increasing intensity of interaction; Zhabotinski reaction; concentrations see Fig. 1b. Residence time 13.8 min. Temperature 36.2°C. Rate of mixing first cell 615 r.p.m., second cell 500 r.p.m. Period of oscillations without coupling; first cell T_{p1}=41.6s, second cell T_{p2}=31.1s. Table: I/II ratio of periods of oscillations in the first and second cell, respectively; φ phase shift between the maxima of oscillations of the cells (first with respect to second).

2D			3D			4D			5D		
I	II	φ	I	II	φ	I	II	φ	I	II	φ
1	1	-30	1	2	0	2	3	0	2	3	0
3	4	+50	4	5	+40	6	6	+60	10	9	+10
5	4	+50	6	4	0	7	7	+80	1	1	0
5	4	+45	5	6	+40	7	4	+20	1	1	0
5	3	+20	8	6	+40	7	6	0			
5	4	+20	7	4	-40	10	9	0			
5	4	+20									

3. Results of simulation. To compare experimental results with model predictions in the case of such a complex dynamic behavior, we need to know the solutions of the appropriate model equations over sufficiently large intervals of the characteristic parameters. We are using a combination of analytical and numerical techniques (e.g. continuation techniques) for construction of the bifurcation diagrams;[21] the techniques are tested on simple models first. Dynamic behavior for particular interesting characteristic values of parameters is then tested by numerical simulation.[11] We shall further discuss several results of such simulations for the system of two identical reaction cells, where equations describing Brussellator model show in single cells simple behavior (either unique stable solution or unstable singularity surrounded by a stable limit cycle) and diffusion coupling in two cells causes rich dynamic behavior including aperiodic sequences of concentrations.

Aperiodic ("turbulent", "chaotic") behavior of concentration trajectories in two coupled oscillators was demonstrated on a model, constructed for that purpose, by Rössler.[13] Similar behavior of the forced Brussellator was studied numerically by Tomita and Kai.[14] Kennedy and Aris[15] have used one of the Rössler's models to discuss development of various regimes including aperiodic ones in reaction-diffusion systems. (MacDonald[16] has discussed some examples and implications of the observations of aperiodic modes in coupled oscillators.

Two coupled reaction-diffusion cells with Brussellator kinetic schema can be described by the set of four coupled autonomous ODE's ($\dot{}$=d/dt) :

$$
\begin{aligned}
\dot{x}_1 &= A - (B+1)x_1 + x_1^2 y_1 + D_1 (x_2 - x_1) \\
\dot{y}_1 &= Bx_1 - x_1^2 y_1 + D_2 (y_2 - y_1) \\
\dot{x}_2 &= A - (B+1)x_2 + x_2^2 y_2 + D_1 (x_1 - x_2) \\
\dot{y}_2 &= Bx_2 - x_2^2 y_2 + D_2 (y_1 - y_2)
\end{aligned}
\qquad (1)
$$

Here x_k, y_k are concentration variables in the k-th cell and D_j transport coefficients different for both components. We shall consider $p=D_1/D_2=\text{const}$.

The location of bifurcation and limit points and primary and secondary[22] Hopf bifurcations for the values of $A=2$, $p=0.1$ are shown in the bifurcation diagram as a function of the values of parameters B and D_1 in Fig.2. As primary we denote here those bifurcations, which occur from the homogeneous stationary solution of (1), (i.e. from $x_1=x_2=A, y_1=y_2=B/A$), as secondary then those bifurcations, which occur from the nonhomogeneous stationary solutions. Numbers m–n denote total number (m) and number of stable (n) stationary solutions.

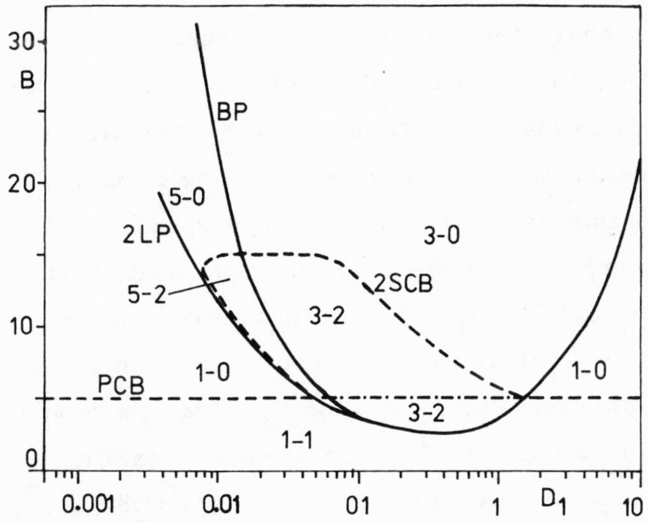

Figure 2) Bifurcation diagram in the plane (B, D_1) for the Brussellator model in two coupled cells; $A=2$, $p=0.1$. Full lines-bifurcations through real eigenvalue; dashed lines-complex (Hopf) bifurcations from stable solution; dashed and dot lines-complex (Hopf) bifurcations from unstable solution; BP – branching point, LP-limit point, PCB-primary complex bifurcation, SCB-secondary complex bifurcation; n-m:n is total number of stationary solutions, m is the number of stable stationary solutions.

Solutions bifurcating from the periodic solutions are not considered in this diagram. Continuation techniques are used for construction of similar and more detailed diagrams.[12,17] As can be inferred from the diagram, one homogeneous solution ($x_1=x_2$, $y_1=y_2$) and four nonhomogeneous solutions (mirror symmetric couples) can occur. Hopf bifurcations indicate the appearance of various time dependent regimes.

Let us follow the dynamic behavior of solutions for the values of parameters A=2, p =0.1, B=5.9 in dependence on the parameter D_1 by numerical simulation. Initial conditions t=0 : $x_1=2.1$, $y_1=2.9$, $x_2=1.9$, $y_2=3.0$ are used in all simulations. For $D_1>D_1^+\doteq 0.9$ (D_1^+ corresponds the secondary Hopf bifurcation point) stable nonhomogeneous periodic solution coexists with stable homogeneous ($x_1=x_2$, $y_1=y_2$) periodic solution. Further increase of D_1 causes first an increase of the amplitude of the nonhomogeneous solution and then the loss of its stability. Stable solutions with more complex characteristics appear. On Fig. 3 the dependence of the values at subsequent maxima of the variable x_2 on D_1 is shown. The behavior is similar as that known from the studies of difference equations ("pitchfork bifurcations"). The time courses of the variables for individual regimes are shown in Figs. 4a-f. Full line corresponds to x_2 and dashed line to x_1. The course of periodic concentration oscillations in both cells is synchronized. For $D_1=1.16$ we can observe one-point cycle oscillations, (Fig.4a) for $D_1=1.18$ two-point cycle oscillations (Fig.4b), for $D_1=1.19$ four-point cycle (Fig.4c) and for $D_1=1.925$ eight-point cycle (Fig.4d) oscillations. The periods of consecutively existing solutions are 4, 8.5, 19, 33.8, i.e. approximately in the ratio 1:2:4:8 (or 2^n, n=0,1,2,3). We have not been able to locate any solutions with higher periods, as for $D_1=1.935$ already aperiodic solution (bifurcation into the invariant torus) sets in. (See Fig. 4e for $D_1=1.22$.) In the aperiodic region we can observe repeating of the profiles after odd number of oscillations. The regions with the "period three" are shown in Fig. 4e. The aperiodic regime here in principle corresponds to the wandering motion of the concentration trajectory between unstable periodic trajectories (perhaps of the increasing complexity). The periodicities of those trajectories then can determine

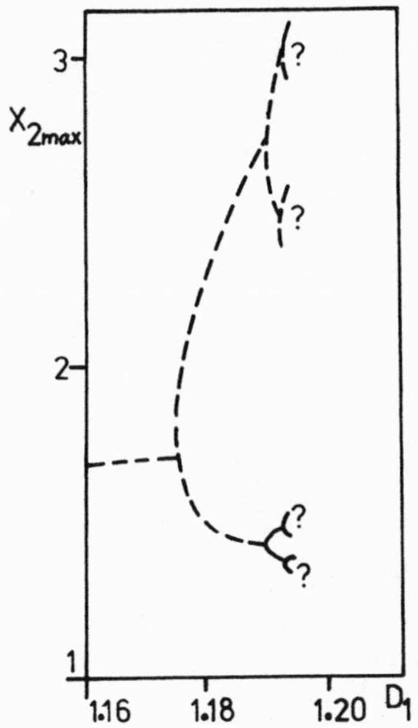

the instantaneous spectral characteristics of the solutions on invariant torus. For sufficiently high value of D_1 the aperiodic solution becomes unstable and homogeneous periodic solution (cf. Fig.4f) sets in.

Figure 3) Dependence of x_{2max} on D_1; Two coupled cells, Brussellator model, A=2, B=5.9, p=0.1; ? aperiodic regime.

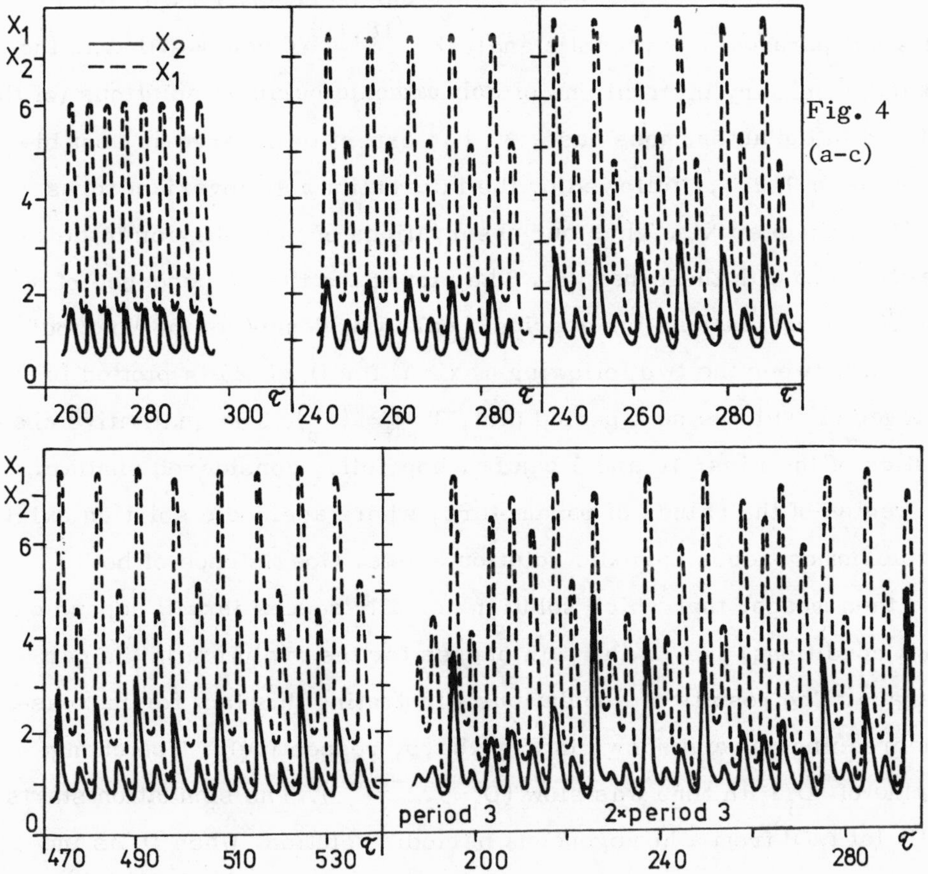

Fig. 4 (a-c)

(d-e)

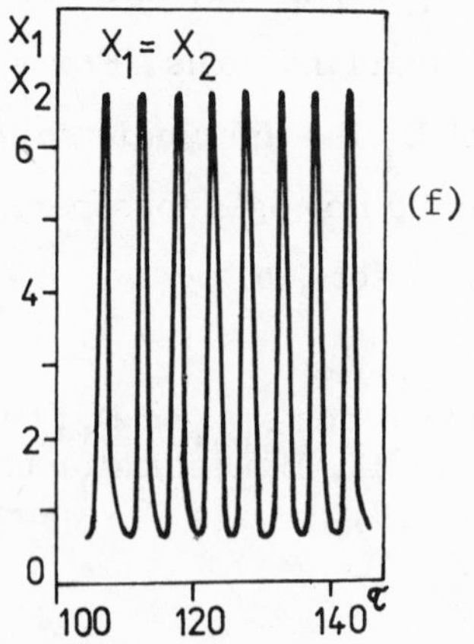

Figure 4) Dependence of the concentrations x_1, x_2 on time.
Two coupled cells, Brussellator model, A=2, B=5.9, p=0.1.
a) D_1=1.16, b) D_1=1.18, c) D_1=1.19, d) D_1=1.1925,
e) D_1=1.22, f) D_1=1.27.

The results of simulation discussed above can be compared with the results of small parameter asymptotic analysis.[(18,19)] We can see, that in the case of the secondary bifurcations of subharmonic periodic solutions (of the Hopf type) in the autonomous case, we can expect to observe at most bifurcations for n=1,2,3,4 followed by the bifurcation into invariant torus (cf. ref (11), Chapter XI). Thus our observations of the behavior in large are in agreement with these results. The plot of the time dependence of the "period" of oscillations for aperiodic case (we denote as the "period" the distance between the two following maxima) for D_1=1.25 is plotted in Fig.5, together with the next period plot, $T_{n+1}=f(T_n)$. The qualitative characteristics of the Figs. 1c and 5 can be, hopefully, considered similar.

In the region of the values of parameters, where aperiodic solution exists, also stable homogeneous periodic solution exists. Coexistence of both solutions then means, that which solution actually settles in will depend on the initial conditions, i.e. on the path chosen for the evolution in the parametric space. The results of simulations illustrating this fact for increasing or decreasing D_1 are shown in Figs.6a,b, respectively. The change of the value of D_1 with time was slow ($D_1=5.2^{-t/50}$). The simulation starts originally (at t=0) from a homogeneous periodic solution. When D_1 is increasing, homogeneous periodic solution changes into inhomogeneous stationary solution, then inhomogeneous periodic solution appears, this further bifurcates into aperiodic regime (we have here passed through the small region (in the parameter space) of the subharmonic bifurcations) and finally homogeneous periodic solution again results. When D_1 is decreased, homogeneous periodic solution does not evolve into an aperiodic solution but changes directly into nonhomogeneous stationary solution and then into nonhomogeneous periodic solution, which for sufficiently small D_1 again evolves into homogeneous periodic solution. Hence here two interesting regions of the values of the parameter D_1 exist; one, where homogeneous periodic solution coexists with the inhomogeneous one and the other, where homogeneous periodic solution coexists with aperiodic solution (solution on an invariant torus). The dimensions of the regions of coexistence of particular solutions depend on the value of the parameter B, e.g. for B=5.5 the region is smaller than for B=5.9. The domain of attraction of the homogeneous periodic solution appears to be small, hence larger perturbations will cause transitions to other solutions, i.e. either nonhomogeneous periodic solution or to aperiodic solution.

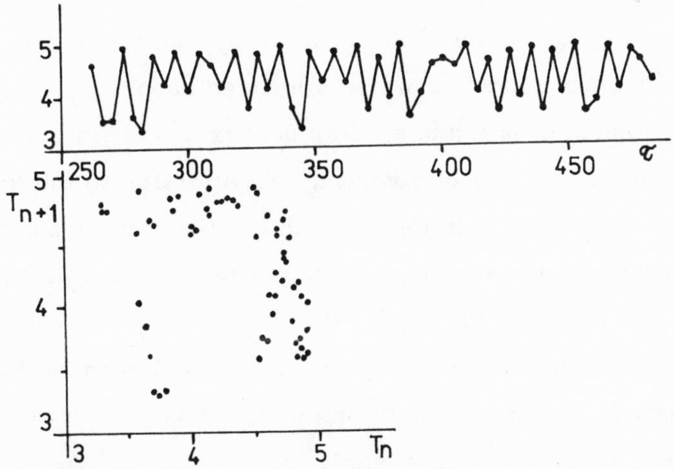

Figure 5) The time course of the oscillation period (time interval between two following maxima in the concentration of x_2). Two coupled cells, Brussellator model, A=2, B=5.9, p=O.1, D_1=1.25.

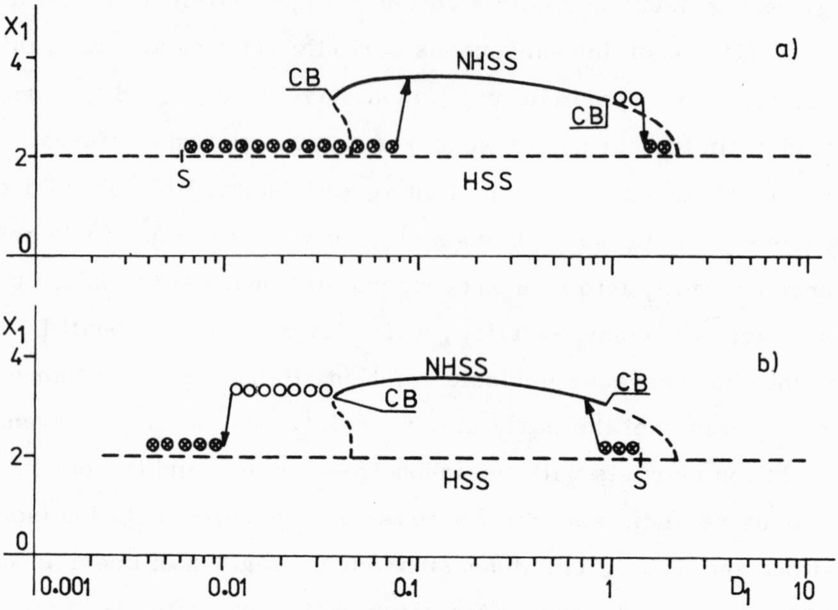

Figure 6) Schematic pictures of the evolution of stationary and periodic solutions with slow change of D_1. Two coupled cells, Brusselator model, A=2, B=6.0, p=0.1
a) increase of D_1, b) decrease of D_1.
s - initial state, o - nonhomogeneous limit cycle. ⊗ homogeneous limit cycle, NHSS - nonhomogeneous stationary state, HSS - homogeneous stationary state. CB - complex bifurcation point.

4. Conclusions. Comparison of the results of simulations with a number of examples of a similar character which appeared recently in the literature and with our experimental data and similar data from the fluid dynamical systems suggests the necessity of directing our attention to the questions of observability of the above solutions. As the regions of parameters where transitory type of solutions exist appear to be small and domains of attractions of some solutions are also very limited, the level and properties of the external noise can in actual experiment crucially determine the properties of the phenomena observed. Questions about observability (and/or relative probabilities of individual regimes), the results of averaging in such a multiple frequency nonconservative systems etc. can be answered only, if we shall formulate corresponding stochastic analogs of (1) and will study them in situations, where the effects of noise with a given amplitude and

spectral properties will be included. Asymptotic techniques for the case of small amplitude random perturbations are now being developed.[20]

REFERENCES

(1) M.ASHKENAZI, M.G. OTHMER, Spatial patterns in coupled biochemical oscillators, J. Math. Biology 5(1978), pp. 305-350.

(2) A. SHAPIRO, F.J.M. HORN, On the possibility of sustained oscillations, multiple steady states and asymmetric steady states in multicell reaction systems, Math Biosciences, 44(1979), pp. 19-39.

(3) M.G. OTHMER, L.E. SCRIVEN, Instability and dynamic pattern in cellular networks, J. Theor. Biol., 32(1971), pp. 507.

(4) V. TORRE, Synchronization of nonlinear biochemical oscillators coupled by diffusion, Biol. Cyber., 17(1975), pp. 137.

(5) M.MAREK, I. STUCHL, Synchronization in two interacting oscillatory systems, Biophys. Chem., 3(1975), pp. 241.

(6) I. PRIGOGINE, R. LEFEVER, Symmetry breaking instabilities in dissipative systems, J. Chem. Phys., 48(1968), pp. 1695-1700.

(7) T. PAVLIDIS, Biological Oscillators: Their Mathematical Analysis, Academic Press, New York, 1973.

(8) S. SMALE, A mathematical model of two cells via Turing's Equations, in: Lectures on Mathematics in the Life Sciences 7, (J.Cowan, ed.) Providence: Amer. Math. Soc. 1975.

(9) M. MAREK, Dissipative structures in chemical systems - Theory and experiment. Synergetics; Far from Equilibrium. Eds. A.Pacault and C. Vidal, Springer Verlag, New York, 1979.

(10) J. KRETBA, Well mixed continuous flow stirred cell with periodic changes of inlet concentrations or temperature, Thesis, Prague Institute of Chemical Technology, 1978.

(11) M. FILIPOVA, Cascade of Reactors With Mutual Mass Exchange, Thesis, Prague Institute of Chemical Technology, 1978.

(12) M. KUBICEK, M. MAREK, Evaluation of limit and bifurcation points for algebraic equations and nonlinear boundary value problems, Appl. Math. and Computation, 5(1979).

(13) O.E. RÖSSLER, Chemical turbulence: Chaos in a small reaction-diffusion system. Z.f. Naturforsch. 31a, (1976), pp. 1168-1172.

(14) K. TOMITA, T. KAI, Stroboscopic phase portrait and strange attractors, Physics Letters 66A, 91 (1978).

(15) C.R. KENNEDY, R. ARIS, Bifurcation of a model diffusion-reaction system, This meeting.

(16) N. MACDONALD, Coupled oscillators in chaotic modes, Nature, 274 (1978), pp. 847

(17) I.SCHREIBER, Dynamic Behaviour of the System of Reaction-diffusion Cells With Mutual Mass Exchange, Thesis, Prague Institute of Chemical Technology, 1979.

(18) G. IOOSS, Topics in Bifurcation of Maps and Applications, Lecture Notes, Math. Studies, North Holland 1979.

(19) G. IOOSS, D.D. JOSEPH, Elementary Stability and Bifurcation Theory, Lecture Notes, University of Minnesota, 1979.

(20) A.D. WENTZEL, M.I. FREIDLIN, Fluctuation in Dynamic Systems- Effects of Small Random Perturbations, Nauka, Moscow, USSR, 1979.

(21) P. HUSTAK, M.KUBICEK, I. MAREK, M. MAREK, Bifurcation in reaction - diffusion systems, in Theory of nonlinear operators, Academie-Verlag-Berlin (1978), pp. 117-128.

(22) M. KUBICEK, Algorithm for evaluation of complex bifurcation points in ordinary differential equations, SIAM J.Appl. Math. 33(1980) in press.

SECTION NINE
LARGE SCALE AND DISTRIBUTED SYSTEMS

Optimization of Distributed Parameter Structures with Repeated Eigenvalues

Edward J. Haug*

Abstract. It has recently been realized that optimum structures; e.g., minimum weight structures whose vibration frequencies and buckling loads are above given lower limits, often have repeated eigenvalues. In this paper, structures whose designs are described by vector parameters and by continuous distributions of material $u(x)$ over a domain $\Omega \subset R^n$, $n=1,2$, or 3, and whose states are governed by matrix equations or by elliptic boundary-value problems are considered. Recent eigenvalue differentiability results for such systems obtained by Rousselet and the author are used with abstract optimization theory to obtain necessary conditions for optimality of a class of structural problems that includes beams, columns, membranes, plates and elastic solids. Technical results involving bifurcation of repeated eigenvalues are obtained as part of the analysis, which may be of interest in their own right.

Examples of Optimization Problems Leading to Repeated Eigenvalues. To see clearly that repeated eigenvalues may systematically arise in optimization problems, consider first a simple two degree of freedom spring-mass system of Fig. 1. The eigenvalue problem for vibration of the rigid body is, with $I = m\ell^2/12$,

$$\begin{bmatrix} 4k_1 + k_2 & k_2 \\ k_2 & 4k_1 + k_2 \end{bmatrix} y = P \begin{bmatrix} 2 & 1 \\ 1 & 2 \end{bmatrix} y \tag{1}$$

where $P = 2m\omega^2/3$ and ω is natural frequency.

*The University of Iowa, Materials Division, College of Engineering, Iowa City, Iowa 52242. Research supported by National Science Foundation Project No. ENG 77-19967.

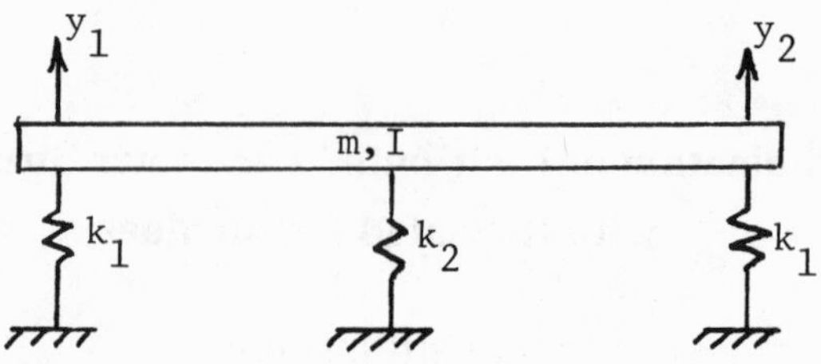

Figure 1. Spring-Mass System.

The optimal design problem treated is to choose the design variables k_1 and k_2 to minimize weight, taken here as

(2) $$\psi_0 = c_1k_1 + c_2k_2$$

where c_i are known, subject to the condition that the eigenvalues P_1 and P_2 of Eq. 1 are no lower than a given constant $P_0 > 0$ and that $k_i \geq 0$, i=1,2.

The eigenvalues of Eq. 1 can easily be calculated as $P_1 = (4k_1 + 2k_2)/3$ and $P_2 = k_1$. Thus, the optimization problem is to find nonnegative k_1 and k_2 to minimize ψ_0 of Eq. 2, subject to the constraints

(3) $$\begin{cases} \psi_1 = P_0 - (4k_1 + 2k_2)/3 \leq 0 \\ \psi_2 = P_0 - 4k_1 \leq 0 \end{cases}$$

Since this is a linear programming problem in two variables, it can be solved by inspection, using the plot of feasible region shown in Fig. 2. The dashed line through vertex A is a level line of the cost function if $c_1/c_2 > 2$, in which case point $A(P_0/4, P_0)$ with a repeated eigenvalue is the optimum. The dashed line through vertex B is a level line of the cost function if $c_1/c_2 < 2$, in which case point $B(3P_0/4, 0)$ with a simple eigenvalue is the optimum. For $c_1/c_2 = 2$ any point on the line between A and B is optimum.

This elementary example shows that for some values of the parameters of the problem a simple eigenvalue occurs at the optimum design and for others a repeated eigenvalue occurs at the optimum design. Thus one does not know a priori whether a repeated eigenvalue will occur

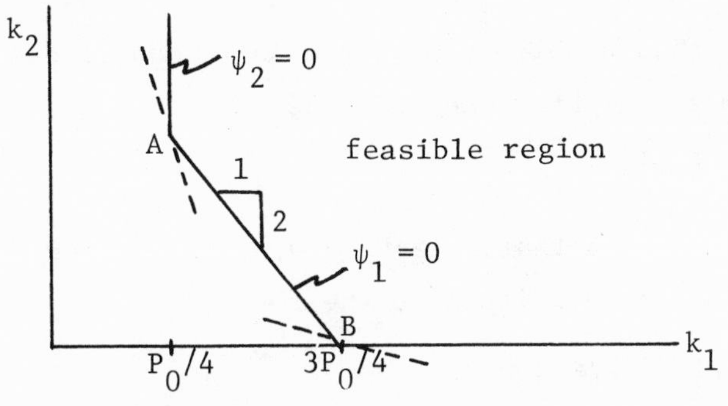

Figure 2. Feasible Domain.

at an optimal design, but the example clearly illustrates that repeated eigenvalue may in some cases arise.

As a second elementary example, consider the compound column of Fig. 3. The bar to the left is rigid and the bar whose top is free to slide along the rigid bar has stiffness EI, with $I = \alpha A^2$, A the cross sectional area of the flexible bar.

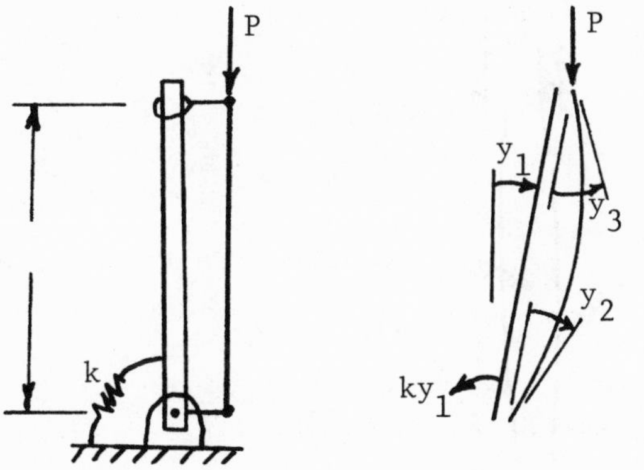

Figure 3. Compound Column.

The eigenvalue problem for the buckling load P is

$$(4) \qquad \begin{bmatrix} \frac{\ell k}{2E\alpha A^2} & 0 & 0 \\ 0 & 2 & -1 \\ 0 & -1 & 2 \end{bmatrix} y = P \frac{\ell^2}{60E\alpha A^2} \begin{bmatrix} 30 & 0 & 0 \\ 0 & 4 & 1 \\ 0 & 1 & 4 \end{bmatrix} y$$

and the eigenvalues may be found as $P_1 = k/\ell$, $P_2 = 12E\alpha A^2/\ell^2$, and

$P_3 = 60E\alpha A^2/\ell^2$. Note that $P_3 > P_2$.

The optimal design problem considered is to choose design variables k and A to minimize weight, taken here as

$$(5) \qquad \psi_0 = a_1 k + a_2 A$$

subject to the condition that $P_i \geq P_0 > 0$, i=1,2, since $P_3 > P_2$. These constraints can be written in the form

$$(6) \qquad \begin{cases} \psi_1 = P_0 - k/\ell \leq 0 \\ \psi_2 = P_0^{1/2} - [12E\alpha]^{1/2}\, A/\ell \leq 0 \end{cases}$$

The problem of choosing k and A to minimize ψ_0 of Eq. 5, subject to the constraints of Eq. 6 is a linear programming problem with only one vertex at the repeated eigenvalue $\psi_1 = \psi_2 = 0$. Thus, for this elementary problem the optimum always leads to a repeated eigenvalue.

As an example of a distributed parameter structural problem, consider the column with variable cross-sectional area u(x) of Fig. 4.

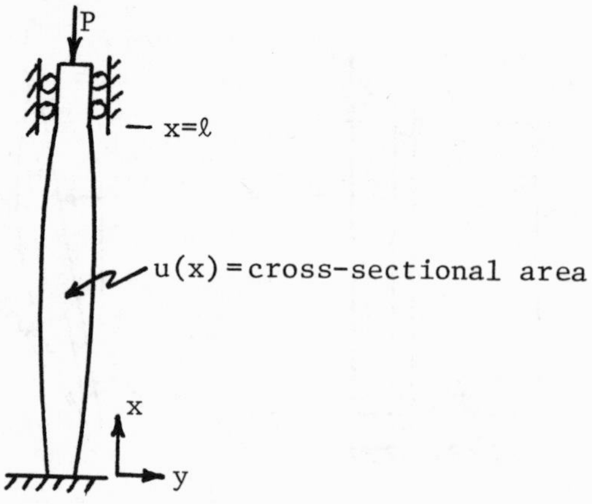

Figure 4. Column.

To minimize volume of structural material, one wishes to find $u \in L^\infty(0,1)$ to minimize

$$(7) \qquad \psi_0 = \int_0^\ell u(x)\,dx$$

subject to a lower bound on the buckling loads

(8) $\psi_i = P_0 - P_i \leq 0, \quad i=1,2$

where the buckling loads P_i are eigenvalues of

(9) $$\begin{cases} A_y \equiv (E\alpha u^2(x)y'')'' = Py'' \equiv PBy \\ y(0) = y'(0) = y(\ell) = y'(\ell) = 0 \end{cases}$$

where $' \equiv \frac{d}{dx}$, E and α are material and geometrical constants, and

(10) $\phi \equiv u_0 - u(x) \leq 0 \quad$ a.e. in $[0,\ell]$

It has recently been shown [1] that one should expect that $\psi_1 = \psi_2 = 0$ in Eq. 8 at the optimum design for some values of $u_0 > 0$ in Eq. 10. This prototype structural optimization problem is shown in Refs. 2 and 3 to be typical of structural optimization problems involving plates and planar elasticity, where the operators A and B are elliptic differential operators. Since $u \in L^\infty(0,\ell)$, it is clear that distributional solutions of the operator equation of Eq. 9 are required. In variational notation, this is

(11) $$\begin{cases} a_u(y,v) \equiv \int_0^\ell E\alpha u^2 y''v''dx = P \int_0^\ell y'v'dx \equiv Pb_u(y,v) \\ y,v \in H_0^2(0,\ell) \end{cases}$$

Directional Derivatives of Eigenvalues. For broad classes of structures, including the column above, it is shown in Ref. 2 that the operators are strongly elliptic, the bilinear forms a_u and b_u are Fréchet differentiable with respect to u, and their differentials in a direction h, $a^{(1)}_{u,h}$ and $b^{(1)}_{u,h}$, are relatively bounded by a_u and b_u, respectively. It is shown in Ref. 2 that if the eigenvalue P is simple, with eigenfunction y normalized by $b_u(y,y) = 1$, then P is Fréchet differentiable with respect to u, and its Fréchet differential in direction h, $P^{(1)}_{u,h}$ is

(12) $P^{(1)}_{u,h} = a^{(1)}_{u,h}(y,y) - Pb^{(1)}_{u,h}(y,y)$

which in the case of the column is

(13) $$P^{(1)}_{u,h} = \int_0^{\ell} 2E\alpha uh(y'')^2 dx$$

If, on the other hand, $P = P_1 = \ldots = P_m$ is an m-fold eigenvalue, with eigenfunctions $y_i(x)$ orthonormalized so that $b_u(y_i, y_j) = \delta_{ij}$, then it is shown in Ref. 3 that P_i, as functions of u, have at most Gateaux differentials $P^{(1)}_{u,h,i}$, which are eigenvalues of the matrix

(14) $$M = \left[a^{(1)}_{u,h} (y_i, y_j) - P\, b^{(1)}_{u,h} (y_i, y_j) \right]$$

Since the eigenvalues of M depend on h, they define the bifurcation of the repeated eigenvalue P into possibly simple eigenvalues as u is changed to y + h, for h small.

In the case of the column with a pair of repeated roots and b_u -orthonormal eigenfunctions y_1 and y_2,

(15) $$M = 2E\alpha \begin{bmatrix} \int_0^{\ell} uh(y_1'')^2 dx & \int_0^{\ell} uhy_1''y_2'' dx \\ \int_0^{\ell} uh\ y_1''y_2''\ dx & \int_0^{\ell} uh(y_2'')^2 dx \end{bmatrix}$$

<u>Necessary Conditions of Optimality</u>. In the case of a simple eigenvalue, the constraint functional of Eq. 8 is differentiable and one may define a convex cone K of feasible directions, for example, in the case of the column associated with Eq. 10 as

(16) $$K = \{h \in L^{\infty}:\ h = \gamma(v - u) \text{ for } v \text{ such that } u_0 - v \leq 0 \text{ and } \gamma > 0\}$$

Now, an optimality condition is that [4, p.83]

(17) $$\psi^{(1)}_{0h} + \mu\, \psi^{(1)}_{1h} \geq 0$$

for some $\mu \geq 0$, and all $h \in K$. In the case of the column, this is

(18) $$\int_0^{\ell} [1 - \mu 2E\alpha\ u(y'')^2] h\ dx \geq 0$$

The situation is more complicated in case of a repeated root. If one writes the constraint of Eq. 8 in the form $\psi_1 = \max_i (P_0 - P_i) = P_0 - \min_i P_i \leq 0$, then the directional derivative of ψ_1 is the smallest eigenvalue of the matrix of Eq. 14. But this is just

$$(19) \qquad \psi_{1h}^{(1)} = - \min_{\substack{\xi \in R^m \\ \xi^T\xi=1}} \xi^T M \xi$$

For fixed ξ, $\xi^T M \xi$ is linear in h, since M is linear in h, so $\min_\xi \ \xi^T Mh \xi$ is concave. Thus, the directional derivative $\psi_{1h}^{(1)}$ of Eq. 19 is convex and an optimality condition for the repeated eigenvalue problem is Eq. 17 [4, p.83], with $\psi_{1h}^{(1)}$ from Eq. 19, i.e.,

$$(20) \qquad \psi_{0h}^{(1)} - \mu \min_{\substack{\xi \in R^m \\ \xi^T\xi=1}} \sum_{i,j=1}^{m} \left[\xi_i \xi_j \left(a_{u,h}^{(1)} (y_i, y_j) - P \, b_{u,h}^{(1)} (y_i, y_j) \right) \right] \geq 0$$

for all $h \in K$. Since for fixed h, $a_{u,h}^{(1)}$ and $b_{u,h}^{(1)}$ are bilinear forms on H_0^2, if one defines $z = \sum_{i=1}^{m} \xi_i y_i$, then

$$(21) \qquad \sum_{i,j=1}^{m} \left[\xi_i \xi_j \left(a_{u,h}^{(1)} (y_i, y_j) - P \, b_{u,h}^{(1)} (y_i, y_j) \right) \right] = a_{u,h}^{(1)} (z,z) - P \, b_{u,h}^{(1)} (z,z)$$

Further, the space of eigenfunctions associated with P is $N_u = \{ z \in H_0^2 : z = \sum_{i=1}^{m} \xi_i y_i, \ \xi_i \text{ real} \}$. The subset of N_u such that $\xi^T\xi = 1$ may now be characterized by the condition

$$b_u(z,z) = \sum_{i,j=1}^{m} \xi_i \xi_j \, b_u(y_i, y_j) = \sum_{i,j=1}^{m} \xi_i \xi_j \, \delta_{ij} = \xi^T \xi = 1$$

since the y_i were selected to be b_u-orthonormal. Thus, Eq. 20 may be written

$$(22) \qquad \psi_{0h}^{(1)} - \mu \min_{\substack{z \in N_u \\ b_u(z,z)=1}} \left(a_{u,h}^{(1)} (z,z) - P \, b_{u,h}^{(1)} (z,z) \right) \geq 0$$

for all $h \in K$.

In the case of the column with a twice repeated root, Eq. 20 is

$$\int_0^\ell h\,dx - \mu 2E\alpha \min_{\substack{\xi \in R^2 \\ \xi^T\xi=1}} \int_0^\ell u(\xi_1 y_1'' + \xi_2 y_2'')^2\, h\,dx \geq 0 \tag{23}$$

for all $h \in K$, where y_1 and y_2 is any b_u -orthonormal pair of eigenfunctions of Eq. 11 associated with P. Use of this condition for determination of an optimum design u appears to be nontrivial.

It is of interest to verify the optimality criterion of Eq. 22 for the simple spring-mass system of Fig. 1, at the solution with repeated root; $k_1 = P_0/4$, $k_2 = P_0$. The minimum in the second term of Eq. 22 is the smaller of the eigenvalues of the matrix M, which are $4h_1/3 + 2h_2/3$ and $4h_1$, so Eq. 22 is, using the differential of ψ_0 of Eq. 2,

$$c_1h_1 + c_2h_2 - \mu \min\{4h_1/3 + 2h_2/3,\ 4h_1\} \geq 0 \tag{24}$$

Consider first the case $4h_1 = 4h_1/3 + 2h_2/3$; i.e., $h_2 = h4_1$. Then Eq. 24 is

$$c_1h_1 + 4c_2h_1 \geq \mu 4h_1$$

for all $h_1 \in R^1$. Thus, equality must hold and

$$\mu = c_1/4 + c_2 \geq 0 \tag{25}$$

Consider second the case $4h_1 > 4h_1/3 + 2h_2/3$; i.e., $h_2 < 4h_1$. Then Eq. 24 is

$$c_1h_1 + c_2h_2 - \mu(4h_1/3 + 2h_2/3) \geq 0$$

Using μ from Eq. 25, this may be written

$$(c_1 - 2c_2)h_2 \leq (c_1 - 2c_2)4h_1$$

Since this must hold for all h_i with $h_2 < 4h_1$, it is necessary that $c_1 - 2c_2 > 0$, or

$$c_1/c_2 \geq 2 \tag{26}$$

Finally, consider the case $4h_1 < 4h_1/3 + 2h_2/3$; i.e., $h_2 > 4h_1$. Then Eq. 24, with μ from Eq. 25 becomes

$$c_2h_2 \geq c_2(4h_1)$$

Since this must hold for all h_i with $h_2 > 4h_1$, it is necessary that

(27) $$c_2 \geq 0$$

The necessary conditions thus reduce to the requirements of Eqs. 25 and 27, which simply require that $c_i \geq 0$, i=1,2, and Eq. 26, which is precisely the condition determined in the first section of the paper for a repeated eigenvalue at the optimum. Thus, the necessary condition of Eq. 22 is sharp for this problem; i.e., it is both necessary and sufficient for an optimum design with repeated eigenvalues.

Connections with Bifurcation Theory. In addition to the results noted in the text on bifurcation of a repeated eigenvalue into branches of perhaps simple eigenvalues, an important connection with bifurcation theory in nonlinear differential equations of structures arises due to the occurance of repeated eigenvalues at an optimum design. It is known in the theory of buckling of structures [5] that qualitatively different nonlinear behavior occurs for simple and repeated eigenvalues (buckling loads). A structure with a single buckling mode at load P (simple eigenvalue) may be "imperfection insensitive"; i.e., if imperfections are present in the structure, then the structure may be loaded up to and in excess of the load P and a stable equilibrium state will occur. If, on the other hand, the structure is optimized and multiple buckling modes exist at a load P (repeated eigenvalue), the structure may be "imperfection sensitive". That is, if imperfections are present in the structure then as the applied load approaches P from below, an unstable configuration may occur and the structure may collapse.

In Ref. 5, Thompson and Hunt refer to the tendency toward coalescence of eigenvalues at an optimum design as a "danger of optimization". Indeed, it should be viewed as a danger of optimization that is carried out using only a linear model of structural behavior. As shown by the foregoing examples, repeated eigenvalues do in fact occur at optimum designs that are constructed using linear structural models. This suggests that to avoid the imperfection sensitive behavior, a nonlinear model of the structure may have to be employed in the design optimization. Thus, one would have to include constraints on nonlinear bifurcation behavior for structures with repeated eigenvalues.

REFERENCES

[1] Olhoff, N., and Rasmussen, S.H., "On Single and Bimodal Optimum Buckling Loads of Clamped Columns", International Journal of Solids and Structures, Vol. 13, 1977, pp.605-614.

[2] Haug, E.J., and Rousselet, B., "Design Sensitivity Analysis in Structural Mechanics I: Static Response Variation", to appear, Journal of Structural Mechanics, 1980.

[3] Haug, E.J., and Rousselet, B., "Design Sensitivity Analysis in Structural Mechanics II: Eigenvalue Variations", to appear, Journal of Structural Mechanics, 1980.

[4] Pshenichnyi, B.N., Necessary Conditions for an Extremum, Marcel Dekker, New York, 1971.

[5] Thompson, J.M.T., and Hunt, G.W., A General Theory of Elastic Stability, Wiley, London, 1973.

A General Limit Cycle Analysis Method for Multivariable Systems

James H. Taylor*

Abstract. The sinusoidal-input describing function (SIDF) technique is a well-known approach for studying limit cycle phenomena in nonlinear systems with one nonlinearity [1,2]. In recent years, a number of extensions of the SIDF method have been developed to permit the analysis of systems containing more than one nonlinearity. In most cases, the nonlinear system models that can be treated by such extensions have been quite restrictive (limited to a few nonlinearities, or to certain specific configurations). Furthermore, some results involve only conservative conditions for limit cycle avoidance, rather than actual limit cycle conditions. The technique described in this paper removes all constraints: Systems described by a general state vector differential equation, with any number of nonlinearities, may be analyzed. In addition, the nonlinearities may be multi-input, and bias effects can be treated.

The general SIDF approach was first fully developed in [3], and its power and use were illustrated by application to a highly nonlinear model of a tactical aircraft in a medium angle-of-attack flight regime [4,5]. Some problems associated with direct simulation (especially "obscuring modes" and the initial condition problem) were also discussed in [5]. This presentation highlights the basic results from [5], and treats a new application (bifurcations in a two-mode panel flutter model) in detail.

1. Introduction. The study of limit cycle (LC) conditions in nonlinear systems is a problem of considerable interest in engineering. An approach to LC analysis that has gained widespread acceptance is the frequency domain/sinusoidal-input describing function (SIDF) method [1,2]. This technique, as it was first developed for systems with one dominant nonlinearity, involved formulating the system in

*School of Mechanical and Aerospace Engineering, Oklahoma State University, Stillwater, Oklahoma 74078. This work was sponsored by the U.S. Office of Naval Research Vehicle Technology Program, under Contract Number N00014-75-C-0432.

the following form:

$$\dot{\underline{x}} = F\underline{x} + \underline{g}\upsilon$$
(1) $$\upsilon = -\phi(\sigma)$$
$$\sigma = \underline{h}^T\underline{x} + \kappa\upsilon$$

where $\underline{x}$ is an n-dimensional state vector. There is thus one single-input/single-output (SISO) nonlinearity, $\phi(\sigma)$, and linear dynamics of arbitrary order that may be represented by the SISO transfer function (in Laplace transform notation) $W(s) = \underline{h}^T(sI-F)^{-1}\underline{g} + \kappa$.

It is then <u>assumed</u> that the input σ may be essentially sinusoidal, e.g., $\sigma = a \cos \omega t$, and the output approximation

(2) $$\phi(\sigma) \cong \text{Re}\,[\psi_1 \exp(i\omega t)]$$
$$\triangleq \text{Re}\,[n_1(a) * a \exp(i\omega t)]$$

is made*. The fourier coefficient† ψ_1 (and thus the "gain" n_1) is generally complex unless $\phi(\sigma)$ is single valued; the real and imaginary parts of ψ_1 represent the in-phase (cosine) and quadrature (-sine) fundamental components of $\phi(a \cos \omega t)$, respectively. The so-called <u>describing function</u> $n_1(a)$ in (2) is "amplitude dependent", thus retaining a basic property of a nonlinear operation. By the principle of harmonic balance, the assumed oscillation -- if it is to exist -- must result in a linearized system with pure imaginary eigenvalues,

$$|i\omega - F + n_1\,\underline{gh}^T| = 0$$

for some value of ω, or by elementary matrix operations

(3) $$W(i\omega) = -\,1/n_1(a)$$

Condition (3) is easy to verify using the polar or Nyquist plot of $W(i\omega)$ [1,2]; in addition, the LC amplitude a is determined in the process.

---

*If $\phi(\sigma)$ is not odd ($\phi(-\sigma) \neq -\phi(\sigma)$), a constant term ("bias" or "D.C. value") must occur in (2); such cases present no difficulty [1,2], but are omitted to simplify the discussion.

†The usual definition of an SIDF is that $n_1(a)$ is chosen to minimize the mean square error between $f(a \cos \omega t)$ and $\text{Re}\,[n_1(a) * a \exp(i\omega t)]$; thus a $n_1(a)$ is the first fourier coefficient [1,2].

It is generally well-understood that SIDF analysis as outlined above is only approximate, so caution is always recommended in its use. The standard caveats that $W(i\omega)$ should be "low pass to attenuate higher harmonics" and that $\phi(\sigma)$ should be "well-behaved" (so that the first harmonic in (2) is dominant) indicate that the analyst has to be familiar with the system behavior, by direct experience or by simulation. Given an appreciation of these warnings, SIDF LC analysis has proven to be a very powerful engineering tool.

The utility of SIDF analysis for systems with one significant SISO nonlinearity as outlined above has naturally resulted in a number of attempts to generalize the technique to the multiple-nonlinearity case. In most cases that preceded [3], only SISO nonlinearities were considered, and bias effects (either due to constant inputs or to "rectification" caused by nonlinear effects) were excluded. Also, special model configurations were often assumed. The earlier results are discussed more fully in [5]. The LC analysis approach described in this paper removes all restrictions with respect to model configuration, nonlinearity type, or the presence of biases.

2. The General SIDF Limit Cycle Analysis Method. The most general system model considered here is

$$\dot{\underline{x}} = \underline{f}\,(\underline{x},\underline{u}) \tag{4}$$

when $\underline{x}$ is an n-dimensional state vector and $\underline{u}$ is an n-dimensional input vector. Assuming that $\underline{u}$ is a vector of constants, denoted $\underline{u}_o$, it is desired to determine if (4) may exhibit LC behavior.

As before, we assume that the state variables are nearly sinusoidal,

$$\underline{x} \cong \underline{x}_c + \mathrm{Re}\,(\underline{a}\,\exp\,(i\omega t)] \tag{5}$$

where $\underline{a}$ is a complex amplitude vector and $\underline{x}_c$ is the state vector center value (which is not a singularity, or solution to $\underline{f}\,(\underline{x}_o,\underline{u}_o) = \underline{0}$ unless the nonlinearities satisfy certain stringent symmetry conditions with respect to $\underline{x}_o$). Then we again assume that higher harmonics are negligible, to make the approximation

(6) $\underline{f}(\underline{x},\underline{u}_o) \cong \underline{f}_{DF}(\underline{u}_o,\underline{x}_c,\underline{a}) + \text{Re}\ [F_{DF}(\underline{u}_o,\underline{x}_c,\underline{a})\ \underline{a}\ \exp\ (i\omega t)]$

The real vector $\underline{f}_{DF}$ and the (generally complex) matrix F_{DF} are obtained by taking the fourier expansions of the elements of $\underline{f}(\underline{x}_c + \text{Re}\ \underline{a}\ \exp (i\omega t), \underline{u}_o)$, and provide the quasi-linear or describing function representation of the nonlinear dynamic relation. The assumed limit cycle exists if $\underline{x}_c$ and $\underline{a}$ can be found so that

(7)
(i) $\underline{f}_{DF}(\underline{u}_o,\underline{x}_c,\underline{a}) = \underline{0}$

(ii) $[i\omega I - F_{DF}(\underline{u}_o,\underline{x}_c,\underline{a})]\underline{a} = 0,\ \underline{a} \neq \underline{0}$

(F_{DF} has a pair of pure imaginary eigenvalues, and $\underline{a}$ is the corresponding eigenvector.)

The nonlinear algebraic equations (7) are generally difficult to solve. An iterative method, based on successive approximation, has been used successfully for a ninth-order, highly nonlinear DE [5].

3. Illustrations.

Example 1. The general SIDF representation of a multi-input nonlinearity is illustrated as follows:

$$
\begin{aligned}
f_5(x) = x_1x_2^3 &\cong [x_{c1}\ x_{c2}^3 + \tfrac{3}{2}\ x_{c2}\ (x_{c1}\ r_{22} + x_{c2}\ r_{12}) + \tfrac{3}{8}\ r_{12}\ r_{22}] \\
&+ [x_{c2}^3 + \tfrac{3}{4}\ x_{c2}\ r_{22}]\ \text{Re}\ [a_1\ \exp\ (i\omega t)] \\
&+ [3x_{c1}\ x_{c2}^2 + \tfrac{3}{4}\ x_{c1}\ r_{22} + \tfrac{3}{2}\ x_{c2}\ r_{12}]\ \text{Re}\ [a_2\ \exp\ i\omega t)] \\
&\triangleq f_{5DF} + f_{5,1}\ \text{Re}\ [a_1\ \exp\ (i\omega t)] + f_{5,2}\ \text{Re}\ [a_2\ \exp\ i\omega t)]
\end{aligned}
$$

where, denoting the conjugate of a_j by a_j^*,

$$r_{ij} = \text{Re}\ [a_i a_j^*] \qquad i,j = 1,2$$

The above result is obtained by substituting for $\underline{x}$ using (5), applying trigonometric identities and discarding the higher harmonic forms. The quantity f_{5DF} is the (hypothetical) fifth element of $\underline{f}_{DF}$, and $f_{5,1}$, $f_{5,2}$ become entries of F_{DF}. By contrast, if Taylor series or "small-signal" linearization is used, the approximation is

$$
\begin{aligned}
f_5(x) = x_1x_2^3 &\cong x_{o1}\ x_{o2}^3 + x_{o2}^3\ \text{Re}\ [a_1\ \exp\ (i\omega t)] \\
&+ 3x_{o1}\ x_{o2}^2\ \text{Re}\ [a_2\ \exp\ (i\omega t)]
\end{aligned}
$$

While this representation is much simpler, it is only realistic when a_1 and a_2 are small.

Example 2. The following second-order differential equation has been derived to describe the local behavior of solutions to a two-mode panel flutter model [6,7]:

$$\ddot{x} + (\alpha + x^2)\,\dot{x} + (\beta + x^2)\,x = 0 \tag{8}$$

Heuristically, it appears that limit cycles may occur for α negative (so that damping is negative for small values of x but positive for large values). Observe also that there are three singularities if β is negative: $x_o = 0, \pm\sqrt{-\beta}$. The corresponding state vector DE is

$$\dot{\underline{x}} = \begin{bmatrix} \dot{x} \\ \ddot{x} \end{bmatrix} = \begin{bmatrix} 0 & 1 \\ -\beta & -\alpha \end{bmatrix} \underline{x} - \begin{bmatrix} 0 \\ x_1^2(x_1 + x_2) \end{bmatrix} \tag{9}$$

The SIDF assumption is that

$$x_1 = x = x_c + a_1 \cos \omega t$$

$$x_2 = \dot{x} = -a_1\omega \sin \omega t$$

(From the relation $x_2 = \dot{x}_1$, it is clear that x_2 has no center value, and that $a_2 = i\omega a_1$ in (5)). Therefore, the combined nonlinearity in (9) is quasi-linearized to be

$$\begin{aligned} x_1^2(x_1 + x_2) &= (x_c + a_1 \cos \omega t)^2\ (x_c + a_1 \cos \omega t - a_1\ \omega \sin\omega t) \\ &\cong (x_c^3 + \tfrac{3}{2} a_1^2\, x_c) + (3x_c^2 + \tfrac{3}{4} a_1^2)\ a_1 \cos \omega t \\ &\qquad + (x_c^2 + \tfrac{1}{4} a_1^2)(-a_1\ \omega \sin \omega t) \end{aligned}$$

Therefore, the conditions of (7) require that

$$\underline{f}_{DF} = \begin{bmatrix} 0 \\ -x_c\ (\beta + x_c^2 + \tfrac{3}{2} a_1^2) \end{bmatrix} = \underline{0} \tag{10}$$

$$F_{DF} = \begin{bmatrix} 0 & 1 \\ -(\beta + 3x_c^2 + \tfrac{3}{4} a_1^2) & -(\alpha + x_c^2 + \tfrac{1}{4} a_1^2) \end{bmatrix} = \begin{bmatrix} 0 & 1 \\ -\omega^2 & 0 \end{bmatrix} \tag{11}$$

Relation (10) shows two possibilities:

(12) Case 1: $x_c = 0$ $\rightarrow a_1 = 2\sqrt{-\alpha}$

$$\omega = \sqrt{\beta - 3\alpha}$$

As predicted, $\alpha < 0$ is required for an LC to exist centered about the origin. The second parameter must satisfy $\beta > 3\alpha$, so β can take on any positive value but cannot be more negative than 3α.

(13) Case 2: $x_c = \pm\sqrt{\frac{\beta - 6\alpha}{5}}$ $\rightarrow a_1 = 2\sqrt{\frac{\alpha - \beta}{5}}$

$$\omega = \sqrt{\beta - 3\alpha}$$

For Case 2 limit cycles to exist, it is necessary that $3\alpha < \beta < \alpha$, so again limit cycles cannot exist unless $\alpha < 0$. One additional constraint must be imposed: $|x_c| > a_1$ must hold or the two limit cycles will overlap. This condition reduces the permitted range of β to $2\alpha < \beta < \alpha$.

One final condition should be investigated: for $2\alpha < \beta < \alpha$, the case 2 LC's must lie inside the case 1 LC: from (12) and (13) this is true if

$$\frac{\beta - 6x}{5} + \frac{4(\alpha - \beta)}{5} < -4\alpha$$

$$-2\alpha - 3\beta < -20\alpha$$

which is indeed satisfied over the range $2\alpha < \beta < \alpha$.

The stability of the case 1 LC can be determined as follows: If $a_1^2 = -4\alpha - \varepsilon < -4\alpha$, then F_{DF} is

$$F_{DF} = \begin{bmatrix} 0 & 1 \\ -(\beta - 3\alpha - \frac{3}{4}\varepsilon) & \frac{1}{4}\varepsilon \end{bmatrix}$$

which for $\varepsilon > 0$ has slightly unstable eigenvalues. Thus a trajectory just inside the LC will grow, indicating that the case 1 LC is stable. A similar analysis of the case 2 LC is more complicated, and thus omitted.

Another viewpoint is provided by the traditional singularity analysis approach (refer to [8]), which involves linearization about $x = 0$ and (if $\beta < 0$) $x = \pm\sqrt{-\beta}$. The linearized F-matrices and singularity characterizations for $\alpha < 0$ are given as follows:

$$\underline{x = 0} \quad \dot{F} = \begin{bmatrix} 0 & 1 \\ -\beta & -\alpha \end{bmatrix} \quad \begin{array}{rcl} \beta < 0 & \rightarrow & \text{saddle} \\ 0 < \beta < \frac{1}{4}\alpha^2 & \rightarrow & \text{unstable node} \\ \beta > \frac{1}{4}\alpha^2 & \rightarrow & \text{unstable focus} \end{array}$$

$$\underline{x = \pm\sqrt{-\beta}} \quad F = \begin{bmatrix} 0 & 1 \\ 2\beta & \beta-\alpha \end{bmatrix} \quad \begin{array}{rcl} \beta_1 < \beta < 0 & \rightarrow & \text{unstable node} \\ \alpha < \beta < \beta_1 & \rightarrow & \text{unstable focus} \\ \beta = \alpha & \rightarrow & \text{center} \\ \beta_2 < \beta < \alpha & \rightarrow & \text{stable focus} \\ \beta < \beta_2 & \rightarrow & \text{stable node} \end{array}$$

where

$$\beta_1 = (\alpha - 4) + 2\sqrt{4 - 2\alpha}$$

$$\beta_2 = (\alpha - 4) - 2\sqrt{4 - 2\alpha}$$

The LC analysis and singularity analysis are completely consistent for $\alpha < 0$, $\beta > 2\alpha$. For all $\beta > 0$, the single singularity is unstable, and for $\alpha < \beta < 0$, the three singularities are unstable, so in both cases the predicted existence of a single stable LC is reasonable. For $\beta = \alpha$, the existence of two center singularities at $x_o = \pm\sqrt{-\beta}$ is in exact accordance with the condition $\beta < \alpha$ for two interior limit cycles to exist, with centers $x_c \cong \pm\sqrt{-\beta}$. The only range of β which seems to give rise to contradictory results is $3\alpha < \beta < 2\alpha$, where the disappearance of the two inner LC's is not consistent with the stable nature of the singularities at $x_o = \pm\sqrt{-\beta}$ and the continuing presence of a large stable LC centered about the origin . The seemingly anomalous result that the SIDF analysis predicts the existence of two overlapping LC's for $3\alpha < \beta < 2\alpha$ might suggest that there may in fact be a single "peanut-shaped" LC inside the large stable LC -- but such a conclusion would only be an intuitive speculation. Since the conjectured inner limit cycle would be quite distinctly nonsinusoidal, it would be necessary to include higher harmonics (e.g., $\underline{x} = \underline{x}_c + \text{Re}\,[\underline{a}_1 \exp(i\omega t)] +$

Re [$\underline{a}_3$ exp (3 iωt)]) in the SIDF analysis in order to reveal its presence. Such an assumption gives rise to substantially more complicated LC existence conditions, so it is not pursued here.

In the terminology of bifurcation theory, we observe that the SIDF analysis indicates the following:

- Bifurcation from a single stable singularity at $x = 0$ to a single stable LC centered about $x = 0$ for $\beta > 0$, α passing from positive to negative,
- Bifurcation from one stable LC enclosing three unstable singularities to one stable LC enclosing two unstable LC's and a saddle for $\alpha < 0$, β passing from greater than α to less than α,
- Disappearance of the two inner LC's for $\alpha < 0$, $\beta < 2\alpha$,
- Disappearance of all limit cycles for $\beta < 3\alpha$.

One quite simple analysis has revealed a great deal of the rich variety of behavior that the DE can exhibit.

5. Conclusion. The basic result in Section 2 shows that there are no inherent restrictions to the generality of the SIDF approach to studying limit cycle conditions. Very complicated high-order and highly nonlinear systems of differential equations have been treated using appropriate computer algorithms for solving nonlinear algebraic equations [5]. A second-order example of significant complexity has been treated in detail, illustrating the power of this method.

REFERENCES

[1] A. GELB and W. E. VANDER VELDE, Multiple-Input Describing Functions and Nonlinear System Design, McGraw-Hill, New York, 1968

[2] D. P. ATHERTON, Nonlinear Control Engineering, Van Nostrand Reinhold, New York, 1975.

[3] J. H. TAYLOR, An algorithmic state-space/describing function technique for limit cycle analysis, IOM, The Analytic Sciences Corp. (TASC), Reading, MA, April 1975. (Also issued as TASC TIM-612-1 to the Office of Naval Research, Oct. 1975).

[4] J. H. TAYLOR, et al, High angle of attack stability and control, Office of Naval Research Report ONR-CR215-237-1, April 1976.

[5] J. H. TAYLOR, A new algorithmic limit cycle analysis method for multivariable systems, IFAC Symp. on Multivariable Systems, U. of New Brunswick, Fredericton N.B. Canada, July 1977.

[6] P. J. HOLMES and J. E. MARSDEN, Bifurcations to divergence and flutter in flow induced oscillations - an infinite dimensional analysis, Automatica 14(1978), pp. 367-384.

[7] J. CARR, Applications of Center Manifold Theory, Lefchetz Center for Dynamical Systems, Div. of Appl. Math, Brown U., Providence, R.I., 1979.

[8] W. J. CUNNINGHAM, Nonlinear Analysis, McGraw-Hill, New York, 1958.